Microeconometrics with R

This book is about doing microeconometrics, defined by Cameron and Trivedi as "the analysis of individual-level data on the economic behavior of individuals or firms using regression methods applied to cross-section and panel data" with R. Microeconometrics became increasingly popular in the last decades, thanks to the availability of many individual data sets and to the development of computer performance.

R appeared in the late nineties as a clone of S. It became increasingly popular among statisticians, especially in fields where S was widely used. Twenty years ago, using R for doing econometrics analysis required a lot of programming because a lot of core methods of econometrics were not available in R. Nowadays, most of the basic methods described in the book are available in contributed packages. Moreover, the set of packages called the tidyverse developed by RStudio (now Posit) for all the basic tasks of an applied statistician (importing, tidying, transforming and visualizing data sets) makes the use of R faster and easier. The book uses extensively specialized econometrics packages and the tidyverse, and it seeks to demonstrate that the adoption of R as the primary software for an econometrician is a relevant choice.

The first part of the book is devoted to the ordinary least squares estimator. Matrix algebra is progressively introduced in this part, and special attention is paid to the interpretation of the estimated coefficients. The second part goes beyond the basic OLS estimator by testing the hypothesis on which this estimator is based and providing more complex estimators relevant when some of these hypotheses are violated. Finally, the third part of the book presents specific estimators devoted to "special" responses, e.g., count, binomial or duration data.

Key Features:

- Many applications using data sets of recent academic works are developed
- Testing and estimation procedures using the programming framework of R and specialized packages are presented
- Two companion packages (micsr and micsr.data), containing respectively functions implementing some estimation and testing procedures not available in other contributed packages and data sets used in the book, are provided

Yves Croissant is professor of Economics at the University Lumière Lyon-2. His main research interests are microeconometrics and transport economics.

Chapman & Hall/CRC
The R Series

Recently Published Titles

A Criminologist's Guide to R: Crime by the Numbers
Jacob Kaplan

Analyzing US Census Data: Methods, Maps, and Models in R
Kyle Walker

ANOVA and Mixed Models: A Short Introduction Using R
Lukas Meier

Tidy Finance with R
Christoph Scheuch, Stefan Voigt, and Patrick Weiss

Deep Learning and Scientific Computing with R torch
Sigrid Keydana

Model-Based Clustering, Classification, and Density Estimation Using mclust in R
Lucca Scrucca, Chris Fraley, T. Brendan Murphy, and Adrian E. Raftery

Spatial Data Science: With Applications in R
Edzer Pebesma and Roger Bivand

Modern Data Visualization with R
Robert Kabacoff

Learn R: As a Language, Second Edition
Pedro J. Aphalo

Spatial Analysis in Geology Using R
Pedro M. Nogueira

Analyzing Baseball Data with R, Third Edition
Jim Albert, Benjamin S. Baumer and Max Marchi

Geocomputation with R, Second Edition
Robin Lovelace, Jakub Nowosad, Jannes Muenchow

For more information about this series, please visit: https://www.crcpress.com/Chapman-HallCRC-The-R-Series/book-series/CRCTHERSER

Microeconometrics with R

Yves Croissant

CRC Press is an imprint of the
Taylor & Francis Group, an **informa** business
A CHAPMAN & HALL BOOK

Designed cover image: © Yves Croissant

First edition published 2025
by CRC Press
2385 NW Executive Center Drive, Suite 320, Boca Raton FL 33431

and by CRC Press
4 Park Square, Milton Park, Abingdon, Oxon, OX14 4RN

CRC Press is an imprint of Taylor & Francis Group, LLC

ISBN: 978-0-367-55446-0 (hbk)
ISBN: 978-0-367-56992-1 (pbk)
ISBN: 978-1-003-10026-3 (ebk)

DOI: 10.1201/9781003100263

Typeset in Latin Modern font
by KnowledgeWorks Global Ltd.

Publisher's note: This book has been prepared from camera-ready copy provided by the authors.

The most important date in a man's life is that of his father's death.

Georges Simenon [1]

Le Fils, 1957

[1] Cited by Pierre Assouline, Simenon : biographie, Paris : Julliard, 1992

Table of contents

List of Figures **xiii**

List of Tables **xvii**

Preface **xix**

R and RStudio . . . xix
R packages . . . xx
Data sets . . . xxi
The micsr package . . . xxi

I Ordinary least squares estimator **1**

1 Simple linear regression model **5**

1.1 Conditional expectation and covariance . . . 5
1.1.1 Linear models: two examples . . . 6
1.1.2 Linear model, conditional expectation and covariance . . . 8
1.2 Model and data set . . . 10
1.3 Computation of the OLS estimator . . . 14
1.4 Geometry of least squares, variance decomposition and coefficient of determination . . . 17
1.4.1 Vectors, variance and covariance . . . 17
1.4.2 Geometry of least squares . . . 17
1.4.3 Variance decomposition and the R^2 . . . 18
1.5 Computation with R . . . 20
1.6 Data generator process and simulations . . . 22
1.6.1 Data generator process . . . 23
1.6.2 Random numbers and simulations . . . 25

2 Statistical properties of the simple linear estimator **33**

2.1 Exact properties of the OLS estimator . . . 33
2.1.1 Errors have 0 expected value . . . 34
2.1.2 Conditional expectation of the errors is 0 . . . 34
2.1.3 Estimator for the variance of the OLS estimator . . . 36
2.1.4 OLS estimator is BLUE . . . 45
2.2 Asymptotic properties of the estimator . . . 49
2.2.1 Convergence in probability . . . 49
2.2.2 Convergence in distribution: central-limit theorem . . . 52
2.2.3 Simulations . . . 54
2.3 Confidence interval and tests . . . 56
2.3.1 Testing hypothesis . . . 56
2.3.2 Confidence interval . . . 59
2.3.3 Exact distribution, the Student distribution . . . 60

2.3.4 Inference with R . . . 60
2.3.5 Delta method . . . 62
2.3.6 Confidence interval for the prediction . . . 63

3 Multiple regression model **67**
3.1 Model and data set . . . 67
3.1.1 Structural model . . . 67
3.1.2 Data set . . . 69
3.2 Computation of the OLS estimator . . . 71
3.3 Geometry of least squares . . . 75
3.3.1 Geometry of the multiple regression model . . . 75
3.3.2 Frisch-Waugh theorem . . . 76
3.4 Computation with R . . . 78
3.4.1 Computation using matrix algebra . . . 78
3.4.2 Efficient computation: QR decomposition . . . 80
3.5 Properties of the estimators . . . 81
3.5.1 Unbiasedness of the OLS estimator . . . 81
3.5.2 Variance of the OLS estimator . . . 82
3.5.3 The OLS estimator is BLUE . . . 84
3.5.4 Asymptotic properties of the OLS estimator . . . 85
3.5.5 The coefficient of determination . . . 85
3.6 Confidence interval and test . . . 86
3.6.1 Simple confidence interval and test . . . 86
3.6.2 Joint confidence interval and test of joint hypothesis . . . 88
3.6.3 The three tests . . . 92
3.6.4 Computation of the three tests . . . 93
3.6.5 Testing that all the slopes are 0 . . . 96
3.7 System estimation and constrained least squares . . . 98
3.7.1 System of equations . . . 99
3.7.2 Constrained least squares . . . 103

4 Interpretation of the Coefficients **105**
4.1 Numerical covariate . . . 106
4.1.1 Response and covariate in level . . . 106
4.1.2 Covariate in logarithm . . . 108
4.1.3 Response in logarithm . . . 109
4.1.4 Response and covariate in log . . . 111
4.1.5 Integer covariate . . . 112
4.2 Categorical covariate . . . 112
4.2.1 Dichotomic variable . . . 113
4.2.2 Polytomic covariate . . . 114
4.3 Several covariates . . . 117
4.3.1 Separate effects . . . 117
4.3.2 Multiplicative effects . . . 118
4.3.3 Polynomials . . . 120
4.4 Marginal effects . . . 122
4.4.1 Computation of the marginal effects with one covariate . . . 123
4.4.2 General computation of marginal effects . . . 126

II Beyond the OLS estimator 129

5 Maximum likelihood estimator 133
5.1 ML estimation of a unique parameter . . . 133
5.1.1 Computation of the ML estimator for the Poisson distribution . . . 133
5.1.2 Computation of the ML estimator for the exponential distribution . 136
5.1.3 Properties of the ML estimator . . . 138
5.1.4 Computation of the variance for the Poisson and the exponential distribution . . . 142
5.2 ML estimation in the general case . . . 145
5.2.1 Computation and properties of the ML estimator . . . 145
5.2.2 Computation of the estimators for the exponential distribution . . . 147
5.2.3 Linear gaussian model . . . 150
5.2.4 Transformation of the response . . . 151
5.3 Tests . . . 156
5.3.1 Tests for nested models: the three classical tests . . . 156
5.3.2 Conditional moment test . . . 163
5.3.3 Tests for non-nested models . . . 168

6 Non-spherical disturbances 173
6.1 Situations where the errors are non-spherical . . . 173
6.1.1 Heteroskedasticity . . . 174
6.1.2 Correlation of the errors . . . 174
6.1.3 System of equations . . . 177
6.2 Testing for non-spherical disturbances . . . 177
6.2.1 Testing for heteroskedasticity . . . 178
6.2.2 Testing for individual effects . . . 180
6.2.3 System of equations . . . 183
6.3 Robust inference . . . 184
6.3.1 Simple linear model . . . 185
6.3.2 Multiple linear model . . . 188
6.4 Generalized least squares estimator . . . 192
6.4.1 General formulation of the GLS estimator . . . 192
6.4.2 Weighted least squares . . . 193
6.4.3 Error component model . . . 194
6.4.4 Seemingly unrelated regression . . . 199

7 Endogeneity 203
7.1 Sources of endogeneity . . . 203
7.1.1 Errors in variables . . . 203
7.1.2 Omitted variable bias . . . 204
7.1.3 Simultaneity bias . . . 206
7.2 Simple instrumental variable estimator . . . 209
7.2.1 Computation of the simple instrumental variable estimator . . . 209
7.2.2 Small sample properties of the **IV** estimator . . . 211
7.2.3 An example: Segregation effects on urban poverty and inequality . . 216
7.2.4 Wald estimator . . . 217
7.3 General IV estimator . . . 221
7.3.1 Computation of the estimator . . . 221
7.3.2 An example: long-term effects of slave trade . . . 223
7.4 Three-stage least squares . . . 226

7.4.1 Computation of the three-stage least square estimator 227
7.4.2 An example: the watermelon market 228
7.5 Fixed effects model 231
7.5.1 Computation of the fixed effects estimator 231
7.5.2 Application: Mincer earning function using a sample of twins 233
7.5.3 Application: Testing Tobin's Q theory of investment using panel data 234
7.6 Specification tests 236
7.6.1 Hausman test 236
7.6.2 Weak instruments 237
7.6.3 Sargan test 237
7.6.4 Individual effects 237
7.6.5 Panel application: Testing Tobin's Q theory of investment 238
7.6.6 Instrumental variable application: slave trade 238

8 Treatment effect 241
8.1 Randomized experiment 242
8.2 Instrumental variable estimator 246
8.3 Regression discontinuity 249
8.3.1 Sharp and Fuzzy 250
8.3.2 Plotting the discontinuity 251
8.3.3 Computing the effect of the treatment 253
8.3.4 Manipulation test 255
8.4 Difference-in-differences 256
8.5 Matching 259
8.6 Synthetic control 267

9 Spatial econometrics 277
9.1 Simple features 277
9.1.1 Structure of a simple feature 277
9.1.2 Computation on sf objects 281
9.2 Two examples 286
9.2.1 Agglomeration economies and diseconomies 286
9.2.2 Solow model 288
9.3 Spatial correlation 292
9.3.1 Contiguity and weights 292
9.3.2 Tests for spatial correlation 295
9.3.3 Local spatial correlation 296
9.4 Spatial econometrics 299
9.4.1 Spatial models and tests 299
9.4.2 Application to the growth model 301

III Special responses 307

10 Binomial models 311
10.1 Functional form and the linear-probability, probit and logit model 311
10.2 Structural models for binomial responses 316
10.2.1 Latent variable and index function 316
10.2.2 Random utility model 316
10.3 Binomial model as a generalized linear model 317
10.3.1 Generalized linear models 317
10.3.2 Estimation with `stats::glm` 319

10.4 Model estimation, evaluation and testing . . . 322
10.4.1 Estimation . . . 322
10.4.2 Evaluation . . . 325
10.4.3 Testing . . . 329
10.5 Endogeneity . . . 332
10.5.1 Maximum likelihood estimation . . . 333
10.5.2 Two-step estimator . . . 333
10.5.3 Minimum χ^2 estimator . . . 335
10.5.4 Application . . . 335
10.6 Ordered models . . . 337

11 Censored and truncated models **343**
11.1 Truncated response, truncated and censored samples . . . 343
11.1.1 Corner solution . . . 343
11.1.2 Data censoring and truncation . . . 345
11.1.3 Sample selection . . . 345
11.1.4 Truncated and censored samples . . . 345
11.2 Tobit-1 model . . . 346
11.2.1 Truncated normal distribution, truncated and censored sample . . . 346
11.2.2 Interpretation of the coefficients . . . 350
11.3 Methods of estimation . . . 352
11.3.1 Non-linear least squares . . . 352
11.3.2 Probit and two-step estimators . . . 353
11.3.3 Maximum Likelihood estimation . . . 353
11.3.4 Semi-parametric estimators . . . 355
11.4 Estimation of the tobit-1 model with R . . . 356
11.4.1 Left-truncated response . . . 357
11.4.2 Right-truncated response . . . 358
11.4.3 Two-sided tobit models . . . 360
11.5 Evaluation and tests . . . 362
11.5.1 Conditional moment tests . . . 362
11.5.2 Endogeneity . . . 363
11.6 Tobit-2 model . . . 366
11.6.1 Two-part models . . . 367
11.6.2 Hurdle models . . . 367
11.6.3 Correlated models . . . 368
11.6.4 Application . . . 370

12 Count data **375**
12.1 Features of count data . . . 375
12.1.1 An empirical survey . . . 375
12.1.2 Analyzing the number of trips . . . 377
12.2 Poisson model . . . 378
12.2.1 Derivation of the Poisson model . . . 378
12.2.2 Interpretation of the coefficients . . . 380
12.2.3 Computation of the variance of the estimator . . . 381
12.2.4 Overdispersion . . . 383
12.3 Overdispersion and excess of zero . . . 386
12.3.1 Mixing model . . . 386
12.3.2 Hurdle and ZIP models . . . 390
12.4 Endogeneity and selection . . . 393

12.4.1 Instrumental variable estimators for count data 393
12.4.2 Sample selection and endogenous switching for count data 396

13 Duration models **401**
13.1 Kaplan-Meier non-parametric estimator 402
13.1.1 Uncensored sample 402
13.1.2 Censored sample 406
13.1.3 Different groups 408
13.2 Parametric and semi-parametric estimators 410
13.2.1 Constant hazard and the exponential distribution 410
13.2.2 Estimation 411
13.2.3 Accelerated failure time model 413
13.2.4 Proportional hazard models 416
13.3 Heterogeneity 420
13.3.1 Detecting heterogeneity 420
13.3.2 Weibull-Gamma model 422
13.4 Other models 423
13.4.1 Multi-state models 423
13.4.2 Grouped data 426
13.4.3 Interval regression 429

14 Discrete choice models **433**
14.1 Data management and model description 433
14.1.1 Data management 433
14.1.2 Model description 437
14.2 Random utility model and multinomial logit model 439
14.2.1 Random utility model 439
14.2.2 Distribution of the error terms 440
14.2.3 IIA property 441
14.2.4 Interpretation 441
14.2.5 Application 442
14.3 Logit models relaxing the iid hypothesis 447
14.3.1 Heteroskedastic logit model 447
14.3.2 Nested logit model 449
14.4 Random parameters (or mixed) logit model 452
14.4.1 Derivation of the model 452
14.4.2 Application 453
14.5 Multinomial probit 458
14.5.1 The model 458
14.5.2 Identification 459
14.5.3 Simulations 460
14.5.4 Application 462

References **465**

Indexes **477**

List of Figures

1.1 Birth weight for non-smoking and smoking mothers 7
1.2 Education and income 7
1.3 Generalized cost for train and plane 11
1.4 Share of rail in function of the threshold time value 12
1.5 Share of rail in function of the threshold time value on a subsample 13
1.6 Model shares with a uniform distribution 13
1.7 Vector algebra 18
1.8 Geometry of the simple regression model 18
1.9 Regression line and residuals 22
1.10 Data generating process 24
1.11 4 different samples 27

2.1 Intercept and the expected value of the error 34
2.2 Education, sex and wage 36
2.3 Size of the error and the precision of the slope estimator 38
2.4 Sample size and precision of the estimator 38
2.5 Variance of x and precision of the estimator 39
2.6 OLS estimator without intercept 47
2.7 Individual slopes 48
2.8 Four samples with χ^2 errors 55
2.9 Empirical distribution of $\hat{\beta}$ for different sample sizes and adjustment by a normal density 55
2.10 Empirical distribution of $\sqrt{N}(\hat{\beta}-\beta)$ for different sample sizes and adjustment by a normal density 56
2.11 Normal distribution and 5% critical value 58
2.12 Critical value and p-value 59
2.13 Confidence and prediction intervals 65
2.14 Predictions for train's model share 66

3.1 Investment rate and demographic growth 70
3.2 Geometry of the multiple regression model 75
3.3 Frisch-Waugh theorem 76
3.4 Ellipse of confidence for the two coefficients 90

4.1 Additive effects of production and region 118
4.2 The Kuznets curve 121

5.1 Empirical distribution and theoretical Poisson probabilities 134
5.2 Log-likelihood curve for a Poisson variable 136
5.3 Empirical distribution and density of the exponential distribution 137
5.4 Log-likelihood curve for an exponential variable 138
5.5 Distribution of the Vuong statistic 171

6.1 Data and regression line for the `urban_gradient` data set 187

7.1 Market equilibrium 207
7.2 Distribution of the IV estimator 215
7.3 Distribution of the IV estimator (zoom) 215
7.4 Income differentials between veterans and non-veterans 221
7.5 Per capita GDP and slave extraction 224

8.1 Boxplot for the test variable between the control and the treatment group . 246
8.2 The four categories of individuals 247
8.3 Discontinuity for probation 252
8.4 Discontinuity for probation using `rdplot` 252
8.5 Placebo test for probation 253
8.6 Density estimation on both sides of the cutoff for the `probation` data set . 256
8.7 Love plot for the `twa` data set 266
8.8 Weights for the units and the predictors 272
8.9 GDP per capita for the Basque Country and its synthetic control 272
8.10 Difference of GDP per capita for the Basque Country and its synthetic control 273
8.11 Plot of the MSPE ratio 274
8.12 Placebos plot 275

9.1 Map of France with its three main cities 280
9.2 Map of American counties 282
9.3 Map of American states 283
9.4 Border states and Texas borders 284
9.5 Counties close to the Texan border 285
9.6 Discontinuity of the mortgage defaults data set 286
9.7 Population growth and initial population in Louisiana 288
9.8 Relation between initial population and population growth in Louisiana . . 288
9.9 Growth and initial GDP, 1960-1995 291
9.10 Queen and rook links for the counties of Louisiana 293
9.11 Moran plot 297
9.12 Map of local Moran statistic 298

10.1 Logistic and normal densities 312
10.2 Fitted probabilities 314
10.3 Histogram of the distribution of the fitted values for the probit mode choice model 328
10.4 Histogram of the distribution of the fitted values for the Airbnb probit model 328
10.5 Probabilities for an ordered model 338
10.6 Marginal effects for an ordered model 338

11.1 Internal and corner solution 344
11.2 Truncated normal distribution 347
11.3 Inverse Mills ratio 348
11.4 Truncated normal distribution for y 349
11.5 OLS bias in a truncated sample 349
11.6 OLS bias in a censored sample 350
11.7 Effect of a change of x 352
11.8 Symmetrically trimmed truncated distribution 355
11.9 Empirical distribution of the response and normal approximation 363

12.1 Unconditional distribution of trips 378
12.2 Rootograms for Poisson models with and without covariates 385
12.3 Squared residuals and fitted values 386
12.4 Rootograms for the Poisson and the Negbin model 390
12.5 Rootograms for the Negbin2, Hurdle and ZIP models 393
12.6 Rootograms for the Negbin Hurdle and ZIP models 394

13.1 Survival curve for the oil data set 404
13.2 Survival curve for the oil data set with a log scale 405
13.3 Survival curves for unemployment duration in the United States 409
13.4 Hazard functions for employment duration 416
13.5 Generalized residuals for the Weibull model 421
13.6 Interval regression 429

List of Tables

5.1 Poisson probabilities for different values of the parameter 135
5.2 Scale elasticity 155

8.1 Balance table for the Afghan girls data set 245

9.1 Results of the growth models 304

10.1 Comparison of the coefficients for the Airbnb data set 315
10.2 Logit and probit models for the mode choice data set 324
10.3 Estimation of the standard deviations of the estimates 325
10.4 IV probit models for the bank data set 337
10.5 Ordered logit models for the financial reform data set 341

11.1 Estimation of charitable giving models 358
11.2 Estimation results of the protection for sale model 365
11.3 Traffic citations 373

12.1 Empirical survey of count data sets 376
12.2 Mixed models for the demand for trips 389
12.3 Cigarette consumption and smoking habits 396
12.4 Poisson and endogenous switching models for alcohol demand 400

13.1 Accelerated failure time model for unemployment duration 416
13.2 Weibull model (AFT and PH) and Cox semi-parametric model for the unemployment data set 420
13.3 Estimation of the grouped duration model for the recall data set 428
13.4 Interval regression for the Kakadu data set 431

Preface

This book is about doing microeconometrics with R. Microeconometrics is defined by Cameron and Trivedi (2005) as "the analysis of individual-level data on the economic behavior of individuals or firms using regression methods applied to cross-section and panel data". We'll use in this book a broader definition of microeconometrics, as some of the empirical examples use countries or regions as a unit of observation. But, nevertheless, the underlying model will have microeconomic foundations, as for example the Solow growth model that is treated at length in Chapter 3 and Chapter 9 and the Kuzets curve presented in Chapter 4. A negative delimitation of what is in this book is that it doesn't contain any analyses of time series.

I used throughout the book some manuals. My main reference was Cameron and Trivedi (2005), but I also used Greene (2018) (especially the online mathematical and statistical appendices that I found particularly useful) and Wooldridge (2010). At an introductory level, I also used Stock and Watson (2015) and Maddala and Lahiri (2009). For some points (geometry of least squares, asymptotic theory, data generating process, computational considerations), I was also inspired by Davidson and MacKinnon (1993, 2004). On more specific subjects, I used also some specialized textbook and surveys: Maddala (1983), Amemiya (1981, 1984) for Chapter 10 and Chapter 11 (binomial and censored models), Cameron and Trivedi (2013) for Chapter 12 (count data), Lancaster (1990) and Kiefer (1988) for Chapter 13 (duration models) and Train (2009) for Chapter 14 (discrete choice models).

R and RStudio

R appeared in the late nineties as a clone of **S** which rapidly became a **GNU** project. It became increasingly popular among statisticians, especially in fields where **S+** was widely used. Its large adoption by econometricians is more recent, especially because of the use by econometricians of other softwares like **SAS**, **GAUSS** and especially nowadays **Stata**, and the lack of some popular estimators and tests for econometrics in **R**. Since then, numerous excellent packages have since been developed that make the adoption of **R** as the primary software for an econometrician a relevant choice. Moreover, **R** has some characteristics that make **R** code concise and clear, compared to other softwares.

RStudio (now **Posit**), founded in 2009 by Joseph J. Allaire, developed an integrated development environment called **RStudio** and a set of **R** packages that changed deeply the way a lot of **R** users (including myself) use **R**. Among them, **quarto** was used to build this book and the set of packages called the **tidyverse** is used extensively for all the basic tasks of an applied statistician which are, citing Wickham, Cetinkaya-Rundel, and Grolemund (2023), importing, tidying, transforming and visualizing data sets. Packages of the

tidyverse, respectively **readr**, **tidyr**, **dplyr** and **ggplot2** enable to perform these tasks using a set of efficient, intuitive and consistent functions.

This book requires basic knowledge of **R** and the **tidyverse**, at the level of the first part of Wickham, Cetinkaya-Rundel, and Grolemund (2023), which is titled "the whole game" and is an excellent introduction for readers of this book that are not **R** users. Moreover, there is a free online version of this book that can be found at https://r4ds.hadley.nz/.

R packages

R has a modular structure. A particular analysis with **R** requires the use of functions that are part of different **packages**. To use these functions, the user can either "attach" the package (using the `library` function) or prefix the name of the function with the name of the package. For example, to use the `waldtest` function of the **lmtest** package, one can either use:

```
library(lmtest)
waldtest(---)
```

or:

```
lmtest::waldtest(---)
```

A relevant strategy is to attach only the packages that are often used in the analysis.

Different kinds of packages should be considered:

- the **base** packages contain the core of **R**. They are included in any distribution of **R** and they are automatically attached, which means that the user can have access directly to the functions they contain. The two most important are **base** and **stats**, which contain respectively the basic general and statistical functions of **R**.
- the **R** distribution also contains 15 **recommended** packages, which are not attached. Among them, we'll use in this book **survival** and **MASS**.
- **contributed** R packages. They are developed by **R** users and are hosted by the Comprehensive R Archive Network (**CRAN**). They can be very easily installed within the RStudio IDE (using the `Packages` tab on the right bottom panel of RStudio). Nowadays, there are more than 20,000 of them.

Twenty years ago, using **R** for econometrics analysis required a lot of programming because a lot of core methods of econometrics were not available neither in the core of **R**, nor in contributed packages. Nowadays, most of the basic methods are available in contributed packages, and the book uses extensively these packages. Some missing methods are implemented in the companion package of the book called **micsr**.

Data sets

This book uses a lot of data sets to illustrate the use of the techniques described in the book. A particular care was given to use original and interesting data sets. Our main source to get data was the Journal of Applied Econometrics data archive maintained by James G. MacKinnon since 1994 and available at http://qed.econ.queensu.ca/jae/[2] and the website of the American Economic Association https://www.aeaweb.org which provides data sets and programs used in articles published in AEA's journal, in particular the American Economic Review and the four American Economic Journals (we particularly used Applied Economics). We also wrote to a lot of authors to ask them whether the data used in their publications was still available and if they could share it with us. Many thanks for those who provided data sets that are now included in the **micsr** and the **micsr.data** packages: John Mullahy (`birthwt`, `cigmales`), Mark Ottoni Wilhelm (`charitable`), Lawrence Katz (`recall`), Joseph Terza (`trips`), Franck Koppelman (`toronto_montreal`), Christopher Skeels and David Drukker (`moffitt`), Winfried Pohlmeier (`doctor_ger`) and Rainer Winkelman (`jobmob`).[3]

The micsr package

The book comes with a companion package called **micsr**. It is available on **CRAN** and is therefore very easy to install. As indicated previously, the book uses extensively the **tidyverse** package and it is therefore recommended to install **tidyverse** before trying to replicate the **R** code contained in the book. **micsr** contains a small subset of the data sets included in the book and some specific methods of estimations and testing procedures for microeconometrics. Data sets used in this book not included in the **micsr** package are available in the **micsr.data** package which is available from **github**. To install it, you should first install the **pak** package and then simply write in the console:

```
pak::pkg_install("ycroissant/micsr.data")
```

Throughout the book, we assume that the reader that is willing to replicate the examples had previously attached the **tidyverse**, **micsr** and **micsr.data** packages, using:

```
library(tidyverse)
library(micsr)
library(micsr.data)
```

[2]This website is still available but is superseded since 2022 by a new website at https://journaldata.zbw.eu/journals/jae.

[3]Franck Vella, Adrian Pagan, Jerry Hausman, Claudia Goldin, Wynand van de Ven, Andrew Jones, Richard J. Smith, Pramila Krishnan, Orley C. Ashenfelter, John Ham, James Adams, John McDonald, Laurence Kotlikoff, Mark Rosenzweig, James Dunvely and Kathryn Graddy kindly answered me that they were unable to share the data I required (most of the time because it was no longer available to them).

micsr provides specific estimation and testing methods that will be presented in the core of the book. It also provides some general purpose functions. Among them, the gaze function will be extensively used throughout the book.

The output of **R** calls is often quite long and takes two much space in a book. For this reason, **micsr** contains a generic function called gaze with methods for a lot of **R** objects. Consider for example a t-test:

```
t.test(extra ~ 1, data = sleep)
```

```

	One Sample t-test

data:  extra
t = 3.413, df = 19, p-value = 0.002918
alternative hypothesis: true mean is not equal to 0
95 percent confidence interval:
 0.5955845 2.4844155
sample estimates:
mean of x 
     1.54 
```

The output is 12 lines long, including three blank lines. gaze writes the main part of the results in just one line:

```
t.test(extra ~ 1, data = sleep) %>% gaze
## t = 3.413, df: 19, pval = 0.003
```

Part I

Ordinary least squares estimator

The ordinary least squares (**OLS**) estimator is the most widely used estimator in econometrics and is often at least the good starting point before exploring more complex tools to perform **regression analysis**.

In a regression analysis, we consider a variable y, called the **response** or the endogenous or the explained variable and the model seeks to explain the variations of this variable from one observation to another. To achieve this goal, a set of other variables x called the **covariates** or explanatory or exogenous variables are introduced. These covariates are not only correlated to the response, but they are assumed to have a causal effect on the response. This means that a variation of x causes a change on the value of y, as the opposite is not true. For example, wages are correlated with education levels. More precisely, a variation of education has a causal effect on wage, which means that wage is the response and education is a covariate.

The dictionary definition of regression is "backward movement, a retreat, a return to an earlier state of development".[4] Regression analysis has nothing to do with this definition. The term was first used in the statistical literature by Francis Galton who studied the relationship between the height of children and the height of parents. His result is that tall parents have tall children, but that "there was a tendency for children's heights to converge toward the average". Galton called this a "regression of children's heights toward the average". Actually, this result is a statistical artifact, called a **regression fallacy**.

Chapter 1 is devoted to the derivation of the simple linear model, which means that we'll consider a unique covariate. Then, Chapter 2 will investigate the statistical properties of the simple linear model. Chapter 3 describes the multiple linear model, used when there is more than one covariate. Finally, Chapter 4 is devoted to the interpretation of the estimators.

Throughout this part, we'll present a real econometric analysis, which has three components:

- a **structural** model, which means that we are going to start from the rational behavior of an individual or of a set of individuals (for example a household that maximizes its utility given its budget constraint), and we'll deduce from this behavior a linear relationship between the response and the covariates, the parameters of this linear relationship being directly linked to the parameters of the structural model.
- a **data set** which is typically presented as a rectangular table for which every line is an observation and every column is a variable. We'll consider tables with one column that contains the response y and a set of K columns that contain the covariates x. We'll denote n the index of the observations, so that $n = 1$ for the first line of the table, $n = 2$ for the second one and $n = N$ for the last one. N is therefore also the number of observations.
- an **estimation method** that is required to compute the estimated values of the unknown parameters as a function of the values of the response and of the covariates on the sample and **tests** that are useful to answer questions like: is the true value of a parameter is equal to 0 or 1?

[4]This paragraph is based on Maddala and Lahiri (2009), pages 59-60 and 102-105.

1

Simple linear regression model

Obviously there will be rarely a deterministic relationship between a response and one or several covariates. Education has a positive causal effect on wage, but there are many other variables that affect wages: sex, ethnicity, experience, abilities, to name a few. Even if a large set of relevant covariates are observed, some are not (this is in particular the case for abilities). Therefore, we won't try to modelize the value of y as a function of covariates x and some unknown parameters γ: $y = f(x, \gamma)$ but, more modestly, the **conditional expectation** of y:

$$\mathrm{E}(y|x) = f(x, \gamma) \tag{1.1}$$

Back to our education / wage model, the conditional expectation is the mean value of the wage in the population for a given value of education. It is therefore a function of x which takes as many values as there are distinct values of x.

In this chapter, we discuss the simplest econometric model, which is the simple linear regression model. This is a **simple** model because there is only one covariate x. This is a **linear** model because the f function is assumed to be linear. Equation 1.1 can then be rewritten as follow:

$$\mathrm{E}(y|x) = \alpha + \beta x \tag{1.2}$$

Section 1.1 details the notion of conditional expectation and its relation with the notion of covariance. Section 1.2 presents the model and the data set that will be used throughout this chapter. Section 1.3 is devoted to the computation of the ordinary least squares estimator. Section 1.4 presents the geometry of least squares. Section 1.5 explains how to compute the ordinary least squares estimator using **R**. Finally Section 1.6 presents the notion of data generator process and explains how simulations can be usefully performed.

1.1 Conditional expectation and covariance

To estimate the two unknown parameters α (the **intercept**) and β (the **slope**), we need a data set that contains different individuals for which x and y are observed. Typically, we'll consider a random sample, which is a small subset of the whole population consisting of a set of individuals randomly drawn from this population.

1.1.1 Linear models: two examples

To understand what the hypothesis of the simple linear model implies, we consider two data sets. The first one is called `birthwt` (Mullahy 1997) and contains birth weights (`birthwt` in ounces) of babies and smoking habits of mothers. The `cigarettes` variable is the number of cigarettes smoked per day during pregnancy (for about 85% of the mothers, it is equal to 0). We first compute a binary variable for smoking mothers:

```
birthwt <- birthwt %>% mutate(smoke = ifelse(cigarettes > 0, 1, 0))
```

and consider a simple linear regression model with `birthwt` as the response and `smoke` as a covariate. As the covariate takes only two values, so does the conditional expectation. In a linear regression model, Equation 1.2 will therefore returns two values, α for $x = 0$ and $\alpha + \beta$ for $x = 1$. α and $\alpha + \beta$ are therefore the expected birth weights for respectively non-smoking and smoking mothers. Note that, because we have as many parameters as values of the covariate, the linear hypothesis is necessarily supported. Natural estimators of the conditional expectations are the conditional sample means, i.e., the average birth weights in the sample for non-smoking and smoking women:

```
cond_means <- birthwt %>%
    group_by(smoke) %>%
    summarise(birthwt = mean(birthwt))
cond_means
```

```
# A tibble: 2 x 2
  smoke birthwt
  <dbl>   <dbl>
1     0    120.
2     1    111.
```

Therefore, our estimation of α is $\hat{\alpha} = 120.1$ and the estimation of $\alpha + \beta$ is 111.1, so that $\hat{\beta} = 111.1 - 120.1 = -8.9$ ounces, which is the estimation of birth weight's loss caused by smoking during pregnancy. This can be illustrated using a scatterplot (see Figure 1.1), with `smoke` on the horizontal axis and `birthwt` on the vertical axis, all the points having only two possible abscissa values, 0 and 1. The conditional means are represented by square points, and we also drew the line that contains these two points. The intercept is therefore the y-value of the square point for $x = 0$ which is $\hat{\alpha} = 120.1$. The slope of the line is the ratio of the vertical and the horizontal distances between the two large points, which are respectively $\hat{\beta}$ and $1 - 0 = 1$ and is therefore equal to $\hat{\beta}$.

Consider now the relationship between education and wage. The `adoptees` data set (Plug 2004) contains the number of years of education `educ` and the annual income `income` (in thousands of US$) for 16481 individuals. We restrict the sample to individuals with an education level between 12 (high school degree) and 15 years (bachelor's degree).

```
adoptees <- adoptees %>%
    filter(educ >= 12, educ <= 15)
```

and we compute the mean income for the four values of education:

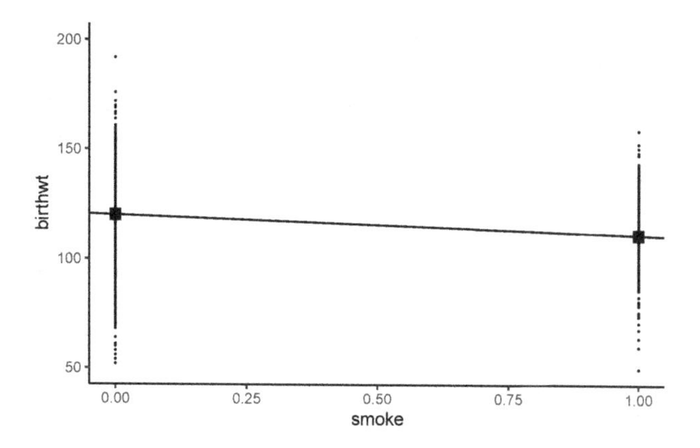

Figure 1.1: Birth weight for non-smoking and smoking mothers

```
adoptees %>% group_by(educ) %>%
    summarise(income = mean(income))
```

```
# A tibble: 4 x 2
   educ income
  <dbl>  <dbl>
1    12   58.9
2    13   68.8
3    14   70.5
4    15   78.4
```

The scatterplot is presented in Figure 1.2:

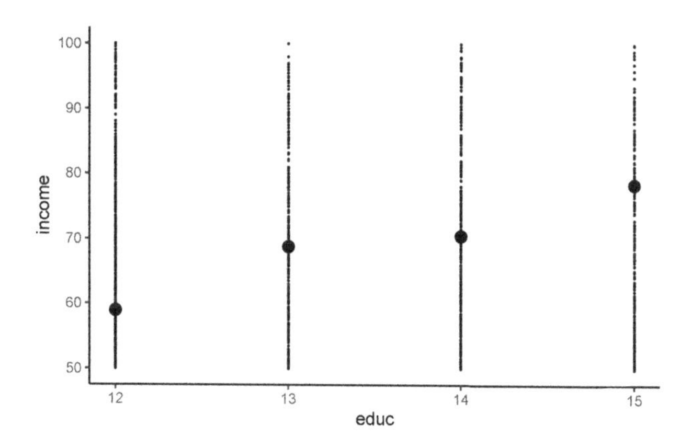

Figure 1.2: Education and income

This time, we have four distinct values of the covariate and we can estimate four conditional sample means, but there are only two parameters to estimate for the simple linear model. Therefore, we can't estimate directly α and β using the conditional means, except in the improbable case where the four conditional means lie on a straight line, which means that, for one additional year of education, the average wage increases exactly by the same amount.

We can see in Figure 1.2 that it is not the case: the increase of income is 9.9 for a 13th year of education, 1.7 for a 14th year and 8 for a 15th year. We therefore need a formal method of estimation which enables us to obtain values of $\hat{\alpha}$ and $\hat{\beta}$. The remaining of this chapter is devoted to the presentation of this estimator, which is called the **ordinary least squares** (**OLS**) estimator.

1.1.2 Linear model, conditional expectation and covariance

For now, it is very important to understand the relation between the conditional expectation and the covariance. The covariance between x and y is, denoting $\mu_x = \mathrm{E}(x)$ and $\mu_y = \mathrm{E}(y)$:

$$\sigma_{xy} = \mbox{cov}(x, y) = \mathrm{E}\left[(x - \mu_x)(y - \mu_y)\right] = \mathrm{E}(xy) - \mu_x\mu_y$$

It is the expectation of the product of the two variables in deviations from their respective expectations or the expectation of the product minus the product of the expectations. The expectation is for the two variables x and y and the law of repeated expectation states that this expectation can be written as:

$$\mbox{cov}(x, y) = \mathrm{E}_x\left[\mathrm{E}_y\left((y - \mu_y)(x - \mu_x) \mid x\right)\right] = \mathrm{E}_x\left[(x - \mu_x)\mathrm{E}_y(y - \mu_y \mid x)\right]$$

Therefore, the covariance between x and y is also the covariance between x and the conditional expectation of y (which is a function of x).

As an example, consider the case where y is the hourly wage and x the education level, which takes the values 0 (highschool), 3 (bachelor) and 5 (graduate), the frequencies for the three levels of education being 0.4, 0.3 and 0.3. Let denote h, b and g these three education levels and assume that $\mathrm{E}(y \mid x) = 8 + 0.5x$, which means that one more year of education increases the expected wage by \$0.5. The expected level of education is $\mu_x = 0.4 \times 0 + 0.3 \times 3 + 0.3 \times 5 = 2.4$. The expected wage for the three levels of education are $\mu_{xh} = 8$, $\mu_{xb} = 9.5$ and $\mu_{xg} = 10.5$. The expected wage is then $\mu_y = 0.4 \times 8 + 0.3 \times 9.5 + 0.3 \times 10.5 = 9.2$. We first compute the first term of the covariance between wage and education for $x = x_h = 0$:

$$\begin{array}{rcl} \mathrm{E}\left[(x - \mu_x)(y - \mu_y) \mid x = x_h\right] &=& \mathrm{E}\left[(x_h - \mu_x)(y - \mu_y) \mid x = x_h\right] \\ &=& (x_h - \mu_x)\mathrm{E}\left[(y - \mu_y) \mid x = x_h\right] \\ &=& (x_h - \mu_x)(\mathrm{E}(y \mid x = x_h) - \mu_y) \\ &=& (0 - 2.4) \times (8 - 9.2) = 2.88 \end{array} \quad (1.3)$$

Similarly, we get $(3 - 2.4) \times (9.5 - 9.2) = 0.18$ for the bachelor level and $(5 - 2.4) \times (10.5 - 9.2) = 3.38$ for the master level. Finally the covariance is obtained as the weight average of these three terms:

$$\sigma_{xy} = 0.4 \times 2.88 + 0.3 \times 0.18 + 0.3 \times 3.38 = 2.22$$

Now consider the case where the conditional expectation of the hourly wage is the same for the three education levels: $\mathrm{E}(y \mid x = x_h) = \mathrm{E}(y \mid x = x_b) = \mathrm{E}(y \mid x = x_m)$. These three conditional expectations are therefore equal to the unconditional expectation μ_y and Equation 1.3 is clearly 0 and so is the covariance. Therefore, a constant conditional expectation of y given x implies that the covariance is 0. Note that the opposite is not true. For example, if $\mu_{yh} = 8$, $\mu_{yb} = 13.2$ and $\mu_{ym} = 6.2$, the relation between the conditional mean of the wage and the level of education is no more linear, and it is not monotonous but inversely

U-shaped (the wage first increases with education and then decreases). The unconditional mean is still: $\mu_y = 0.4 \times 8 + 0.3 \times 13.2 + 0.3 \times 6.2 = 9.2$, but the covariance is then:

$$\sigma_{xy} = 0.4 \times (0-2.4) \times (8-9.2) + 0.3 \times (3-2.4) \times (13.2-9.2) + 0.3 \times (5-2.4) \times (6.2-9.2) = 0$$

Therefore, a 0 covariance doesn't imply that the conditional mean is constant. However, in the special case where the conditional mean is a linear function of the conditional variable, the two properties of 0 covariance and constant conditional means are equivalent.

Consider now the sample counterpart of this theoretical covariance. Consider a sample of 10 individuals for which the levels of education and the hourly wages are:

education	0	0	0	0	3	3	3	5	5	5
wage	8	7	8	9	9	9.5	10	10	10.5	11

The sample means of x and y are : $\bar{x} = 2.4$ and $\bar{y} = 9.2$. The conditional means of y for a given value of x in the sample are simply the means for the three subsamples defined by an education level:

- $x_h = 0 : \bar{y}_h = \frac{8+7+8+9}{4} = 8$,
- $x_b = 3 : \bar{y}_b = \frac{9+9.5+10}{3} = 9.5$,
- $x_m = 5 : \bar{y}_m = \frac{10,10.5,11}{3} = 10.5$,

and the covariance is obtained as the mean of the products of the two variables in deviations from their mean, or equivalently by the difference between the mean of the products and the products of the means:[1]

$$\hat{\sigma}_{xy} = \frac{\sum_{n=1}^N (x_n - \bar{x})(y_n - \bar{y})}{N} = \frac{\sum_{n=1}^N x_n y_n}{N} - \bar{x}\bar{y}$$

The sum of the products is $\sum_n x_n y_n = 243$ and the covariance is then:

$$\hat{\sigma}_{xy} = \frac{243}{10} - 2.4 \times 9.2 = 2.22$$

We also can consider the three subsamples defined by education levels, denoting $n = 1, ..., H$, $n = H+1, ..., H+B$, $n = H+B+1, ..., N$ (with $H = 4$, $B = 3$ and $N = 10$) observations for, respectively, a high-school, a bachelor and a master level. We then have:

$$\hat{\sigma}_{xy} = \frac{\sum_{n=1}^H (x_n - \bar{x})(y_n - \bar{y}) + \sum_{n=H+1}^{H+B} (x_n - \bar{x})(y_n - \bar{y}) + \sum_{n=H+B+1}^{N} (x_n - \bar{x})(y_n - \bar{y})}{N}$$

For example, for $n = 1, ..., H$, the value of x is the same: $x_n = x_h = 0$. Therefore, we can write $\sum_{n=1}^H (x_n - \bar{x})(y_n - \bar{y}) = (x_h - \bar{x}) \sum_{n=1}^H (y_n - \bar{y})$ and more generally:

[1] Note that we use a biased estimator of the covariance as the denominator is N and not $N-1$. The difference is negligible if N is large.

$$\hat{\sigma}_{xy} = \frac{(x_h - \bar{x})\sum_{n=1}^{H}(y_n - \bar{y}) + (x_b - \bar{x})\sum_{n=H+1}^{B}(y_n - \bar{y}) + (x_m - \bar{x})\sum_{n=H+B+1}^{N}(y_n - \bar{y})}{N}$$

Moreover $\frac{\sum_{n=1}^{H}(y_n - \bar{y})}{H} = (\bar{y}_h - \bar{y})$ and then:

$$\hat{\sigma}_{xy} = \frac{(x_h - \bar{x})H(\bar{y}_h - \bar{y}) + (x_b - \bar{x})B(\bar{y}_b - \bar{y}) + (x_m - \bar{x})M(\bar{y}_m - \bar{y})}{N}$$

Denoting $f_h = \frac{H}{N}$, $f_b = \frac{B}{N}$ and $f_m = \frac{M}{N}$ the empirical frequencies of the three education levels in the sample:

$$\hat{\sigma}_{xy} = f_h(x_h - \bar{x})(\bar{y}_h - \bar{y}) + f_b(x_b - \bar{x})(\bar{y}_b - \bar{y}) + (x_m - \bar{x})(\bar{y}_m - \bar{y})$$

Finally, denoting $k = h, b, m$ the education levels:

$$\hat{\sigma}_{xy} = \hat{\sigma}_{x\bar{y}_x} = \sum_k f_k(x_k - \bar{x})(\bar{y}_k - \bar{y})$$

Therefore, the covariance between x and y is also the covariance between x and the mean values of y for given values of x. We have here: $\bar{y}_h = 8$, $\bar{y}_b = 9.5$ and $\bar{y}_m = 10.5$. The covariance is then:

$$\hat{\sigma}_{xy} = 0.4\times(0-2.4)\times(8-9.2)+0.3\times(3-2.4)\times(9.5-9.2)+0.3\times(5-2.4)\times(10.5-9.2) = 2.22$$

Consider now the case where the mean wage is the same for every education level: $\bar{y}_b = \bar{y}_l = \bar{y}_m = \bar{y} = 9.2$. In this case, the covariance is obviously 0 because $\bar{y}_k - \bar{y} = 0\ \forall k$.

1.2 Model and data set

We'll consider in this chapter the question of mode shares for inter-urban transportation. More precisely, considering that a trip can be made using one out of two transport modes (air and rail), how can we modelize the market shares of both modes? We'll use in this section a popular model in transportation economics which is the price-time model. `price_time` contains aggregate data about rail and air transportation between Paris and 13 French towns in 1995, it is reproduced from Bonnel (2004) pp. 364-366.

```
price_time %>% print(n = 3)
```

```
# A tibble: 13 x 7
  town  trafic_rail trafic_air price_rail price_air time_rail time_air
  <chr>       <dbl>      <dbl>      <dbl>     <dbl>     <dbl>    <dbl>
1 Bord~        2005       1400       48.3      82.6       242      165
```

```
2 Brest        471      428      51.6      98.6      308      170
3 Cler~        429      243      32.7      89.4      266      140
# i 10 more rows
```

For the sake of simplicity, we'll use shorter names for the variables:

```
prtime <- price_time %>%
    set_names(c("town", "qr", "qa", "pr", "pa", "tr", "ta"))
```

Variables are prices (`pr` and `pa`) in euros, transport times (`tr` and `ta`) in minutes and thousands of trips (`qf` and `qa`) for the two modes (`r` for rail and `a` for air). We first compute the market shares of rail:

```
prtime <- mutate(prtime, sr = qr / (qr + qa))
prtime %>% pull(sr) %>% summary
##    Min. 1st Qu.  Median    Mean 3rd Qu.    Max.
##   0.136   0.339   0.524   0.555   0.868   0.943
```

Rail's market share exhibits huge variations in the sample, ranging from 14 to 94%. For an individual, the relevant cost of a trip is the generalized cost, which is the sum of the monetary cost and the value of the travel time. Denoting h^i the time value of individual i, in euros per hour, the generalized cost for the two modes are:

$$\left\{ \begin{array}{rcl} c_a^i &=& p_a + h^i t_a/60 \\ c_r^i &=& p_r + h^i t_r/60 \end{array} \right. \tag{1.4}$$

Plane is typically faster and more expensive than train, which means that in the time-value / generalized cost plane, the generalized cost for rail will be represented by a line with a lower intercept (the price of train is lower) and with a higher slope (transport time is higher) than the one that corresponds to air. Generalized cost for both modes and for the two towns of Bordeaux and Nice are presented in Figure 1.3.

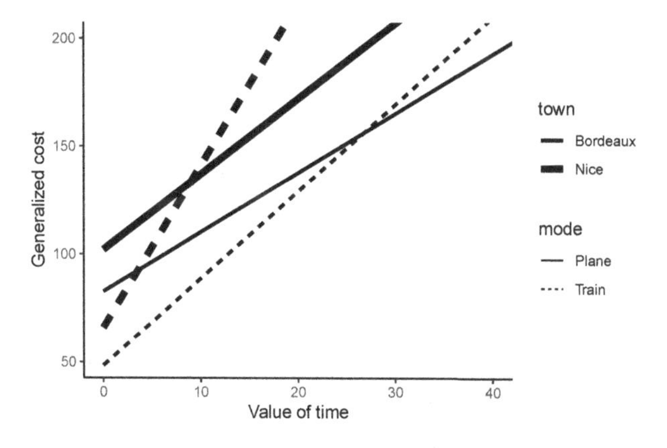

Figure 1.3: Generalized cost for train and plane

Every individual will choose the mode with the lowest generalized cost. For example, i will choose the train if $c_r^i < c_a^i$. This will depend on the individual value of time: an individual with a high value of time will choose the plane as an individual with a lower travel time will choose the train. For given values of prices and travel times, one can compute a value of travel time h^* which equates the generalized costs of the two modes in Equation 1.4:

$$h^* = 60\frac{p_a - p_r}{(t_r - t_a)}$$

For Bordeaux and Nice, these time values are respectively 26.7 and 9 euros per hour. Nice is actually very far from Paris and only people with a very low value of time would spend 7.5 hours in the train instead of taking a plane. We now compute this threshold value for every city:

```
prtime <- mutate(prtime,  h = (pa - pr) / ( (tr - ta) / 60) )
prtime %>% pull(h) %>% summary
##     Min. 1st Qu.  Median    Mean 3rd Qu.    Max.
##      9.0    15.9    20.4    59.0    67.2   314.3
```

There are huge variations of the threshold value of time, as it ranges from 9 (Nice) to 314 (Nantes) euros per hour. Before considering a theoretical model that links the market share of train with the threshold value of time, let's have a first glance at Figure 1.4 of this relationship using a scatterplot.

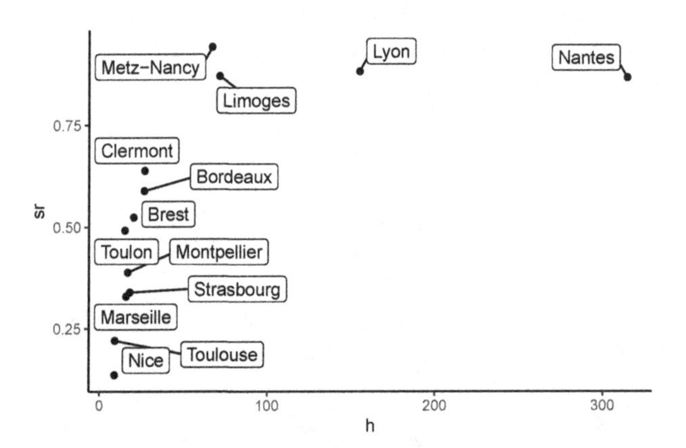

Figure 1.4: Share of rail in function of the threshold time value

The relationship between the threshold value of time and rail's market share seems approximately linear, except for cities where the market share of train is very high (more than 75%). For now, we'll remove these four cities from the sample and plot on Figure 1.5 the scatterplot for this restricted sample.[2]

```
prtime <- filter(prtime, sr < 0.75)
```

Now, we consider the distribution of the values of time. If h follows a given distribution between a and b, train's market share is the share of the population for which the value of time is between a and h^* (and plane's market share is the share of the population for which the value of time is between h^* and b). The simplest probability distribution is the

[2]This is obviously a very raw method to remove the non-linearity. Real time-price models use a much relevant solution, i.e., a transformation of the two variables in order to deal with the non-linearity.

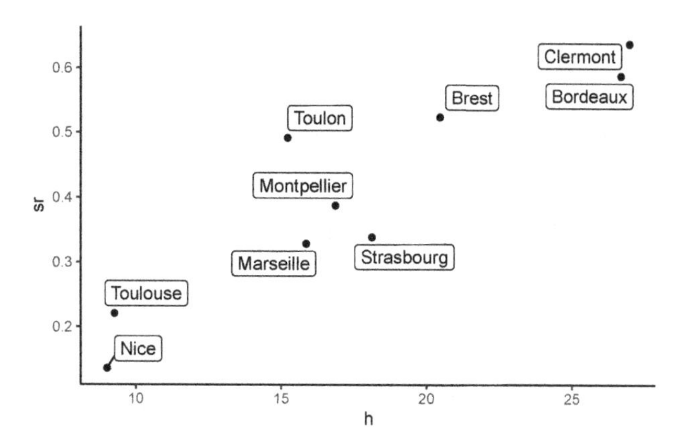

Figure 1.5: Share of rail in function of the threshold time value on a subsample

uniform distribution, which is defined by a constant density equal to $\frac{1}{b-a}$ between a and b. It is represented in Figure 1.6. The area of the rectangle of width $[a, b]$ and height $[0, \frac{1}{b-a}]$ is 100%, because the whole population has a time value between a and b. This rectangle has two components:

- a first rectangle of width $[a, h^*]$ which includes people for which time value is below h^* and therefore take the train,
- a second rectangle of width $[h^*, b]$ which includes people for which time value is higher than h^* and therefore take the plane.

Stated differently: $s_f = \frac{h^*-a}{b-a} = -\frac{a}{b-a} + \frac{1}{b-a}h^*$ and this model therefore predicts a linear relationship between h^* and s_f, the intercept being $-\frac{a}{b-a}$ and the slope $\frac{1}{b-a}$. Of course, rail's market share depends on other variables than the threshold value of time, so that the linear relationship concerns the conditional expectation of rail's market share. With $y = s_f$ and $x = h^*$, we therefore have a linear model of the form: $\mathrm{E}(y|x) = \alpha + \beta x$. Moreover, the two parameters to be estimated α and β are functions of the structural parameters of the model a and b which are the minimum and the maximum of the value of time.

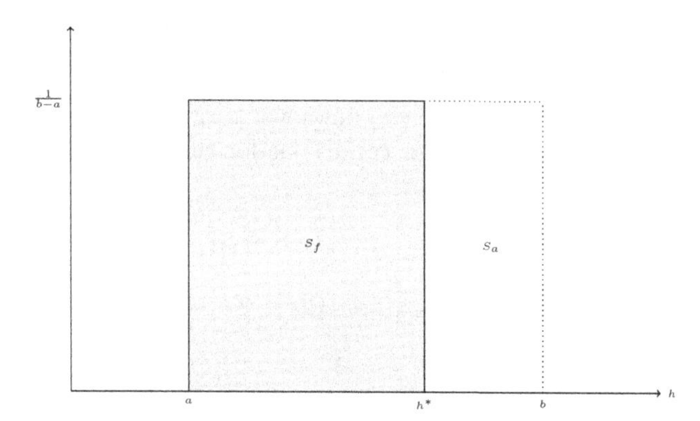

Figure 1.6: Model shares with a uniform distribution

1.3 Computation of the OLS estimator

The model we seek to estimate is: $\mathrm{E}(y_n \mid x_n) = \alpha + \beta x_n$. The difference between the observed value of y and its conditional expectation is called the **error** for observation n:

$$y_n - \mathrm{E}(y_n \mid x_n) = \epsilon_n$$

The linear regression model can therefore be rewritten as:

$$y_n = \mathrm{E}(y_n \mid x_n) + \epsilon_n = \alpha + \beta x_n + \epsilon_n$$

ϵ_n is the error for observation n when the values of the unknown parameters (α, β) are set to their true values (α_o, β_o). For given values of (α, β), obtained using an estimation method, ϵ_n will be called the **residual** for observation n. The residual for an observation is therefore the vertical distance between the point for observation n and the regression line.

We seek to draw a straight line that is as closest as possible to all the points of our sample as in Figure 1.5. In the simple linear regression model, the distance between a point and the line is defined by the vertical distance, which is the residual for this observation. For the whole sample, we need to aggregate this individual measures of distance. Summing them is not an issue, as there are positive and negative values of the residuals and the sum may be very close to zero even if the individual residuals are very high in absolute values. One solution would be to use the sum of the absolute values of the residuals,[3] but with the OLS estimator, we'll consider the sum of the squares of the residuals (also called the residual sum of squares (**RSS**). Taking the squares, as taking the absolute values, removes the sign of the individual residuals and it results in an estimator which has nice mathematical and statistical properties. We'll therefore consider a function f which depends on the value of the response and the covariate in the sample (two vectors x and y of length N) and on two unknown parameters α and β, respectively the intercept and the slope of the regression line:

$$f(\alpha, \beta | x, y) = \sum_{n=1}^{N} (y_n - \alpha - \beta x_n)^2$$

Note that we write f as a function of the two unknown parameters conditional on the values of x and y for a given sample. First-order conditions for the minimization of f are:

$$\left\{ \begin{array}{rcl} \frac{\partial f}{\partial \alpha} &=& -2\sum_{n=1}^N (y_n - \alpha - \beta x_n) = 0 \\ \frac{\partial f}{\partial \beta} &=& -2\sum_{n=1}^N x_n (y_n - \alpha - \beta x_n) = 0 \end{array} \right. \quad (1.5)$$

Or, dividing by -2:

$$\sum_{n=1}^{N} (y_n - \alpha - \beta x_n) = \sum_{n=1}^{N} \epsilon_n = 0 \quad (1.6)$$

[3]This estimator is called the least absolute deviations estimator.

$$\sum_{n=1}^{N} x_n (y_n - \alpha - \beta x_n) = \sum_{n=1}^{N} x_n \epsilon_n = 0 \tag{1.7}$$

Equation 1.6 indicates that the sum (or the mean) of the residuals in the sample is 0. Dividing this expression by N also implies that, denoting $\bar{y}$ and $\bar{x}$ the sample means of the response and of the covariate:

$$\bar{y} = \alpha + \beta\bar{x}, \tag{1.8}$$

which means that the sample mean is on the regression line. Denoting $\hat{\epsilon}_n$ the residuals of the OLS estimator, Equation 1.7 states that $\sum_n x_n \hat{\epsilon}/N = 0$, i.e., that the average cross-product of the covariate and the residual is 0. But, as the sample mean of the residuals $\bar{\hat{\epsilon}}$ is 0, this expression is also the covariance between the covariate and the residuals:

$$\hat{\sigma}_{x\hat{\epsilon}} = \frac{\sum_{n=1}^{N}(x_n - \bar{x})(\hat{\epsilon}_n - \bar{\hat{\epsilon}})}{N} = \frac{\sum_{n=1}^{N} x_n \hat{\epsilon}_n}{N} - \bar{x}\bar{\hat{\epsilon}} = \frac{\sum_{n=1}^{N} x_n \hat{\epsilon}_n}{N} = 0$$

which means that the regression line is such that there is no correlation between the covariate and the residuals in the sample. Subtracting $\bar{y} - \alpha - \beta\bar{x}$, which is 0 (see Equation 1.8) from Equation 1.7, one gets:

$$\sum_{n=1}^{N} x_n \left[(y_n - \bar{y}) - \beta (x_n - \bar{x})\right] = 0$$

Moreover, $\sum_{n=1}^{N} \bar{x} \left[(y_n - \bar{y}) - \beta (x_n - \bar{x})\right] = 0$ and so:

$$\sum_{n=1}^{N} (x_n - \bar{x}) \left[(y_n - \bar{y}) - \beta (x_n - \bar{x})\right] = 0 \tag{1.9}$$

Solving for β, we finally get the estimator of the slope:

$$\hat{\beta} = \frac{\sum_{n=1}^{N}(x_n - \bar{x})(y_n - \bar{y})}{\sum_{n=1}^{N}(x_n - \bar{x})^2} = \frac{S_{xy}}{S_{xx}} = \frac{\hat{\sigma}_{xy}}{\hat{\sigma}_x^2} = \hat{\rho}_{xy}\frac{\hat{\sigma}_y}{\hat{\sigma}_x} \tag{1.10}$$

Equation 1.10 gives three formulations for this estimator:

- the first indicates that it is the ratio of the covariation of x and y: $S_{xy} = \sum_{n=1}^{N}(x_n - \bar{x})(y_n - \bar{y})$ and the variation of x: $S_{xx} = \sum_{n=1}^{N}(x_n - \bar{x})^2$,
- the second is obtained by dividing both sides of the ratio by the sample size, so that the estimator is now the ratio of the sample covariance between x and y and the sample variance of x,
- the third is obtained by introducing the coefficient of correlation between x and y: $\hat{\rho}_{xy} = \frac{\hat{\sigma}_{xy}}{\hat{\sigma}_x \hat{\sigma}_y}$, so that the estimator is also expressed as the product of the coefficient of correlation and the ratio of the standard deviations.

This last formulation is particularly intuitive: $\hat{\rho}_{xy}$ is a pure measure of the correlation between the covariate and the response. This number has no unit and lies in the $-1/+1$ interval, a value of -1 $(+1)$ indicating a perfect negative (positive) correlation and the value of 0 no correlation. The ratio of the standard deviations gives the relevant unit to the slope, which is the unit of y divided by the unit of x. With this estimator of the slope in hand, we easily get the estimator of the intercept using Equation 1.8:

$$\hat{\alpha} = \bar{y} - \hat{\beta}\bar{x} \tag{1.11}$$

Consider now that prior to the estimation, we sustracted from the covariate and the response their sample means. We then used $\tilde{y}_n = y_n - \bar{y}$ and $\tilde{x}_n = x_n - \bar{x}$ as the response and the covariate. In this case, as the mean of these two transformed variables are zero, the intercept is 0 (from Equation 1.11) and the slope is simply $\sum \tilde{y}_n \tilde{x}_n / \sum \tilde{x}_n^2$.

Once the parameters of the regression line have been computed, one can define the **fitted value** for observation n as the value returned by the regression line for x_n, which is:

$$\hat{y}_n = \hat{\alpha} + \hat{\beta} x_n$$

By definition, we have for the fitted model: $y_n = \hat{y}_n + \hat{\epsilon}_n$. Note that, for the "true" model, we have $y_n = \mathrm{E}(y|x = x_n) + \epsilon_n$, so that the residuals are an estimation of the errors and the fitted values $\hat{y}_n$ are an estimation of the conditional expectations of y. Moreover, denoting $\bar{\hat{y}}$ the sample mean of the fitted values:

$$\frac{\sum_n (\hat{y}_n - \bar{\hat{y}})(\hat{\epsilon}_n - \bar{\hat{\epsilon}})}{N} = \frac{\sum_n \hat{y}_n \hat{\epsilon}_n}{N} - \bar{\hat{y}}\bar{\hat{\epsilon}} = \frac{\sum_n \hat{y}_n \hat{\epsilon}_n}{N} = \frac{\sum_n (\hat{\alpha} + \hat{\beta} x_n)\hat{\epsilon}_n}{N} = 0$$

Therefore, there is no correlation between $\hat{y}$ and $\hat{\epsilon}$, which is clear as $\hat{y}$ is a linear function of x and x is uncorrelated with $\hat{\epsilon}$. $(\hat{\alpha}, \hat{\beta})$ is an optimum of the objective function. To check that this optimum is a minimum, we have to compute the second derivatives. From Equation 1.5, we get: $\frac{\partial^2 f}{\partial \alpha^2} = 2N$, $\frac{\partial^2 f}{\partial \beta^2} = 2\sum_n x_n^2$ and $\frac{\partial^2 f}{\partial \alpha \partial \beta} = 2\sum_n x_n$. We need, for a maximum, positive direct second derivatives, which is obviously the case, and also a positive determinant of the matrix of second derivatives:

$$D = \frac{\partial^2 f}{\partial \alpha^2}\frac{\partial^2 f}{\partial \beta^2} - \left(\frac{\partial^2 f}{\partial \alpha \partial \beta}\right)^2 = 4N\sum_n x_n^2 - 4\left(\sum_n x_n\right)^2 = 4N^2\left(\frac{\sum_n x_n^2}{N} - \bar{x}^2\right) > 0$$

which is the case as the term in brackets is the variance of x and is therefore positive.

1.4 Geometry of least squares, variance decomposition and coefficient of determination

The OLS estimator relies on variances and covariances of several observable (y and x) and computed ($\hat{y}$, $\hat{\epsilon}$) variables. Its properties can be nicely illustrated using vector algebra, each variable being represented by a vector, and by plotting these vectors.[4]

1.4.1 Vectors, variance and covariance

Every variable z used in a regression is a vector of $\mathcal{R}_N$, i.e., a set of N real values: $z^\top = (z_1, z_2, \ldots, z_N)$. The length (or norm) of the vector is: $\|z\| = \sqrt{\sum_{n=1}^N z_n^2}$. Remember that the OLS estimator can always be computed with data measured in deviations from their sample mean. Then, $\|z\|^2/N$ is the variance of the variable, or the norm of the vector is $\sqrt{N}$ times the standard deviation of the corresponding variable. The inner (or scalar) product of two vectors is denoted by $z^\top w = w^\top z = \sum_{n=1}^N z_n w_n$ (note that the inner product is commutative). For corresponding variables expressed in deviations from their respective means it is, up to the $1/N$ factor, the covariance between the two variables. Denoting θ the angle formed by the two vectors, we also have:[5] $z^\top w = \cos\theta \|z\| \|w\|$.

Consider as an example: $x = (4, 3)$, $z = (4.5, 6)$ and $w = (-6, 4.5)$. The three vectors are plotted in Figure 1.7. The norm of x is $\|x\| = \sqrt{4^2 + 3^2} = 5$. Similarly, $\|z\| = 7.5$ and $\|w\| = 7.5$. The cosinus of the angle formed by x and z with the horizontal axis is $\cos\theta_x = 4/5 = 0.8$ and $\cos\theta_z = 4.5/7.5 = 0.6$. The angle formed by x and z is therefore: $\theta = \arccos 0.6 - \arccos 0.8 = 0.284$, with $\cos 0.284 = 0.96$. We can then check that $z^\top x = 4 \times 4.5 + 3 \times 6 = 36$, which is equal to: $\cos\theta\|z\|\|w\| = 0.96 \times 7.5 \times 5$. As the absolute value of $\cos\theta$ is necessarily lower than or equal to 1, the inner product of two vectors is lower than the product of the norms of the two vectors, and $\cos\theta = \frac{x^\top z}{\|x\|\|z\|}$ is the ratio of the covariance between x and z and the product of their standard deviations, which is the coefficient of correlation between the two underlying variables x and z. Consider now z and w. Their inner product is: $z^\top x = 4.5 \times -6 + 6 \times 4.5 = 0$. This is because z and w are two orthogonal vectors, which means that the two underlying variables are uncorrelated.

1.4.2 Geometry of least squares

The geometry of the simple linear regression model is represented in Figure 1.8. With $N = 2$, x and y are two vectors in a plane. For the "true" model, the y vector is the sum of two vectors: βx (which is the conditional expectation of y) and ϵ, which is the vector of errors. For the estimated model, y is the sum of the fitted values $\hat{y} = \hat{\beta}x$ and the residuals $\hat{\epsilon}$. Using the OLS estimator, we must minimize the sum of squares of the residuals, i.e., the norm of the $\hat{\epsilon}$ vector. Obviously, this implies that $\hat{\epsilon}$ should be orthogonal to $\hat{y}$ and therefore to x, which implies that the residuals are uncorrelated to the fitted values (and to the covariate) in the sample. Note also that, except in the unlikely case where $\hat{\beta} = \beta$, $\|\hat{\epsilon}\| < \|\epsilon\|$ which means that the residuals have a smaller variance than the errors. Note finally that what

[4]For a detailed presentation of the geometry of least squares, see Davidson and MacKinnon (2004), chapter 2.

[5]See Davidson and MacKinnon (2004), p. 48.

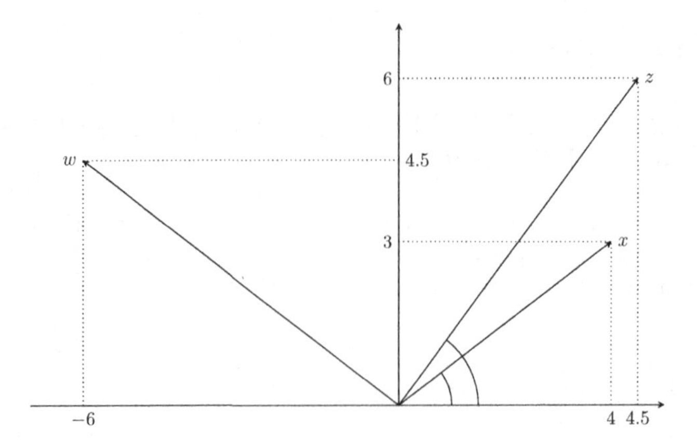

Figure 1.7: Vector algebra

determines $\hat{y}$ and $\hat{\epsilon}$ is not x per se, but the subspace defined by it, in our case the horizontal straight line. For example, consider the regression of y on $z = 0.5x$; we would then obtain exactly the same values for $\hat{y}$ and $\hat{\epsilon}$, the only difference being that the estimator of β would be two times larger.

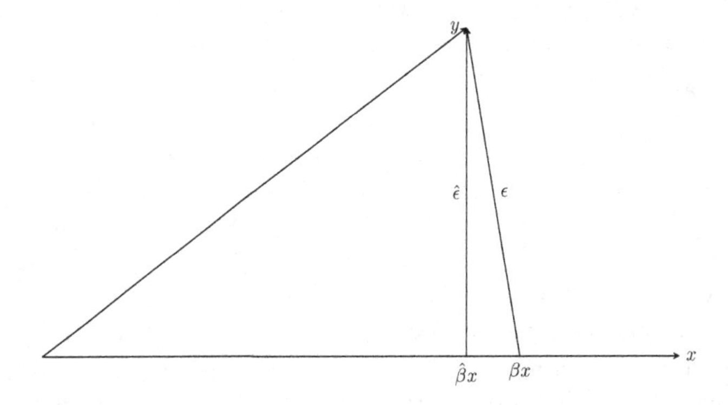

Figure 1.8: Geometry of the simple regression model

1.4.3 Variance decomposition and the R^2

For one observation n, we have:

$$y_n - \bar{y} = (y_n - \hat{y}_n) + (\hat{y}_n - \bar{y}) \tag{1.12}$$

The difference between y for individual n and the sample mean is therefore the sum of:

- a residual variation: $(y_n - \hat{y}_n) = \hat{\epsilon}_n$,
- an explained variation: $\hat{y}_n - \bar{y} = \hat{\beta}(x_n - \bar{x})$.

Taking the square of Equation 1.12 and summing for all n, we get:

$$\begin{array}{rcl}\sum_{n=1}^N (y_n - \bar{y})^2 &=& \sum_{n=1}^N \left(\hat{\epsilon}_n + \hat{\beta}(x_n - \bar{x})\right)^2 \\ &=& \sum_{n=1}^N \hat{\epsilon}_n^2 + \hat{\beta}^2 \sum_{n=1}^N (x_n - \bar{x})^2 \\ &+& 2\hat{\beta}\sum_{n=1}^N \hat{\epsilon}_n x_n - 2\hat{\beta}\bar{x}\sum_{n=1}^N \hat{\epsilon}_n\end{array}$$

But $\sum_{n=1}^N \hat{\epsilon}_n x_n = 0$ (Equation 1.7) and $\sum_{n=1}^N \hat{\epsilon}_n = 0$ (Equation 1.6), so that:

$$\sum_{n=1}^N (y_n - \bar{y})^2 = \hat{\beta}^2 \sum_{n=1}^N (x_n - \bar{x})^2 + \sum_{n=1}^N \hat{\epsilon}_n^2 \quad (1.13)$$

This equation indicates that the total sum of squares (TSS) of the response equals the sum of the explained sum of squares (ESS) and of the residual sum of squares (RSS). This latter term is also called the **deviance**, and it is the objective function for the OLS estimator. Dividing by N, we also have on the left-hand size the variance of y and on the right-hand side the sum of the variances of $\hat{y}$ and $\hat{\epsilon}$. This is the formula of the **variance decomposition** of the response. It can be easily understood using Figure 1.8. It is clear from this figure that $y = \hat{y} + \hat{\epsilon}$ and that $\hat{y}$ and $\hat{\epsilon}$ are orthogonal. Therefore, applying the Pythagorean theorem, we have $\|y\|^2 = \|\hat{y}\|^2 + \|\hat{\epsilon}\|^2$. Up to the $1/N$ factor, if the response is measured in deviation from its sample mean, we have on the left the total variance of y and on the right the sum of the variances of $\hat{y}$ and $\hat{\epsilon}$. Note that it is a unique feature of the ordinary least squares estimator. Any other estimator will generally result with a vector of residuals which won't be orthogonal to $\hat{y}$ (or to x) and therefore Equation 1.13 won't apply.

The coefficient of determination, denoted by R^2, measures the share of the variance of the response which is explained by the model, or one minus the share of the residual variation. We then have:

$$R^2 = \frac{\hat{\beta}^2 \sum_{n=1}^N (x_n - \bar{x})^2}{\sum_{n=1}^N (y_n - \bar{y})^2} = 1 - \frac{\sum_{n=1}^N \hat{\epsilon}_n^2}{\sum_{n=1}^N (y_n - \bar{y})^2} \quad (1.14)$$

using Equation 1.10, we finally get:

$$R^2 = \left[\frac{\sum_{n=1}^N (x_n - \bar{x})(y_n - \bar{y})}{\sum_{n=1}^N (x_n - \bar{x})^2}\right]^2 \frac{\sum_{n=1}^N (x_n - \bar{x})^2}{\sum_{n=1}^N (y_n - \bar{y})^2} = \left[\frac{\hat{\sigma}_{xy}}{\hat{\sigma}_x^2}\right]^2 \left[\frac{\hat{\sigma}_x^2}{\hat{\sigma}_y^2}\right] = \frac{\hat{\sigma}_{xy}^2}{\hat{\sigma}_x^2 \hat{\sigma}_y^2} = \hat{\rho}_{xy}^2$$

R^2 is therefore simply the square of the coefficient of correlation between x and y. We have seen previously that, denoting θ the angle formed by two vectors, $\cos\theta$ is the coefficient of correlation between the two underlying variables (if they are measured in deviations from their means). Therefore, in Figure 1.8, R^2 is represented by the square of the cosine of the angle formed by the two vectors $\hat{y}$ and y. As this angle tends to 0 (the two vectors point almost in the same direction), R^2 tends to 1. On the contrary, if this angle tends to $\pi/2$, the two vectors become almost orthogonal, and R^2 tends to 0.

1.5 Computation with R

The slope of the regression line can easily be computed "by hand", using any of the formula indicated in Equation 1.10, using the `dplyr::summarise` function. We first compute the variations of x and y and the covariation of x and y, the two standard deviations and the coefficient of correlation between x and y.

```
stats <- prtime %>%
    summarise(N = nrow(prtime), xb = mean(h), yb = mean(sr),
              Sxx = sum( (h - xb) ^ 2), Syy = sum( (sr - yb) ^ 2),
              Sxy = sum( (h - xb) * (sr - yb)),
              sx = sqrt(Sxx / N), sy = sqrt(Syy / N),
              sxy = Sxy / N, rxy = sxy / (sx * sy))
```

We can then compute the estimator of the slope using any of the three formulas:

```
hbeta <- stats$Sxy / stats$Sxx
hbeta
## [1] 0.02393
stats$sxy / stats$sx ^ 2
## [1] 0.02393
stats$rxy * stats$sy / stats$sx
## [1] 0.02393
```

The estimation of the intercept is obtained using Equation 1.11:

```
halpha <- stats$yb - hbeta * stats$xb
halpha
## [1] -0.01548
```

Much more simply, the OLS estimator can be obtained using the `lm` function (for linear model). It is a very important function in **R**, not only because it implements efficiently the most important estimator used in econometrics, but also because **R** functions that implement other estimators often mimic the features of the `lm` function. Therefore, once one is at ease with using the `lm` function, using other estimating function of **R** will be straightforward. `lm` is a function that has many arguments, but the first two are fundamental and almost mandatory:[6]

- `formula` is a symbolic description of the model to be estimated,
- `data` is a data frame that contains the variables used in the formula.

Here, our formula writes `sr ~ h`, which means `sr` as a function of `h`. The data frame is `prtime`. The result of the `lm` function may be directly printed (the result is then lost), or saved in an object, which can be later printed or manipulated:

[6] Actually, the second argument `data` is not mandatory because estimation can be performed without using a data frame.

```
lm(sr ~ h, prtime)
```

```
Call:
lm(formula = sr ~ h, data = prtime)

Coefficients:
(Intercept)            h
    -0.0155       0.0239
```

```
pxt <- lm(sr ~ h, prtime)
```

writing directly `pxt` is like writing `print(pxt)`, the side effect is to print a short description of the results, namely a remind of the function call and the name and the values of the fitted coefficients. `lm` returns an object of class `lm` which is a list of 12 elements; their names can be retrieved using the `names` function:

```
names(pxt)
```

```
 [1] "coefficients"  "residuals"     "effects"       "rank"
 [5] "fitted.values" "assign"        "qr"            "df.residual"
 [9] "xlevels"       "call"          "terms"         "model"
```

An element of this list can be extracted using the `$` operator. For example:

```
pxt$coefficients
## (Intercept)           h
##    -0.01548     0.02393
```

returns the named vector of coefficients. `pxt$residuals` and `pxt$fitted.values` return two vectors of length N containing the residuals and the fitted values. However, it is not advised to use the `$` operator to retrieve the elements of a fitted model. Specific functions, called extractors, should be used instead. For example, to retrieve the coefficients, the residuals and the fitted values as previously, we would use:

```
coef(pxt)
resid(pxt)
fitted(pxt)
```

There are other functions that extract important information about the model, as the number of observation (`nobs`) and the sum of square residuals (`deviance`):

```
nobs(pxt)
## [1] 9
deviance(pxt)
## [1] 0.03606
```

The results of the estimation are presented in Figure 1.9; we've added to the scatterplot:

- the sample mean, indicated by a large circle,
- the regression line,
- the residuals, represented by arrows: upward arrows represent positive residuals (e.g., Brest and Toulon), and downward arrows represent negative residuals (e.g., Strasbourg and Nice).

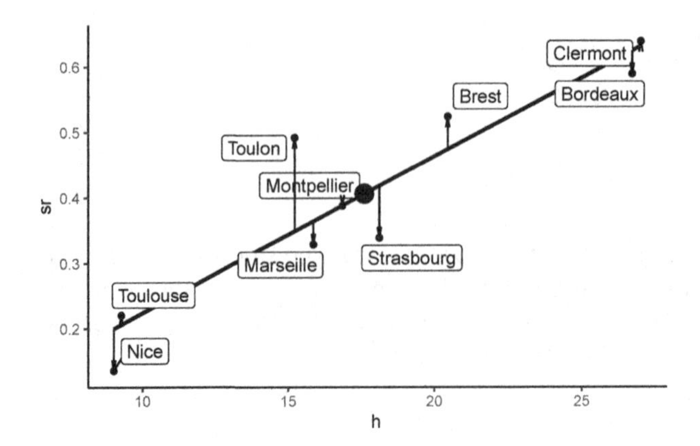

Figure 1.9: Regression line and residuals

Individual coefficients can be extracted using the `[` operator. As `coef` returns a named vector, one can either indicate the position or the name of the coefficient to be extracted:

```
int <- unname(coef(pxt)[1])
slope <- unname(coef(pxt)["h"])
c(int, slope)
## [1] -0.01548  0.02393
```

Note that we use the `unname` function in order to remove the name of the extracted coefficient. Once the intercept (`int`) and the slope (`slope`) are extracted, the structural parameters can be retrieved, as the intercept is $\alpha = -\frac{a}{b-a}$ and the slope $\beta = \frac{1}{b-a}$. Therefore, $a = -\frac{\alpha}{\beta}$ and $b = \frac{1}{\beta} + a$. We finally get:

```
ahat = - int / slope
bhat = 1 / slope + ahat
c(ahat, bhat)
## [1]  0.647 42.431
```

Time values therefore lie between 0.65 and 42.43 euros per hour. The mean (and median) time value is the mean of the two extreme values, which is 21.54 euros per hour.

1.6 Data generator process and simulations

Inferential statistics rely on the notions of population and sample. The population is a large and exhaustive set of observations, and a sample is a small subset of observations drawn in this population. These notions are relevant in medical sciences and often in economic studies. For example, the first data set we used concerned smoking habits and birth weight. The population of interest is all American pregnant women in 1988, and a sample of 1388 of them was drawn from this population. The second data set concerned the relation between education and wage. The population was American labor force in 1992, and a sample of 16,481 workers was drawn from this population. On the contrary, our third data set doesn't fit with these notions of population and sample. The sample consists of major towns in France that are connected on a regular basis by rail and air to Paris. It contains 13 observations, which are not 13 observations randomly drawn from a large set of cities, but which are more or less all the relevant cities.

1.6.1 Data generator process

An interesting alternative is the notion of **data generator process** (**DGP**). It describes how the data are assumed to have been generated. We assume in the linear regression model that:

- $\mathrm{E}(y|x = x_n) = \alpha + \beta x_n$: the expected value of y for a given value of x is a linear function of x,
- $y_n = \mathrm{E}(y|x = x_n) + \epsilon_n$: the observed value of y_n is obtained by adding to the conditional expectation of y for $x = x_n$ a random variable ϵ called the error.

From Equation 1.10, we can write the estimator of the slope as:

$$\hat{\beta} = \frac{\sum_{n=1}^N (x_n - \bar{x})(y_n - \bar{y})}{\sum_{n=1}^N (x_n - \bar{x})^2} = \sum_{n=1}^N \frac{(x_n - \bar{x})}{\sum_{n=1}^N (x_n - \bar{x})^2} y_n = \sum_n c_n y_n \tag{1.15}$$

with $c_n = \frac{x_n - \bar{x}}{\sum_{n=1}^N (x_n - \bar{x})^2}$. The OLS estimator is therefore a linear estimator, i.e., a linear combination of the values of y. The coefficients of this linear combination c_n are such that $\sum_{n=1}^N c_n = 0$ and that:

$$\sum_{n=1}^N c_n^2 = \frac{\sum_{n=1}^N (x_n - \bar{x})^2}{\left(\sum_{n=1}^N (x_n - \bar{x})^2\right)^2} = \frac{1}{\sum_{n=1}^N (x_n - \bar{x})^2} = \frac{1}{N\hat{\sigma}_x^2}$$

Replacing y_n by $\alpha + \beta x_n + \epsilon_n$, we then express $\hat{\beta}$ as a function of ϵ_n:

$$\hat{\beta} = \sum_{n=1}^N c_n(\alpha + \beta x_n + \epsilon_n) = \alpha \sum_{n=1}^N c_n + \beta \sum_{n=1}^N \frac{x_n(x_n - \bar{x})}{\sum_{n=1}^N (x_n - \bar{x})^2} + \sum_{n=1}^N c_n \epsilon_n$$

As $\sum_{n=1}^N x_n(x_n - \bar{x}) = \sum_{n=1}^N (x_n - \bar{x})^2$ and $\sum_{n=1}^N c_n = 0$, we finally get:

$$\hat{\beta} = \beta + \sum_{n=1}^N c_n \epsilon_n \tag{1.16}$$

The deviation of the estimator of the slope of the OLS regression line $\hat{\beta}$ from the true value β is therefore a linear combination of the N errors. Consider our sample used to estimate the price-time model. From a DGP perspective, this sample has been generated using the formula: $y = \alpha + \beta x + \epsilon$. Consider now that the "true" values of α and β are -0.2 and 0.032, we can in this case compute the vector of errors for our sample ($\epsilon = y - \alpha - \beta x$):

```
alpha <- - 0.2
beta <- 0.032
y <- prtime %>% pull(sr)
x <- prtime %>% pull(h)
eps = y - alpha - beta * x
```

We then compute the OLS estimator, and we retrieve $\hat{\epsilon}$ and $\hat{y}$.

```
z <- lm(y ~ x)
hat_eps <- z %>% residuals
hat_y <- z %>% fitted
```

The observed data set, in the DGP perspective, is represented in Figure 1.10.

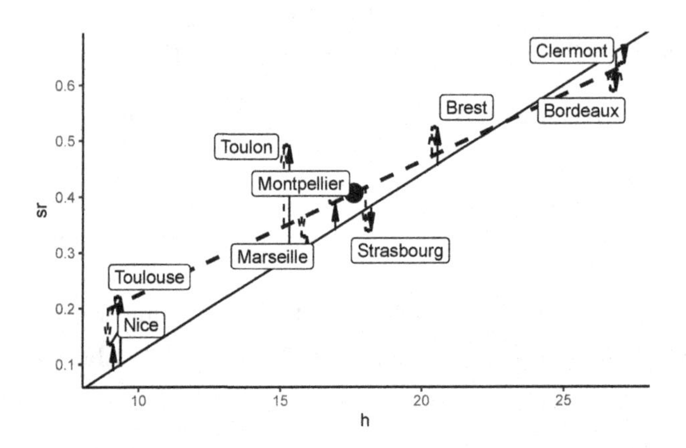

Figure 1.10: Data generating process

The "true" model is represented by the plain line. The errors are represented by the plain arrows (positive errors for upward arrows, negative errors for downward arrows). Each value of y_n is the sum of the conditional expectation of y for the value of x: $E(y|x = x_n) = \alpha + \beta x_n$ (the value returned by the plain line for the given value of x) and the error ϵ_n represented by the plain arrow. For our specific sample, we have a specific vector of ϵ_n, which means a specific set of points and a specific regression line, the dashed line on the figure. Each value of y_n is then also the sum of the fitted value (the one returned by the regression line for $x = x_n$) and the residual, represented by a dashed arrow.

We can check that the residuals sum to 0 and are uncorrelated with the covariate and the fitted values:

```
sum(hat_eps)
## [1] 1.735e-18
sum(hat_eps * x)
```

```
## [1] 3.556e-16
sum(hat_eps * hat_y)
## [1] 3.903e-18
```

which is not the case for the errors:

```
sum(eps)
## [1] 0.3818
sum(eps * x)
## [1] 4.036
sum(eps * hat_y)
## [1] 0.09069
```

We can also check that the residuals are "smaller" than the errors by computing their standard deviations:

```
sd(eps)
## [1] 0.08496
sd(hat_eps)
## [1] 0.06713
```

1.6.2 Random numbers and simulations

Now consider an other fictive sample for the same values of x, i.e., consider that the values of the threshold values of time are the same, but that the other factors that influence rail's shares (the ϵ) are different. To generate such a fictive sample, we need to generate the ϵ vector, using a function that generates random numbers.[7] With **R**, these functions have a name composed of the letter `r` (for random) and the abbreviated name of the statistical distribution: for example `runif` draws numbers in a uniform distribution (by default within a $0-1$ range) and `rnorm` draws numbers in a normal distribution (by default with zero expectation and unit standard deviation). These functions have a mandatory argument which is the number of draws. For example, to get five numbers drawn from a standard normal distribution:

```
rnorm(5)
## [1]  0.01875 -0.18425 -1.37133 -0.59917  0.29455
```

Using the same command once more, we get a completely different sequence:

```
rnorm(5)
## [1]  0.3898 -1.2081 -0.3637 -1.6267 -0.2565
```

As stated previously, the `rnorm` function doesn't draw random numbers but computes a sequence of numbers that looks like a random sequence. Imagine that `rnorm` actually computes a sequence of thousands of numbers, what is obtained using `rnorm(5)` is 5 consecutive numbers in this sequence, for example from the 5107th to the 5111th number. The position of the first element is called the **seed**, and it can be set to an integer using the `set.seed`

[7] More precisely, a function that generates a sequence of numbers that looks like random numbers.

function. Using the same seed while starting a simulation, we would then get exactly the same pseudo-random numbers each time and therefore the same results:

```
set.seed(7L)
rnorm(5)
## [1]  2.2872 -1.1968 -0.6943 -0.4123 -0.9707
rnorm(5)
## [1] -0.9473  0.7481 -0.1170  0.1527  2.1900
set.seed(7L)
rnorm(5)
## [1]  2.2872 -1.1968 -0.6943 -0.4123 -0.9707
rnorm(5)
## [1] -0.9473  0.7481 -0.1170  0.1527  2.1900
```

The DGP is completely described by specifying the distribution of ϵ. We'll consider here a normal distribution with mean 0 and standard deviation $\sigma_\epsilon = 0.08$. A pseudo-random sample can then be constructed as follow:

```
N <- prtime %>% nrow
x <- prtime %>% pull(h)
alpha <- - 0.2 ; beta <- 0.032 ; seps <- 0.08
eps <- rnorm(N, sd = seps)
y <- alpha + beta * x + eps
asmpl <- tibble(y, x)
```

which gives the following OLS estimates:

```
lm(y ~ x, asmpl) %>% coef
## (Intercept)           x
##     -0.25218     0.03841
```

The notion of DGP enables to perform simulations, that have two purposes:

- the first (and the most important in practice) is that, in some situations, it is impossible to get analytical results for the properties of an estimator, and those properties can in this case be obtained using simulations,
- the second is pedagogical, as theoretical results can be confirmed and illustrated using simulations.

A simulation consists of creating a large number of samples, computing some interesting numbers for every sample and calculating some relevant statistics for these numbers. For example, we'll compute the slope of the OLS estimate for every sample, and we'll calculate the mean and the standard deviation for this slope. A minimal simulation is presented in Figure 1.11, where we present four different samples obtained for four different sets of errors.

For every sample, there is a specific set of errors (arrows) and therefore specific data points and regression lines. The estimated intercepts range from -0.212 to -0.015 and the slopes from 0.024 to 0.032. The important point is that β is, in practice, an unknown fixed parameter. $\hat{\beta}$ depends on the N observations of the sample and therefore the N values of the errors.

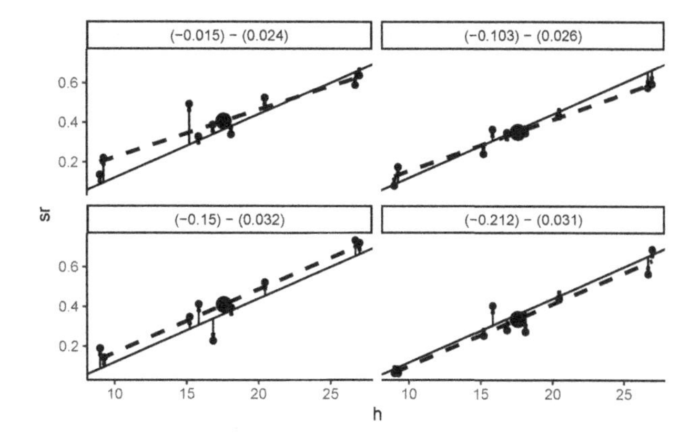

Figure 1.11: 4 different samples

Each sample is characterized by a specific vector of errors and therefore by a different value of the estimator.

More generally, we denote R the number of replications and we construct a data frame with $R \times N$ lines. The columns are the 13 values of x (fixed in this simulation), the vector ϵ drawn in a normal distribution with a standard deviation equal to 0.08 and $y = \alpha + \beta x + \epsilon$, with $\alpha = 0.2$ and $\beta = 0.032$ as previously. Finally we add a variable `id` that is a vector containing integers from 1 to R repeated 13 times that will be used to identify the sample:

```
alpha <- - 0.2 ; beta <- 0.032 ; seps <- 0.08
R <- 1E03 ; N <- length(x)
datas <- tibble(id = rep(1:R, each = N),
                x = rep(x, R),
                eps = rnorm(R * N, sd = seps),
                y = alpha + beta * x + eps)
```

Running regressions on every sample requires to use `lm(y ~ x)` not on the whole data set, but on every subset defined by a value of `id`. The `lm` function has a `subset` argument that must be a logical expression used to select only a subset of the data frame. For example, to get the fitted model for the first sample, the logical expression is `id == 1` and one can use:

```
lm(y ~ x, data = datas, subset = id == 1)
```

The slope can also be calculated using Equation 1.15, which shows that the OLS estimator is a linear combination of the values of the response:

```
datas %>% filter(id == 1) %>%
  mutate(cn = (x - mean(x)) / sum((x - mean(x)) ^ 2)) %>%
  summarise(slope = sum(cn * y)) %>%
  pull
## [1] 0.03287
```

To get values of the estimator for every sample, we have to loop on this command for `id` from 1 to `R`. A good practice in **R** is to avoid the use of explicit loops whenever it's possible.

The `group_by` / `summarise` couple of functions from the `dplyr` package can be used instead.

```
results <- datas  %>%
  group_by(id) %>%
  mutate(cn = (x - mean(x)) / sum((x - mean(x)) ^ 2)) %>%
  summarise(slope = sum(cn * y), intercept = mean(y) - slope * mean(x))
results
```

```
# A tibble: 1,000 x 3
     id  slope intercept
  <int>  <dbl>     <dbl>
1     1 0.0329    -0.166
2     2 0.0309    -0.181
3     3 0.0348    -0.213
4     4 0.0267    -0.129
5     5 0.0270    -0.127
# i 995 more rows
```

The result is now a tibble with `R` lines, as the computation of the OLS estimator is performed for every sample. As we now have a large number of values of $\hat{\beta}$, we can analyze its statistical properties, for example its mean and its standard deviation for the 1000 random samples:

```
results %>% summarise(mean = mean(slope), sd = sd(slope))
```

```
# A tibble: 1 x 2
    mean      sd
   <dbl>   <dbl>
1 0.0320 0.00437
```

We'll see in Chapter 2 that the expected value of $\hat{\beta}$ is $\beta = 0.032$ (we'll say that the OLS estimator is unbiased) and that its standard deviation is $\sigma_\epsilon/\sigma_x/\sqrt{N} = 0.0041$. The mean and standard deviations for our 1000 values of $\hat{\beta}$ are estimates of these two theoretical values. We can see that the mean of $\hat{\beta}$ is actually equal to 0.032 and that the standard deviation equals 0.00437, which is close to the theoretical value.

A more general method to perform simulation is to use list-columns of data frames. To conduct such an analysis step by step, we start with a very small example, with $R = 2$ and $N = 3$.

```
R <- 2
N <- 3
small_data <- tibble(id = rep(1:R, each = N),
                     x = rep(x[1:N], R),
                     eps = rnorm(R * N, sd = seps),
                     y = alpha + beta * x + eps)
small_data
```

```
# A tibble: 6 x 4
```

```
     id     x     eps     y
  <int> <dbl>   <dbl> <dbl>
1     1  26.7 -0.0624 0.593
2     1  20.5  0.0243 0.479
3     1  27.0  0.111  0.775
4     2  26.7 -0.0124 0.643
5     2  20.5 -0.0356 0.419
# i 1 more row
```

Starting with `small_data`, we use `tidy::nest` to nest the data frame using the `id` variable (`.by = id`); the name of the results is by default `data`, but it can be customized using the `.key` argument:

```
datalst <- small_data %>% nest(.by = id, .key = "smpl")
datalst
```

```
# A tibble: 2 x 2
     id smpl
  <int> <list>
1     1 <tibble [3 x 3]>
2     2 <tibble [3 x 3]>
```

Each line now corresponds to a value of `id` and `smpl` is a list column that contains a tibble,

```
datalst %>% pull(smpl)
```

```
[[1]]
# A tibble: 3 x 3
      x     eps     y
  <dbl>   <dbl> <dbl>
1  26.7 -0.0624 0.593
2  20.5  0.0243 0.479
3  27.0  0.111  0.775

[[2]]
# A tibble: 3 x 3
      x     eps     y
  <dbl>   <dbl> <dbl>
1  26.7 -0.0124 0.643
2  20.5 -0.0356 0.419
3  27.0  0.141  0.805
```

We now write a function that takes as argument a data frame containing the x and y columns and returns a one-line tibble that contains the fitted intercept and slope:

```
ols <- function(s){
  x <- s$x ; y <- s$y
  tibble(slope = sum((x - mean(x)) * (y - mean(y))) /
```

```
            sum( ( x - mean(x)) ^ 2),
          intercept = mean(y) - slope * mean(x))
}
```

This function can be applied for a given sample:

```
ols(small_data[1:N, ])
```

```
# A tibble: 1 x 2
   slope intercept
   <dbl>     <dbl>
1 0.0329    -0.198
```

But, as the `smpl` column of `datalst` is a list, the `ols` function can be applied to all the elements of the list, which can be done using the `purrr::map` function which takes as argument a list and a function:

```
datalst %>% pull(smpl) %>% map(ols)
```

```
[[1]]
# A tibble: 1 x 2
   slope intercept
   <dbl>     <dbl>
1 0.0329    -0.198

[[2]]
# A tibble: 1 x 2
   slope intercept
   <dbl>     <dbl>
1 0.0483    -0.573
```

and returns a list. But this can more easily be done using `mutate` or `transmute`, so that the result is a tibble:

```
results <- datalst %>% transmute(smpl = map(smpl, ols))
results
```

```
# A tibble: 2 x 1
  smpl
  <list>
1 <tibble [1 x 2]>
2 <tibble [1 x 2]>
```

Finally, `tidyr::unnest` expands the list column containing tibbles into rows and columns.

```
results %>% unnest(cols = smpl)
```

```
# A tibble: 2 x 2
   slope intercept
   <dbl>     <dbl>
1 0.0329    -0.198
2 0.0483    -0.573
```

which returns a standard tibble with one row for each draw and one column for each computed statistic. Going back to the full example with $R = 1000$, we get:

```
datas %>% nest(.by = id, .key = "smpl") %>%
  transmute(smpl = map(smpl, ols)) %>%
  unnest(cols = smpl) %>%
  summarise(mean = mean(slope), sd = sd(slope))
```

```
# A tibble: 1 x 2
    mean      sd
   <dbl>   <dbl>
1 0.0320 0.00437
```

2

Statistical properties of the simple linear estimator

To analyze the statistical properties of the OLS estimator, we use Equation 1.16 that indicates that the difference between the estimated slope and the true value is a linear combination of the errors:

$$\hat{\beta} = \beta + \sum_{n=1}^{N} c_n \epsilon_n, \text{ with } c_n = (x_n - \bar{x})/S_{xx} \tag{2.1}$$

The properties of $\hat{\beta}$ are therefore directly deduced from those of ϵ. We'll consider two sets of properties:

- **exact** properties that apply whatever the size of the sample is (Section 2.1),
- **asymptotic** properties that indicate approximate results, the approximation being better and better as the sample size grows (Section 2.2).

Section 2.3 explains how these properties can be used to construct confidence intervals and tests.

2.1 Exact properties of the OLS estimator

The OLS estimator is a random variable, for which we observe one value, obtained with a given sample. The exact properties of the OLS estimator concern:

- its expected value: if the true value is β_o, what is the expected value of $\hat{\beta}$, β_o or another value?
- its variance (or standard deviation): is the variance small (the estimation is precise) or large?

The computation of the expected value indicates the presence or absence of a **bias**. Therefore, we check here whether there is a systematic error (called the bias) while performing the estimation. The variance indicates the **efficiency** (or the precision) of the estimator. It measures the amount of the **sampling error**, i.e., the average distance between the value of the estimator and its expected value.

To analyze the properties of the OLS estimator, we'll make different hypotheses and we'll see that if these hypotheses are satisfied, the OLS is the best (most efficient) linear unbiased estimator. To illustrate the results of this chapter, we'll use the price-time model, with the same GDP as previously: $\alpha = -0.2$, $\beta = 0.032$ and $\sigma_\epsilon = 0.08$ and we'll consider different departures from this reference case.

2.1.1 Errors have 0 expected value

The reference model being $y_n = \alpha + \beta x_n + \epsilon_n$ with $\mathrm{E}(\epsilon_n) = 0$, consider the alternative model: $y_n = \gamma + \beta x_n + \eta_n$, for which the slope is the same and the error term is η_n, with $\mathrm{E}(\eta) = \mu_\eta \neq 0$. We have therefore: $y_n = \alpha + \beta x_n + (\eta_n + \gamma - \alpha)$ or: $\eta_n = \epsilon_n + \alpha - \gamma$ and finally: $\mathrm{E}(\eta) = \mu_\eta = \mathrm{E}(\epsilon) + \alpha - \gamma = \alpha - \gamma$.

Therefore, the alternative model is: $y_n = \gamma + \mu_\gamma + \beta x_n + \epsilon$, which is the same model as the initial model with α replaced by $\gamma + \mu_\gamma$. Therefore, it is impossible to discriminate between the initial and the alternative model, as what can be estimated is the sum of the intercept and the expected value of the errors $(\gamma + \mu_\gamma)$ and the two elements of this sum can't be estimated separately. This is an illustration of a very important problem in econometrics called **identification** that we'll encounter in subsequent chapters. We can say here that γ and μ_γ are not identified, but that their sum is. Therefore, we can set one of the two parameters to any value. For example, we can simply set $\mu_\gamma = 0$, i.e., suppose that the expected value of the errors is 0 and the other parameter γ became identified, i.e., it can be estimated using the data.

Figure 2.1 illustrates the "reference" model (plain line and ϵ_n represented by plain vectors) and the alternative model (dashed line and η_n represented by dashed vectors).

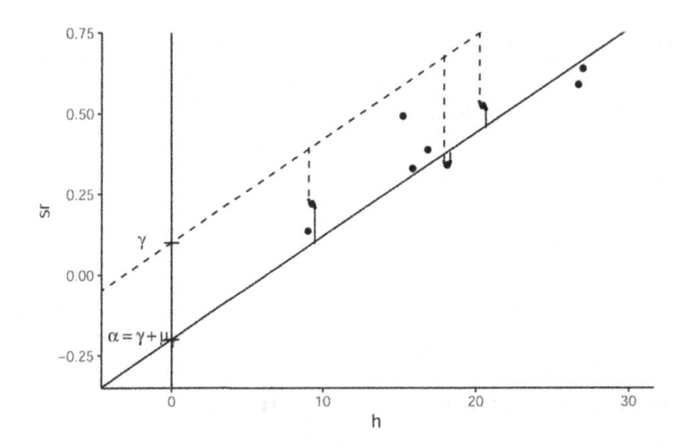

Figure 2.1: Intercept and the expected value of the error

2.1.2 Conditional expectation of the errors is 0

As we have seen, the hypothesis that $\mathrm{E}(\epsilon) = 0$ can always be stated if the model contains an intercept. On the contrary, the hypothesis that the expected value of ϵ conditional on x is 0 ($\mathrm{E}(\epsilon|x) = 0$) is much more problematic and the violation of this hypothesis has dramatic consequences for the OLS estimator. It is important to understand that this condition actually implies that there is no correlation between the error and the covariate. Starting with the expression of the covariance between the error and the covariate: $\mathrm{cov}(x, \epsilon) = \mathrm{E}\left((x - \mu_x)(\epsilon - \mu_\epsilon)\right)$, with μ_x and μ_ϵ the expected values of x and ϵ, we can rewrite this covariance using conditional expectation, using the rule of repeated expectations:

$$
\begin{array}{rcl}
\mathrm{cov}(x, \epsilon) & = & \mathrm{E}_x\left[\mathrm{E}\left((x - \mu_x)(\epsilon - \mu_\epsilon)|x\right)\right] \\
& = & \mathrm{E}_x\left[(x - \mu_x)\left(\mathrm{E}(\epsilon - \mu_\epsilon)|x\right)\right] \\
& = & \mathrm{E}_x\left[(x - \mu_x)\mathrm{E}(\epsilon|x)\right]
\end{array}
$$

The covariance between x and ϵ is therefore equal to the covariance between x and the conditional expectation of ϵ. If $\mathrm{E}(\epsilon|x)$ is a constant (equal to μ_ϵ, the unconditional expectation), the covariance is:

$$\mathrm{cov}(x, \epsilon) = \mathrm{E}\left[(x - \mu_x)\mu_\epsilon\right] = \mu_\epsilon \mathrm{E}\left[x - \mu_x\right] = 0$$

Therefore, a constant conditional expectation of ϵ (not necessarily 0 but we have seen previously than we can safely suppose that it is 0) implies that the covariance between the errors and the covariate is 0 or, stated differently, that the errors are uncorrelated with the covariate. From Equation 2.1, the conditional expectation of the estimator is:

$$\mathrm{E}(\hat{\beta} \mid x) = \beta + \sum_{n=1}^{N} \mathrm{E}(c_n \epsilon_n \mid x_n) = \beta + \sum_{n=1}^{N} c_n \mathrm{E}(\epsilon_n \mid x_n) \qquad (2.2)$$

If the conditional expectation of the errors is constant ($\mathrm{E}(\epsilon_n \mid x_n) = \mu_\epsilon$), $\sum_{n=1}^N c_n \mathrm{E}(\epsilon_n|x) = \mu_\epsilon \sum_{n=1}^N c_n = 0$ as $\sum_n c_n = 0$, so that $\mathrm{E}(\hat{\beta}|x) = \beta$, which means that the expected value of the estimator is the true value. In this case, the estimator is **unbiased**. Therefore, the hypothesis of constant conditional expectation of the errors is crucial.

It is very important to understand why, in practice, this hypothesis may be violated. As an illustration, consider the wage / education model. It is well documented in a lot of countries that, for a given value of education, women earn less than men on average. This means that the conditional expectation of wage is lower for women or, graphically, that in a scatterplot, points for women will be in general below the line which indicates the conditional expectation of wages and that points for men will be above this line. To see whether this can induce a bias in the OLS estimator, rewrite Equation 2.2 as:

$$\mathrm{E}(\hat{\beta} \mid x) = \beta + \frac{\sum_{n=1}^N (x_n - \bar{x})\mathrm{E}(\epsilon_n|x_n)}{\sum_{n=1}^N (x_n - \bar{x})^2}$$

The second term is the ratio of the covariance between ϵ and x and the variance of x. If not zero, this ratio is the bias of the OLS estimator. In our example, the key question is whether being a male is correlated with education:

- with no correlation, the OLS estimator is unbiased,
- with a negative correlation, the OLS estimator is downward biased,
- with a positive correlation, the OLS estimator is upward biased.

These three situations are depicted in Figure 2.2. The common feature of the three rows of the figure is that women (depicted by circles) are in general below the expected value line (plain line) and men (depicted by triangles) are above. For row 1 of the figure, considering the horizontal position of the points, women and men are approximately uniformly disposed, which indicates the absence of correlation between education and being a male. The regression (dashed) line is then very close to the expected value line, the OLS estimator is unbiased. For row 2 of the figure, males have in general a lower level of education than females. There is therefore a negative correlation between education and being a man and the consequence is that the OLS estimator is downward biased. Finally, for row 3 of the figure, the correlation is positive and the OLS estimator is upward biased.

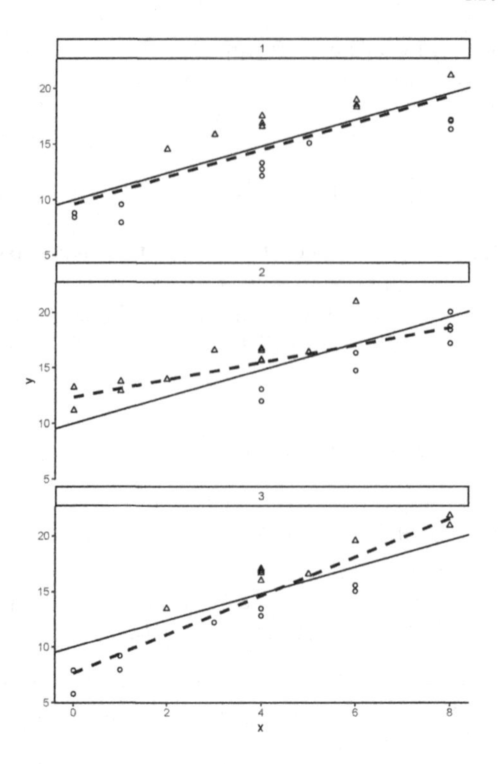

Figure 2.2: Education, sex and wage

Consider the latter case in details. Women have a lower wage than men for two reasons: they are less educated and, for a given value of education, they receive a lower wage than males. Increasing the education level from, say, 4 to 5 years will have two effects on the expected wage:

- direct positive effect of education on wage,
- indirect positive effect: as being a man is positively correlated with education, considering a higher level of education, we'll get a subpopulation with a higher share of males, and therefore a higher wage.

The OLS estimator estimates the sum of these two effects and is therefore in this case upward biased.

2.1.3 Estimator for the variance of the OLS estimator

Consider now the conditional variance of the OLS estimator:

$$\begin{aligned}
\mathrm{V}(\hat{\beta} \mid x) &= \mathrm{E}\left(\left(\hat{\beta} - \beta\right)^2 \mid x\right) \\
&= \mathrm{E}\left(\left(\sum_{n=1}^N c_n \epsilon_n\right)^2 \mid x\right) \\
&= \frac{1}{S_{xx}^2} \mathrm{E}\left(\left(\sum_{n=1}^N (x_n - \bar{x})\epsilon_n\right)^2 \mid x\right)
\end{aligned}$$

To compute the variance, we therefore have to take the expected value of N^2 terms, N of them being of the form: $(x_n - \bar{x})^2\epsilon_n^2$ and the $N^2 - N$ other of the form: $(x_n - \bar{x})(x_m - \bar{x})\epsilon_n\epsilon_m$. This is best understood by arranging the N^2 terms in a square matrix of dimension N. With $N = 4$, we have[1]:

$$\left(\begin{array}{cccc}
(x_1-\bar{x})^2\mathrm{E}(\epsilon_1^2) & (x_1-\bar{x})(x_2-\bar{x})\mathrm{E}(\epsilon_1\epsilon_2) & (x_1-\bar{x})(x_3-\bar{x})\mathrm{E}(\epsilon_1\epsilon_3) & (x_1-\bar{x})(x_4-\bar{x})\mathrm{E}(\epsilon_1\epsilon_4) \\
(x_2-\bar{x})(x_1-\bar{x})\mathrm{E}(\epsilon_2\epsilon_1) & (x_2-\bar{x})^2\mathrm{E}(\epsilon_2^2) & (x_2-\bar{x})(x_3-\bar{x})\mathrm{E}(\epsilon_2\epsilon_3) & (x_2-\bar{x})(x_4-\bar{x})\mathrm{E}(\epsilon_2\epsilon_4) \\
(x_3-\bar{x})(x_1-\bar{x})\mathrm{E}(\epsilon_3\epsilon_1) & (x_3-\bar{x})(x_2-\bar{x})\mathrm{E}(\epsilon_3\epsilon_2) & (x_3-\bar{x})^2\mathrm{E}(\epsilon_3^2) & (x_3-\bar{x})(x_4-\bar{x})\mathrm{E}(\epsilon_3\epsilon_4) \\
(x_4-\bar{x})(x_1-\bar{x})\mathrm{E}(\epsilon_4\epsilon_1) & (x_4-\bar{x})(x_2-\bar{x})\mathrm{E}(\epsilon_4\epsilon_2) & (x_4-\bar{x})(x_3-\bar{x})\mathrm{E}(\epsilon_4\epsilon_3) & (x_4-\bar{x})^2\mathrm{E}(\epsilon_4^2)
\end{array}\right)$$

2.1.3.1 Uncorrelation and homoskedasticity

$\mathrm{V}(\hat{\beta} \mid x)$ is obtained by taking the sum of these N^2 terms, N terms depending on conditional variances ($\mathrm{E}(\epsilon_n^2 \mid x_n) = \mathrm{V}(\epsilon_n \mid x_n)$) and $N \times (N-1)$ on conditional covariances ($\mathrm{E}(\epsilon_n\epsilon_m \mid x_n, x_m) = \mathrm{cov}(\epsilon_n, \epsilon_m \mid x_n, x_m)$). The resulting estimator has a very simple form if two hypothesis are made:

- the errors are **homoskedastic**, which means that their variances don't depend on x, the N terms that contain the conditional variances are then equal to $(x_n - \bar{x})^2\sigma_\epsilon^2$,
- the errors are **uncorrelated**, the $N \times (N-1)$ terms that involve the covariance are then equal to 0.

With these two hypotheses in hand, only the diagonal terms are not zero and their sum is $\sigma_\epsilon^2 \sum_{n=1}^N (x_n - \bar{x})^2 = \sigma_\epsilon^2 S_{xx}$, which finally leads to the simplified formula of the variance of $\hat{\beta}$:

$$\mathrm{V}(\hat{\beta} \mid x) = \sigma_{\hat{\beta}}^2 = \frac{\sigma_\epsilon^2 S_{xx}}{S_{xx}^2} = \frac{\sigma_\epsilon^2}{S_{xx}} = \frac{\sigma_\epsilon^2}{N\hat{\sigma}_x^2} \tag{2.3}$$

Note that this is the "true" variance of $\hat{\beta}$ if the two hypotheses are satisfied, and that it can't be computed as it depends on the unknown parameter σ_ϵ. The square root of Equation 2.3 is the standard deviation of β, is measured in the same unit as β and is commonly called the **standard error** of $\hat{\beta}$. It is therefore a convenient indicator of the precision of the estimator:

$$\sigma_{\hat{\beta}} = \frac{\sigma_\epsilon}{\sqrt{N}\hat{\sigma}_x} \tag{2.4}$$

It is clear from Equation 2.4 that the precision of the estimator depends on three components which will be described in details in the next subsection.

2.1.3.2 Determinants of the precision of the OLS estimator

First consider the "size" of the error, measured by its standard deviation σ_ϵ. Figure 2.3 presents a scatterplot for six samples which use the same DGP, except that samples on the second line are generated with a smaller value of σ_ϵ. The "true model" ($\alpha + \beta x$) is represented by a plain line and the regression line is dashed. Obviously, the estimation is much more precise on the second line of Figure 2.3, because of small-sized errors.

[1]We don't explicitly indicate that the expected values are conditional on x to save space.

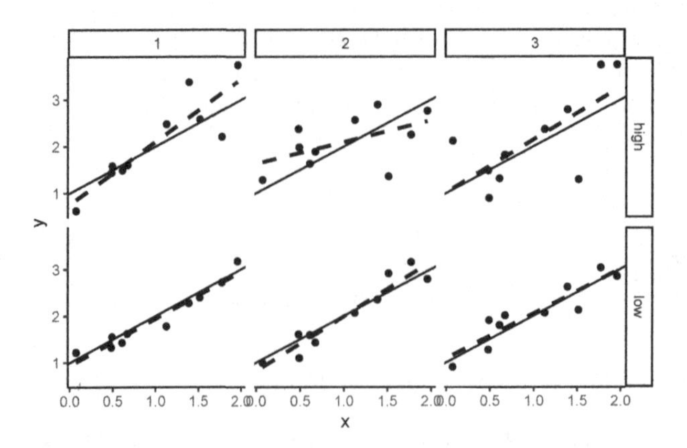

Figure 2.3: Size of the error and the precision of the slope estimator

Next consider the sample size. In Figure 2.4, we take the same value of σ_ϵ as in row 1 of Figure 2.3, but we increase the sample size to 40 for the samples of the second line. In large samples (second line in Figure 2.4) the slope is very precisely estimated, which means that the value of $\hat{\beta}$ is almost the same from a sample to another. On the contrary, with a small sample size (first line in Figure 2.4), the slopes of the regression lines are very different for the three samples, which indicates that the standard error of $\hat{\beta}$ is high (or that the estimator is imprecise).

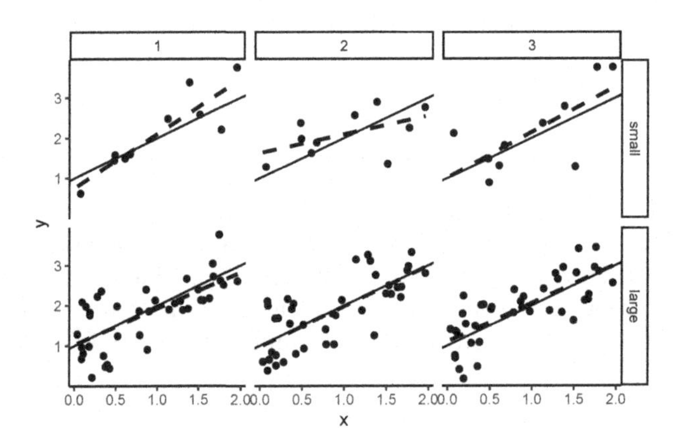

Figure 2.4: Sample size and precision of the estimator

Finally, in Figure 2.5, we consider a variation of the variance of x. For samples on the second line, the variance of x is much smaller than for samples on the first line. The larger the variance of x is, the more precise is the estimator of the slope. Obviously, it is difficult to estimate the effect of education on wage if all the individuals in the sample have almost the same level of education. Consider the extreme case of no variation of x in a sample; in this case it is impossible to estimate the effect of x, as all the observations are characterized by the same value of x.

All these results can be illustrated by simulations, using the `price_time` data set. For convenience, we replicate in the following code the operations we performed on this data set in Chapter 1:

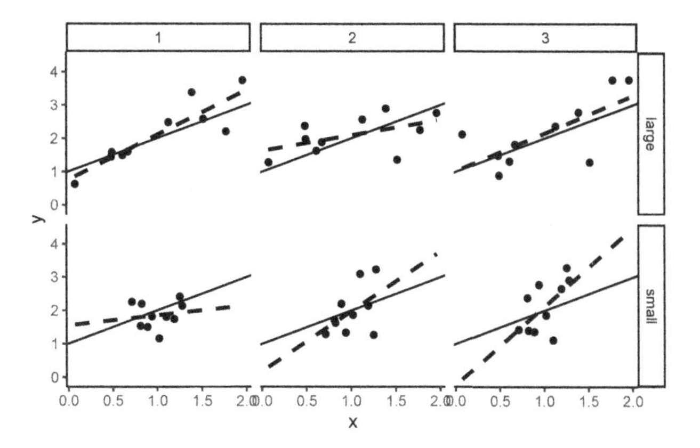

Figure 2.5: Variance of x and precision of the estimator

```
prtime <- price_time %>%
  set_names(c("town", "qr", "qa", "pr", "pa", "tr", "ta")) %>%
  mutate( sr = qr / (qr + qa),
          h = (pa - pr) / ( (tr - ta) / 60)) %>%
  filter(sr < 0.75)
```

We start with our reference case ($\sigma_\epsilon = 0.08$, $N = 9$ and x is the vector of the threshold value of time for the nine selected cities of the `prtime` data set, with a sample standard deviation $\hat{\sigma}_x = 6.085$).

```
alpha <- - 0.2 ; beta <- 0.032
seps <- 0.08
x <- pull(prtime, h)
N <- length(x)
```

We generate $R = 100$ samples using each time the same vector of covariate (x) and drawing the errors in a normal distribution:

```
R <- 100
dataref <- tibble(smpls = rep(1:R, each = N),
                  x     = rep(x, R),
                  eps   = rnorm(R * N, sd = seps),
                  y     = alpha + beta * x + eps)
```

To illustrate the influence of σ_ϵ on the precision of the estimator, we take a value of $\sigma_\epsilon = 0.04$, i.e., we divide σ_ϵ by 2 compared to the reference case:

```
dataseps <- tibble(smpls = rep(1:R, each = N),
                   x     = rep(x, R),
                   eps   = rnorm(R * N, sd = seps / 2),
                   y     = alpha + beta * x + eps)
```

Next, we increase the sample size to $N = 36$, i.e., we multiply the sample size by 4. More specifically, for every sample each value of x is repeated four times:

```
N <- length(x) * 4
xN <- rep(x, 4)
datasN <- tibble(smpls = rep(1:R, each = N),
                 x     = rep(xN, R),
                 eps   = rnorm(R * N, sd = seps),
                 y     = alpha + beta * x + eps)
```

Finally, we increase the variation of x, simply by multiplying all the values by 2. In this case, the standard deviation of x is also multiplied by 2.

```
N <- length(x)
xv <- x * 2
datasvx <- tibble(smpls = rep(1:R, each = N),
                  x     = rep(xv, R),
                  eps   = rnorm(R * N, sd= seps),
                  y     = alpha + beta * x + eps)
```

The standard deviation for the reference case is:

$$\hat{\sigma}_{\hat{\beta}} = \frac{\sigma_\epsilon}{\sqrt{N}\hat{\sigma}_x} = \frac{0.08}{\sqrt{9} \times 6.085} = 0.0044$$

which is very close to the standard deviation of $\hat{\beta}$ for our $R = 100$ samples:

```
dataref %>% group_by(smpls) %>%
  summarise(slope = sum( (x - mean(x)) * (y - mean(y)) ) /
              sum( (x - mean(x)) ^ 2)) %>%
    summarise(mean = mean(slope), sd = sd(slope))
```

```
# A tibble: 1 x 2
    mean      sd
   <dbl>   <dbl>
1 0.0316 0.00420
```

Note that we used twice the `summarise` function. The first time, it is used with `group_by` so that a tibble with `R` lines is returned, containing the values of the estimator `slope`. The second time, a one-line tibble is returned containing the mean and the standard deviation of the `R` values of the estimator.

When σ_ϵ is divided by 2 (from 0.08 to 0.04), the standard deviation of $\hat{\beta}$ should also be divided by 2 (from 0.00438 to 0.00219), which is approximately the value of the standard deviation of the values of $\hat{\beta}$ for our $R = 100$ samples:

```
dataseps %>% group_by(smpls) %>%
  summarise(hbeta = sum( (x - mean(x)) * (y - mean(y)) ) /
              sum( (x - mean(x)) ^ 2)) %>%
  summarise(mean = mean(hbeta), sd = sd(hbeta))
```

```
# A tibble: 1 x 2
    mean      sd
   <dbl>   <dbl>
1 0.0318 0.00222
```

When the sample size is multiplied by 4, $\hat{\sigma}_\beta$ should also be divided by 2:

```
datasN %>% group_by(smpls) %>%
  summarise(hbeta = sum( (x - mean(x)) * (y - mean(y)) ) /
              sum( (x - mean(x)) ^ 2)) %>%
  summarise(mean = mean(hbeta), sd = sd(hbeta))
```

```
# A tibble: 1 x 2
    mean      sd
   <dbl>   <dbl>
1 0.0323 0.00191
```

Finally, when every value of x is multiplied by 2, x's standard deviation is also multiplied by 2 and $\hat{\sigma}_\beta$ should be divided by 2:

```
datasvx %>% group_by(smpls) %>%
  summarise(hbeta = sum( (x - mean(x)) * (y - mean(y)) ) /
              sum( (x - mean(x)) ^ 2)) %>%
  summarise(mean = mean(hbeta), sd = sd(hbeta))
```

```
# A tibble: 1 x 2
    mean      sd
   <dbl>   <dbl>
1 0.0324 0.00247
```

2.1.3.3 Variance of $\hat{\alpha}$ and covariance between $\hat{\alpha}$ and $\hat{\beta}$

To get the variance of the estimator of the intercept, consider the "true" and the fitted model for one observation:

$$\begin{cases} y_n &= \alpha + \beta x_n + \epsilon_n \\ y_n &= \hat{\alpha} + \hat{\beta} x_n + \hat{\epsilon}_n \end{cases} \tag{2.5}$$

Equating the two expressions in Equation 2.5, we get:

$$(\hat{\alpha} - \alpha) + (\hat{\beta} - \beta)x_n + (\hat{\epsilon}_n - \epsilon_n) = 0 \tag{2.6}$$

Summing Equation 2.6 for the whole sample and dividing by N:[2]

$$(\hat{\alpha} - \alpha) + (\hat{\beta} - \beta)\bar{x} - \bar{\epsilon} = 0 \tag{2.7}$$

[2]Note that the mean of the residuals is 0.

Finally, subtracting Equation 2.7 from Equation 2.6:

$$(\hat{\beta} - \beta)(x_n - \bar{x}) + \hat{\epsilon}_n - (\epsilon_n - \bar{\epsilon}) = 0 \tag{2.8}$$

From Equation 2.7, the variance of $\hat{\alpha}$ is the expected value of the square of:

$$\begin{array}{rcl}(\hat{\alpha} - \alpha) & = & -(\hat{\beta} - \beta)\bar{x} + \bar{\epsilon} \\ & = & -\sum_n c_n \epsilon_n \bar{x} + \bar{\epsilon} \\ & = & -\sum_n \left(\bar{x} c_n - \frac{1}{N}\right) \epsilon_n \end{array}$$

With the hypothesis of homoskedastic and uncorrelated errors, the variance simplifies to:

$$\sigma_{\hat{\alpha}}^2 = \sigma_\epsilon^2 \sum_n \left(\bar{x}^2 c_n^2 + \frac{1}{N^2} - \frac{2\bar{x}}{N} c_n\right)$$

As $\sum_n c_n = 0$ and $\sum_n c_n^2 = \frac{1}{N\hat{\sigma}_x^2}$, we finally get:

$$\sigma_{\hat{\alpha}}^2 = \frac{\sigma_\epsilon^2}{N\hat{\sigma}_x^2}(\hat{\sigma}_x^2 + \bar{x}^2)$$

Finally, to get the covariance between the slope and the intercept, we take the product of the two estimators in deviation from their expected values.

$$(\hat{\alpha} - \alpha)(\hat{\beta} - \beta) = -\left[\sum_n \left(\bar{x} c_n - \frac{1}{N}\right) \epsilon_n\right] \left[\sum_n c_n \epsilon_n\right]$$

Taking the expected value, we get:

$$\hat{\sigma}_{\hat{\alpha}\hat{\beta}} = -\sigma_\epsilon^2 \sum_n \left(\bar{x} c_n^2 - \frac{1}{N} c_n\right) = -\bar{x} \frac{\sigma_\epsilon^2}{N\hat{\sigma}_x^2}$$

We can then compactly write the variances and the covariance of the OLS estimator in matrix form:

$$\begin{pmatrix} \sigma_{\hat{\alpha}}^2 & \sigma_{\hat{\alpha}\hat{\beta}} \\ \sigma_{\hat{\alpha}\hat{\beta}} & \sigma_{\hat{\beta}}^2 \end{pmatrix} = \frac{\sigma_\epsilon^2}{N\hat{\sigma}_x^2} \begin{pmatrix} \bar{x}^2 + \hat{\sigma}_x^2 & -\bar{x} \\ -\bar{x} & 1 \end{pmatrix} \tag{2.9}$$

2.1.3.4 Estimation of the variance of the errors

The standard deviation of the OLS estimator can't be computed because it depends on an unknown parameter σ_ϵ. To get an estimation of $\sigma_{\hat{\beta}}$, we therefore need to estimate first σ_ϵ. If the error were observed, a natural estimator would be obtained by computing the empirical variance of the errors in the sample: $\frac{1}{N}\sum_{n=1}^N (\epsilon_n - \bar{\epsilon})^2$. As the errors are not observed, this estimator cannot be computed, but a feasible estimator is obtained by replacing the unobserved errors by the residuals:

$$\hat{\sigma}_\epsilon^2 = \frac{\sum_{n=1}^N \hat{\epsilon}_n^2}{N}$$

To analyze the properties of this estimator, we first compute the variance of one residual. From Equation 2.8, a residual can be written as:

$$\hat{\epsilon}_n = \epsilon_n - \bar{\epsilon} - (\hat{\beta} - \beta)(x_n - \bar{x}) = \epsilon_n - \frac{1}{N}\sum_n \epsilon_n - (x_n - \bar{x})\sum_n c_n \epsilon_n$$

Taking the expected value of the square of this expression and noting that $\mathrm{E}\left(\bar{\epsilon}(\hat{\beta} - \beta)(x_n - \bar{x})\right) = \frac{1}{N}(x_n - \bar{x})\sigma_\epsilon^2 \sum_n c_n = 0$, we get the following variance:

$$\mathrm{V}(\hat{\epsilon}_n) = \sigma_\epsilon^2 + \frac{1}{N}\sigma_\epsilon^2 + \frac{(x_n - \bar{x})^2}{S_{xx}}\sigma_\epsilon^2 - 2\frac{1}{N}\sigma_\epsilon^2 - 2\frac{(x_n - \bar{x})^2}{S_{xx}}\sigma_\epsilon^2$$

Re-arranging terms:

$$\sigma_{\hat{\epsilon}_n}^2 = \sigma_\epsilon^2\left(1 - \frac{1}{N} - \frac{(x_n - \bar{x})^2}{S_{xx}}\right)$$

Note that $\sigma_{\hat{\epsilon}_n} < \sigma_\epsilon$, which means that residuals are on average "smaller" than errors; this is a direct consequence of the fact that we minimize the sum of the squares of the residuals (see Section 1.4.2). Summing for all the observations, we get the expected value of the sum of squares residuals:

$$\mathrm{E}\left(\sum_{n=1}^N \hat{\epsilon}_n^2\right) = \sigma_\epsilon^2(N-2)$$

Therefore, the previously computed estimator of the variance of the errors $\hat{\sigma}_\epsilon^2$ is biased:

$$\mathrm{E}(\hat{\sigma}_\epsilon^2) = \frac{\mathrm{E}\left(\sum_{n=1}^N \hat{\epsilon}_n^2\right)}{N} = \sigma_\epsilon^2\frac{N-2}{N}$$

More precisely, $\hat{\sigma}_\epsilon$ is downward biased, by a factor of $\sqrt{\frac{N-2}{N}}$. For example, for $N = 10, 20, 100$, we get $\sqrt{\frac{N-2}{N}} = 0.89, 0.95, 0.99$, which means a 11, 5, 1% downward bias for the estimated standard deviation. As the factor $\frac{N-2}{N}$ tends to 1 for $N \rightarrow +\infty$, the bias will be negligible for large samples, but can be severe in small samples. We'll from now denote $\dot{\sigma}_\epsilon$ the unbiased estimator:

$$\dot{\sigma}_\epsilon = \sqrt{\frac{N}{N-2}}\hat{\sigma}_\epsilon$$

$\dot{\sigma}_\epsilon$ is often called the **residual standard error**.

2.1.3.5 Exact distribution of the OLS estimator with normal errors

If the distribution of the errors is normal, as the OLS estimator is a linear combination of the errors, its exact distribution is also normal. Therefore:

$$\hat{\beta}_N \sim \mathcal{N}\left(\beta, \frac{\sigma_\epsilon}{\sqrt{N}\hat{\sigma}_x}\right)$$

Subtracting the mean and dividing by the standard deviation, we get a standard normal deviate:

$$\frac{\hat{\beta} - \beta}{\sigma_\epsilon / (\sqrt{N}\hat{\sigma}_x)} \sim \mathcal{N}(0, 1)$$

σ_ϵ is unknown; replacing it by its unbiased estimator $\dot{\sigma}_\epsilon$ induces an increase of the uncertainty and the distribution changes from a normal to a Student t distribution:

$$\frac{\hat{\beta} - \beta}{\dot{\sigma}_\epsilon / (\sqrt{N}\hat{\sigma}_x)} \sim t_{N-2}$$

The Student distribution is symmetric, has fatter tails than the normal distribution and converges in distribution to the normal distribution. Actually, it is worth considering the Student and not the normal distribution as an approximation only for small-sized samples. For example, for sample sizes of 10, 20, 50 and 100, 5% critical values for a Student are 2.31, 2.1, 2.01 and 1.98, as the critical value is 1.96 for the normal distribution.

2.1.3.6 Computation of the variance of the OLS estimator with R

We go back to the price-time model estimated in the previous chapter. Remember that the fitted model (called `pxt`) was obtained with the following code:

```
prtime <- price_time %>%
  set_names(c("town", "qr", "qa", "pr", "pa", "tr", "ta")) %>%
  mutate( sr = qr / (qr + qa),
          h = (pa - pr) / ( (tr - ta) / 60)) %>%
  filter(sr < 0.75)
pxt <- lm(sr ~ h, prtime)
```

We first compute "by hand" the standard error of the OLS estimator and we then use the relevant methods for `lm` objects to do so. We first extract the x vector, its length N, its mean $\bar{x}$ and its sample standard deviation $\hat{\sigma}_x$:

```
x <- pull(prtime, h)
N <- length(x)
bx <- mean(x)
sx <- sqrt(mean((x - bx) ^ 2))
```

We then get the sum of square residuals, and the residual standard error:

```
heps <- resid(pxt)
SSR <- sum(heps ^ 2)
seps <- sqrt(SSR / (N - 2))
seps
## [1] 0.07177
```

which finally leads to the estimators of the standard deviation of the OLS coefficients:

```
sbeta <- seps / (sqrt(N) * sx)
salpha <- sqrt(bx ^ 2 + sx ^ 2) * seps / (sqrt(N) * sx)
c(salpha, sbeta)
## [1] 0.073262 0.003931
```

All this information can be retrieved easily with **R** using specific functions. To get the sample size and the number of degrees of freedom (which is, in the simple linear model, $N-2$), we have:

```
nobs(pxt)
## [1] 9
df.residual(pxt)
## [1] 7
```

$\dot{\sigma}_\epsilon$ is computed using:

```
sigma(pxt)
## [1] 0.07177
```

The matrix of variance-covariance of the estimators is obtained using the `vcov` function:

```
vcov(pxt)
##             (Intercept)          h
## (Intercept)   0.0053673 -2.722e-04
## h            -0.0002722  1.546e-05
```

To get the standard deviations of the intercept and the slope estimators, we first extract the diagonal elements of this matrix and we next take the square roots of the values:

```
pxt %>% vcov %>% diag %>% sqrt
## (Intercept)           h
##    0.073262    0.003931
```

More simply, the `micsr::stder` function can be used:

```
pxt %>% stder
```

2.1.4 OLS estimator is BLUE

We have seen previously that the OLS estimator is a linear estimator (i.e., it is a linear combination of the N values of y for the sample):

$$\hat{\beta} = \sum_{n=1}^N \left(\frac{(x_n - \bar{x})}{\sum_{n=1}^N (x_n - \bar{x})^2} \right) y_n = \sum_{n=1}^N c_n y_n$$

Moreover, we have seen that if $\mathrm{E}(\epsilon \mid x) = 0$, it is unbiased and, with the hypothesis of homoskedastic and uncorrelated errors, we have established that its variance is: $\sigma_\epsilon^2 / S_{xx}$. We'll show in this subsection that among all the linear unbiased estimators, the OLS estimator is the one with the smallest variance. For these reasons, the OLS estimator is the **best linear unbiased estimator** (**BLUE**).

2.1.4.1 Comparing OLS with other linear unbiased estimators

Consider another linear estimator, with weights a_n:

$$\tilde{\beta} = \sum_{n=1}^N a_n y_n$$

Replacing y_n by $\alpha + \beta x_n + \epsilon_n$, we have:

$$\tilde{\beta} = \sum_{n=1}^N a_n(\alpha + \beta x_n + \epsilon_n) = \alpha \sum_{n=1}^N a_n + \beta \sum_{n=1}^N a_n x_n + \sum_{n=1}^N a_n \epsilon_n$$

Therefore, for any unbiased estimator, one must have $\sum_{n=1}^N a_n = 0$ and $\sum_{n=1}^N a_n x_n = 1$. We then have: $\tilde{\beta} - \beta = \sum_{n=1}^N a_n \epsilon_n$ and the variance of $\tilde{\beta}$ is:

$$\sigma_{\tilde{\beta}}^2 = \mathrm{E}\left(\left[\sum_{n=1}^N a_n \epsilon_n\right]^2\right) = \sigma_\epsilon^2 \sum_{n=1}^N a_n^2$$

defining $d_n = a_n - c_n$, we have:

$$\sum_{n=1}^N a_n^2 = \sum_{n=1}^N (c_n + d_n)^2 = \sum_{n=1}^N c_n^2 + \sum_{n=1}^N d_n^2 + 2\sum_{n=1}^N d_n c_n$$

But the last term is 0 because:

$$\begin{array}{rcl}
\sum_{n=1}^N d_n c_n & = & \sum_{n=1}^N (a_n - c_n) c_n \\
& = & \frac{1}{S_{xx}} \sum_{n=1}^N a_n x_n - \frac{1}{S_{xx}} \bar{x} \sum_{n=1} a_n - \sum_{n=1}^N c_n^2 \\
& = & 0
\end{array}$$

so that $\sum_{n=1}^N a_n^2 = \sum_{n=1}^N c_n^2 + \sum_{n=1}^N d_n^2$ and:

$$\sigma_{\tilde{\beta}}^2 = \sigma_\epsilon^2 \sum_{n=1}^N a_n^2 = \sigma_\epsilon^2 \left(\sum_{n=1}^N c_n^2 + \sum_{n=1}^N d_n^2\right) = \sigma_{\hat{\beta}}^2 + \sigma_\epsilon^2 \sum_{n=1}^N d_n^2$$

Therefore, $\sigma_{\tilde{\beta}}^2 > \sigma_{\hat{\beta}}^2$, which means that the OLS estimator is **BLUE**, i.e., it is, among all the unbiased linear estimators, the one with the lower variance.

2.1.4.2 Practical example

Consider as an example the price-time model. The model we have previously estimated is:

```
pxt <- lm(sr ~ h, prtime)
```

Consider now the same model without intercept ($\alpha = 0$). As $\alpha = -a/(b-a)$, a and b being respectively the minimal and the maximum time value, $\alpha = 0$ implies that the minimal time value is 0. To fit the model that imposes this hypothesis, we need to fit the same model without intercept. In **R**, this is performed using either `- 1` or `+ 0` in the formula :

```
pxt2 <- lm(sr ~ h - 1, prtime)
pxt2 <- lm(sr ~ h + 0, prtime)
```

The same model can also be estimated by updating the previous fitted model `pxt`, using the `update` function which takes as first argument the model we wish to update:

```
pxt2 <- update(pxt, . ~ . + 0)
pxt2 <- update(pxt, . ~ . - 1)
```

The formula is updated using `.`, which means the same thing as in the initial model. Therefore, `. ~ .` means the initial formula and we remove the intercept by either "adding" `0` or "subtracting" `1`.

The fitted model is presented in Figure 2.6.

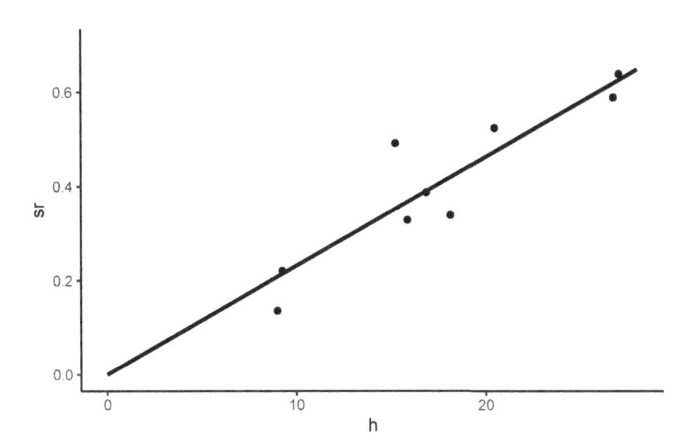

Figure 2.6: OLS estimator without intercept

For this model without intercept, the formula for the slope is:

$$\hat{\beta} = \frac{\sum_{n=1}^{N} x_n y_n}{\sum_{n=1}^{N} x_n^2}$$

Or, replacing y_n by $\beta x_n + \epsilon_n$:

$$\hat{\beta} = \beta + \frac{\sum_{n=1}^N x_n \epsilon_n}{\sum_{n=1}^N x_n^2}$$

for which the variance is:

$$\sigma^2_{\hat{\beta}} = \frac{\sigma^2_\epsilon}{\sum_{n=1}^N x_n^2} = \frac{\sigma^2_\epsilon}{N A_{x^2}}$$

where A_{x^2} is the arithmetic mean of the squares of x.

An alternative estimation method consists of drawing lines from every point to the origin, as illustrated in Figure 2.7, and to estimate β by the arithmetic mean of the N slopes, which are y_n/x_n. Formally, we have:

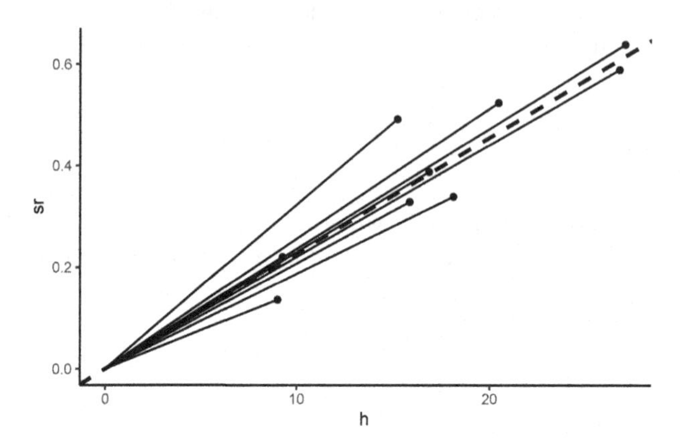

Figure 2.7: Individual slopes

$$\tilde{\beta} = \frac{1}{N}\sum_{n=1}^N \frac{y_n}{x_n}$$

This is a linear estimator, with weights $a_n = \frac{1}{N}\frac{1}{x_n}$. Replacing y_n by $\beta x_n + \epsilon_n$, we get :

$$\tilde{\beta} = \beta + \frac{1}{N}\sum_{n=1}^N \frac{\epsilon_n}{x_n}$$

This linear estimator is therefore unbiased. Its variance is:

$$\sigma_{\tilde{\beta}} = \frac{\sigma^2_\epsilon}{N^2}\sum_{n=1}^N \frac{1}{x_n^2} = \frac{\sigma^2_\epsilon}{N H_{x^2}}$$

where $H_{x^2} = \frac{N}{\sum_{n=1}^N \frac{1}{x_n^2}}$ is the harmonic mean of x^2. As the harmonic mean is always lower than the arithmetic mean, $\sigma_{\tilde{\beta}} > \sigma_{\hat{\beta}}$ and therefore $\tilde{\beta}$ is less precise than $\hat{\beta}$. The value of this alternative estimator can be computed as follow:

```
slope <- prtime %>% transmute(slope = sr / h) %>%
    summarise(slope = mean(slope)) %>% pull(slope)
slope
## [1] 0.02277
```

Once the estimator is computed, we can calculate $\hat{\sigma}_\epsilon$ and $\hat{\sigma}_{\tilde{\beta}}$ and, as intermediate results, the arithmetic and the harmonic means of x^2:

```
reg2 <- prtime %>%
  mutate(resid = sr - slope * h) %>%
  summarise(seps = sqrt(sum(resid ^  2) / 8),
            H = 9 / sum( 1 / h ^ 2),
            A = sum(h ^ 2) / 9,
            sdtilde = seps / sqrt(N) / sqrt(H))
reg2
## # A tibble: 1 x 4
##      seps     H     A sdtilde
##     <dbl> <dbl> <dbl>   <dbl>
## 1 0.0678  204.  347. 0.00158
```

We check that the harmonic mean (204) is lower than the arithmetic mean (347). Comparing with the OLS results:

```
pxt2 %>% stder
##        h
## 0.001205
```

we confirm that the OLS estimator has a lower standard error than the alternative estimator.

2.2 Asymptotic properties of the estimator

Asymptotic properties of an estimator deal with the behavior of this estimator when the sample size increases without bound. Compared to exact properties which are true and hold whatever the sample size is, asymptotic properties are approximations, the better the larger the sample size is. Two notions of convergence, that rely on two fundamental theorems are used:

- the convergence in probability, based on the **law of large numbers**,
- the convergence in distribution, based on the **central-limit** theorem.

2.2.1 Convergence in probability

We consider an estimator as a sequence of random numbers, indexed by the size of the sample on which it has been estimated: $\left\{\hat{\beta}_N\right\}$. This sequence converges in probability to a constant θ if:

$$\lim_{N\to\infty} \mathrm{P}(\mid \hat{\beta}_N - \theta \mid > \nu) = 0 \ \forall \nu$$

This is denoted by: $\hat{\beta}_N \xrightarrow{p} \theta$ or $\mathrm{plim}\,\hat{\beta} = \theta$. Convergence in probability implies convergence in mean square, which is defined by:

$$\lim_{N\to+\infty} \mathrm{E}\left((\hat{\beta}_N - \theta)^2\right) = 0$$

and means that:

$$\begin{cases} \lim_{N\to+\infty} \mathrm{E}(\hat{\beta}_N) = \theta \\ \lim_{N\to+\infty} V(\hat{\beta}_N) = 0 \end{cases}$$

If an estimator converges in mean square to its true value β, we'll write $\hat{\beta}_N \xrightarrow{\text{m.s.}} \beta$ and we'll also use $\mathrm{plim}\,\hat{\beta}_N = \beta$, as convergence in mean squares implies convergence in probability.[3] We'll also say in this case that the estimator is consistent. Note that, on the opposite, an estimator may be inconsistent for two reasons:

- the estimator doesn't converge in probability to any value,
- the estimator converges in probability to $\theta \neq \beta$.

The consistency of an estimator shouldn't be confused with the property of unbiasedness, even if we often encounter estimators which are unbiased *and* consistent:

- unbiasedness is an exact property (true or false whatever the sample size), and it refers only to the expected value of the estimator and doesn't say anything about its variance,
- consistency is an asymptotic property, which implies a limit for the expected value (β) and for the variance (0) of the estimator.

Therefore, an unbiased estimator can be inconsistent and, conversely, a consistent estimator can be biased. Consider for example that we have a random sample of N observations of a variable x which has a mean and a variance equal respectively to μ and σ^2. A natural estimator of μ is the arithmetic mean: $\bar{x}_N = \frac{1}{N}\sum_{n=1}^N x_n$, with expected value and variance:

$$\begin{cases} \mathrm{E}(\bar{x}_N) & = & \mathrm{E}\left(\frac{1}{N}\sum_{n=1}^N x_n\right) = \frac{1}{N}\sum_{n=1}^N \mathrm{E}(x_n) = \frac{1}{N}\sum_{n=1}^N \mu = \mu \\ \mathrm{V}(\bar{x}_N) & = & \mathrm{V}\left(\frac{1}{N}\sum_{n=1}^N x_n\right) = \frac{1}{N^2}\sum_{n=1}^N \mathrm{V}(x_n) = \frac{1}{N^2}\sum_{n=1}^N \sigma^2 = \frac{\sigma^2}{N} \end{cases}$$

This estimator is unbiased and consistent (the variance tends to 0 and the expected value is equal to the population mean μ). Consider now two alternative estimators.[4] The first one is:

[3]See for example Amemiya (1985), pages 89-90.

[4]These two estimators are inspired by Davidson and MacKinnon (2004), page 97, and Davidson and MacKinnon (1993), pages 123-124.

$$\dot{x}_N = \frac{1}{N-1}\sum_{n=1}^{N} x_n$$

Its first two moments can easily be obtained by writing $\dot{x}_N$ as a function of $\bar{x}_N$: $\dot{x}_N = \frac{N}{N-1}\bar{x}_N$, so that $\mathrm{E}(\dot{x}_N) = \frac{N}{N-1}\mu$ and $\mathrm{V}(\dot{x}_N) = \left(\frac{N}{N-1}\right)^2 \frac{\sigma^2}{N}$. The estimator is upward biased, by a multiplicative factor of $\frac{N}{N-1}$. The bias is severe in small samples (for example 25% if N is equal to 5), but becomes negligible as N grows. As the variance tends to 0 and the expected value to μ, $\dot{x}_N$ is consistent.

The second estimator is:

$$\tilde{x}_N = \frac{1}{2}x_1 + \frac{1}{2}\frac{1}{N-1}\sum_{n=2}^{N} x_n$$

It consists of first taking the mean for the whole sample except the first observation and then taking the simple average between it and the first observation: $\tilde{x}_N = \frac{1}{2}x_1 + \frac{1}{2}\bar{x}_{N-1}$. It is unbiased, as:

$$\mathrm{E}(\tilde{x}_N) = \frac{1}{2}\mathrm{E}(x_1) + \frac{1}{2}\mathrm{E}(\bar{x}_{N-1}) = \frac{1}{2}\mu + \frac{1}{2}\mu = \mu$$

and the variance is:

$$\mathrm{V}(\tilde{x}_N) = \frac{1}{4}\mathrm{V}(x_1) + \frac{1}{4}\mathrm{V}(\bar{x}_{N-1}) = \frac{1}{4}\sigma^2 + \frac{1}{4}\frac{\sigma^2}{N-1}$$

which tends to $\frac{1}{4}\sigma^2$ as $N \rightarrow +\infty$. Therefore, this unbiased estimator is not consistent. The problem is that the weight of the first observation is constant and, therefore, the value obtained for x_1 influences the estimator, whatever the size of the sample.

The OLS estimator writes:

$$\hat{\beta}_N = \beta + \frac{\sum_{n=1}^N (x_n - \bar{x})\epsilon_n}{N\hat{\sigma}_x^2} = \beta + \frac{\hat{\sigma}_{x\epsilon}}{\hat{\sigma}_x^2} \tag{2.10}$$

where $\bar{x}$, $\hat{\sigma}_x^2$ and $\hat{\sigma}_{x\epsilon}$ are the sample estimates of the population mean and variance of x and of the covariance between x and ϵ.[5] As the sample size increases, these three estimators converge to their population counterpart, namely $\mu_x = \mathrm{E}(x)$, $\sigma_x^2 = \mathrm{V}(x)$ and $\sigma_{x\epsilon} = \mathrm{cov}(x, \epsilon)$. We therefore have:

$$\mathrm{plim}\,\hat{\beta}_N = \beta + \frac{\sigma_{x\epsilon}}{\sigma_x^2} = \theta$$

θ equals β, in which case the estimator is consistent, if x is uncorrelated in the population with ϵ ($\sigma_{x\epsilon} = 0$).

[5] Note that the numerator of Equation 2.10 is not exactly the sample covariance, which is $\sum_n (x_n - \bar{x})(\epsilon_n - \bar{\epsilon})/N$.

2.2.2 Convergence in distribution: central-limit theorem

When $N \rightarrow +\infty$, $\hat{\beta}_N$ has a degenerate distribution, as it converges to a constant (which is β if the estimator is consistent) as its variance tends to 0. In this subsection, we seek to analyze the shape of the distribution of $\hat{\beta}$ as the sample size grows. We therefore need to consider a transformation of $\hat{\beta}_N$ which has a constant variance, and we'll see that it is $\sqrt{N}(\hat{\beta} - \beta)$. Starting again with the equation that relates $\hat{\beta}_N$ to the errors and defining $w_n = \frac{x_n - \bar{x}}{\sqrt{N}\hat{\sigma}_x}$, we have:

$$\hat{\beta}_N = \beta + \sum_{n=1}^{N} c_n \epsilon_n = \beta + \frac{\sum_{n=1}^{N} w_n \epsilon_n}{\sqrt{N}\hat{\sigma}_x}$$

Note that w_n sums to 0 (as c_n), but that $\sum_{n=1}^{N} w_n^2 = 1$. Subtracting β and multiplying by $\sqrt{N}$, we get:

$$z = \sqrt{N}(\hat{\beta}_N - \beta) = \sum_{n=1}^{N} w_n \frac{\epsilon_n}{\hat{\sigma}_x}$$

The distribution of $\sqrt{N}(\hat{\beta}_N - \beta)$ is the distribution of a linear combination of N random deviates $\epsilon_n/\hat{\sigma}_x$, with an unknown distribution, a 0 expected value and a standard deviation equal to $\sigma_\epsilon/\hat{\sigma}_x$. The first two moments of $z = \sqrt{N}(\hat{\beta}_N - \beta)$ don't depend on N ($\mbox{E}(z) = 0$ and $\mbox{V}(z) = \sum_n w_n^2 \sigma_\epsilon^2/\hat{\sigma}_x^2 = \sigma_\epsilon^2/\hat{\sigma}_x^2$). As N tends to infinity, the distribution of $\sqrt{N}(\hat{\beta}_N - \beta)$ still has a 0 expected value and a standard deviation equals to $\sigma_\epsilon/\hat{\sigma}_x$. The central-limit theorem states that the distribution of $\sqrt{N}(\hat{\beta}_N - \beta)$ **converges in distribution** to a normal distribution as N tends to infinity, whatever the distribution of ϵ. This is denoted by:

$$\sqrt{N}(\hat{\beta}_N - \beta) \xrightarrow{d} \mathcal{N}\left(0, \frac{\sigma_\epsilon}{\hat{\sigma}_x}\right)$$

Stated differently, the **asymptotic distribution** of $\hat{\beta}$ is a normal distribution with an expected value equal to β and a standard deviation equal to $\frac{\sigma_\epsilon}{\sqrt{N}\hat{\sigma}_x}$:

$$\hat{\beta}_N \overset{a}{\sim} \mathcal{N}\left(\beta, \frac{\sigma_\epsilon}{\sqrt{N}\hat{\sigma}_x}\right)$$

To illustrate the strength of the central-limit theorem, consider the simple case of the arithmetic mean of N independent random numbers with an expected value equal to 0 and a standard deviation equal to 1: $\bar{x}_n = \frac{\sum_{n=1} x_n}{N}$.[6] $\bar{x}_n$ has a 0 expected value and a variance equal to $1/N$. As we already now, $\bar{x}_n$ converges in probability to 0 and has therefore a degenerate distribution. Consider now:

$$z_N = \sqrt{N}\bar{x}_n = \frac{\sum_{n=1} x_n}{\sqrt{N}}$$

The expected value of z_N is still 0, but its standard deviation is now 1. Its third moment is:

[6] Inspired by Davidson and MacKinnon (1993), pages 126-127.

$$E(z_N^3) = \frac{\mathrm{E}\left((\sum_{n=1}^N x_n)^3\right)}{N^{3/2}}$$

Developing the sum for $N = 3$, we have:

$$\left(\sum_{n=1}^3 x_n\right)^3 = (x_1^2 + x_2^2 + x_3^2 + 2x_1x_2 + 2x_1x_3 + 2x_2x_3)(x_1 + x_2 + x_3)$$

Taking the expected value of this sum, we get terms like:

- $\mathrm{E}(x_n x_m^2) = \mathrm{E}(x_n)\mathrm{V}(x_m^2) = 0 \times 1 = 0$,
- $\mathrm{E}(x_n, x_m, x_l) = \mathrm{E}(x_n)\mathrm{E}(x_m)\mathrm{E}(x_l) = 0 \times 0 \times 0 = 0$,
- $\mathrm{E}(x_n^3) = \mu_3$.

Therefore, only the last category of terms remains while taking the expected value of the sum. As we have N of them, the third moment of z_N is therefore:

$$E(z_N^3) = \frac{N\mu_3}{N^{3/2}} = \frac{\mu_3}{\sqrt{N}}$$

Therefore, as N tends to infinity, $E(z_N^3)$ tends to 0, whatever the value of μ_3.

Consider now the fourth moment:

$$\begin{aligned}\left(\sum_{n=1}^3 x_n\right)^4 &= (x_1^2 + x_2^2 + x_3^2 + 2x_1x_2 + 2x_1x_3 + 2x_2x_3) \\ &\times (x_1^2 + x_2^2 + x_3^2 + 2x_1x_2 + 2x_1x_3 + 2x_2x_3)\end{aligned}$$

- terms like $x_n x_m^3$, $x_n x_m x_l^2$ and $x_n x_m x_l x_p$ have zero expected values,
- terms like $x_n^2 x_m^2$ have an expected value of $1 \times 1 = 1$. For $N = 3$, there are 18 of them and, more generally, for a given value of N, there are $3N(N-1)$ of them,
- terms like x_n^4 have an expected value of μ_4 and there are N of them.

Therefore:

$$E(z_N^4) = \frac{3N(N-1) + N\mu_4}{N^2} = 3\frac{N-1}{N} + \frac{\mu_4}{N}$$

which tends to 3, as N tends to infinity.

Therefore, for "large" N, the distribution of z_N doesn't depend on the shape parameters of x_n (μ_3 and μ_4) and its third and fourth moments tend to 0 and 3, which are the corresponding values for a normal distribution. These reasonings can easily be extended to higher moments, the general conclusion being that, when N tends to infinity, all the moments of z_N tend to those of the normal distribution. The asymptotic distribution of z_N is therefore normal and doesn't depend on the characteristics of the distribution of x_N.

2.2.3 Simulations

The law of large numbers and the central-limit theorem can be interestingly illustrated using simulations. Consider errors that follow a standardized chi-square distribution with one degree of freedom. Remember that a chi-squared with one degree of freedom is simply the square of a standard normal deviate: $x = z^2$. We thus have $\mathrm{E}(x) = \mathrm{E}(z^2) = \mathrm{V}(z) = 1$ and:

$$\mathrm{V}(x) = \mathrm{E}\left((x-1)^2\right) = \mathrm{E}\left(z^4\right) - 1 = 3 - 1 = 2$$

Therefore, $v = \frac{x-1}{\sqrt{2}}$ has a zero expected value and a standard deviation equal to 1. One can show that its third and fourth centered moments are $2\sqrt{2}$ and 15. Therefore, the distribution is:

- highly asymmetric, i.e., it has a long tail on the right side of the distribution, and a negative median, which is lower than the mean (equal to zero),
- highly leptokurtic, the fourth moment (15) is much larger than the value of 3 of the normal distribution; it has therefore a much higher mode and fatter tails than a normal distribution.

Now, going back to the `prtime` data, we generate a sample using the following DGP:

$$y_n = \alpha + \beta x_n + \epsilon_n$$

with $\alpha = -0.2$, $\beta = 0.032$ and: $\epsilon_n = \sigma_\epsilon \frac{z_n^2 - 1}{\sqrt{2}}$, where $\sigma_\epsilon = 0.08$ and z_n is a random draw on a standard normal distribution.

```
set.seed(1)
x <- prtime %>% pull(h)
N <- length(x)
asmpl <- tibble(h = x,
                eps = seps * (rnorm(N) ^ 2 - 1) / sqrt(2),
                sr = alpha + beta * h + eps)
v <- lm(sr ~ h, asmpl)
v %>% residuals %>% round(2)
##     1     2     3     4     5     6     7     8     9
## -0.01 -0.03  0.01  0.10 -0.03  0.00 -0.02 -0.01 -0.02
v %>% residuals %>% sum
## [1] -1.041e-17
```

The sum of the residuals is still equal to zero, but we can see in Figure 2.8, where a scatterplot is drawn for four random samples that the distribution of the errors (and therefore the distribution of the residuals) is highly asymmetric (we have only a couple of positive values, some of them being very large).

We then generate a large number of samples and for each of them, we compute the estimator of the slope and we plot the empirical distribution of $\hat{\beta}$ using a histogram. We consider different sample sizes; we use the "repeated in fixed sample" hypothesis,[7] i.e., we increase

[7] See Davidson and MacKinnon (1993), pp. 116-117 for details.

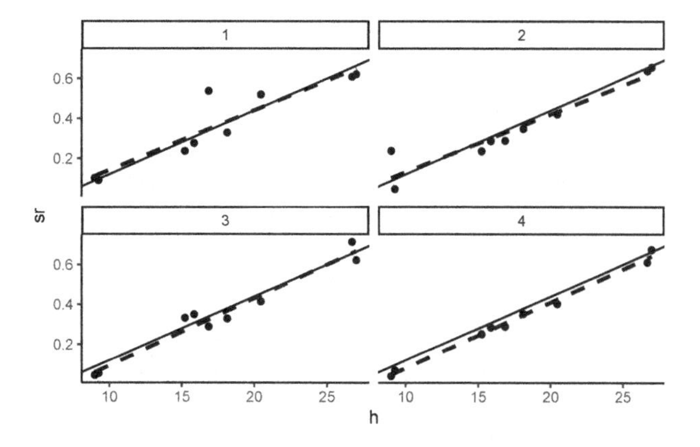

Figure 2.8: Four samples with χ^2 errors

the size of the sample by duplicating the same values of x. The histograms are presented in Figure 2.9, along with the normal density curve. The distribution of $\hat{\beta}$ is centered on β, whatever the sample size, which illustrates the fact that the estimator is unbiased. As the sample size is growing, we can see two changes in the shape of the histogram:

- it is more and more concentrated around the mean value of $\hat{\beta}$, which is due to the fact that the standard deviation of $\hat{\beta}$ is inversely proportional to sample size,
- the adjustment by the normal density curve is very bad in small samples; especially, the distribution of the estimator is highly leptokurtic, but the adjustment gets much better for larger samples.

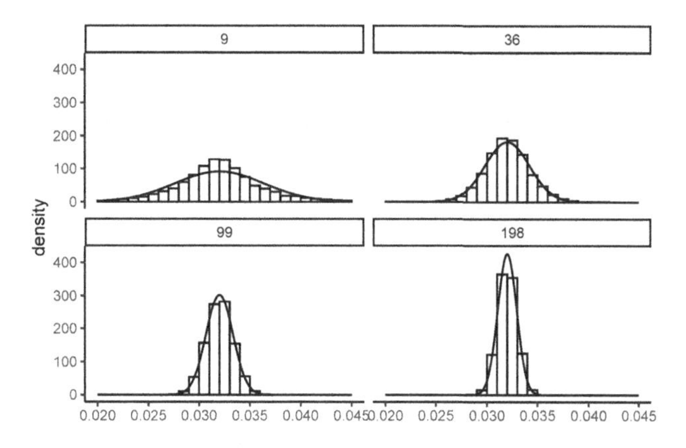

Figure 2.9: Empirical distribution of $\hat{\beta}$ for different sample sizes and adjustment by a normal density

Next, we plot in Figure 2.10 the distribution of $\sqrt{N}(\hat{\beta} - \beta)$, which has constant mean and standard deviation (respectively 0 and $\sigma_\epsilon / \sigma_x$). Therefore, only the shape of the distribution changes when the sample size increases. We can see more precisely on this figure the strength of the central-limit theorem, even for errors that follow a distribution very different from the normal.

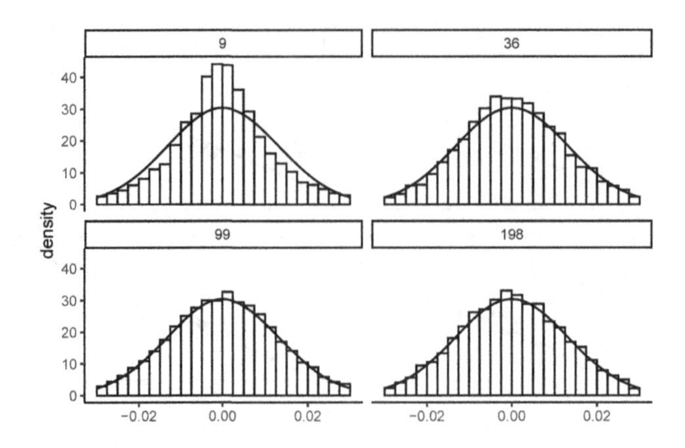

Figure 2.10: Empirical distribution of $\sqrt{N}(\hat{\beta}-\beta)$ for different sample sizes and adjustment by a normal density

2.3 Confidence interval and tests

With the set of hypotheses we have made concerning the errors of the model, the distribution of the estimator is completely defined by:

$$\hat{\beta}_N \overset{a}{\sim} \mathcal{N}\left(\beta, \frac{\sigma_\epsilon}{\sqrt{N}\hat{\sigma}_x}\right)$$

where $\overset{a}{\sim}$ means that the normal distribution is asymptotic and is actually a very good approximation if the sample size is large enough (which is the case in general in microeconometrics studies), whatever the distribution of the errors. Moreover, as $\hat{\beta}_N = \beta + \sum_{n=1}^N c_n \epsilon_n$ (the estimator is a linear combination of the errors), if the errors are normal, then the distribution of $\hat{\beta}$ is **exactly normal** (see Section 2.1.3.5). Removing from $\hat{\beta}$ its expected value and dividing by its standard deviation, we get a standard normal variable:

$$\frac{\hat{\beta}_N - \beta}{\sigma_{\hat{\beta}_N}} = \frac{\sqrt{N}\hat{\sigma}_x}{\sigma_\epsilon}(\hat{\beta}_N - \beta) \overset{a}{\sim} \mathcal{N}(0,1)$$

This result enables to perform two tasks:

- testing hypothesis,
- constructing a confidence interval, either for the coefficients or for the predictions of the model.

2.3.1 Testing hypothesis

We want to test the hypothesis that $\text{H}_0 : \beta = \beta_0$, the alternative hypothesis being $\text{H}_1 : \beta \neq \beta_0$. Denote $z_{\alpha/2}$ the critical value of a standard normal distribution at the α% error level. It is defined by: $\text{P}(\mid z \mid > z_{\alpha/2}) = \alpha$ or:

$$\mathrm{P}(\mid z \mid \leq z_{\alpha/2}) = 1 - \alpha \tag{2.11}$$

Consider for example $\alpha = 5\%$. To obtain the critical value, the `qnorm` function can be used, which takes a probability p as argument and returns a quantile q. By default, it returns the value such that $\mathrm{P}(z < q) = p$, but the value of $\mathrm{P}(z > q) = p$ is returned if the `lower.tail` argument is set to `FALSE`:

```
qnorm(0.025)
## [1] -1.96
qnorm(0.975)
## [1] 1.96
qnorm(0.025, lower.tail = FALSE)
## [1] 1.96
```

In this case, the critical value is 1.96, which means that, drawing in a standard normal distribution, one gets on average 95% of values lower, in absolute values, than 1.96. The preceding command indicates respectively that:

- 2.5% of the values of a normal distribution are lower than −1.96,
- 97.5% of the values of a normal distribution are lower than 1.96,
- 2.5% of the values of a normal distribution are greater than 1.96.

The 5% critical value is presented in Figure 2.11. If H_0 is true, $(\hat{\beta}_N - \beta_0)/\sigma_{\hat{\beta}_N}$ is a draw in a standard normal distribution and we therefore should have an absolute value lower than 1.96, 95% of the time. Obviously, $\hat{\beta}$ will almost never be exactly equal to β_0, even if H_0 is true because of sampling error. We have therefore the following decision rule, say at the 95% confidence level:

- if the absolute value of the computed statistic $(\hat{\beta}_N - \beta_0)/\sigma_{\hat{\beta}_N}$ is greater than the critical value, we'll say that the difference between $\hat{\beta}$ and β_0 is too large to be caused by sampling error; we therefore reject the hypothesis,
- if the absolute value of the computed statistic $(\hat{\beta}_N - \beta_0)/\sigma_{\hat{\beta}_N}$ is lower than the critical value, we'll say that the difference between $\hat{\beta}$ and β_0 is small enough to be caused by sampling error; we therefore don't reject the hypothesis.

Consider as an example $\hat{\beta} = 3.46$, $\beta_0 = 4$ and $\sigma_\epsilon = 0.3$. The computed statistic is $\frac{3.46-4}{0.3} = -1.8$.

```
hbeta <- 3.46 ; betao <- 4 ; shbeta <- 0.3
stat <- (hbeta - betao) / shbeta
stat
## [1] -1.8
```

It is lower, in absolute value, than 1.96; we therefore don't reject the null hypothesis at the 5% error level.

A more general tool is the **probability value**. It is the probability of drawing a value at least as large as the one we obtained (in absolute value) if the hypothesis is true. It is given by:

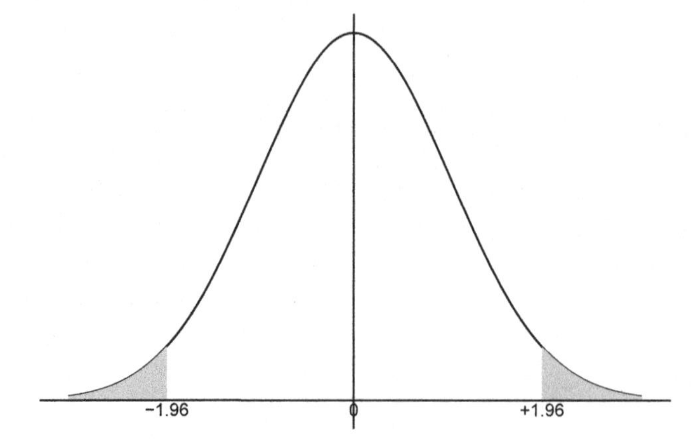

Figure 2.11: Normal distribution and 5% critical value

$$p = 2\left[1 - \Phi\left(\left|\frac{\hat{\beta} - \beta_0}{\sigma_{\hat{\beta}}}\right|\right)\right]$$

Probability values are computed using the `pnorm` function, which takes as argument a value of the variable (q) and computes the probability for a given value of its argument. The default behavior of `pnorm` is to return $p = \mbox{P}(z < q)$, but the upper tail, given by $\mbox{P}(z > x)$ is returned by setting the `lower.tail` argument to `FALSE`.

```
pnorm(stat)
## [1] 0.03593
pnorm(abs(stat))
## [1] 0.9641
1 - pnorm(abs(stat))
## [1] 0.03593
pnorm(abs(stat), lower.tail = FALSE)
## [1] 0.03593
2 * pnorm(abs(stat), lower.tail = FALSE)
## [1] 0.07186
```

The computed statistics can be of both signs, so the last formula is the most robust: first take the absolute value of the statistic, then compute the upper tail for a normal distribution and finally multiply it by 2. The p-value is greater than 5%; therefore, the hypothesis is not rejected at the 5%. The interest of the p-value is that, once it is computed, it is very easy to get the decision, whatever the error level (and even whatever the distribution). The 5-10% critical values and the p-value are represented in Figure 2.12. The absolute value of the statistic is 1.80, the critical values at the 5 and 10% are 1.96 and 1.64. Then:

- the absolute value of the statistic being lower than the 5% critical value; the hypothesis is not rejected at the 5% level,
- the absolute value of the statistic being greater than the 10% critical value; the hypothesis is rejected at the 10% level.

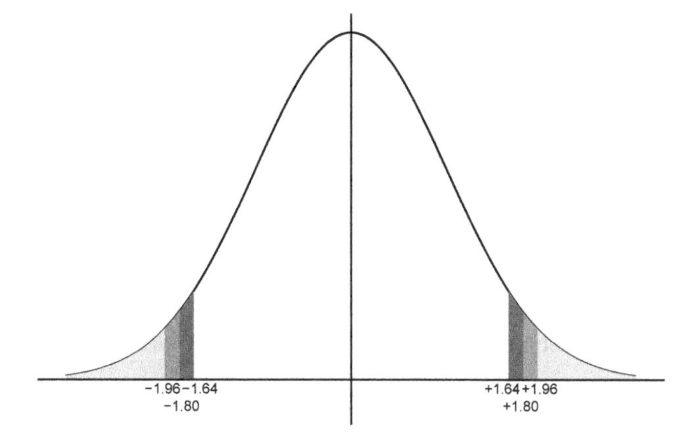

Figure 2.12: Critical value and p-value

The p-value is equal to 7.2%:

- the p-value is greater than 5%; the hypothesis is not rejected at the 5% level,
- the p-value is lower than 10%; the hypothesis is rejected at the 10% level.

2.3.2 Confidence interval

Knowing the distribution of the estimator enables one to go beyond the point estimation of the unknown parameter and to introduce the uncertainty by giving an interval of values which contains the real value of the unknown parameter with a given confidence. This is called a **confidence interval**. To obtain it, we start with Equation 2.11:

$$\mathrm{P}\left(\left|\frac{\hat{\beta}-\beta}{\sigma_{\hat{\beta}}}\right| < z_{\alpha/2}\right) = 1-\alpha$$

Developing this expression, we get:

$$\mathrm{P}\left(\hat{\beta} - \sigma_{\hat{\beta}} z_{\alpha/2} < \beta < \hat{\beta} + \sigma_{\hat{\beta}} z_{\alpha/2}\right) = 1-\alpha$$

which gives, in our example:

```
ic <- round(hbeta + c(-1, 1) * 1.96 * shbeta, 3)
ic
## [1] 2.872 4.048
```

This confidence interval indicates that there is a probability of 95% that the true value of β is between 2.872 and 4.048.

2.3.3 Exact distribution, the Student distribution

In real settings, $\sigma_{\hat{\beta}} = \frac{\sigma_\epsilon}{\sqrt{N}\hat{\sigma}_x}$ is unknown because σ_ϵ is an unknown parameter. Replacing σ_ϵ by the unbiased estimator of $\dot{\sigma}_\epsilon$ we get $\dot{\sigma}_{\hat{\beta}}$, the standard error of the estimation of the slope:

$$\dot{\sigma}_{\hat{\beta}} = \frac{\dot{\sigma}_\epsilon}{\sqrt{N}\hat{\sigma}_x}$$

As σ_ϵ is estimated, some more noise is added, so that the distribution of $\hat{\beta}$ is no longer a normal, but a Student t with $N-2$ degrees of freedom (see Section 2.1.3.5):

$$\frac{\hat{\beta}_N - \beta}{\dot{\sigma}_{\hat{\beta}_N}} = \frac{\sqrt{N}\hat{\sigma}_x}{\dot{\sigma}_\epsilon}(\hat{\beta}_N - \beta) \sim t_{N-2}$$

The Student distribution has a 0 expected value and a variance equal to $\frac{N-2}{N-4}$, which tends to 1 for large N. Moreover, the Student distribution converges in distribution to a normal distribution. Therefore, for large N, the same inference as the one presented for known σ_ϵ can be applied, using the normal distribution as a good approximation. For small samples, however, critical values of the Student distribution should be used. The relevant 95% critical values are computed below for numbers of degrees of freedom equal to 5, 10, 50, 100 and 1000:

```
qt(0.025, df = c(5, 10, 50, 100, 1000), lower.tail = FALSE) %>% round(3)
```

```
[1] 2.571 2.228 2.009 1.984 1.962
```

Therefore, the normal distribution can be safely used if the sample has at least a few hundreds of observations.

2.3.4 Inference with R

R has different functions that extract information about the statistical properties of the fitted model. To illustrate their use, we use once again the price-time model:

```
pxt <- lm(sr ~ h, prtime)
```

Detailed results of the model are computed using the `summary` method for `lm` objects:

```
spxt <- summary(pxt)
```

which returns an object of class `summary.lm`. Moreover, `summary.lm` prints nicely. It is therefore customary to use `summary` without storing the result in an object but only to visualize the detailed results of the fitted model:

```
summary(pxt)
```

```
Call:
lm(formula = sr ~ h, data = prtime)

Residuals:
     Min       1Q   Median       3Q      Max
-0.07938 -0.03535 -0.00034  0.01430  0.14291

Coefficients:
            Estimate Std. Error t value Pr(>|t|)
(Intercept) -0.01548    0.07326   -0.21   0.8386
h            0.02393    0.00393    6.09   0.0005 ***
---
Signif. codes:  0 '***' 0.001 '**' 0.01 '*' 0.05 '.' 0.1 ' ' 1

Residual standard error: 0.0718 on 7 degrees of freedom
Multiple R-squared:  0.841, Adjusted R-squared:  0.818
F-statistic: 37.1 on 1 and 7 DF,  p-value: 0.000497
```

The output first indicates the "call", i.e., the function that has been used to estimate the model. Then, the distribution of the residuals is summarized using the **five numbers** (the range `min` and `max`, the two quartiles and the median).[8] Note that the mean is not indicated, as it is necessarily 0 for a model fitted by OLS. Next, the table of coefficients is printed, containing:

- the names of the effects,
- the value of the estimates ($\hat{\beta}$),
- their standard errors ($\dot{\sigma}_{\hat{\beta}}$),
- the Student statistic which is the ratio of the previous two columns and is a special case of the test statistic $(\hat{\beta} - \beta_0)/\dot{\sigma}_{\hat{\beta}}$ where $\beta_0 = 0$,
- the probability value of this statistic.

Thes kinds of tests are often considered as tests of significance of the corresponding covariate. If the hypothesis that $\beta_0 = 0$ is rejected, we would say that the coefficient is "significant", which means more precisely that it is significantly different from 0. As we have a very small sample, it is worth considering the critical value of a Student instead of a normal distribution. We get here:

```
cv <- qt(0.025, df = df.residual(pxt), lower.tail = FALSE)
cv
## [1] 2.365
```

- for the intercept, the t statistic is much lower than the critical value and the probability value is far greater than 5%; therefore, the hypothesis that $\alpha = 0$ is not rejected,
- for the slope, the t statistic is much higher than the critical value and the probability value is far lower than 5%; therefore, the hypothesis that $\beta = 0$ is rejected.

[8]Any series can be summarized this way using the `fivenum` function.

This table of coefficients is a matrix that is stored in the `summary.lm` object with the `coefficients` name. As such, it can be extracted using `spxt$coefficients` or using the `coef` method of `summary.lm`: `coef(spxt)`. Finally, the printed output ends with some general indicators (often **GOF** for goodness-of-fit indicators) as the residual standard error ($\dot{\sigma}_\epsilon$, which can be extracted using the `sigma` function), two measures of the coefficient of determination, and the F statistic that is relevant for the multiple regression model.

The `confint` function computes the confidence interval for the coefficients:

```
confint(pxt, level = 0.9)
```

```
                  5 %    95 %
(Intercept) -0.15428 0.12332
h            0.01648 0.03138
```

we set here the `level` argument to `0.9` (the default value being `0.95`) and the results indicate that there is a 90% probability that the true value of the slope is between 0.016 and 0.031.

2.3.5 Delta method

It's often the case that the parameters of interest are not the fitted parameters, but some functions of them. In the price-time model, the fitted parameters are α and β, but the structural parameters (the lower and higher values of the travel time) are a and b:

$$\left\{\begin{array}{rcl} a &=& F^a(\alpha,\beta) = -\frac{\alpha}{\beta} \\ b &=& F^b(\alpha,\beta) = \frac{1-\alpha}{\beta} \end{array}\right. \tag{2.12}$$

The structural parameters are easily retrieved using Equation 2.12. The so-called delta method can be used to compute their standard deviations. Denoting f the first derivatives of F, we write a first-order Taylor expansion for F^a and F^b:

$$\left\{\begin{array}{rcl} a &=& F^a(\alpha_0,\beta_0) + (\alpha-\alpha_0)f^a_\alpha(\alpha_0,\beta_0) + (\beta-\beta_0)f^a_\beta(\alpha_0,\beta_0) \\ b &=& F^b(\alpha_0,\beta_0) + (\alpha-\alpha_0)f^b_\alpha(\alpha_0,\beta_0) + (\beta-\beta_0)f^b_\beta(\alpha_0,\beta_0) \end{array}\right.$$

So that the variances of the fitted structural parameters are approximately:

$$\left\{\begin{array}{rcl} \hat{\sigma}^2_{\hat{a}} &=& f^a_\alpha(\alpha_0,\beta_0)^2\hat{\sigma}^2_{\hat{\alpha}} + f^a_\beta(\alpha_0,\beta_0)^2\hat{\sigma}^2_{\hat{\beta}} + 2f^a_\alpha(\alpha_0,\beta_0)f^a_\beta(\alpha_0,\beta_0)\hat{\sigma}_{\hat{\alpha}\hat{\beta}} \\ \hat{\sigma}^2_{\hat{b}} &=& f^b_\alpha(\alpha_0,\beta_0)^2\hat{\sigma}^2_{\hat{\alpha}} + f^b_\beta(\alpha_0,\beta_0)^2\hat{\sigma}^2_{\hat{\beta}} + 2f^b_\alpha(\alpha_0,\beta_0)f^b_\beta(\alpha_0,\beta_0)\hat{\sigma}_{\hat{\alpha}\hat{\beta}} \end{array}\right.$$

Replacing (α_0,β_0) by $(\hat{\alpha},\hat{\beta})$ and using the formulas for the variances and covariance of $\hat{\alpha}$ and $\hat{\beta}$ given in Equation 2.9, we get:

$$\left\{\begin{array}{rcl} \hat{\sigma}_{\hat{a}} &=& \frac{\dot{\sigma}_\epsilon}{\sqrt{N}\hat{\sigma}_x}\frac{1}{\hat{\beta}}\sqrt{\hat{\sigma}^2_x + \left(\bar{x}+\frac{\hat{\alpha}}{\hat{\beta}}\right)^2} \\ \hat{\sigma}_{\hat{b}} &=& \frac{\dot{\sigma}_\epsilon}{\sqrt{N}\hat{\sigma}_x}\frac{1}{\hat{\beta}}\sqrt{\hat{\sigma}^2_x + \left(\bar{x}-\frac{1-\hat{\alpha}}{\hat{\beta}}\right)^2} \end{array}\right.$$

```
bx <- mean(x)
sx <- sqrt(mean( (x - bx) ^ 2))
halpha <- coef(pxt)[1] %>% unname
hbeta <- coef(pxt)[2] %>% unname
hseps <- sigma(pxt)
ab <- c(- halpha / hbeta, (1 - halpha) / hbeta)
sab <- hseps / sx / sqrt(nobs(pxt)) / hbeta *
    sqrt(c(sx ^ 2 + (bx + halpha / hbeta) ^ 2,
           sx ^ 2 + (bx - (1 - halpha) / hbeta) ^ 2)
        )
ab
## [1]  0.647 42.431
sab
## [1] 2.961 4.197
```

which finally leads to the 95% confidence interval:

```
matrix(ab, 2, 2) + matrix(c(- sab, sab), 2) * cv
```

```
        [,1]   [,2]
[1,] -6.354  7.648
[2,] 32.505 52.356
```

There is therefore a 95% probability that the maximum value of time is between 32.5 and 52.4 euros and the hypotheses that the minimum value of time is 0 is not rejected.

2.3.6 Confidence interval for the prediction

Once the model is estimated, a prediction for every observation can be computed using the formula of the conditional expectation of y for $x = x_n$, which is:

$$\mathrm{E}(y \mid x = x_n) = \alpha + \beta x_n + \mathrm{E}(\epsilon \mid x = x_n) = \alpha + \beta x_n$$

As $\hat{\alpha}$ and $\hat{\beta}$ are unbiased estimators of α and β, $\hat{y}_n = \hat{\alpha} + \hat{\beta} x_n$ is an unbiased estimator of $\mathrm{E}(y \mid x = x_n)$. Applying the formula for the variance of a sum, we have: $\mathrm{V}(\hat{y}) = \mathrm{V}(\hat{\alpha}) + x_n^2 \mathrm{V}(\hat{\beta}) + 2x_n \mathrm{cov}(\hat{\alpha}, \hat{\beta})$. Using Equation 2.9, we get:

$$\sigma_{\hat{y}_n}^2 = \frac{\sigma_\epsilon^2}{N\sigma_x^2}\left(\sigma_x^2 + (x_n - \bar{x})^2\right)$$

which finally leads to the formula of the standard deviation of the predictions:

$$\sigma_{\hat{y}_n} = \frac{\sigma_\epsilon}{\sqrt{N}}\sqrt{1 + \frac{(x_n - \bar{x})^2}{\hat{\sigma}_x^2}}$$

$\sigma_{\hat{y}_n}$ increases with the deviation of x_n from the sample mean. Moreover, $\sigma_{\hat{y}}$ tends to 0 when N tends to infinity, which means that $\hat{y}_n$ is a consistent estimator of $\mathrm{E}(y \mid x = x_n)$.

Consider now the standard deviation of $y_n = \mathrm{E}(y \mid x = x_n) + \epsilon_n$. To the variation due to the estimation of α and β, we have to add the one associated with ϵ_n. Therefore, the variance of y_n is the sum of $\sigma^2_{\hat{y}_n}$ and σ^2_ϵ and therefore its standard deviation is:

$$\sigma_{y_n} = \sqrt{\sigma^2_{\hat{y}_n} + \sigma^2_\epsilon} = \frac{\sigma_\epsilon}{\sqrt{N}}\sqrt{1 + \frac{(x_n - \bar{x})^2}{\hat{\sigma}^2_x} + N}$$

Note that when $N \rightarrow \infty$, σ_{y_n}, contrary Note that $\sigma_{\hat{y}_n}$ tends to σ_ϵ and not to 0.

A **confidence interval** for $\hat{y}_n$ is obtained by adding and subtracting to the point estimator the estimated standard deviation $\sigma_{\hat{y}_n}$ times the critical value (here a Student t with $N-2 = 7$ degrees of freedom). A **prediction interval** for y_n is obtained the same way, but using σ_{y_n} instead of $\sigma_{\hat{y}_n}$. The following code computes the two standard deviations and the relevant limits of the confidence / prediction intervals:

```
mux <- mean(prtime$h)
N <- nobs(pxt)
sx2 <- sum( (prtime$h - mux) ^ 2) / N
tcv <- qt(0.975, df = df.residual(pxt))
prtime <- prtime %>%
    mutate(fitted = fitted(pxt),
           sehy = sigma(pxt) / sqrt(N) *
             sqrt( 1 + (prtime$h - mux) ^ 2 / sx2),
           sey = sigma(pxt) / sqrt(N) *
             sqrt( 1 + (prtime$h - mux) ^ 2 / sx2 + N),
           lowhy = fitted - tcv * sehy, uphy  = fitted + tcv * sehy,
           lowy  = fitted - tcv *sey, upy   = fitted + tcv * sey)
prtime %>% select(fitted, lowhy, uphy, lowy, upy) %>% head(3)
```

```
# A tibble: 3 x 5
  fitted lowhy  uphy  lowy   upy
   <dbl> <dbl> <dbl> <dbl> <dbl>
1  0.624 0.522 0.726 0.426 0.822
2  0.474 0.412 0.536 0.293 0.655
3  0.631 0.527 0.735 0.432 0.830
```

These values can also be obtained with the `predict` function, with the `interval` argument set to `"confidence"` or `"prediction"`:

```
predict(pxt, interval = "confidence")
predict(pxt, interval = "prediction")
```

In Figure 2.13, the two intervals are represented:

- for the confidence interval, we use `geom_smooth` with `method = "lm"` and the default `TRUE` value for the `se` argument; in this case, we have a gray zone which figures the confidence interval (by default at the 95% level, but another level can be used by setting the `level` argument to the desired level),
- for the confidence interval, we use `geom_errorbar` that draws vertical segments, which represent here the limits of the confidence interval we have computed.

```
prtime %>% ggplot(aes(h, sr)) + geom_point() +
  geom_smooth(method = "lm", color = "black") +
    geom_errorbar(aes(ymin = lowy, ymax = upy))
```

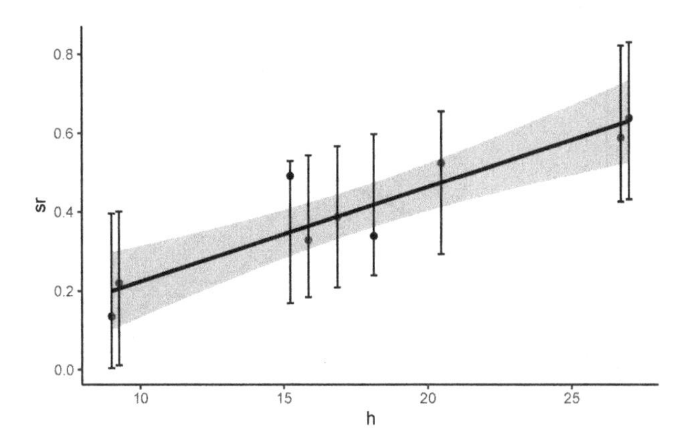

Figure 2.13: Confidence and prediction intervals

As an example, consider trips from Bordeaux to Paris. Reported transport time is 242 minutes, which is approximately 4 hours. The high speed track, opened in 2018 reduces this transport time to a minimum of 2 hours and 6 minutes. We consider 3 hours as the mean transport time, and we consider the average price to be 75 euros. Assuming that the conditions on the air transport market are unchanged, what prediction can we make about the change of the modal share of rail? We first construct a tibble called `bordeaux` with two lines: the first contains the actual features of the Paris-Bordeaux trip, and the second the new features.

```
bordeaux <- prtime %>% filter(town == "Bordeaux") %>%
  select(town, pr, pa, tr, ta) %>%
  add_row(town = "BordeauxSim", pr = 75,
          pa = 82.6, tr = 180, ta = 165) %>%
    mutate(h = (pa - pr) / ( (tr - ta) / 60))
```

The prediction of train's modal share is obtained using the `predict` function with the `new` argument which is a data frame containing the values of the covariates for which we want to compute predictions (this is the `bordeaux` table in our example):

```
prd <- predict(pxt, new = bordeaux, interval = "confidence") %>% as_tibble
prd
```

```
# A tibble: 2 x 3
    fit   lwr   upr
  <dbl> <dbl> <dbl>
1 0.624 0.522 0.726
2 0.712 0.580 0.844
```

The model predicts that the train's share increases from 0.624 to 0.712, but the confidence intervals are quite large and overlap. Present and predicted market shares are represented by a triangle and by a circle in Figure 2.14, along with the confidence intervals.

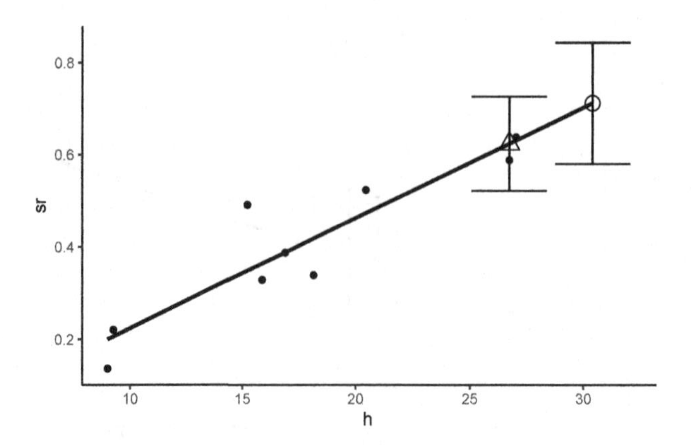

Figure 2.14: Predictions for train's model share

3

Multiple regression model

In this chapter, we'll analyze the computation and the properties of the OLS estimator when the number of covariates (K) is at least 2. Actually, we'll analyze in depth the case when $K = 2$ (generalizing from $K = 2$ to $K > 2$ being quite simple) and, compared to the previous two chapters, we'll insist on two important points:

- the use of matrix algebra, which makes the computation of the estimator and the analysis of its properties elegant and compact,
- the correlation between the covariates; actually we'll show that the multiple and the simple linear models are different only if there exists such a correlation.

We'll roughly follow the same plan as for the simple linear model: Section 3.1 presents the structural model and the data set that we'll use throughout the chapter, Section 3.2 the computation of the estimator, Section 3.3 the geometry of the multiple linear model, Section 3.4 the computation of the estimator with **R**, Section 3.5 its statistical properties and Section 3.6 the inference methods (confidence interval and tests). Finally, Section 3.7 presents system estimation and the constrained least squares estimator.

3.1 Model and data set

To illustrate the multiple regression model, we'll use the example of the estimation of a model explaining economic growth, using a cross-section of countries.

3.1.1 Structural model

One of the most popular growth models in the economic literature is the Solow-Swan model. The production Y (or more precisely the added value or GDP) is performed using two production factors, labor L and capital K. Physical labor is transformed in effective labor using a term called A. A is time-varying (typically increasing) and therefore represents the effect of technical progress which increases the productivity of labor. The functional form of the production function is a Cobb-Douglas:

$$Y(t) = K(t)^{\kappa} \left[A(t)L(t)\right]^{1-\kappa} \tag{3.1}$$

Coefficients for capital and labor are respectively κ and $1 - \kappa$. They represent the elasticity of the production respective to each factor, but also the share of each factor in the national income (κ is therefore the share of profits and $1 - \kappa$ the share of wages).

Each variable is a continuous function of time t. We'll denote, for each variable $\dot{V} = \frac{dV}{dt}$ the derivative with respect to time. Finally, we'll denote $y(t) = \frac{Y(t)}{A(t)L(t)}$ and $k(t) = \frac{Y(t)}{A(t)L(t)}$ production and capital per unit of effective labor. We therefore have:

$$y(t) = k(t)^\kappa \tag{3.2}$$

We'll hereafter omit (t) to make the notation less cluttered. Variation of capital is investment less depreciation. We assume that investment equals savings and that a constant percentage of income (i) is saved every year. The depreciation rate is denoted by δ. We then have:

$$\dot{K} = \frac{dK}{dt} = iY - \delta K$$

The growth of the capital stock per unit of effective labor k is then:

$$\dot{k} = \frac{d\frac{K}{AL}}{dt} = \frac{\dot{K}AL - (A\dot{L} + \dot{A}L)K}{A^2L^2} = \frac{\dot{K}}{AL} - \frac{K}{AL}\left(\frac{\dot{A}}{A} + \frac{\dot{L}}{L}\right) = (iy - \delta k) - k\left(\frac{\dot{A}}{A} + \frac{\dot{L}}{L}\right)$$

Denoting $n = \frac{\dot{L}}{L}$ and $g = \frac{\dot{A}}{A}$ the growth rates of L and A, i.e., the demographic growth rate and the technological progress rate, we finally have:

$$\dot{k}(t) = iy(t) - (n + g + \delta)k(t) = ik(t)^\kappa - (n + g + \delta)k(t)$$

At the steady state, the growth rate of capital per unit of effective labor is 0. Solving $\dot{k}(t) = 0$ we get the steady state value of $k(t)$, denoted by k^*:

$$k^* = \left(\frac{i}{n + g + \delta}\right)^{\frac{1}{1-\kappa}}$$

Or:

$$\left(\frac{K}{Y}\right)^* = \frac{k^*}{y^*} = k^{*(1-\kappa)} = \frac{i}{n + g + \delta}$$

From Equation 3.1, the production per capita is:

$$\frac{Y(t)}{L(t)} = A(t)k(t)^\kappa$$

Replacing $k(t)$ by k^*, $A(t)$ by $A(0)e^{gt}$ and taking logs, we get:

$$\ln\frac{Y(t)}{L(t)} = \ln A(0) + gt + \frac{\kappa}{1-\kappa}\ln i - \frac{\kappa}{1-\kappa}\ln(n + g + \delta)$$

Finally, let's denote $\ln A(0) = a + \epsilon$. ϵ is an error term which represents the initial dispersion between countries in terms of initial value of technical progress. With $C = a + gt$, and $v = \ln(n + g + \delta)$, the linear model that we wish to estimate is finally:

$$\ln \frac{Y}{L} = C + \frac{\kappa}{1-\kappa} \ln i - \frac{\kappa}{1-\kappa} \ln v + \epsilon \tag{3.3}$$

We therefore get a multiple regression model for which the response is the log of GPD per capita, ($\ln \frac{Y}{L}$) and the two covariates are $\ln i_n$ (the saving rate) and $\ln v_n$ (the sum of the demographic growth, the technical progress and the depreciation rates). Moreover, the structural model imposes some restrictions on the coefficients that can be tested. The two slopes are, in terms of the structural parameters of the theoretical model: $\beta_i = \frac{\kappa}{1-\kappa}$ and $\beta_v = -\frac{\kappa}{1-\kappa}$. Moreover, κ is the elasticity of the GDP with the capital and also the share of profits in GDP. A common approximation for the value of this parameter is about 1/3, which implies: $\beta_i = -\beta_v = \frac{\kappa}{1-\kappa} = 0.5$.

Mankiw, Romer, and Weil (1992) proposed a generalization of the Solow-Swan model that includes human capital, denoted by H. The production function is now:

$$Y(t) = K(t)^\kappa H(t)^\lambda \left[A(t)L(t)\right]^{1-\kappa-\lambda}$$

λ is the share of human capital in the GDP and the share of labor is now $(1-\kappa-\lambda)$. The model is very similar to the one we previously developed. We first compute the growth rate of physical and human capital ($\dot{k}$ and $\dot{h}$) per unit of effective labor, we set these two growth rates to 0 to get the stocks of physical and human capital at the steady state per unit of effective labor (k^* and h^*) and we introduce these two values in the production function to get:

$$\ln \frac{Y}{L} = C + \frac{\kappa}{1-\kappa-\lambda} \ln i + \frac{\lambda}{1-\kappa-\lambda} \ln e - \frac{\kappa+\lambda}{1-\kappa-\lambda} \ln v + \epsilon \tag{3.4}$$

where e is the per capita level of human capital. The model now contains three covariates and three slopes (β_i, β_e and β_v). Moreover, the structural model implies a structural restriction ($\beta_i + \beta_e + \beta_v = 0$) that is testable.

3.1.2 Data set

`growth` contains the data used by Mankiw, Romer, and Weil (1992). It consists of 121 countries for 1985.

```
growth %>% print(n = 3)
```

```
# A tibble: 121 x 9
  country group  gdp60 gdp85 gdpgwth popgwth   inv school  growth
  <chr>   <fct>  <dbl> <dbl>   <dbl>   <dbl> <dbl>  <dbl>   <dbl>
1 Algeria other   2485  4371     4.8   0.026 0.241  0.045  0.565
2 Angola  lqdata  1588  1171     0.8   0.021 0.058  0.018 -0.305
3 Benin   lqdata  1116  1071     2.2   0.024 0.108  0.018 -0.0412
# i 118 more rows
```

This data set contains a variable called `group` which enables the selection of subsamples. The modalities of this variable are:

- "oil": for countries whose most part of the GDP is linked to oil extraction,
- "oecd": for OECD countries,
- "lqdata": for countries with low quality data,
- "other": for other countries.

The variables used in the following regressions are per capita GDP in 1985 (gdp85), investment rate (inv) and growth rate of the population (popgwth). To get the variable denoted by v in the previous section, we need to add to the growth rate of the population the technical progress rate and the rate of depreciation. As these two variables are difficult to measure consistently, the authors assume that they don't exhibit any cross-country variation and that they sum to 5%. Therefore, v equals popgwth + 0.05.

We first investigate the relationship between the two covariates: inv and popgwth. Figure 3.1 presents the scatterplot, the size of the points being proportional to GDP per capita.

```
growth %>% ggplot(aes(popgwth, inv)) +
    geom_point(aes(size = gdp85, shape = group)) + stat_ellipse() +
    geom_smooth(color = "black")
```

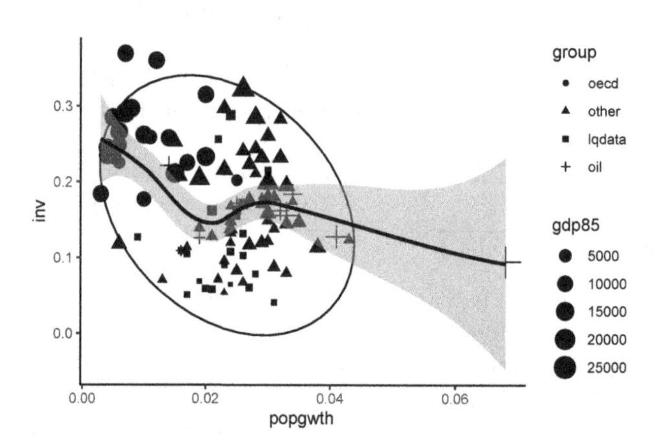

Figure 3.1: Investment rate and demographic growth

There is a weak negative correlation between the two variables and rich countries are in general characterized by a low demographic growth rate and a high investment rate. We also remark that there is an outlier, Kuwait, which has a very high demographic growth rate. We then compute the variable v and we rename inv and school in i and e as in Equation 3.4:

```
growth <- mutate(growth, v = popgwth + 0.05) %>%
  rename(i = inv, e = school)
```

3.2 Computation of the OLS estimator

In this section, we present the computation of the OLS estimator with a number of covariates $K \geq 2$. We start with the case where $K = 2$, for which it is possible to compute the estimators using roughly the same notations as the one used for the simple linear regression model. Then we'll perform the computation in the general case using matrix algebra.

For one observation, denoting x_{n1} and x_{n2} the two covariates for observation n, the model is:

$$y_n = \alpha + \beta_1 x_{n1} + \beta_2 x_{n2} + \epsilon_n$$

Each observation is now a point in the 3-D space defined by (x_1, x_2, y) and $\gamma^\top = (\alpha, \beta_1, \beta_2)$ are the coordinates of a plane that returns the expected value of y for a given value of x_1 and x_2. The residual sum of squares is:

$$f(\alpha, \beta) = \sum_{n=1}^{N} \left(y_n - \alpha - \beta_1 x_{n1} - \beta_2 x_{n2}\right)^2$$

which leads to the following three first-order conditions:

$$\left\{\begin{array}{rcl} \frac{\partial f}{\partial \alpha} &=& -2\sum_{n=1}^{N}\left(y_n - \alpha - \beta_1 x_{n1} - \beta_2 x_{n2}\right) = 0 \\ \frac{\partial f}{\partial \beta_1} &=& -2\sum_{n=1}^{N} x_{n1}\left(y_n - \alpha - \beta_1 x_{n1} - \beta_2 x_{n2}\right) = 0 \\ \frac{\partial f}{\partial \beta_2} &=& -2\sum_{n=1}^{N} x_{n2}\left(y_n - \alpha - \beta_1 x_{n1} - \beta_2 x_{n2}\right) = 0 \end{array}\right. \quad (3.5)$$

Dividing the first line of Equation 3.5 by the sample size, we get:

$$\bar{y} - \alpha - \beta_1 \bar{x}_1 - \beta_2 \bar{x}_2 = 0 \quad (3.6)$$

which means that the sample mean is on the regression plane, or that the sum of the residuals is zero. For the last two lines of Equation 3.5, the terms in parentheses are the residual for one observation and as its mean is 0, they indicate that the sample covariance between the residuals and both covariates should be 0. Therefore, we get exactly the same conditions as the one obtained for the simple linear regression model. Subtracting Equation 3.6 from the last two lines of Equation 3.5, we get:

$$\left\{\begin{array}{l} \sum_{n=1}^{N} x_{n1}\left[(y_n - \bar{y}) - \beta_1(x_{n1} - \bar{x}_1) - \beta_2(x_{n2} - \bar{x}_2)\right] = 0 \\ \sum_{n=1}^{N} x_{n2}\left[(y_n - \bar{y}) - \beta_1(x_{n1} - \bar{x}_1) - \beta_2(x_{n2} - \bar{x}_2)\right] = 0 \end{array}\right.$$

or replacing x_{nk} by $x_{nk} - \bar{x}_k$ and developing terms:

$$\left\{\begin{array}{rcl} \sum_{n=1}^{N}(x_{n1} - \bar{x}_1)(y_n - \bar{y}) &=& \beta_1 \sum_{n=1}^{N}(x_{n1} - \bar{x}_1)^2 + \beta_2 \sum_{n=1}^{N}(x_{n1} - \bar{x}_1)(x_{n2} - \bar{x}_2) \\ \sum_{n=1}^{N}(x_{n2} - \bar{x}_2)(y_n - \bar{y}) &=& \beta_1 \sum_{n=1}^{N}(x_{n1} - \bar{x}_1)(x_{n2} - \bar{x}_2) + \beta_2 \sum_{n=1}^{N}(x_{n2} - \bar{x}_2)^2 \end{array}\right. \quad (3.7)$$

We therefore have a system of two linear equations with two unknown parameters (β_1 and β_2) that could be solved, for example, by substitution. However, the use of matrix algebra enables to solve such a problem in a much simpler way. Denote:

$$X = \begin{pmatrix} x_{11} & x_{12} \\ x_{21} & x_{22} \\ \vdots & \vdots \\ x_{N1} & x_{N2} \end{pmatrix} \quad y = \begin{pmatrix} y_1 \\ y_2 \\ \vdots \\ y_N \end{pmatrix} \quad \beta = \begin{pmatrix} \beta_1 \\ \beta_2 \end{pmatrix}$$

X is a matrix with two columns (K in the general case) and N lines (the number of observations) and y is a vector of length N. We define $\tilde{I} = I - J/N$ where I is a $N \times N$ identity matrix and J a $N \times N$ matrix of ones. For example, for $N = 3$:

$$\tilde{I} = I - J/N = \begin{pmatrix} 1 & 0 & 0 \\ 0 & 1 & 0 \\ 0 & 0 & 1 \end{pmatrix} - \begin{pmatrix} \frac{1}{3} & \frac{1}{3} & \frac{1}{3} \\ \frac{1}{3} & \frac{1}{3} & \frac{1}{3} \\ \frac{1}{3} & \frac{1}{3} & \frac{1}{3} \end{pmatrix}$$

Premultiplying a vector z of length N by J/N, we get a vector of length N containing the sample mean of z $\bar{z}$ repeated N times. As premultiplying z by the identity matrix returns z, premultiplying z by $\tilde{I}$ returns a vector of length N containing the N values of z in difference from the sample mean. Note that $\tilde{I}$ is **idempotent**, which means that $\tilde{I} \times \tilde{I}$. It can be checked by using direct multiplication, but also by reminding that premultiplying a vector by $\tilde{I}$ removes the sample mean from the values of the vector. The transformed vector then has a zero mean, so that applying the same premultiplication one more time will leave it unchanged. Therefore, $\tilde{I}(\tilde{I}z) = \tilde{I}z$. Denoting $\tilde{z} = \tilde{I}z$, we get:

$$\tilde{X} = \tilde{I}X = \begin{pmatrix} x_{11} - \bar{x}_1 & x_{12} - \bar{x}_2 \\ x_{21} - \bar{x}_1 & x_{22} - \bar{x}_2 \\ \vdots & \vdots \\ x_{N1} - \bar{x}_1 & x_{N2} - \bar{x}_2 \end{pmatrix}, \quad \tilde{y} = \tilde{I}y = \begin{pmatrix} y_1 - \bar{y} \\ y_2 - \bar{y} \\ \vdots \\ y_N - \bar{y} \end{pmatrix}$$

Then:

$$\tilde{X}^\top \tilde{X} = \begin{pmatrix} \sum_{n=1}^N (x_{n1} - \bar{x}_1)^2 & \sum_{n=1}^N (x_{n1} - \bar{x}_1)(x_{n2} - \bar{x}_2) \\ \sum_{n=1}^N (x_{n1} - \bar{x}_1)(x_{n2} - \bar{x}_2) & \sum_{n=1}^N (x_{n2} - \bar{x}_2)^2 \end{pmatrix} = \begin{pmatrix} S_{11} & S_{12} \\ S_{12} & S_{22} \end{pmatrix}$$

and

$$\tilde{X}^\top \tilde{y} = \begin{pmatrix} \sum_{n=1}^N (x_{n1} - \bar{x}_1)(y_n - \bar{y}) \\ \sum_{n=1}^N (x_{n2} - \bar{x}_2)(y_n - \bar{y}) \end{pmatrix} = \begin{pmatrix} S_{1y} \\ S_{2y} \end{pmatrix}$$

S_{kk} is the total variation of x_k, S_{kl} the covariation of x_k and x_l and S_{ky} the covariation between covariate k and the response. Note that the quantities S_{kk} and S_{ky} were already present in the simple linear model as S_{xx} and S_{xy} (there are now two of them). The new term is S_{kl}, which measures the correlation between the two covariates.

Equation 3.7 can then be written in matrix form as:

$$\tilde{X}^\top \tilde{y} = \tilde{X}^\top \tilde{X} \beta$$

And the OLS estimator is obtained by premultiplying both sides of the equation by the inverse of $\tilde{X}^\top \tilde{X}$:

$$\hat{\beta} = \left(\tilde{X}^\top \tilde{X}\right)^{-1} \tilde{X}^\top \tilde{y} \tag{3.8}$$

Note that premultiplying the vector $\tilde{X}^\top \tilde{y}$ by the inverse of $\tilde{X}^\top \tilde{X}$ is a natural extension of the computation we've performed for the simple linear model, which consisted of dividing S_{xy} by S_{xx}. To understand this formula, we write $\tilde{X}^\top \tilde{X}$ and $\tilde{X}^\top \tilde{y}$ as:

$$\tilde{X}^\top \tilde{X} = \begin{pmatrix} S_{11} & S_{12} \\ S_{12} & S_{22} \end{pmatrix} = N \begin{pmatrix} \hat{\sigma}_1^2 & \hat{\sigma}_{12} \\ \hat{\sigma}_{12} & \hat{\sigma}_2^2 \end{pmatrix} = N\hat{\sigma}_1\hat{\sigma}_2 \begin{pmatrix} \frac{\hat{\sigma}_1}{\hat{\sigma}_2} & \hat{\rho}_{12} \\ \hat{\rho}_{12} & \frac{\hat{\sigma}_2}{\hat{\sigma}_1} \end{pmatrix} \tag{3.9}$$

and

$$X^\top y = \begin{pmatrix} S_{1y} \\ S_{2y} \end{pmatrix} = N \begin{pmatrix} \hat{\sigma}_{1y} \\ \hat{\sigma}_{2y} \end{pmatrix} = N\hat{\sigma}_y \begin{pmatrix} \hat{\sigma}_1\hat{\rho}_{1y} \\ \hat{\sigma}_2\hat{\rho}_{2y} \end{pmatrix}$$

- the first formulation uses the total sample variations / covariations,
- the second one divides every term by N to obtain sample variances and covariances,
- the third one divides the covariances by the product of the standard deviations to get sample coefficients of correlation.

To compute the estimator, we need to compute the inverse of $\tilde{X}^\top \tilde{X}$, which is:

$$\left(\tilde{X}^\top \tilde{X}\right)^{-1} = \frac{\begin{pmatrix} S_{22} & -S_{12} \\ -S_{12} & S_{11} \end{pmatrix}}{S_{11}S_{22} - S_{12}^2} = \frac{\begin{pmatrix} \hat{\sigma}_2^2 & -\hat{\sigma}_{12} \\ -\hat{\sigma}_{12} & \hat{\sigma}_1^2 \end{pmatrix}}{N(\hat{\sigma}_1^2\hat{\sigma}_2^2 - \hat{\sigma}_{12}^2)} = \frac{\begin{pmatrix} \frac{\hat{\sigma}_2}{\hat{\sigma}_1} & -\hat{\rho}_{12} \\ -\hat{\rho}_{12} & \frac{\hat{\sigma}_1}{\hat{\sigma}_2} \end{pmatrix}}{N\hat{\sigma}_1\hat{\sigma}_2(1 - \hat{\rho}_{12}^2)} \tag{3.10}$$

Equation 3.8 finally gives:

$$\left\{ \begin{array}{rcl} \hat{\beta}_1 &=& \frac{S_{22}S_{1y} - S_{12}S_{2y}}{S_{11}S_{22} - S_{12}^2} = \frac{\hat{\sigma}_2^2\hat{\sigma}_{1y} - \hat{\sigma}_{12}\hat{\sigma}_{2y}}{\hat{\sigma}_1^2\hat{\sigma}_2^2 - \hat{\sigma}_{12}^2} = \frac{\hat{\rho}_{1y} - \hat{\rho}_{12}\hat{\rho}_{2y}}{1 - \hat{\rho}_{12}^2} \frac{\hat{\sigma}_y}{\hat{\sigma}_1} \\ \hat{\beta}_2 &=& \frac{S_{11}S_{2y} - S_{12}S_{1y}}{S_{11}S_{22} - S_{12}^2} = \frac{\hat{\sigma}_1^2\hat{\sigma}_{2y} - \hat{\sigma}_{12}\hat{\sigma}_{1y}}{\hat{\sigma}_1^2\hat{\sigma}_2^2 - \hat{\sigma}_{12}^2} = \frac{\hat{\rho}_{2y} - \hat{\rho}_{12}\hat{\rho}_{1y}}{1 - \hat{\rho}_{12}^2} \frac{\hat{\sigma}_y}{\hat{\sigma}_2} \end{array} \right.$$

If the two covariates are uncorrelated in the sample ($S_{12} = \hat{\sigma}_{12} = \hat{\rho}_{12} = 0$), we have:

$$\left\{ \begin{array}{rcl} \hat{\beta}_1 &=& \frac{S_{1y}}{S_{11}} = \frac{\hat{\sigma}_{1y}}{\hat{\sigma}_1^2} = \hat{\rho}_{1y}\frac{\hat{\sigma}_y}{\hat{\sigma}_1} \\ \hat{\beta}_2 &=& \frac{S_{2y}}{S_{22}} = \frac{\hat{\sigma}_{2y}}{\hat{\sigma}_2^2} = \hat{\rho}_{2y}\frac{\hat{\sigma}_y}{\hat{\sigma}_2} \end{array} \right.$$

which is exactly the same formula that we had for the unique slope in the case of the simple regression model. This means that, if x_1 and x_2 are uncorrelated in the sample, regressing y on x_1 or on x_1 and x_2 leads exactly to the same estimator for the slope of x_1.

The general formula for $\hat{\beta}_1$ in term of the estimator of the simple linear model $\hat{\beta}_1^S$ (Equation 1.10) is:

$$\hat{\beta}_1 = \frac{\hat{\rho}_{1y} - \hat{\rho}_{12}\hat{\rho}_{2y}}{1-\hat{\rho}_{12}^2}\frac{\hat{\sigma}_y}{\hat{\sigma}_1} = \hat{\beta}_1^s \frac{1 - \frac{\hat{\rho}_{12}\hat{\rho}_{2y}}{\hat{\rho}_{1y}}}{1-\hat{\rho}_{12}^2} = \hat{\beta}_1^s \frac{1 - \hat{\rho}_{12}^2 \frac{\hat{\rho}_{2y}}{\hat{\rho}_{12}\hat{\rho}_{1y}}}{1-\hat{\rho}_{12}^2} \tag{3.11}$$

We have $\mid \hat{\rho}_{12}\hat{\rho}_{1y} \mid \leq \mid \hat{\rho}_{2y} \mid$, or $\left| \frac{\hat{\rho}_{2y}}{\hat{\rho}_{12}\hat{\rho}_{1y}} \right| \geq 1$.

Consider the case where the two covariates are positively correlated with the response. As an example, consider the wage equation with x_1 education and x_2 a dummy for males. Two cases should then be analyzed:

- the two covariates are positively correlated (males are more educated than females on average). In this case $\hat{\rho}_{2y}/(\hat{\rho}_{12}\hat{\rho}_{1y}) > 1$, and the numerator of Equation 3.11 is lower than $1 - \hat{\rho}_{12}^2$, so that $\hat{\beta}_1 < \hat{\beta}_1^S$. $\hat{\beta}_1^S$ is upward biased because it estimates the sum of the positive direct effect of education on wage and a positive indirect effect (more education leads to a subpopulation with a higher share of males and therefore higher wages),
- the two covariates are negatively correlated (males are less educated than females on average). In this case, $\hat{\rho}_{2y}/(\hat{\rho}_{12}\hat{\rho}_{1y}) < 0$ and the numerator of Equation 3.11 is greater than 1, so that $\hat{\beta}_1 > \hat{\beta}_1^S$. $\hat{\beta}_1^S$ is downward biased because it estimates the sum of the positive direct effect of education on wage and a negative indirect effect (more education leads to a subpopulation with a lower share of males and therefore lower wages).

The general derivation of the OLS estimator can be performed using matrix algebra, denoting j_N a vector of 1 of length N, $Z = (j_N, X)$ a vector formed by binding a vector of 1 to the matrix of covariates and $\gamma^\top = (\alpha, \beta^\top)$ the vector of parameters obtained by adding the intercept α to the vector of slopes:

$$f(\gamma) = (y - Z\gamma)^\top(y - Z\gamma) = y^\top y + \gamma^\top Z^\top Z\gamma - 2\gamma^\top Z^\top y$$

The $K+1$ first-order conditions are:

$$\frac{\partial f}{\partial \gamma} = 2Z^\top Z\gamma - 2Z^\top y = -2Z^\top(y - Z\gamma) = 0$$

The last expression indicates that all the columns of Z are orthogonal to the vector of residuals $y - Z\gamma$. Solving for γ, we get:

$$\hat{\gamma} = (Z^\top Z)^{-1}Z^\top y \tag{3.12}$$

The matrix of second derivatives is:

$$\frac{\partial^2 f}{\partial\gamma\partial^\top\gamma} = 2Z^\top Z$$

It is a positive semi-definite matrix, so that $\hat{\gamma}$ is a minimum of f. Comparing Equation 3.12 and Equation 3.8, we can see that the formula of the OLS is the same with different matrices (respectively $\tilde{X}$ and Z), the first equation returning $\hat{\beta}$ and the second one $\hat{\gamma}^\top = (\hat{\alpha}, \hat{\beta}^\top)$.

3.3 Geometry of least squares

3.3.1 Geometry of the multiple regression model

The geometry of the multiple regression model is presented in Figure 3.2.

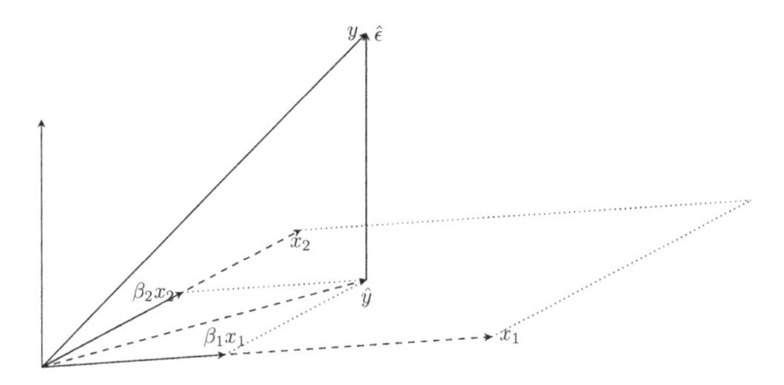

Figure 3.2: Geometry of the multiple regression model

We now have $N = 3$ and $K = 2$ and therefore each variable is a vector in the 3-D space. As the two covariates x_1 and x_2 are linearly independent, they span a subspace of dimension 2, which is a plane in this 3-D space. $\hat{y}$ is the orthogonal projection of y on this subspace and $\hat{\epsilon}$, is the vector that links $\hat{y}$ and y. $\hat{\epsilon}$ is therefore the projection of y on the complement to the subspace defined by (x_1, x_2), which is a straight line orthogonal to the plane spanned by (x_1, x_2). Therefore $\hat{\epsilon}$ is orthogonal to x_1 and to x_2, which means that the residuals are uncorrelated with the two covariates. The decomposition of y on the sum of two orthogonal vectors $\hat{\epsilon}$ and $\hat{y}$ doesn't depend on the two variables x_1 and x_2 per se, but on the subspace spanned by x_1 and x_2. This means that any couple of independent linear combination of x_1 and x_2 will leads to the same subspace as the one defined by x_1 and x_2 and therefore to the same residuals and the same fitted values.

More formally, as $\hat{\beta} = (X^\top X)^{-1}X^\top y$, we have $\hat{y} = X\hat{\beta} = X(X^\top X)^{-1}X^\top y = P_X y$. P is sometimes called the "hat" matrix, as it "puts a hat" on y. This matrix transforms the vector of response on a vector of prediction. As $\hat{\epsilon} = y - \hat{y}$, we also have $\hat{\epsilon} = y - P_X y = (I - P_X)y = M_X y$. We therefore consider two matrices P_X and M_X:

$$\left\{ \begin{array}{lcl} P_X & = & X(X^\top X)^{-1}X^\top \\ M_X & = & I - X(X^\top X)^{-1}X^\top \end{array} \right. \tag{3.13}$$

that are square and symmetric of dimension $N \times N$, which means that they are in practice large matrices, and are therefore never computed in practice. However, they have very interesting analytical features. First, they are idempotent, which means that $P_X \times P_X = P_X$ and $M_X \times M_X = M_X$. This means that while premultiplying a vector by such a matrix, this vector is projected in a subspace. For example, premultiplying y by P_X gives $\hat{y}$, the vector of fitted values. It is a linear combination of x_1 and x_2 and therefore belongs to the subspace spanned by x_1 and x_2. Therefore, $P_X\hat{y}$ obviously equal $\hat{y}$ and therefore $P_X \times P_X = P_X$. Except for the identity matrix, idempotent matrices are not full rank. Their rank can be easily computed using the fact that the rank of a matrix is equal to its trace (the sum of the diagonal elements) and that the trace of a product of matrices is invariant to any permutation of the matrices: $\text{tr}ABC = \text{tr}BCA = \text{tr}CAB$.

For a regression with an intercept, the model matrix Z has $K+1$ column, the first one being a column of one. In this case, the rank of P_Z and M_Z are: $\mbox{rank}\, P_Z = \mbox{tr}\, P_Z = \mbox{tr}\, Z(Z^\top Z)^{-1}Z^\top = \mbox{tr}\,(Z^\top Z)^{-1}Z^\top Z = \mbox{tr}\, I_{K+1} = K+1$ and $\mbox{rank}\, M_Z = \mbox{tr}\,(I_N - P_X) = \mbox{tr}\, I_N - \mbox{tr}\, P_Z = N - K - 1$.

Finally, the two matrices are orthogonal: $P_Z M_Z = P_Z(I - P_Z) = P_Z - P_Z = 0$, which means that they perform the projection of a vector on two orthogonal subspaces.

Getting back to Figure 3.2, P_X project y on the 2-D subspace (a plane) spanned by x_1 and x_2 and M_X project y on a 1-D subspace (the straight line orthogonal to the previous plane). M_X and P_X perform therefore an orthogonal decomposition of y in $\hat{y}$ and $\hat{\epsilon}$, which means that $\hat{y} + \hat{\epsilon} = y$ and that $\hat{y}^\top \hat{\epsilon} = 0$.

3.3.2 Frisch-Waugh theorem

Consider the regression of y on a set of regressors X which, for some reasons, is separated in two subsets X_1 and X_2. Suppose that we are only interested in the coefficients β_2 associated with X_2. The Frisch-Waugh theorem states that the same estimator $\hat{\beta}_2$ is obtained:

- by regressing y on X_1 and X_2,
- by first regressing y and each column of X_2 on X_1, then taking the residuals $M_1 y$ and $M_1 X_2$ of these regressions and finally regressing $M_1 y$ on $M_1 X_2$.

Figure 3.3 illustrates the Frisch-Waugh theorem.

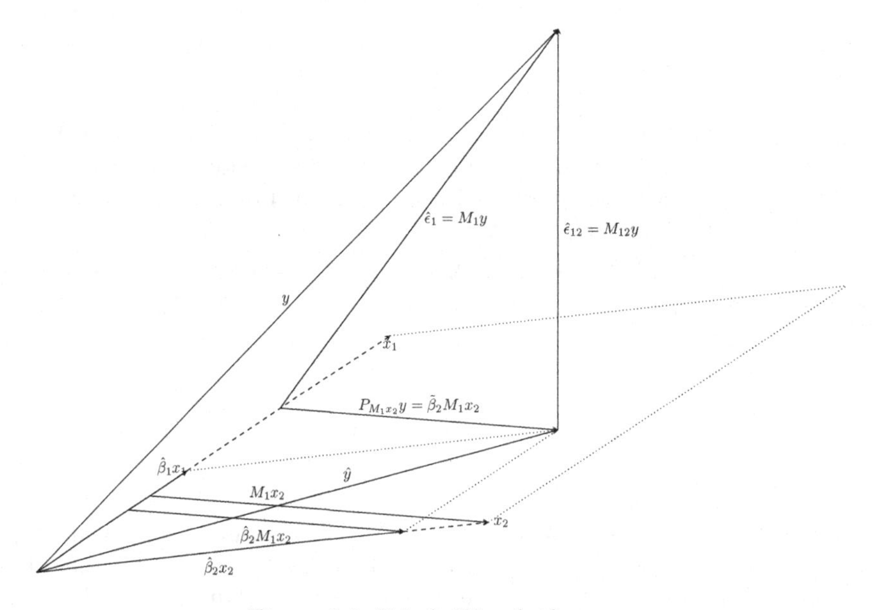

Figure 3.3: Frisch-Waugh theorem

The **first regression** is the regression of y on x_1 and x_2. We then get an orthogonal decomposition of y in the vector of fitted values $\hat{y} = P_{12} y$ and of the residuals $\hat{\epsilon}_{12} = M_{12} y$. We also show in this figure the decomposition of $\hat{y}$ in x_1 and x_2, which is represented by the sum of the two vectors $\hat{\beta}_1 x_1$ and $\hat{\beta}_2 x_2$. $\hat{\beta}_2$ is the estimator of x_2 on this first regression and is represented by the ratio between $\hat{\beta}_2 x_2$ and x_2.

The **second regression** is the regression of M_1y on M_1x_2. M_1x_2 is the residual of the regression of x_2 on x_1. Therefore, this vector lies in the line that is in the plane spanned by x_1 and x_2 and is orthogonal to x_1. M_1y is the residual of y on x_1; it is therefore orthogonal to x_1.

As both M_1y and $M_{12}y$ are orthogonal to x_1, so is the vector that joins those two vectors. Therefore, this vector is parallel to M_1x_2, and it is therefore the fitted value of the second regression (M_1y on M_1x_2), $P_{M_1x_2}y$. It is also equal to $\tilde{\beta}_2M_1x_2$, $\tilde{\beta}_2$ being the estimator of x_2 on the second regression. Note also that $\hat{\epsilon}_{12} = M_{12}y$ is the residual of the second regression, which is therefore the same as the residuals of the first regression.

Finally, consider the regression of $\hat{\beta}_2x_2$ on x_1. The residual of this regression is $\hat{\beta}_2M_1x_2$. As it lies on the plane spanned by x_1 and x_2 and is orthogonal to x_1, it is parallel to $P_{M_1x_2}y = \tilde{\beta}_2M_1x_2$. Moreover, the Frisch-Waugh theorem states that both vectors have the same length and are therefore identical, which means that $\tilde{\beta}_2 = \hat{\beta}_2$ and that the two regressions give identical estimators.

The Frisch-Waugh is easily demonstrated using some geometric arguments. Consider the regression with all the covariates:

$$y = X_1\hat{\beta}_1 + X_2\hat{\beta}_2 + M_{12}y$$

Then, premultiply both sides of the model by M_1:

$$M_1y = M_1X_1\hat{\beta}_1 + M_1X_2\hat{\beta}_2 + M_1M_{12}y$$

$M_1X_1\hat{\beta}_1$ is 0 as $X_1\hat{\beta}_1$ is obviously in the subset spanned by X_1 and therefore its projection on the orthogonal complement is 0. $M_{12}y$ is orthogonal to the subset spanned by X and is therefore also orthogonal to the subset spanned by X_1. Therefore $M_1M_{12}y = M_{12}y$. We therefore have:

$$M_1y = M_1X_2\hat{\beta}_2 + M_{12}y$$

For which the estimation is:

$$\hat{\beta}_2 = (X_2^\top M_1X_2)^{-1}X_2^\top M_1y$$

which is exactly the estimation obtained by regressing M_1y on M_1X_2. We finally note (an important result that will be used in Section 3.6.3), that $\hat{\epsilon}_1 = M_1y$, $\hat{\epsilon}_{12} = M_{12}y$ and $P_{M_1x_2}y$ form a right triangle, $\hat{\epsilon}_1$ being the hypotenuse. Therefore, using the Pythagorean theorem, we have:

$$\| \hat{\epsilon}_{12}^2 \| + \| P_{M_1X_2}y \| = \| \hat{\epsilon}_1^2 \| \tag{3.14}$$

3.4 Computation with R

To estimate the multiple linear model, we use as for the single linear model the `lm` function; the difference being that now, on the right side of the formula, we have several variables (here two), separated by the `+` operator. Actually, formulas have a much richer syntax that includes other operators, for example `*` and `:`. This will be discussed in Chapter 4.

```
slw_tot <- lm(log(gdp85) ~ log(i) + log(v), growth)
slw_tot %>% coef
## (Intercept)      log(i)      log(v)
##      9.6293      1.4780     -0.4573
```

3.4.1 Computation using matrix algebra

The estimator can also be computed "by hand", using matrix algebra. To start, we use the `model.frame` function which, as `lm`, has `formula` and `data` arguments. For pedagogical purposes, we add the `group` variable in the formula.

```
mf <- model.frame(log(gdp85) ~ log(i) + log(v) + group, growth)
head(mf, 3)
##     log(gdp85) log(i) log(v)  group
## 1        8.383 -1.423 -2.577  other
## 2        7.066 -2.847 -2.645 lqdata
## 3        6.976 -2.226 -2.604 lqdata
nrow(mf)
## [1] 107
nrow(growth)
## [1] 121
```

`model.frame` returns a data frame that contains the data required for estimating the model described in the formula. More precisely, it performs three tasks:

- it selects only the columns of the initial data frame that are required for the estimation,
- it transforms these variables if required, here `gdp85`, `i` and `v` are transformed in logarithms and the columns are renamed accordingly,
- it selects only the observations for which there are no missing values for the relevant variables; note here that `growth` has 121 rows and `mf` only 107 rows.

Several interesting elements of the model can be extracted from the model frame. The Z matrix is obtained using the `model.matrix` function, which also uses a `formula/data` interface, the data being the model frame `mf`:

```
Z <- model.matrix(log(gdp85) ~ log(i) + log(v) + group, mf)
head(Z, 3)
```

```
(Intercept) log(i) log(v) groupother grouplqdata groupoil
```

```
1            1 -1.423 -2.577         1          0          0
2            1 -2.847 -2.645         0          1          0
3            1 -2.226 -2.604         0          1          0
```

Note that the model matrix includes an intercept[1] and the `group` variable, which is a categorical variable is transformed into a set of dummy variables. More precisely, a dummy variable is created for all the modalities except the first one (`"oecd"`). The response is obtained using the `model.response` function:

```
y <- model.response(mf)
head(y, 3)
##     1     2     3
## 8.383 7.066 6.976
```

Once the model matrix and the response vector are created, the estimator can easily be computed using matrix operators provided by `R`. In particular:

- `%*%` is the matrix product operator (`*` performs an element per element product),
- `t()` transposes a matrix,
- `solve()` solves a linear system of equation or computes the inverse of a matrix,
- `crossprod` takes the inner products of two matrices (or of one matrix and a vector).

The most straightforward formula to get the OLS estimator is:

```
solve(t(Z) %*% Z) %*% t(Z) %*% y
```

```
                [,1]
(Intercept) 15.0991
log(i)       0.8332
log(v)       1.6089
groupother  -1.4135
grouplqdata -2.1321
groupoil    -0.8888
```

But `crossprod` is more efficient, $A^\top B$ being obtained using `crossprod(A, B)` and $A^\top A$ is either `crossprod(A, A)` or `crossprod(A)`. Therefore, $Z^\top Z$ and $Z^\top y$ are respectively obtained using:

```
crossprod(Z)
crossprod(Z, y)
```

Moreover, `solve` can be used to solve a system of linear equations: `solve(A, z)` compute the vector w such that $Aw = z$, which is $w = A^{-1}z$. Therefore, the OLS estimator can be computed using the more efficient and compact following code:

```
solve(crossprod(Z), crossprod(Z, y))
```

[1]As seen in Section 2.1.4.2; to remove it, one has to use either `+ 0` or `- 1` in the formula.

3.4.2 Efficient computation: QR decomposition

The efficient method used by `lm` to compute the OLS estimate is the QR decomposition:[2] the matrix of covariates Z can be written as the product of an orthonormal matrix Q of dimension $N \times (K+1)$ (therefore $Q^\top Q = I$) and an upper triangular matrix R of dimension $(K+1) \times (K+1)$. Then, the linear model can be written as:

$$y = Z\gamma + \epsilon = QR\gamma + \epsilon = Q\delta + \epsilon$$

with $\delta = R\gamma$. With Q as the matrix of covariates, the OLS estimator is $\hat{\delta} = (Q^\top Q)^{-1} Q^\top y = Q^\top y$, which is obtained without matrix inversion. Then $\hat{y} = Q\hat{\gamma} = QQ^\top y$ and $\hat{\epsilon} = (I - QQ^\top)y$. With Z as the matrix of covariates, the OLS estimator is $\hat{\gamma} = R^{-1}\delta$. The inverse of a triangular matrix can be very easily obtained with a very high numerical accuracy. Moreover, $\hat{\gamma}$ can be obtained without inverting R. With $K = 2$, we have:

$$\begin{pmatrix} \delta_0 \\ \delta_1 \\ \delta_2 \end{pmatrix} = \begin{pmatrix} r_{11} & r_{12} & r_{13} \\ 0 & r_{22} & r_{23} \\ 0 & 0 & r_{33} \end{pmatrix} \begin{pmatrix} \alpha \\ \beta_1 \\ \beta_2 \end{pmatrix} = \begin{pmatrix} r_{11}\alpha + r_{12}\beta_1 + r_{13}\beta_2 \\ r_{22}\beta_1 + r_{23}\beta_2 \\ r_{33}\beta_2 \end{pmatrix}$$

which can be solved recursively:

- $\beta_2 = \delta_2 / r_{33}$,
- $\beta_1 = (\delta_1 - r_{23}\beta_2)/r_{22}$,
- $\alpha = (\delta_0 - r_{12}\beta_1 - r_{13}\beta_2)/r_{11}$.

Finally, $(X^\top X)^{-1} = (R^\top Q^\top QR)^{-1} = (R^\top R)^{-1} = R^{-1}R^{-1\top}$.

We illustrate the computation of the OLS estimator using the QR decomposition and using the previously estimated growth model, without the `group` covariate. The QR decomposition is performed using the `qr` function:

```
Z <- model.matrix(log(gdp85) ~ log(i) + log(v), mf)
qrZ <- qr(Z)
```

`qr` returns an object of class `qr`, and the `qr.R` and `qr.Q` function can be used to retrieve the two matrices. We check that R is upper triangular and that $Q^\top Q = I$:

```
R <- qr.R(qrZ)
Q <- qr.Q(qrZ)
crossprod(Q)
```

```
          [,1]      [,2]      [,3]
[1,] 1.000e+00 3.849e-17 8.088e-17
[2,] 3.849e-17 1.000e+00 2.971e-17
[3,] 8.088e-17 2.971e-17 1.000e+00
```

```
R
```

[2]See Davidson and MacKinnon (1993), section 1.5, pp. 25-31.

```
  (Intercept) log(i)  log(v)
1      -10.34 19.193 27.1993
2        0.00  5.084 -0.4102
3        0.00  0.000  1.3600
```

$\hat{\delta}$ is then obtained as the cross-products of Q and y:

```
hdelta <- crossprod(Q, y) %>% drop
hdelta
## [1] -83.678   7.702  -0.622
```

and $\hat{\gamma}$ is obtained by solving recursively $\hat{\delta} = R\hat{\gamma}$:

```
beta_2 <- hdelta[3] / R[3, 3]
beta_1 <- (hdelta[2] - R[2, 3] * beta_2) / R[2, 2]
alpha <- (hdelta[1] - R[1, 2] * beta_1  - R[1, 3] * beta_2) / R[1, 1]
c(alpha, beta_1, beta_2)
## [1]  9.6293  1.4780 -0.4573
```

3.5 Properties of the estimators

In this section, we'll briefly analyze the statistical properties of the OLS estimator with more than one covariate. Most of these properties are similar to the one we have described in Chapter 2. They'll be presented in this section using matrix algebra.

3.5.1 Unbiasedness of the OLS estimator

The vector of slopes can be written as a linear combination of the vector of response and then of vector of the errors:

$$\begin{array}{rcl}\hat{\beta} &=& (X^\top \tilde{I} X)^{-1} X^\top \tilde{I} y \\ &=& (X^\top \tilde{I} X)^{-1} X^\top \tilde{I}(X\beta + \epsilon) \\ &=& \beta + (X^\top \tilde{I} X)^{-1} X^\top \tilde{I}\epsilon \\ &=& \beta + (\tilde{X}^\top \tilde{X})^{-1} \tilde{X}^\top \epsilon\end{array} \tag{3.15}$$

The expected value of $\hat{\beta}$ conditional on X is:

$$\mbox{E}(\hat{\beta} \mid X) = \beta + (\tilde{X}^\top \tilde{X})^{-1} \tilde{X}^\top \mbox{E}(\epsilon \mid X)$$

The unbiasedness condition is therefore that $\mathrm{E}(\epsilon \mid X) = 0$, which is a direct generalization of the result obtained for the simple linear regression model, namely ϵ has a constant expected value (that can be set to 0 without any restriction) whatever the value of the covariates. It implies also that the population covariance between the errors and any of the covariates is 0.

3.5.2 Variance of the OLS estimator

The variance of $\hat{\beta}$ is now a matrix of variances and covariances:

$$\mathrm{V}(\hat{\beta} \mid X) = \mathrm{E}\left[(\hat{\beta} - \beta)(\hat{\beta} - \beta)^\top \mid X\right]$$

Using Equation 3.15:

$$\mathrm{V}(\hat{\beta} \mid X) = \mathrm{E}\left[(\tilde{X}^\top \tilde{X})^{-1}\tilde{X}^\top \epsilon\epsilon^\top \tilde{X}(\tilde{X}^\top \tilde{X})^{-1} \mid X\right]$$

$$\mathrm{V}(\hat{\beta} \mid X) = \frac{1}{N}\left(\frac{1}{N}\tilde{X}^\top \tilde{X}\right)^{-1}\left[\frac{1}{N}\mathrm{E}(\tilde{X}^\top \epsilon\epsilon^\top \tilde{X} \mid X)\right]\left(\frac{1}{N}\tilde{X}^\top \tilde{X}\right)^{-1} \quad (3.16)$$

This is a **sandwich** formula, the **meat**: $\frac{1}{N}\mathrm{E}\left(X'\bar{I}\epsilon\epsilon^\top \bar{I}X \mid X\right)$ being surrounded by two slices of **bread**: $\left(\frac{1}{N}\tilde{X}^\top \tilde{X}\right)^{-1}$. Note that the two matrices are square and of dimension K. The bread is just the inverse of the covariance matrix of the covariates. The meat is the variance of the score vector, i.e., the vector of the first-order conditions. For $K = 2$, it is the expected value of:

$$\frac{1}{N}\begin{pmatrix} \left(\sum_{n=1}^N (x_{n1} - \bar{x}_1)\epsilon_n)\right)^2 & \left(\sum_{n=1}^N (x_{n1} - \bar{x}_1)\epsilon_n)\right)\left(\sum_{n=1}^N (x_{n2} - \bar{x}_2)\epsilon_n)\right) \\ \left(\sum_{n=1}^N (x_{n1} - \bar{x}_1)\epsilon_n)\right)\left(\sum_{n=1}^N (x_{n2} - \bar{x}_2)\epsilon_n)\right) & \left(\sum_{n=1}^N (x_{n2} - \bar{x}_2)\epsilon_n)\right)^2 \end{pmatrix}$$

which is a generalization of the single regression case where the "meat" reduces to the scalar $\left(\sum_{n=1}^N (x_n - \bar{x})\epsilon_n)\right)^2$. As for the simple regression model, the formula of the variance simplifies with the hypothesis that the errors are homoskedastic ($\mathrm{E}(\epsilon_n^2 \mid x) = \sigma_\epsilon^2$) and uncorrelated ($\mathrm{E}(\epsilon_n\epsilon_m \mid x) = 0 \; \forall \; m \neq n$). In this case, the meat reduces to $\sigma_\epsilon^2 \frac{1}{N}\tilde{X}^\top \tilde{X}$, i.e., up to a scalar to the matrix of covariance of the covariates, and Equation 3.16 becomes:

$$V(\hat{\beta}) = \sigma_\epsilon^2(\tilde{X}^\top \tilde{X})^{-1} \quad (3.17)$$

which can be rewritten, using Equation 3.10:

$$\mathrm{V}(\hat{\beta}) = \frac{\sigma_\epsilon^2}{N\hat{\sigma}_1\hat{\sigma}_2(1 - \hat{\rho}_{12}^2)}\begin{pmatrix} \frac{\hat{\sigma}_2}{\hat{\sigma}_1} & -\hat{\rho}_{12} \\ -\hat{\rho}_{12} & \frac{\hat{\sigma}_1}{\hat{\sigma}_2} \end{pmatrix}$$

from which we get:

$$\sigma_{\hat{\beta}_k} = \frac{\sigma_\epsilon}{\sqrt{N}\hat{\sigma}_k\sqrt{1-\hat{\rho}_{12}^2}} \text{ for } k=1,2,\ \hat{\sigma}_{\hat{\beta}_1\hat{\beta}_2} = -\frac{\hat{\rho}_{12}\sigma_\epsilon^2}{N\hat{\sigma}_1\hat{\sigma}_2(1-\hat{\rho}_{12}^2)} \text{ and } \hat{\rho}_{\hat{\beta}_1\hat{\beta}_2} = -\hat{\rho}_{12} \quad (3.18)$$

First remark that if $\hat{\rho}_{12} = 0$, which means that the two covariates are uncorrelated, the formula for the standard deviation of a slope in the multiple regression model reduces to the formula of the single regression model, which means that the standard deviation is proportional to:

- the standard deviation of the error,
- the inverse of the standard deviation of the corresponding covariate,
- the inverse of the square root of the sample size.

When the two covariates are correlated, the last term $1/\sqrt{1-\hat{\rho}_{12}^2}$ is added and inflates the standard deviation. This means that the more the covariates are correlated (whatever the sign of the correlation) the larger is the standard deviation of the slope. The intuition is that, if the two covariates are highly correlated, it is difficult to estimate precisely the separate effect of each of them.

σ_ϵ^2 being unknown, $\sigma^2_{\hat{\beta}}$ can't be computed. If the error were observed, a natural estimator of σ_ϵ^2 would be $\sum_{n=1}^N(\epsilon_n - \bar{\epsilon})^2/N$. As the errors are unknown, one can use the residuals instead, which are related to the errors by the relation: $\hat{\epsilon} = M_Z y = M_Z(Z\gamma + \epsilon) = M_Z\epsilon$, the last equality standing because $Z\gamma$ is a vector of the subspace defined by the columns of Z and therefore $M_Z Z\gamma = 0$. Therefore, we have: $\hat{\epsilon}^\top\hat{\epsilon} = \epsilon^\top M_Z\epsilon$; as it is a scalar, it is equal to its trace. Using the rule of permutation, we get: $\hat{\epsilon}^\top\hat{\epsilon} = \epsilon^\top M_Z \epsilon = \mbox{tr}\, M_Z\epsilon\epsilon^\top$. With spherical disturbances, we have $\mbox{E}(\hat{\epsilon}^\top\hat{\epsilon}) = \mbox{tr}\, M_Z\sigma_\epsilon^2 I = \sigma_\epsilon^2 \mbox{tr} M_Z = (N-K-1)\sigma_\epsilon^2$. Therefore, an unbiased estimator of σ_ϵ^2 is:

$$\dot{\sigma}_\epsilon^2 = \frac{\hat{\epsilon}^\top\hat{\epsilon}}{N-K-1} = \frac{\sum_{n=1}^N \hat{\epsilon}_n^2}{N-K-1} \quad (3.19)$$

and replacing σ_ϵ^2 by $\dot{\sigma}_\epsilon^2$ in Equation 3.17, we get an unbiased estimator of the covariance matrix of the estimators:

$$\hat{V}(\hat{\beta}) = \dot{\sigma}_\epsilon^2(\tilde{X}^\top\tilde{X})^{-1} \quad (3.20)$$

We now go back to the estimation of the growth model. As in Mankiw, Romer, and Weil (1992), we use a restricted sample by excluding countries for which most of the GDP is linked to oil extraction and those with low quality data.

```
growth_sub <- growth %>% filter(! group %in% c("oil", "lqdata"))
slw_tot <- lm(log(gdp85) ~ log(i) + log(v), growth_sub)
```

The covariance matrix of the estimators is obtained using the `vcov` function:

```
vcov(slw_tot)
```

```
            (Intercept)  log(i)  log(v)
(Intercept)      2.3811 0.13554 0.80956
log(i)           0.1355 0.02922 0.03213
log(v)           0.8096 0.03213 0.28501
```

The `summary` method computes detailed results of the regression, in particular the table of coefficients, a matrix that can be extracted using the `coef` method:

```
slw_tot %>% summary %>% coef
```

```
            Estimate Std. Error t value  Pr(>|t|)
(Intercept)    5.346     1.5431   3.464 8.990e-04
log(i)         1.318     0.1709   7.708 5.383e-11
log(v)        -2.017     0.5339  -3.778 3.224e-04
```

3.5.3 The OLS estimator is BLUE

The demonstration made in Section 2.1.4 for the simple linear model can easily be extended to the multiple regression model. The OLS estimator is: $\hat{\gamma} = (Z^\top Z)^{-1}Z^\top y = \gamma + (Z^\top Z)^{-1}Z^\top\epsilon$. Consider another linear estimator:

$$\tilde{\gamma} = Ay = \left[(Z^\top Z)^{-1}Z^\top + D\right] y = (I + DZ)\gamma + \left[(Z^\top Z)^{-1}Z^\top + D\right]\epsilon$$

The unbiasedness of $\tilde{\gamma}$ implies that $DZ = 0$, so that:

$$\tilde{\gamma} - \gamma = \left[(Z^\top Z)^{-1}Z^\top + D\right]\epsilon = (\hat{\gamma} - \gamma) + D\epsilon = (\hat{\gamma} - \gamma) + Dy$$

because $DZ = 0$ implies that $D\epsilon = Dy$. Therefore, $\tilde{\gamma} - \hat{\gamma} = Dy$. The covariance between the OLS estimator and the vector of differences of the two estimators is:

$$\mathrm{E}\left[(\tilde{\gamma} - \hat{\gamma})(\hat{\gamma} - \gamma)^\top\right] = \mathrm{E}\left[D\epsilon\epsilon^\top Z(Z^\top Z)^{-1}\right]$$

with spherical disturbances, this reduces to:

$$\mathrm{E}\left[(\tilde{\gamma} - \hat{\gamma})(\hat{\gamma} - \gamma)^\top\right] = \sigma_\epsilon^2 DZ(Z^\top Z)^{-1} = 0$$

Therefore, we can write the variance of $\tilde{\gamma}$ as:

$$\mathrm{V}(\tilde{\gamma}) = \mathrm{V}\left[\hat{\gamma} + (\tilde{\gamma} - \hat{\gamma})\right] = \mathrm{V}(\hat{\gamma}) + \mathrm{V}(Dy)$$

as $\mathrm{V}(Dy)$ is a covariance matrix, it is semi-definite positive and, therefore, the difference between the variance matrix of any unbiased linear estimator and the OLS estimator is a semi-definite positive matrix, which means that the OLS estimator is the most efficient linear unbiased estimator.

3.5.4 Asymptotic properties of the OLS estimator

Asymptotic properties of the multiple regression model are direct extensions of those we have seen for the simple regression model. With $\mbox{E}(\hat{\beta}) = \beta$ and $\mbox{V}(\hat{\beta}) = \frac{\sigma_\epsilon^2}{N}\left(\frac{1}{N}\tilde{X}^\top X\right)^{-1}$, the OLS estimator is consistent ($\mbox{plim}\,\hat{\beta} = \beta$) if the covariance matrix of the covariates $\frac{1}{N}\tilde{X}^\top \tilde{X}$ converges to a finite matrix. The central-limit theorem implies that:

$$\sqrt{N}(\hat{\beta}_N - \beta) \xrightarrow{d} \mathcal{N}\left(0, \sigma_\epsilon^2\left(\frac{1}{N}\tilde{X}^\top\tilde{X}\right)^{-1}\right)$$

or

$$\hat{\beta}_N \overset{a}{\sim} \mathcal{N}\left(\beta, \frac{\sigma_\epsilon^2}{N}\left(\frac{1}{N}\tilde{X}^\top\tilde{X}\right)^{-1}\right)$$

3.5.5 The coefficient of determination

From Equation 1.14, one way to write the coefficient of determination for the simple linear regression model is:

$$R^2 = 1 - \frac{\sum_{n=1}^N \hat{\epsilon}_n^2}{\sum_{n=1}^N (y_n - \bar{y})^2} = 1 - \frac{\hat{\sigma}_\epsilon^2}{\hat{\sigma}_y^2}$$

The last term is obtained by dividing the residual sum of squares and the total variation of y by the sample size N. This gives rise to biased estimators of the variances of the errors and of the response. Consider now the unbiased estimators: the numerator should then be divided by $N - K - 1$ and $\dot{\sigma}_\epsilon^2$ is obtained. Similarly, the denominator should be divided by $N - 1$ to obtain an unbiased estimation of the variance of the response. We then obtain the **adjusted coefficient of determination**:

$$\bar{R}^2 = 1 - \frac{\dot{\sigma}_\epsilon^2}{\dot{\sigma}_y^2} = 1 - \frac{\sum_{n=1}^N \hat{\epsilon}_n^2/(N-K-1)}{\sum_{n=1}^N (y_n - \bar{y})^2/(N-1)} = 1 - \frac{N-1}{N-K-1}\frac{\sum_{n=1}^N \hat{\epsilon}_n^2}{\sum_{n=1}^N (y_n - \bar{y})^2}$$

R^2 necessarily increases when one more covariate is added to the regression because, even if this covariate is irrelevant, its sample correlation with the response will never be exactly 0 and therefore the sum of square will decrease. This is not the case with $\bar{R}^2$ because $\dot{\sigma}_\epsilon^2$ is the ratio of two terms which both increase when a covariate is added. Note also that $\bar{R}^2$ is not necessarily positive.

3.6 Confidence interval and test

In Section 2.3, we have seen how to compute a confidence interval for a single parameter and how to perform a test for one hypothesis. The confidence interval was a segment, i.e., a range of values that contains the true value of the parameter with a given probability, and the tests were performed using a normal or a Student distribution. Of course, the same kind of analysis can be performed with a multiple regression. But, in this latter case:

- a confidence interval can also be computed for several coefficients one at a time,
- tests of multiple hypotheses can be performed, using a chi-squared or a Fisher distribution.

To illustrate these points, we'll use the Solow model. Remember that the model to estimate is:

$$\ln y = \alpha + \beta_i \ln i + \beta_v \ln v + \epsilon$$

where y is the per capita gdp, i the investment rate and v the sum of the labor force growth rate, the depreciation rate and the technical progress rate. Moreover, the relation between β_i and β_v and the structural parameter κ, which is the share of profits in the GDP is $\beta_i = -\beta_v = \kappa/(1-\kappa)$. We'll analyze the confidence interval for the couple of coefficients (β_i, β_v) and we'll test two hypotheses:

- the first is imposed by the model; we must have $\beta_i + \beta_v = 0$,
- the second corresponds to a reasonable value of the share of profits, which is approximately one-third; therefore, we'll test the hypothesis that $\kappa = 1/3$, which implies that $\beta_i = 0.5$.

Note that these two hypotheses imply that $\beta_v = -0.5$.

3.6.1 Simple confidence interval and test

The asymptotic distribution of the estimators is a multivariate normal distribution.[3] The distribution of one estimator (say $\hat{\beta}_1$)[4] is a univariate normal distribution with, for the two covariates case, a standard deviation equal to: $\sigma_{\hat{\beta}_1} = \frac{\sigma_\epsilon}{\sqrt{N}\hat{\sigma}_1\sqrt{1-\hat{\rho}_{12}^2}}$. Therefore $(\hat{\beta}_1 - \beta_1)/\sigma_{\hat{\beta}_1}$ follows (exactly if ϵ is normal) a standard normal distribution. Confidence interval and tests for a coefficient are therefore computed exactly as for the case of the simple linear regression model. In particular, using Equation 3.18 and Equation 3.19, we get:

$$t_k = \frac{\hat{\beta}_k - \beta_k}{\hat{\sigma}_{\hat{\beta}_k}} = \frac{\hat{\beta}_k - \beta_k}{\frac{\dot{\sigma}_\epsilon}{\sqrt{N}\hat{\sigma}_1\sqrt{1-\hat{\rho}_{12}^2}}} \tag{3.21}$$

that follows exactly a Student distribution with $N-K-1$ degrees of freedom if the errors

[3] If the errors are normal, the exact distribution of the estimators is normal (see Section 2.2.2).

[4] Or of a linear combination of several estimators.

are normal and asymptotically a normal distribution whatever the distribution of the errors. Therefore, $1-\alpha$ confidence intervals are: $\hat{\beta}_k \pm \mbox{cv}_{1-\alpha/2}\dot{\sigma}_{\hat{\beta}_k}$ where $\mbox{cv}_{1-\alpha/2}$ is the critical value of either a Student or a normal distribution. For a given hypothesis: $H_0: \beta_k = \beta_{k0}$, $t_{k0} = (\hat{\beta}_k - \beta_{k0})/\hat{\sigma}_{\hat{\beta}_k}$ is a draw on a normal or a Student distribution if H_0 is true.

Remember that "reasonable" values of the two slopes in our growth model should be $+/-0.5$. We check whether these values are in the confidence intervals, using the `confint` function:

```
confint(slw_tot, level = 0.95)
```

```
              2.5 % 97.5 %
(Intercept)  2.2698  8.422
log(i)       0.9768  1.658
log(v)      -3.0814 -0.953
```

0.5 and -0.5 are not in the 95% confidence interval for respectively $\hat{\beta}_i$ and $\hat{\beta}_v$. The hypotheses that the true values of the parameters are 0 is very easy to compute, as they are based on the t statistics that are routinely returned by the `summary` method for `lm` objects. But, in our example, the hypothesis that $\beta_i = 0$ is the hypothesis that the share of profits is 0, which is of little interest. More interestingly, we could check the hypothesis that the coefficients are equal to ± 0.5. In this case, we can "manually" compute the test statistics by extracting the relevant elements in the matrix returned by `coef(summary(x))`:

```
v <- slw_tot %>% summary %>% coef
(v[2:3, 1] - c(0.5, - 0.5)) / v[2:3, 2]
## log(i) log(v)
##  4.783 -2.842
```

which confirms that both hypotheses are rejected. The same kind of linear hypothesis on one coefficient can be simpler tested by using a different parametrization of the same model. The trick is to write the model in such a way that a subset of coefficients are 0 if the hypothesis is true. For example, in order to test the hypothesis that $\beta_i = 0.5$, we must have in the model $(\beta_i - 0.5)\ln i$; therefore, the term $-0.5 \ln i$ is added on the right side of the formula and should therefore be added also on the left side. Adding also $0.5 \ln v$ on both sides of the formula, we finally get:

$$(\ln y - 0.5 \ln i + 0.5 \ln v) = \alpha + (\beta_i - 0.5)\ln i + (\beta_v + 0.5)\ln v + \epsilon$$

and the two slopes are now equal to 0 under H_0.

```
slw_totb <- lm(log(gdp85 * sqrt(v / i) ) ~ log(i) + log(v), growth_sub)
slw_totb %>% summary %>% coef
```

```
            Estimate Std. Error t value  Pr(>|t|)
(Intercept)   5.3459     1.5431   3.464 8.990e-04
log(i)        0.8176     0.1709   4.783 8.924e-06
log(v)       -1.5172     0.5339  -2.842 5.830e-03
```

which gives exactly the same values for the t statistics.

A simple hypothesis may also concern a linear combination of several coefficients and not the value of one coefficient. For example, the structural growth model implies that $\beta_i + \beta_v = 0$. If the hypothesis is true, $\mbox{E}(\hat{\beta}_i + \hat{\beta}_v) = 0$, the variance being $\mbox{V}(\hat{\beta}_i + \hat{\beta}_v) = \hat{\sigma}^2_{\hat{\beta}_1} + \hat{\sigma}^2_{\hat{\beta}_2} + 2\hat{\sigma}_{\hat{\beta}_1\hat{\beta}_2}$, the statistic can then be computed by extracting the relevant elements of the covariance matrix of the fitted model:

```
v <- vcov(slw_tot)
v_sum <- v[2, 2] + v[3, 3] + 2 * v[2, 3]
stat_sum <- (coef(slw_tot)[2] + coef(slw_tot)[3]) %>% unname
t_sum <- stat_sum / sqrt(v_sum) %>% unname
pval_sum <- 2 * pt(abs(t_sum), df = df.residual(slw_tot),
                   lower.tail = FALSE)
c(stat = stat_sum, t = t_sum, pv = pval_sum)
##     stat        t       pv
## -0.6996 -1.1372  0.2592
```

The hypothesis is therefore not rejected, even at the 10% level. Once again, such a linear hypothesis can be more easily tested using a different parametrization. Introducing in the model the term $(\beta_i + \beta_v) \ln i$, subtracting $\beta_v \ln_i$ and rearranging terms:

$$\begin{array}{rcl} \ln y &=& \alpha + (\beta_i + \beta_v) \ln i + \beta_v \ln v - \beta_v \ln i + \epsilon \\ &=& \alpha + (\beta_i + \beta_v) \ln i + \beta_v \ln \frac{v}{i} + \epsilon \end{array}$$

We then have a model for which the two covariates are now $\ln i$ and $\ln v/i$, the hypothesis being that the coefficient associated to $\ln i$ is equal to 0.

```
slw_totc <- lm(log(gdp85) ~ log(i) + log(v / i), growth_sub)
slw_totc %>% summary %>% coef
```

```
            Estimate Std. Error t value  Pr(>|t|)
(Intercept)   5.3459     1.5431   3.464 0.0008990
log(i)       -0.6996     0.6152  -1.137 0.2592063
log(v/i)     -2.0172     0.5339  -3.778 0.0003224
```

3.6.2 Joint confidence interval and test of joint hypothesis

3.6.2.1 Joint confidence interval

We now consider the computation of confidence interval for more than one parameter. The distribution of $\hat{\beta}$ is:

$$\hat{\beta} \overset{a}{\sim} \mathcal{N}\left(\beta, \frac{\sigma_\epsilon^2}{N}\left(\frac{1}{N}\tilde{X}^\top\tilde{X}\right)^{-1}\right)$$

In the simple regression model, subtracting the expected value and dividing by the standard deviation, we get a standard normal deviate. Taking the square, we get a χ^2 with 1 degree of freedom. If the K slopes are uncorrelated, $\sum_{k=1}^K (\hat{\beta}_k - \beta_{k0})^2/\sigma^2_{\hat{\beta}_k}$ is a χ^2 with K degrees of freedom. If the slopes are correlated, this correlation should be "corrected"; more precisely, a quadratic form of the vector of slopes in deviation from its expectation with the inverse of its variance should be computed:

$$q_K = (\hat{\beta} - \beta)^\top \mathrm{V}(\hat{\beta})^{-1}(\hat{\beta} - \beta) = (\hat{\beta} - \beta)^\top \frac{N}{\sigma_\epsilon^2}\left(\frac{1}{N}\tilde{X}^\top \tilde{X}\right)(\hat{\beta} - \beta) \sim \chi^2_K$$

For $K = 2$, we have:

$$q_2 = \left(\hat{\beta}_1 - \beta_1, \hat{\beta}_2 - \beta_2\right) \frac{N}{\sigma_\epsilon^2} \begin{pmatrix} \hat{\sigma}_1^2 & \hat{\sigma}_{12} \\ \hat{\sigma}_{12} & \hat{\sigma}_2^2 \end{pmatrix} \begin{pmatrix} \hat{\beta}_1 - \beta_1 \\ \hat{\beta}_2 - \beta_2 \end{pmatrix}$$

$$q_2 = \frac{N}{\sigma_\epsilon^2}\left[\hat{\sigma}_1^2(\hat{\beta}_1 - \beta_1)^2 + \hat{\sigma}_2^2(\hat{\beta}_2 - \beta_2)^2 + 2\hat{\sigma}_{12}(\hat{\beta}_1 - \beta_1)(\hat{\beta}_2 - \beta_2)\right] \quad (3.22)$$

Using Equation 3.21, this expression can also be rewritten in terms of the Student statistics t_k:

$$q_2 = \frac{1}{1 - \hat{\rho}_{12}^2}\left(t_1^2 + t_2^2 + 2\hat{\rho}_{12}t_1t_2\right) = \frac{1}{1 - \hat{\rho}_{\hat{\beta}_1\hat{\beta}_2}^2}\left(t_1^2 + t_2^2 - 2\hat{\rho}_{\hat{\beta}_1\hat{\beta}_2}t_1t_2\right) \quad (3.23)$$

the last equality resulting from the fact that the coefficient of correlation of two slopes is the opposite of the coefficient of correlation of the corresponding covariates (see Equation 3.18). Therefore, a $(1 - \alpha)$ confidence interval for a couple of coefficients (β_1, β_2) is the set of values for which Equation 3.23 is lower than the critical value for a χ^2 with 2 degrees of freedom, which is, for example, 5.99 for two degrees of freedom and $1 - \alpha = 0.95$. Equating Equation 3.23 to this critical value, we get the equation of an ellipse, with two particular nested cases:

- if $\hat{\rho}_{12} = 0$, i.e., if the two covariates and therefore the two coefficients are uncorrelated, the expression reduces to the sum of squares of the t statistics, which is an ellipse with vertical and horizontal tangents,
- if $\hat{\rho}_{12} = 0$ and $\hat{\sigma}_1 = \hat{\sigma}_2$, the expression reduces further to the equation of a circle.

Equation 3.23 can't be computed as it depends on σ_ϵ which is unknown. Replacing σ_ϵ by $\dot{\sigma}_\epsilon$, we get:

$$\hat{q}_2 = \frac{1}{1 - \hat{\rho}_{12}^2}\left(\hat{t}_1^2 + \hat{t}_2^2 + 2\hat{\rho}_{12}\hat{t}_1\hat{t}_2\right) \quad (3.24)$$

where $\hat{q}_2$ now follows asymptotically a χ^2 distribution with 2 degrees of freedom. If the errors are normal, dividing by K (here 2), we get an exact Fisher F distribution with 2 and $N - K - 1$ degrees of freedom. In the simple case of no correlation between the covariates, the χ^2 and the F statistics are therefore the sum and the mean of the squares of the t statistics. As for the Student distribution, which converges in distribution to a normal distribution, $K\times$ the F statistic converges in distribution to a χ^2 with K degrees of freedom. For example, with $K = 2$, the critical value for a F distribution with 2 and ∞ degrees of freedom is half the corresponding χ^2 value with 2 degrees of freedom (5.99 for the 95% confidence level), which is 2.996. The confidence interval is represented in Figure 3.4; the circle point is the point estimation, the vertical and horizontal segments are the separate confidence intervals for both coefficients at the 95% level. We've also added a diamond point that corresponds to the hypothesis $\kappa = 1/3$, which implies that $\beta_i = -\beta_v = \frac{\kappa}{1-\kappa} = 0.5$. Finally, we add a line with a slope equal to -1 and an intercept equal to 0 which corresponds to the hypothesis that $\beta_i = -\beta_v$.

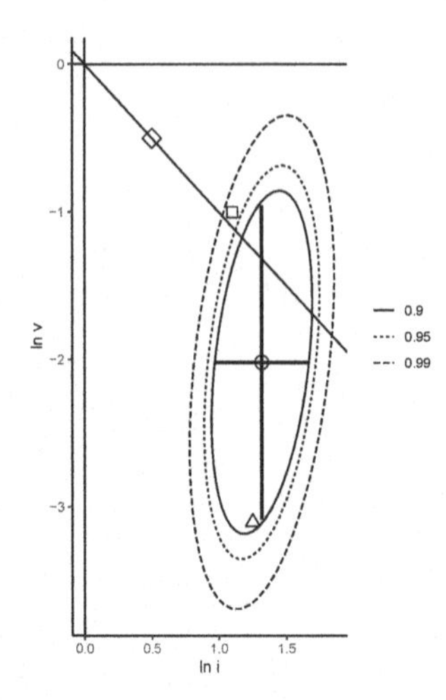

Figure 3.4: Ellipse of confidence for the two coefficients

The confidence ellipse is a "fat sausage"[5] with the long part of the sausage slightly oriented in the lower-left/upper-right direction. This is because, as we have seen previously (see Figure 3.1), the two covariates exhibit a small negative correlation, which implies a small positive correlations between $\hat{\beta}_i$ and $\hat{\beta}_v$. The ellipse is also "higher" than "wide", because v has a smaller variance than i and therefore $\hat{\beta}_v$ has a larger variance than $\hat{\beta}_i$. Note the difference between a set of two simple hypotheses and a joint hypothesis. Consider for example:

- $(\beta_i = 1.1, \beta_v = -1)$ represented by a square point: both simple hypotheses are not rejected (the two values are in the unidimensional confidence interval), but the joint hypothesis is rejected, as the corresponding square point is outside the 95% confidence interval ellipse,
- $(\beta_i = 1.25, \beta_v = -3.1)$ represented by a triangle point: the simple hypothesis $\beta_i = 1.25$ is not rejected, the simple hypothesis $\beta_b = -3.1$ is rejected, but the joint hypothesis is not rejected, the triangle point being inside the 95% confidence interval ellipse.

The hypothesis that the two coefficients sum to 0 is not rejected as some points of the straight line that figures this hypothesis are in the confidence ellipse. Concerning the hypothesis that $\beta_i = 0.5$ and $\beta_v = -0.5$, the two simple hypotheses and the joint hypothesis are rejected at the 95% confidence level; the two estimates are neither in the segments representing the simple confidence interval nor inside the ellipse figuring the joint confidence interval.

[5]Stock and Watson (2015), p. 235.

3.6.2.2 Joint hypothesis

To test a joint hypothesis for the values of a couple of parameters (β_{10}, β_{20}), we have just seen that we can simply check whether the corresponding point is inside or outside the confidence interval ellipse. We can also compute the statistic given in Equation 3.24 for $(\beta_1 = \beta_{10}, \beta_2 = \beta_{20})$ and compare it to the critical value.

The statistic is computed using the elements of the matrix returned by `coef(summary(x))`, which contains in particular the estimations and their standard deviations. We first compute the t statistics corresponding to the two simple hypothesis $\beta_i = 0.5$ and $\beta_v = -0.5$:

```
sc <- slw_tot %>% summary %>% coef
t_i <- (sc[2, 1] - 0.5) / sc[2, 2]
t_v <- (sc[3, 1] + 0.5) / sc[3, 2]
```

We then apply the simplified formula, which is the sum or the mean of the squares of the t statistics:

```
St2 <- t_i ^ 2 + t_v ^ 2
Mt2 <- (t_i ^ 2 + t_v ^ 2) / 2
c(St2, Mt2)
## [1] 30.95 15.47
```

for which the distributions under H_0 are respectively χ^2 and F, with the following critical values:

```
qchisq(0.95, df = 2)
## [1] 5.991
qf(0.95, df1 = 2, df2 = df.residual(slw_tot))
## [1] 3.124
```

The critical values being much smaller than the computed statistics, the joint hypothesis is clearly rejected. The exact formula corrects the correlation between the two estimators using the coefficient of correlation between the two covariates. We compute this coefficient using the data frame of the fitted model, which is obtained using the `model.frame` function. Using the initial data frame `growth` wouldn't give the correct value, as the estimation is not performed on the full data set because of missing data and because the estimation is performed on a subsample with some groups of countries that are excluded:

```
mf <- model.frame(slw_tot)
r_iv <- summarise(mf, r = cor(`log(i)`, `log(v)`)) %>% pull(r)
r_iv
## [1] -0.352
```

Note that the names of the covariates are not regular names as they contain parentheses, therefore they should be surrounded by the ` sign. The coefficient of correlation between the two covariates is -0.352. We obtain:

```
Fstat <- (t_i ^ 2 + t_v ^ 2 + 2 * r_iv * t_i * t_v) / (1 - r_iv ^ 2) / 2
```

```
Fstat
## [1] 23.13
```

The statistic (23.125) is slightly greater than the approximative value (15.475) which was previously computed; we therefore reject once again the null hypothesis.

3.6.3 The three tests

In the previous subsection, we started with a general (unconstrained) model, we constructed a confidence ellipse for the two parameters, and we were able to test a set of hypotheses, either by checking whether the values of the parameters corresponding to the hypotheses were inside the confidence ellipse, or by computing the value of the statistic for the tested values of the parameters. Actually, this testing principle, based on the unconstrained model, is just one way of testing a hypothesis. The geometry of least squares and the Frisch-Waugh theorem highlights the fact that any set of hypotheses can be tested using the fact that this set of hypotheses gives rise to two models: a **constrained model**, which imposes the hypotheses and an **unconstrained model**, which doesn't impose the hypotheses. The same test can be performed using the constrained model, the unconstrained model or both, which give rise to three test principles:[6]

- **Wald** test is based only on the unconstrained model,
- **lagrange multiplier** or **score** test is based on the constrained model,
- **likelihood ratio** test is based on the comparison between the two models.

A set of J linear hypotheses is written as:

$$R\gamma = q$$

where R is a matrix of dimension $J \times (K+1)$ and q is a vector of length J. J, the number of hypotheses, is necessarily lower or equal to K. Actually, a set of J hypotheses can always be rewritten as a model of the form $y = X_1\beta_1 + X_2\beta_2 + \epsilon$, the hypothesis being $\mathrm{H}_0 : \beta_2 = 0$. In this setting, the three tests are easily constructed using the Frisch-Waugh theorem, with $H_0 : \beta_2 = 0$.

3.6.3.1 Wald test

The Wald test is based on the unconstrained model, for which a vector of slopes $\hat{\beta}_2$ is estimated. Using the Frisch-Waugh theorem, this vector can be obtained as the regression of the residuals of y on X_1 ($M_1 y$) on the residuals of every column of X_2 on X_1 ($M_1 X_2$). Then, $\hat{\beta}_2 = (X_2^\top M_1 X_2)^{-1} X_2^\top M_1 y$, with expected value and variance equal to β_2 (0 under H_0) and $\mathrm{V}(\hat{\beta}_2) = \sigma_\epsilon^2 (X_2^\top M_1 X_2)^{-1}$. Convergence in distribution implies that:

$$\hat{\beta}_2 \overset{a}{\sim} \mathcal{N}\left(0, \mathrm{V}(\hat{\beta}_2)\right)$$

[6]This section is largely based on Davidson and MacKinnon (1993), section 3.6, pp. 88-94. The three "classical" tests are often understood as tests suitable for models estimated by maximum likelihood (see Section 5.3.1). Davidson and MacKinnon (1993) advocate the presentation of the three test principles for other estimators, including the linear regression model.

The distribution of the quadratic form of a centered vector of normal random variables of length J with the inverse of its covariance matrix is a χ^2 with J degrees of freedom:

$$\text{wald} = \hat{\beta}_2^\top \text{V}(\hat{\beta}_2)^{-1}\hat{\beta}_2 = \frac{\hat{\beta}_2^\top (X_2^\top M_1 X_2)\hat{\beta}_2}{\sigma_\epsilon^2}$$

which is also, replacing $\hat{\beta}_2$ by its expression:

$$\text{wald} = \frac{y^\top M_1 X_2 (X_2^\top M_1 X_2)^{-1} X_2^\top M_1 y}{\sigma_\epsilon^2} = \frac{y^\top P_{M_1 X_2} y}{\sigma_\epsilon^2}$$

3.6.3.2 Lagrange multiplier or score test

Consider the constrained model, which imposes $\beta_2 = 0$. It is therefore obtained by regressing y on X_1 only, and the vector of residuals is $\hat{\epsilon}_1 = M_1 y$. The idea of this test is that, if H_0 is true, $X_2^\top \hat{\epsilon}_1$ should be close to zero, $\hat{\epsilon}_1$ being "almost" orthogonal to the subspace spanned by X_2. Therefore, we consider the vector $X_2^\top \hat{\epsilon}_1 = X_2^\top M_1 y$, which, under H_0, should have a 0 expected value. The variance of this vector is: $\sigma_\epsilon^2 (X_2^\top M_1 X_2)^{-1}$ so that, applying the central-limit theorem:

$$X_2^\top \hat{\epsilon}_1 \equiv X_2^\top M_1 y \overset{a}{\sim} \mathcal{N}\left(0, \sigma_\epsilon^2 X_2^\top M_1 X_2\right)$$

The statistic is obtained, as previously, by computing the quadratic form:

$$\text{score} = y^\top M_1 X_2 \left(\sigma_\epsilon^2 X_2^\top M_1 X_2\right)^{-1} X_2^\top M_1 y = \frac{y^\top M_1 X_2 (X_2^\top M_1 X_2)^{-1} X_2^\top M_1 y}{\sigma_\epsilon^2} = \frac{y^\top P_{M_1 X_2} y}{\sigma_\epsilon^2}$$

3.6.3.3 Likelihood ratio test

The likelihood ratio test is based on the comparison of the objective function (the sum of square residuals) for the constrained and the unconstrained model. Remember, from Equation 3.14, that:

$$\| \hat{\epsilon}_{12}^2 \| + \| P_{M_1 X_2} y \| = \| \hat{\epsilon}_1^2 \|$$

The first term on the left and the term on the right are residual sums of squares (respectively for the unconstrained and the constrained model). Therefore, the likelihood ratio test is based on: $\text{SSR}_c - \text{SSR}_{nc} = \|P_{M_1 X_2} y\|^2 = y^\top P_{M_1 X_2} y$. Dividing by σ_ϵ^2, we get exactly the same statistic as previously.

3.6.4 Computation of the three tests

Consider the augmented Solow model: $y = \alpha + \beta_i i + \beta_v v + \beta_e e + \epsilon$, with $\beta_i = \frac{\kappa}{1-\kappa-\lambda}$, $\beta_v = -\frac{\kappa+\lambda}{1-\kappa-\lambda}$ and $\beta_e = \frac{\lambda}{1-\kappa-\lambda}$. For convenience, we compute the logarithm of the variables before using `lm`:

```
growth_sub2 <- growth_sub %>%
  mutate(y = log(gdp85), i = log(i), v = log(v), e = log(e))
mrw2 <- lm(y ~ i + v + e, growth_sub2)
```

We consider two hypotheses:

- $\beta_i + \beta_v + \beta_e = 0$, this hypothesis is directly implied by the structural model,
- $\kappa = 1/3$, the share of profits has a reasonable value.

The second hypothesis implies that $\beta_i = \frac{1/3}{2/3-\lambda}$ and $\beta_e = \frac{\lambda}{2/3-\lambda}$, and therefore that $\beta_e = 2\beta_i - 1$. The model can be reparametrized in such way that two slopes are 0 if the two hypotheses are satisfied:

$$y + e - v = \alpha + \beta_i(i + 2e - 3v) + (\beta_e - 2\beta_i + 1)(e - v) + (\beta_i + \beta_e + \beta_v)v$$

Therefore, the unconstrained model can be written as a model with $y+e-v$ as the response and $i + 2e - 3v$, $e - v$ and v as the three covariates and the constrained model as a model with the same response but with $i + 2e - 3$ as the unique covariate.

```
growth_sub2 <- growth_sub2 %>%
  mutate(y2 = y + e - v, i2 = i + 2 * e - 3 * v, e2 = e - v)
mrw_c <-  lm(y2 ~ i2,           growth_sub2)
mrw_nc <- lm(y2 ~ i2 + e2 + v, growth_sub2)
coef_nc <- mrw_nc %>% summary %>% coef
coef_nc
```

```
            Estimate Std. Error t value  Pr(>|t|)
(Intercept)  7.79131     1.1924  6.5340 8.301e-09
i2           0.70037     0.1506  4.6510 1.488e-05
e2           0.32981     0.3611  0.9133 3.642e-01
v           -0.06886     0.4654 -0.1480 8.828e-01
```

The two hypotheses can be tested one by one, as the test is that:

- the slope of $e2 = e - v$ equals 0 for the hypothesis that $\kappa = 1/3$,
- the slope of v equals 0 for the hypothesis that $\beta_i + \beta_e + \beta_v = 0$.

Both hypotheses are not rejected, even at the 10% level. To test the joint hypothesis, we can first use the approximate formula which is the sum or the mean of the squares of the t statistics (respectively χ^2 with 2 degrees of freedom and F with 2 and 71 degrees of freedom):

```
appr_chisq <- coef_nc[3, 3] ^ 2 + coef_nc[4, 3] ^ 2
appr_f <- appr_chisq / 2
c(appr_chisq, appr_f)
## [1] 0.856 0.428
pchisq(appr_chisq, df = 2, lower.tail = FALSE)
## [1] 0.6518
```

```
pf(appr_f, df1 = 2, df2 = 71, lower.tail = FALSE)
## [1] 0.6535
```

Based on this approximation, the joint hypothesis is clearly not rejected. We now turn to the computation of the statistics, using the three test principles.

3.6.4.1 Wald test

Considering the initial unconstrained model, for which the formula is $y \sim i + v + e$, the set of the two hypotheses can be written in matrix form as:

$$\begin{pmatrix} 0 & 2 & 0 & -1 \\ 0 & 1 & 1 & 1 \end{pmatrix} \begin{pmatrix} \alpha \\ \beta_i \\ \beta_v \\ \beta_e \end{pmatrix} = \begin{pmatrix} 1 \\ 0 \end{pmatrix}$$

The matrix R and the vector q are created in **R**:

```
R <- matrix(0, 2, 4) ; q <- c(1, 0)
R[1, 2] <- 2 ; R[1, 4] <- - 1 ; R[2, 2:4] <- 1
```

We then have $H_0 : R\gamma - q = 0$ and, for the fitted unconstrained model, we get: $R\hat{\gamma} - q$. Under H_0 the expected value of this vector is 0 and its estimated variance is $R^\top \hat{V}_{\hat{\gamma}} R = \hat{\sigma}_\epsilon^2 R^\top (Z^\top Z)^{-1} R$. Then:

$$(R\hat{\gamma} - q)^\top \left[R\hat{V}_{\hat{\gamma}} R^\top\right]^{-1} (R\hat{\gamma} - q) \tag{3.25}$$

is asymptotically a χ^2 with $J = 2$ degrees of freedom. Dividing by J, we get a F statistic with 2 and 71 degrees of freedom.

```
t(R %*% coef(mrw2) - q) %*%
  solve(R %*% vcov(mrw2) %*% t(R)) %*%
  (R %*% coef(mrw2) - q) / 2
##        [,1]
## [1,] 0.4233
```

`car::linearHypothesis` performs Wald tests with a nice syntax: the first argument is a fitted model, and the second one is a vector of characters which contains the character representation of the hypothesis:

```
car::linearHypothesis(mrw2, c("i + v + e = 0", "e = 2 * i - 1")) %>% gaze
## F = 0.423, df: 2-71, pval = 0.657
```

3.6.4.2 Likelihood ratio test

The computation of the likelihood ratio statistic is very simple once the two models have been estimated. The residual sums of squares of the two models are extracted using the `deviance` method, and we divide the difference of the two sums of squares by $\dot{\sigma}_\epsilon^2$ (the `sigma` method is used to extract the residual standard error) and by 2 to get an F statistic.

```
(deviance(mrw_c) - deviance(mrw2))/ sigma(mrw2) ^ 2 / 2
## [1] 0.4233
```

3.6.4.3 Score test

For the score test, we consider the reparametrized model and we define Z_1 as a matrix containing a column of one and $i + 2e - 3v$ and X_2 as a matrix containing $e - v$ and v.

```
Z1 <- model.matrix(~ i2, model.frame(mrw_nc))
X2 <- model.matrix(~ e2 + v - 1, model.frame(mrw_nc))
```

The test is based on the vector $X_2^\top \hat{\epsilon}_c$, where $\hat{\epsilon}_c$ is the vector of the residuals for the constrained model. Under H_0, the expected value of $X_2^\top \hat{\epsilon}_c$ is 0 and its variance is $\sigma_\epsilon^2 X_2^\top M_1 X_2$. $M_1 X_2$ is a matrix of residuals of all the columns of X_2 on Z_1.

```
ec <- resid(mrw_c)
M1X2 <- resid(lm(X2 ~ Z1))
```

Note that we've used `lm` with only a formula argument. In this case, the response and the covariates are vectors or matrices and not columns of a tibble. Note also that the left-hand side of a formula is a matrix and not a vector; in this case, each column is supposed to be a response and `lm` fit as many models than there are columns in this matrix. Then, the `resid` method no longer returns a vector, but a matrix, each column being a vector of residuals for one of the fitted models.

The statistic is then $\hat{\epsilon}_c^\top X_2 \left[X_2^\top M_1 X_2\right]^{-1} X_2 \hat{\epsilon}_c / \sigma_\epsilon^2$ and it is computed using an estimator of σ_ϵ^2, and dividing by $J = 2$ to get an F statistic.

```
t(crossprod(X2, ec)) %*%
  solve(crossprod(M1X2)) %*%
  crossprod(X2, ec) / sigma(mrw_c) ^ 2 / 2
##        [,1]
## [1,] 0.4301
```

The statistic is slightly different from the one computed previously, the difference being only due to the fact that the estimation of σ_ϵ is based, for the score test, on the constrained model.

3.6.5 Testing that all the slopes are 0

The test that all the slopes are 0 is routinely reported by software performing OLS estimation. It can be computed using any of the three test principles, but the likelihood ratio test is particularly appealing in this context, the constrained model being a model with only an intercept: $y_n = \alpha + \epsilon_n$. In this case, $\hat{\alpha} = \bar{y}$ and $\hat{\epsilon} = (y_n - \bar{y})$. Therefore, the residual sum of squares for the constrained model is $S_{yy} = \sum_n (y_n - \bar{y})^2$ and is also denoted by TSS for total sum of squares. The statistic is then: $(\text{TSS} - \text{RSS})/\sigma_\epsilon^2 \sim \chi^2_K$, which is a χ^2 with K degrees of freedom if the hypothesis that all the slopes are 0 is true. To compute this statistic, σ_ϵ^2 has to be estimated. A natural estimator is $\text{RSS}/(N-K-1)$ but, if H_0 is true, $\text{TSS}/(N-1)$ is also an unbiased estimator. Moreover, dividing by the sample size (N) and not by the number of degrees of freedom leads to biased but consistent estimators. Using the first estimator of σ_ϵ^2 and dividing by K, we get the F statistic with K and $N - K - 1$ degrees of freedom:

$$\frac{\text{TSS} - \text{RSS}}{\text{RSS}} \frac{N-K-1}{K} \sim F_{K,N-K-1}$$

Using TSS/N as an estimator of σ_ϵ^2, we get a very simple statistic that is asymptotically a χ^2 with K degrees of freedom:

$$N\frac{\text{TSS} - \text{RSS}}{\text{TSS}} \overset{a}{\sim} \chi^2_K$$

These two statistics are closely related to the R^2 which is, using this notation, equal to $1 - \text{RSS}/\text{TSS} = (\text{TSS} - \text{RSS})/\text{TSS}$. We can then write the F statistic as:

$$\frac{R^2}{1-R^2}\frac{N-K-1}{K}$$

and the asymptotic χ^2 statistic as NR^2.

There is no easy way to extract the R^2 and the F statistic with **R**. Both are computed by the `summary` method of `lm`:

```
slm <- slw_tot %>% summary
slm %>% names
```

```
 [1] "call"          "terms"         "residuals"     "coefficients"
 [5] "aliased"       "sigma"         "df"            "r.squared"
 [9] "adj.r.squared" "fstatistic"    "cov.unscaled"
```

and can be extracted "manually" from the list returned by `summary` using the `$` operator:

```
slm$r.squared
## [1] 0.5989
slm$adj.r.squared
## [1] 0.5878
slm$fstatistic
## value numdf dendf
## 53.76  2.00 72.00
```

More simply, `micsr:rsq` and `micsr:ftest` can be used:

```
slw_tot %>% rsq
## [1] 0.5989
slw_tot %>% rsq(type = "adj")
## [1] 0.5878
```

```
slw_tot %>% ftest %>% gaze
## F = 53.755, df: 2-72, pval = 0.000
```

The interest of this testing strategy is not limited to the test that all the slopes of a real model are 0. It can also be used to test any set of hypotheses, using reparametrization and the Frisch-Waugh theorem. Consider for example the hypothesis that $\kappa = 0.5$ and $\beta_i + \beta_e + \beta_v = 0$. We have seen previously that, after reparametrization, this corresponds to the model:

```
mrw_nc <- lm(y2 ~ i2 + e2 + v, growth_sub2)
mrw_nc %>% coef
## (Intercept)          i2          e2           v
##     7.79131     0.70037     0.32981    -0.06886
```

with, if the two hypotheses are true, the two slopes associated with `e2` and `v` equal to 0. Now, using the Frisch-Waugh theorem, and denoting $Z_1 = (j, i_2)$ and $X_2 = (e_2, v)$:

```
y2b <- lm(y2 ~ i2, growth_sub2) %>% resid
e2b <- lm(e2 ~ i2, growth_sub2) %>% resid
vb <- lm(v ~ i2, growth_sub2) %>% resid
mrw_ncb <- lm(y2b ~ e2b + vb - 1)
mrw_ncb %>% coef
##      e2b       vb
## 0.32981 -0.06886
```

We get exactly the same estimators as previously for `e2` and `v`, but now the joint hypothesis is that all the slopes of the second model are 0. Therefore, the test is based on the F statistic that is returned by `summary(lm(x))` and doesn't require any further calculus:

```
mrw_ncb %>% ftest %>% gaze
## F = 0.435, df: 2-73, pval = 0.649
```

Actually, a degrees of freedom correction should be performed to get exactly the same results because `mrw_ncb` has $75 - 2 = 73$ degrees of freedom, as the real number of degrees of freedom is $75 - 4 = 71$.

3.7 System estimation and constrained least squares

Very often in economics, the phenomenon under investigation is not well described by a single equation, but by a system of equations. Moreover, there may be inter-equations constraints on the coefficients. It is particularly the case in the field of the microeconometrics of consumption or production. For example, the behavior of a producer is described by a minimum cost equation along with equations of factor demand and the behavior of a consumer is described by a set of demand equations.

3.7.1 System of equations

We consider therefore a system of L equations denoted by $y_l = Z_l\beta_l + \epsilon_l$, with $l = 1 \ldots L$. In matrix form, the system can be written as follows:

$$\begin{pmatrix} y_1 \\ y_2 \\ \vdots \\ y_L \end{pmatrix} = \begin{pmatrix} Z_1 & 0 & \ldots & 0 \\ 0 & Z_2 & \ldots & 0 \\ \vdots & \vdots & \ddots & \vdots \\ 0 & 0 & \ldots & Z_L \end{pmatrix} \begin{pmatrix} \gamma_1 \\ \gamma_2 \\ \vdots \\ \gamma_L \end{pmatrix} + \begin{pmatrix} \epsilon_1 \\ \epsilon_2 \\ \vdots \\ \epsilon_L \end{pmatrix} \tag{3.26}$$

Therefore, the whole system can be estimated directly by stacking the vector of responses and by constructing a block-diagonal matrix of covariates, each block being the matrix of covariates for one equation.

As an example, consider the analysis of production characteristics (returns to scale, elasticities of substitution between pairs of inputs). The modern approach of production analysis consists of first considering the minimum cost function, which depends on the level of production and on input unit prices $C(y, p_1, \ldots p_J)$ and then of deriving the demands for input using Shepard Lemma:

$$\frac{\partial C}{\partial p_j} = x_j(y, p_1, \ldots, p_J)$$

The cost function is obviously homogeneous of degree 1 in input unit prices, which means that, for a given level of input, if all the prices increase proportionally, the quantity of the different inputs are the same and therefore the cost function increases by the same percentage. This writes: $C(y, \lambda p_1, \ldots \lambda p_J) = \lambda C(y, p_1, \ldots, p_J)$ and $x_j(y, \lambda p_1, \ldots, \lambda p_J) = x_j(y, p_1, \ldots p_J)$; the latter relation indicating that the demands for input are homogeneous of degree 0 in input unit prices. Among the different functional forms that have been proposed to estimate the cost function, the translog specification is the most popular. It can be considered as the second-order approximation of a general cost function:

$$\ln C = \alpha + \beta_y \ln y + \frac{1}{2}\beta_{yy} \ln^2 y + \sum_i \beta_i \ln p_i + \frac{1}{2}\sum_i \sum_j \beta_{ij} \ln p_i \ln p_j$$

Using Shephard Lemma, the cost share of input i is the derivative of $\ln C$ with $\ln p_i$.

$$s_i = \frac{\ln C}{\ln p_i} = \frac{p_i x_i}{C} = \beta_i + \beta_{ii} \ln p_i + \frac{1}{2}\sum_{j \neq i}(\beta_{ij} + \beta_{ji}) \ln p_j$$

Homogeneity of degree 1 in input prices implies that the cost shares don't depend on the level of prices. Therefore, $\sum_j^I \beta_{ij} = 0$, or $\beta_{iI} = -\sum_j^{I-1} \beta_{ij}$ and:

$$\begin{aligned}
\textstyle\sum_i^I \sum_j^I \beta_{ij} \ln p_i \ln p_j &= \textstyle\sum_i^I \ln p_i \left[\sum_j^{I-1} \beta_{ij} \ln p_j + \beta_{iJ} \ln p_I\right] \\
&= \textstyle\sum_i^I \ln p_i \sum_j^{I-1} \beta_{ij} \ln \frac{p_j}{p_I} \\
&= \textstyle\sum_i^{I-1} \ln p_i \sum_j^{I-1} \beta_{ij} \ln \frac{p_j}{p_I} + \ln p_I \sum_j^{I-1} \beta_{Ij} \ln \frac{p_j}{p_I} \\
&= \textstyle\sum_i^{I-1} \ln p_i \sum_j^{I-1} \beta_{ij} \ln \frac{p_j}{p_I} + \ln p_I \sum_j^{I-1} \beta_{jI} \ln \frac{p_j}{p_I} \\
&= \textstyle\sum_i^{I-1} \ln p_i \sum_j^{I-1} \beta_{ij} \ln \frac{p_j}{p_I} - \ln p_I \sum_i^{I-1} \sum_j^{I-1} \beta_{ji} \ln \frac{p_j}{p_I} \\
&= \textstyle\sum_i^{I-1} \sum_j^{I-1} \beta_{ij} \ln \frac{p_i}{p_I} \ln \frac{p_j}{p_I}
\end{aligned}$$

Moreover, the cost shares sum to 1 whatever the value of the prices, so that $\sum_i \beta_i = 1$. Therefore, the cost function can be rewritten as:

$$C^* = \alpha + \beta_y \ln y + \frac{1}{2}\beta_{yy} \ln^2 y + \sum_i^{I-1} \beta_i p_i^* + \frac{1}{2}\sum_{i=1}^{I-1} \beta_{ii} {p_i^*}^2 + \sum_{i=1}^{I-1}\sum_{j>i}^{I-1} \beta_{ij} p_i^* p_j^*$$

where $z^* = \ln(z/p_I)$ and the cost shares are:

$$s_i = \beta_i + \sum_{j=1}^{I-1} \beta_{ij} p_j^*$$

Consider the case where $I = 3$. In this case, the complete system of equations is:

$$\left\{ \begin{array}{rcl} C^* &=& \alpha + \beta_y \ln y + \frac{1}{2}\beta_{yy} \ln^2 y + \beta_1 p_1^* + \beta_2 p_2^* + \frac{1}{2}\beta_{11} {p_1^*}^2 + \beta_{12} p_1^* p_2^* + \frac{1}{2}\beta_{22} {p_2^*}^2 \\ s_1 &=& \beta_1 + \beta_{11} p_1^* + \beta_{12} p_2^* \\ s_2 &=& \beta_2 + \beta_{12} p_1^* + \beta_{22} p_2^* \end{array} \right.$$

There are 14 parameters to estimate in total (8 in the cost function and 3 in each of the cost share equations), but there are 6 linear restrictions; for example, the coefficient of p_1^* in the cost equation should be equal to the intercept of the cost share for the first factor.

We estimate the translog cost function with the `apples` data set of Ivaldi et al. (1996) who studied the production cost of apple producers. Farms in this sample produce apples and other fruits (respectively `apples` and `otherprod`). The authors observe the sales of apples and other fruits as well as the quantity of apple produced. Therefore, they are able to compute the unit price of apples. Both sales are divided by this unit price, so that `apples` is measured in apple quantity, and `otherprod` is measured in "equivalent" apple quantities. Therefore, they can be summed in order to have a unique output variable `y`. The expenses in the tree factors are given by `capital`, `labor` and `materials` and the corresponding unit prices are `pc`, `pl` and `pm`. The data set is an unbalanced panel of 173 farms observed for three years (1984, 1985 and 1986). We consider only one year (1985) and, for a reason that will be clear later, we divide all the variables by their sample mean:

```
ap <- apples %>% filter(year == 1985) %>%
    transmute(y = otherprod + apples,
              ct = capital + labor + materials,
              sl = labor / ct, sm = materials / ct,
              pk = log(pc / mean(pc)), pl = log(pl / mean(pl)) - pk,
              pm = log(pm / mean(pm)) - pk, ct = log(ct / mean(ct)) - pk,
              y = log(y / mean(y)), y2 = 0.5 * y ^ 2,
              ct = ct, pl2 = 0.5 * pl ^ 2,
              pm2 = 0.5 * pm ^ 2, plm = pl * pm)
```

We then create three formulas corresponding to the system of three equations (the cost function and the two factor shares):

```
eq_ct <- ct ~ y + y2 + pl + pm + pl2 + plm + pm2
```

```
eq_sl <- sl ~ pl + pm
eq_sm <- sm ~ pl + pm
```

These equations can be estimated one by one using OLS, but in this case, the trans-equations restrictions are ignored. The whole system can also be estimated directly by stacking the three vectors of responses and constructing a block diagonal matrix of covariates, each block being the relevant set of covariates for one equation. We use for this purpose the **Formula** package (Zeileis and Croissant 2010). This package extends usual formulas in two directions: first, several covariates can be indicated on the left- hand side of the formula (using the `+` operator), and several parts can be defined on both sides of the formula, using the `|` sign. For example:

```
y_1 + y_2 | y_3 ~ x_1 + x_2 | x_3 | x_4 + x_5
```

This formula has two sets of responses, the first containing y_1 and y_2, and the second y_3. Three sets of covariates are defined on the right-hand side of the formula.

For our production analysis, we first create a "meta" formula which contains the three responses on the left side and the whole set of covariates on the right side. Then, we extract, using `model.matrix`, the three model matrices for the three equations (`Z_c`, `Z_l` and `Z_m` respectively for the cost, labor share and material share equations). The column names of these matrices are customized using the `nms_cols` function which, for example, turns the original column names of `Z_l` (`(Intercept)`, `pl` and `pm`) to `sl_cst`, `sl_pl` and `sl_pm`. We then construct the block diagonal matrix (using the `Matrix::bdiag` function) and use our customized names:

```
library(Formula)
eq_sys <- Formula(ct + sl + sm ~ y + y2 + pl + pm + pl2 + plm + pm2)
mf <- model.frame(eq_sys, ap)  ; Z_c <- model.matrix(eq_ct, mf)
Z_l <- model.matrix(eq_sl, mf) ; Z_m <- model.matrix(eq_sm, mf)
nms_cols <- function(x, label)
    paste(label, c("cst", colnames(x)[-1]), sep = "_")
nms_c <- nms_cols(Z_c, "cost") ; nms_l <- nms_cols(Z_l, "sl")
nms_m <- nms_cols(Z_m, "sm")
Zs <- Matrix::bdiag(Z_c, Z_l, Z_m) %>% as.matrix
colnames(Zs) <- c(nms_c, nms_l, nms_m)
head(Zs, 2)
```

```
     cost_cst  cost_y cost_y2 cost_pl cost_pm cost_pl2 cost_plm
[1,]        1  0.1661  0.0138 -0.3187 -0.0787  0.05077  0.02508
[2,]        1 -1.2880  0.8294 -0.3480 -0.9062  0.06056  0.31537
     cost_pm2 sl_cst sl_pl sl_pm sm_cst sm_pl sm_pm
[1,] 0.003097      0     0     0      0     0     0
[2,] 0.410581      0     0     0      0     0     0
```

`Formula::model.part` enables to retrieve any part of the model. Here, we want to extract the three responses which are on the only left side of the formula. Therefore, we set `lrs` and `rhs` respectively to 1 and 0. The result is a data frame with three variables `ct`, `sl` and `sm`. We then use `dplyr::pivot_longer`, to stack the three responses in one column. Note the use of the optional argument `cols_vary` that is set to `"slowest"`, so that the elements are stacked columnwise:

```
Y <- model.part(eq_sys, mf, rhs = 0, lhs = 1)
ys <- Y %>% pivot_longer(1:3, cols_vary = "slowest",
                         names_to = "equation", values_to = "response")
print(ys, n = 2)
```

```
# A tibble: 411 x 2
  equation response
  <chr>       <dbl>
1 ct          0.232
2 ct         -1.25
# i 409 more rows
```

The estimation can be performed using the response vector and the matrix of covariates:

```
lm(ys$response ~ Zs - 1)
```

However, nicer input is obtained by constructing a tibble by binding the columns of `ys` and `Zs` and then using the usual formula-data interface.

```
stack_data <- ys %>% bind_cols(Zs)
stack_data %>% print(n = 2)
```

```
# A tibble: 411 x 16
  equation response cost_cst cost_y cost_y2 cost_pl cost_pm cost_pl2
  <chr>       <dbl>    <dbl>  <dbl>   <dbl>   <dbl>   <dbl>    <dbl>
1 ct          0.232        1  0.166  0.0138  -0.319 -0.0787   0.0508
2 ct         -1.25         1 -1.29   0.829   -0.348 -0.906    0.0606
# i 409 more rows
# i 8 more variables: cost_plm <dbl>, cost_pm2 <dbl>, sl_cst <dbl>,
#   sl_pl <dbl>, sl_pm <dbl>, sm_cst <dbl>, sm_pl <dbl>, sm_pm <dbl>
```

In order to avoid having to write the whole long list of covariates, the dot can be used on the right-hand side of the formula, which means in this context all the variables (except the response which is on the left-hand side of the formula). The intercept and the `equation` variable should be omitted from the regression.

```
ols_unconst <- lm(response ~ . - 1 - equation, stack_data)
ols_unconst %>% coef
## cost_cst    cost_y   cost_y2   cost_pl   cost_pm cost_pl2 cost_plm
##  0.01742   0.43634   0.11320   0.39270   0.46545  0.23015 -0.21353
## cost_pm2    sl_cst     sl_pl     sl_pm    sm_cst    sm_pl    sm_pm
##  0.18063   0.49383   0.11957  -0.10219   0.32165 -0.09014  0.10369
```

3.7.2 Constrained least squares

Linear restrictions on the vector of coefficients to be estimated can be represented using a matrix R and a numeric vector q: $R\gamma = q$,[7] where $\gamma^\top = (\gamma_1^\top, \dots, \gamma_L^\top)$ is the stacked vector of the coefficients for the whole system of equations. The OLS estimator is now the solution of a constrained optimization problem. Denoting λ a vector of Lagrange multipliers,[8] and ϵ the errors, the objective function is:

$$L = \epsilon^\top \epsilon + 2\lambda^\top (R\gamma - q)$$

The Lagrangian can also be written as:

$$L = y^\top y - 2\gamma^\top Z^\top y + \gamma^\top Z^\top Z\gamma + 2\lambda(R\gamma - q)$$

The first-order conditions are:

$$\left\{ \begin{array}{rcl} \frac{\partial L}{\partial \gamma} &=& -2Z^\top y + 2Z^\top Z\gamma + 2R^\top \lambda = 0 \\ \frac{\partial L}{\partial \lambda} &=& 2(R\gamma - q) = 0 \end{array} \right.$$

which can also be written in matrix form as:

$$\left(\begin{array}{cc} Z^\top Z & R^\top \\ R & 0 \end{array} \right) \left(\begin{array}{c} \gamma \\ \lambda \end{array} \right) = \left(\begin{array}{c} Z^\top y \\ q \end{array} \right)$$

The constrained OLS estimator can be obtained using the formula for the inverse of a partitioned matrix:[9]

$$\left(\begin{array}{cc} A_{11} & A_{12} \\ A_{21} & A_{22} \end{array} \right)^{-1} = \left(\begin{array}{cc} B_{11} & B_{12} \\ B_{21} & B_{22} \end{array} \right) = \left(\begin{array}{cc} A_{11}^{-1}(I + A_{12}F_2A_{21}A_{11}^{-1}) & -A_{11}^{-1}A_{12}F_2 \\ -F_2A_{21}A_{11}^{-1} & F_2 \end{array} \right)$$

with $F_2 = \left(A_{22} - A_{21}A_{11}^{-1}A_{12}\right)^{-1}$. We have here $F_2 = -\left(R(Z^\top Z)^{-1}R^\top\right)^{-1}$. The constrained estimator is then: $\hat{\gamma}_c = B_{11}Z^\top y + B_{12}q$, with $B_{11} = (Z^\top Z)^{-1}\left(I - R^\top (R(Z^\top Z)^{-1}R^\top)^{-1}R(Z^\top Z)^{-1}\right)$ and $B_{12} = (Z^\top Z)^{-1}R^\top \left(R(Z^\top Z)^{-1}R^\top\right)^{-1}$

The unconstrained estimator being $\hat{\beta}_{nc} = \left(Z^\top Z\right)^{-1} Z^\top y$, we finally get:

$$\hat{\gamma}_c = \hat{\gamma}_{nc} - (Z^\top Z)^{-1}R^\top (R(Z^\top Z)^{-1}R^\top)^{-1}(R\hat{\gamma}_{nc} - q) \tag{3.27}$$

The difference between the constrained and unconstrained estimators is then a linear combination of the excess of the linear constraints of the model evaluated for the unconstrained model. For the system of cost and factor shares for apple production previously described, we have 14 coefficients and 6 restrictions:

[7]See Section 3.6.4.1.
[8]These multipliers are multiplied by 2 in order to simplify the first-order conditions.
[9]See Greene (2018), online appendix p. 1076.

```
R <- matrix(0, nrow = 6, ncol = 14)
R[1, c(4,  9)] <- R[2, c(5, 12)] <- R[3, c(6, 10)] <-
  R[4, c(7, 11)] <- R[5, c(7, 13)] <-
  R[6, c(8, 14)] <- c(1, -1)
```

For example, the first line of `R` returns the difference between the fourth coefficient (`cost_pl`) and the ninth coefficient (`sl_cst`). As the vector q is 0 in our example, this means that the coefficient of p_l^* in the cost equation should equal the intercept in the labor share equation. Applying Equation 3.27, we get:

```
excess <- R %*% coef(ols_unconst)
XpX <- crossprod(model.matrix(ols_unconst))
beta_c <- coef(ols_unconst) -
    drop(solve(XpX) %*% t(R) %*%
         solve(R %*% solve(XpX) %*% t(R)) %*% excess)
beta_c
## cost_cst   cost_y  cost_y2  cost_pl  cost_pm cost_pl2 cost_plm
##  0.02019  0.44485  0.11118  0.49570  0.34208  0.11690 -0.10409
## cost_pm2   sl_cst    sl_pl    sl_pm   sm_cst    sm_pl    sm_pm
##  0.10775  0.49570  0.11690 -0.10409  0.34208 -0.10409  0.10775
```

More simply, `micsr::clm` can be used, which computes the constrained least squares estimator with, as arguments, a `lm` object (the unconstrained model) and `R`, the matrix of restrictions (and optionally a `q` vector):

```
ols_const <- clm(ols_unconst, R)
```

and returns a `lm` object.

Finally, the **systemfit** package (Henningsen and Hamann 2007) is devoted to system estimation and provides a `systemfit` function. Its main arguments are a list of equations and a data frame, but it also has `restrict.matrix` and `restrict.rhs` arguments to provide respectively R and q.

```
systemfit::systemfit(list(cost = eq_ct, labor = eq_sl,
                          materials = eq_sm),
                     data = ap, restrict.matrix = R)
```

The output of `systemfit` is large and is not reproduced here. The full strength of this package will be presented in later chapters, while describing the seemingly unrelated regression (Section 6.4.4) and the three-stage least squares (Section 7.4) estimators.

4

Interpretation of the Coefficients

In the simplest model, a regression with a real numerical response and a real numerical covariate: $y_n = \alpha + \beta x_n + \epsilon_n$; the interpretation of β is simply the constant marginal effect of x_n on y_n or, mathematically, the derivative of y with respect to x. However, linear models are much richer than this simple one, as some covariates may be:

- integer and not real (e.g., the number of children),
- dummy variables (e.g., 0 for a male and 1 for a female),
- categorical variables (e.g., marital status, with modalities married, single, divorced, widower),
- introduced in the regression after a transformation, for example, by taking the logarithm or the square root (this can be also true for the response),
- introduced as a polynomial, for example, not only x is introduced, but also its square,
- jointly introduced with an interaction, i.e., not only x_1 and x_2 are introduced, but also their product.

In all these cases, β is no longer the derivative of y with respect to x and special care is required in the interpretation of the estimated coefficients. Throughout this chapter, we'll use the data set of Costanigro, Mittelhammer, and McCluskey (2009) (called `wine`) who estimate a hedonic price function for wines.

```
wine %>% print(n = 3)
```

```
# A tibble: 9,600 x 11
  price cases score   age region    variety  vintage reserve vineyard
  <dbl> <int> <int> <int> <fct>     <fct>    <fct>   <fct>   <fct>
1  27.1   300    86     3 sonoma    pinot    95      no      no
2  32.1   500    90     3 mendocino cabernet 97      yes     no
3  25.5  1635    84     2 mendocino pinot    other   no      no
# i 9,597 more rows
# i 2 more variables: estate <fct>, class <fct>
```

The data set contains 9600 wines, for which one observes:

- `price`: the price in US$ of 2000 per bottle,
- `cases`: the production in number of cases, a case containing 12 bottles of 75 cl,
- `region`: the region where the wine comes from,
- `vineyard`: equal to `"yes"` if vineyard information is provided and `"no"` otherwise,
- `age`: the age of the wine in years, which is an integer, from 1 to 6,
- `score`: the score given by the Wine Spectator magazine, between 0 and 100.

We divide the score by 100 so that the maximum attainable score is 1:

```
wine <- wine %>% mutate(score = score / 100)
```

Section 4.1 deals with numerical covariates and Section 4.2 with categorical covariates. Section 4.3 introduces interactions between several covariates. Finally, Section 4.4 presents useful tools to compute marginal effects.

4.1 Numerical covariate

We first consider the production as the unique covariate. High quality wines are often produced in small areas, with a low yield; we therefore expect a negative relationship between price and production. We consider two measures of production and price:

- production is measured in cases (x) and in m^3 (w). A case contains 12 bottles of 75 cl, or 9 liters, which is also 9/1000 m^3. Therefore, the number of cases must be multiplied by 0.009 in order to get the production in m^3 ($w = 0.009x$),
- price (y) is measured in US$ of 2000 per bottle. The price index in the United States was equal to 172 in 2000 and to 258 in 2020. Therefore, to get the price in US$ of 2020 (q), we have to multiply the price in US$ of 2000 by $258/172 = 1.5$ ($q = 1.5y$).

We perform these transformations, and we select only the relevant variables:

```
wine2 <- wine %>% select(x = cases, y = price) %>%
  mutate(w = x * 0.009,  q = y * 1.5)
```

4.1.1 Response and covariate in level

We start by a set of simple linear regressions of y / q on x / w:

```
nn_yx <- lm(y ~ x, wine2)
nn_yw <- lm(y ~ w, wine2)
nn_qx <- lm(q ~ x, wine2)
nn_qw <- lm(q ~ w, wine2)
nn_yx %>% coef
## (Intercept)            x
##  31.7118276   -0.0003312
```

The initial model, with y and x is: $y_n = \alpha + \beta_x x_n + \epsilon_n$. y being measured in US$ of 2000, so are α, βx_n and ϵ_n. The intercept indicates the expected value of a bottle of wine when the production is 0, which of course makes no sense. However, we can consider the intercept as the expected value of a bottle for a wine with a very low production. We get here: 31.7. $\beta_x x$ is measured in US$ of 2000 and x in cases, β_x is measured in US$ of 2000 per case. Therefore, when the production increases by one case, the price of a bottle decreases by -3.3×10^{-4} US$ of 2000, i.e., 0.03 cent. The residuals and the fitted values are also measured in US$ of 2000. To see the response, the fitted values and the residuals, we can create a tibble, using the `resid` and the `fitted` functions:

```
tibble(y = wine2$y, heps = resid(nn_yx), hy = fitted(nn_yx))
```

or more simply using the `broom::augment` function, which adds to the model frame of a fitted object different columns, including the residuals (`.resid`) and the fitted values (`.fitted`):

```
library(broom)
nn_yx %>% augment
```

```
# A tibble: 9,600 x 8
      y     x .fitted .resid      .hat .sigma       .cooksd .std.resid
  <dbl> <int>   <dbl>  <dbl>     <dbl>  <dbl>         <dbl>      <dbl>
1  27.1   300    31.6 -4.49  0.000120    19.5 0.00000319       -0.231
2  32.1   500    31.5  0.554 0.000118    19.5 0.0000000479      0.0285
3  25.5  1635    31.2 -5.67  0.000112    19.5 0.00000475       -0.291
4  27.0   800    31.4 -4.41  0.000116    19.5 0.00000299       -0.226
5  26.4  1150    31.3 -4.93  0.000114    19.5 0.00000368       -0.253
# i 9,595 more rows
```

For example, for the first wine in the sample, the observed price is $27.1 and is the sum of:

- the prediction of the model, i.e., the estimation of the expected value of a bottle for this level of production: 31.6,
- the residual, i.e., the difference between the actual and the predicted price, which is negative for the first bottle (−4.49).

We now measure the price in US$ of 2020. This means that the unit of measurement of the response, as well as that of α, $\beta_x x_n$ and ϵ_n is changed. More precisely, all these values are multiplied by 1.5. $\beta_x x_n$ is multiplied by 1.5, but as x_n is unchanged, β_x is multiplied by 1.5.

```
coef(nn_qx)
## (Intercept)           x
##  47.5677414  -0.0004969
coef(nn_qx) / coef(nn_yx)
## (Intercept)           x
##         1.5         1.5
head(resid(nn_qx), 2)
##       1       2
## -6.7387  0.8307
head(resid(nn_qx), 2) / head(resid(nn_yx), 2)
##   1   2
## 1.5 1.5
```

For example, the price of the first bottle is lower than the estimated expected value for this given level of production by an amount of −6.74 US$ of 2020, which is 1.5 times the residual of the initial regression (−4.49).

Consider now that the production is measured in m^3 and the price in US$ of 2000. The reference model can in this case be rewritten as:

$$y_n = \alpha + \beta_x(w_n/0.009) + \epsilon_n = \alpha + (\beta_x/0.009)w_n + \epsilon_n = \alpha + \beta_w w_n + \epsilon_n$$

Compared to the reference model, the three elements of the model are still measured in US$ of 2000, and it is particularly the case for $\beta_w w_n$. As the production is now measured in m^3, it has been multiplied by 0.009, and therefore, the slope is divided by the same amount. The intercept, the predictions and the residuals still have the same unit of measurement (US$ of 2000) and therefore the same values.

```
nn_yw <- lm(y ~ w, wine2)
coef(nn_yw)
## (Intercept)           w
##    31.71183    -0.03681
coef(nn_yx) / coef(nn_yw)
## (Intercept)           x
##       1.000       0.009
```

The price of a bottle of very rare wine is still 31.7 US$ of 2000, and the slope now indicates that when the production increases by 1 m^3, the price of a bottle decreases by -0.037, which is approximately 4 cents.

4.1.2 Covariate in logarithm

We now consider that the covariate is the logarithm of the production: $y_n = \alpha + \beta_x \ln x_n + \epsilon_n$. The response is still measured in US$ of 2000, and therefore so are α, $\beta_x \ln x_n$, ϵ_n, $\hat{y}_n$ and $\hat{\epsilon}_n$.

```
nl_yx <- lm(y ~ log(x), wine2)
coef(nl_yx)
## (Intercept)      log(x)
##      57.008      -3.708
```

The intercept is now the expected value of the price of a bottle for $x = 1$ (and not 0 as previously) and is therefore not particularly meaningful. β_x is now the derivative of y with $\ln x$, which is $\beta_x = \frac{dy}{d\ln x} = \frac{dy}{dx/x}$. Therefore, β_x is the ratio of an absolute variation of y and a relative variation of x. For $dx/x = 1$, which means that the production is multiplied by 2, $dy = -3.71$, which means that the price of the bottle decreases by -3.71 US$ of 2000.

```
nl_yx %>% augment
```

```
# A tibble: 9,600 x 8
      y `log(x)` .fitted .resid     .hat .sigma      .cooksd .std.resid
  <dbl>    <dbl>   <dbl>  <dbl>    <dbl>  <dbl>        <dbl>      <dbl>
1  27.1     5.70    35.9  -8.74 0.000219   19.1 0.0000230       -0.458
2  32.1     6.21    34.0  -1.87 0.000156   19.1 0.000000747     -0.0979
3  25.5     7.40    29.6  -4.07 0.000105   19.1 0.00000240      -0.214
4  27.0     6.68    32.2  -5.18 0.000120   19.1 0.00000443      -0.272
5  26.4     7.05    30.9  -4.48 0.000106   19.1 0.00000293      -0.235
# i 9,595 more rows
```

With this new specification, for the first bottle, the price is lower than the expected value of the price by an amount of -8.74. If the production is measured in m^3, we get, replacing x_n by $w_n/0.009$:

$$y_n = \alpha + \beta_x \ln(w_n/0.009) + \epsilon_n = y_n = (\alpha - \beta_x \ln 0.009) + \beta_x \ln w_n + \epsilon_n$$

```
nl_yw <- lm(y ~ log(w), wine2)
coef(nl_yw)
## (Intercept)      log(w)
##      39.544      -3.708
```

Therefore, the slope is unchanged because it measures the effect of doubling the production on the price, which obviously doesn't depend on the unit of measurement of the production. The intercept is modified by an amount equal to $-\beta_x \ln 0.009$:

```
coef(nl_yw)[1] - coef(nl_yx)[1]
## (Intercept)
##      -17.46
- coef(nl_yw)[2] * log(0.009)
## log(w)
## -17.46
```

Consider now that the price is measured in US$ of 2020:

$$y_n = \frac{q_n}{1.5} = \alpha + \beta_x \ln x_n + \epsilon_n$$

All the elements of the initial model are therefore multiplied by 1.5:

```
nl_qx <- lm(q ~ log(x), wine2)
coef(nl_qx)
## (Intercept)      log(x)
##      85.513      -5.561
coef(nl_qx) / coef(nl_yx)
## (Intercept)      log(x)
##         1.5         1.5
resid(nl_qx)[1:3]
##       1       2       3
## -13.112  -2.801  -6.112
resid(nl_qx)[1:3] / resid(nl_yx)[1:3]
##   1   2   3
## 1.5 1.5 1.5
```

4.1.3 Response in logarithm

Consider now the case when the response is in logarithm. In terms of the response, we then start from a multiplicative model:

$$y_n = e^{\alpha+\beta_x x}(1+\eta_n)$$

where η_n is a multiplicative error. For example, if $\eta_n = -0.2$, it means that for the given wine, the price is 20% lower than the expected value for this level of production. Taking logs, we get:

$$\ln y_n = \alpha + \beta_x x_n + \ln(1+\eta_n) = \alpha + \beta_x x_n + \epsilon_n$$

The error term is now $\epsilon = \ln(1+\eta_n)$. But, if η_n is "small": $\ln(1+\eta_n) \approx \eta_n$. For example, $\ln(1+0.05) = 0.049$ and $\ln(1-0.10) = -0.105$. Therefore, ϵ can be interpreted as a percentage of variation between the actual price and the expected value for the given level of production.

```
ln_yx = lm(log(y) ~ x, wine2)
coef(ln_yx)
## (Intercept)             x
##   3.3076419   -0.0000138
```

The intercept is now the expected value of $\ln y$ when $x = 0$. Taking the exponential of the intercept, we get:

```
exp(coef(ln_yx)[1])
## (Intercept)
##       27.32
```

which is quite close to the value obtained in the reference model (31.71). β_x is now the derivative of $\ln y$ with x:

$$\frac{d\ln y}{dx} = \frac{dy/y}{dx}$$

It is therefore the ratio of a relative variation of y and an absolute variation of x. The results indicate that when the production increases by one case, the relative variation of the price is -10^{-5}, i.e., approximately 0.001%. Consider now the production in m^3:

$$\ln y_n = \alpha + \beta_x(w_n/0.009) + \epsilon_n = \alpha + (\beta_x/0.009)w_n + \epsilon_n$$

The only effect of this change of measurement for the covariate is that, as the covariate is multiplied by 0.009, β_x is divided by the same amount:

```
ln_yw = lm(log(y) ~ w, wine2)
coef(ln_yw)
## (Intercept)            w
##    3.307642    -0.001533
coef(ln_yx) / coef(ln_yw)
## (Intercept)            x
##       1.000        0.009
```

the intercept, the fitted values and the residuals being unchanged. If the response is measured in US$ of 2020, starting from the initial model, we have:

$$\ln y_n = \ln \frac{q_n}{1.5} = \alpha + \beta_x x_n + \epsilon_n$$

which leads to:

$$\ln q_n = (\alpha + \ln 1.5) + \beta_x x_n + \epsilon_n$$

Therefore, the only effect is that the intercept increases by $\ln 1.5$:

```
ln_qw <- lm(log(q) ~ w, wine2)
coef(ln_qw)
## (Intercept)           w
##    3.713107   -0.001533
coef(ln_qw) - coef(ln_yw)
## (Intercept)           w
##   4.055e-01  -6.072e-18
log(1.5)
## [1] 0.4055
```

the slope, the predictions and the residuals being unchanged.

4.1.4 Response and covariate in log

Consider finally the case where both the response and the covariate are in logs:

$$\ln y_n = \beta_0 + \beta_x \ln x_n + \epsilon_n$$

The slope is the derivative of $\ln y$ with $\ln x$:

$$\beta_x = \frac{d \ln y}{d \ln x} = \frac{dy/y}{dx/x}$$

and is therefore a ratio of two relative variations.

```
ll_yx <- lm(log(y) ~ log(x), wine2)
coef(ll_yx)
## (Intercept)      log(x)
##      4.3042     -0.1466
```

When $dx/x = 1$, which means that the production is doubling, the relative variation of the price is -0.147 or -14.7%. Equivalently, when the production increases by 1%, the price decreases by -0.147%; β_x is an elasticity and is a number without unit. Therefore, it is invariant to any linear transformation of the covariate or of the response.

```
ll_yw <- lm(log(y) ~ log(w), wine2)
ll_qx <- lm(log(q) ~ log(x), wine2)
ll_qw <- lm(log(q) ~ log(w), wine2)
rbind(ll_yx = coef(ll_yx), ll_yw = coef(ll_yw),
```

```
        ll_qx = coef(ll_qx), ll_qw = coef(ll_qw))
##          (Intercept)  log(x)
## ll_yx          4.304 -0.1466
## ll_yw          3.614 -0.1466
## ll_qx          4.710 -0.1466
## ll_qw          4.019 -0.1466
```

4.1.5 Integer covariate

Consider the same model: $y_n = \alpha + \beta_x x_n + \epsilon_n$, but x_n is now an integer. In the `wine` data set, `age` is the age of the wine in years, from 1 to 6.

```
lm(price ~ age, wine) %>% coef
## (Intercept)         age
##       9.910       7.288
```

β is still the derivative of y with x, but the computation of the derivative makes little sense when x is an integer, as infinitesimal variations are not relevant in this case. Therefore, instead of a marginal effect, one is interested in a discrete variation of x_n (for example from 3 to 4). If we compute the variation of $\mbox{E}(y \mid x = x_n)$ when x increases from x_n to $x_n + 1$, we get:

$$(\alpha + \beta_x(x_n + 1)) - (\alpha + \beta_x x_n) = \beta_x$$

Therefore, β_x is not only the derivative, but it is also the variation of y caused by an increase of one unit of x. In our example, when the age of the wine increases by one year, its expected price increases by \$7.3.

4.2 Categorical covariate

A categorical variable is a variable that defines a set of modalities, each observations belonging to one modality and only one. The preferred way to deal with categorical variables with **R** is to use a variable type that is called **factor**, the modalities being called **levels**. Factors have at least two levels. A categorical variable with two levels is a **dichotomic** variable, and the factor can be replaced by a dummy variable (a numerical variable for which the only two values are 0 and 1). For example, sex can be stored as a factor with levels male and female, or as a dummy variable with for example 1 for females and 0 for males. When the number of levels is greater than 2, the variable is **polytomic**, and it is advisable to store the variable as a factor. Base **R** proposed several functions to deal with factors, but we'll use the **forcats** packages which is part of the **tidyverse**.

4.2.1 Dichotomic variable

We consider here the "vineyard" variable which has two modalities ("yes" and "no"), and we create a variant of this variable by creating a dummy variable equal to 1 if vineyard is equal to "yes" and 0 otherwise:

```
wine <- wine %>% mutate(vyd = ifelse(vineyard == "yes", 1L, 0L))
```

4.2.1.1 Response in level

We first compute the average price for the two categories:

```
Mwine <- wine %>% group_by(vineyard) %>% summarise(price = mean(price))
Mwine
## # A tibble: 2 x 2
##   vineyard price
##   <fct>    <dbl>
## 1 no        28.9
## 2 yes       34.9
```

The average price difference between the two categories is substantial ($5.97). We now consider the estimation of the linear model, with x_n a dummy variable equal to 1 if the wine has vineyard indications. The effect of x_n varying from 0 to 1 on the conditional expectation of the price is:

$$(\alpha + \beta_x \times 1) - (\alpha + \beta_x \times 0) = \beta_x$$

We then estimate the linear model, using vineyard and vyd:

```
lm(price ~ vyd, wine) %>% coef
## (Intercept)         vyd
##      28.900       5.974
lm(price ~ vineyard, wine) %>% coef
## (Intercept) vineyardyes
##      28.900       5.974
```

The intercept is the average price for wines with no vineyard indication, and β_x is the price increase caused by vineyard indication. Note also that $\alpha + \beta_x$ is the average price for wines with a vineyard indication. Identical results are obtained using either vyd and vineyard. The difference is that in the first case, vyd is directly used in the regression. In the second case, the factor vineyard is replaced in the regression by a dummy variable called vineyardyes equal to 1 when vineyard equals "yes".

Removing the intercept, we get:

```
lm(price ~ vineyard - 1, wine) %>% coef
##  vineyardno vineyardyes
##       28.90       34.87
```

Two dummies are created, one for `vineyard = "yes"` as previously done, but also one for `vineyard = "no"`. The two coefficients are now average prices for the two categories.

4.2.1.2 Response in log

Consider now a regression with the response in log and a dummy covariate: $\ln y_n = \alpha + \beta_x x_n + \epsilon_n$. Consider two wines with the same value of ϵ, but one with $x = 1$ and the other with $x = 0$, and call y^1 and y^0 the respective prices. The difference in log price is:

$$\ln y^1 - \ln y^0 = \beta_x$$

Denoting $\tau_x = \frac{y^1 - y^0}{y^0}$ the variation rate of the price for a wine with vineyard indication compared to a wine without vineyard indication, we get:

$$\ln y^1 - \ln y^0 = \ln(1 + \tau_x) = \beta_x$$

β_x is therefore equal to $\ln(1+\tau)$ and, as seen previously, for a "small" value of τ_x, $\ln(1+\tau_x) \approx \tau_x$. Therefore, β_x is approximately the relative price difference for a wine with and without vineyard indications.

```
log_bin <- lm(log(price) ~ vineyard, wine)
coef(log_bin)
## (Intercept) vineyardyes
##       3.191       0.248
```

$\hat{\beta}_x = 0.25$ is the approximation of the relative difference (25%), as the true relative difference is given by the formula: $\tau = e^{\beta_x} - 1$, which is 0.28. Therefore, $\hat{\beta}_x$ gives quite a good approximation of the relative difference.

4.2.2 Polytomic covariate

Consider now the case of a polytomic covariate. `region` is a factor with numerous modalities:

```
wine %>% pull(region) %>% levels
## [1] "other"        "napavalley" "baycentral" "sonoma"       "southcoast"
## [6] "carneros"     "sierra"       "mendocino"  "washington"
```

To simplify, we keep only the two most important regions (Napa Valley and Sonoma) and merge all the other regions in a `other` category. This operation is easily performed using the `forcats::fct_lump` function:

```
wine <- wine %>% mutate(reg = fct_lump(region, 2))
wine %>% count(reg)
```

```
# A tibble: 3 x 2
  reg            n
```

```
  <fct>       <int>
1 napavalley  2770
2 sonoma      2358
3 Other       4472
```

The reg variable has then three modalities:

```
wine %>% pull(reg) %>% levels
## [1] "napavalley" "sonoma"     "Other"
```

4.2.2.1 Response in level

We first compute the average price for these three modalities:

```
Mwine <- wine %>% group_by(reg) %>% summarise(price = mean(price))
Mwine
```

```
# A tibble: 3 x 2
  reg         price
  <fct>       <dbl>
1 napavalley  41.3
2 sonoma      28.7
3 Other       23.9
```

Consider a linear model, without and with an intercept:

```
lin_1 <- lm(price ~ reg - 1, wine)
lin_2 <- lm(price ~ reg, wine)
lin_1 %>% coef
## regnapavalley      regsonoma       regOther
##         41.28          28.72          23.90
lin_2 %>% coef
## (Intercept)    regsonoma     regOther
##       41.28       -12.55       -17.37
```

Without an intercept, three dummy variables are created and introduced in the regression. With an intercept, only two dummy variables are introduced. We have noted previously that the three levels are napavalley, sonoma and Other. The name of the dummy variables are obtained by merging the name of the variable reg and the name of the level. Note that when an intercept is introduced, the dummy variable corresponding to the first level (here napavalley) is omitted. Without an intercept, the three estimated coefficients are simply the average prices for the three regions. With an intercept, the intercept is the average price for the first level of the variable (the Napa Valley region) and the two other coefficients are the differences between the average price for the two regions and the average price for the reference region (Napa Valley). For example, using the second regression (lin_2), the average price for the Sonoma region can be computed as the sum of the first two coefficients:

```
unname(coef(lin_2)[1] + coef(lin_2)[2])
## [1] 28.72
```

The reference level can be changed by modifying the order of the levels of the factor. This task can easily be performed using the `forcats::fct_relevel` function and indicating, for example, that `"Other"` should be the first level:

```
wine <- mutate(wine, reg = fct_relevel(reg, "Other"))
lin_2b <- lm(price ~ reg, wine)
coef(lin_2b)
##    (Intercept) regnapavalley      regsonoma
##          23.90         17.37           4.82
```

The average price for Sonoma is now equal to the sum of the first coefficient (the intercept which is the average price for "other regions") and the third coefficient (the difference between the average price in Sonoma and "other regions"):

```
unname(coef(lin_2b)[1] + coef(lin_2b)[3])
## [1] 28.72
```

4.2.2.2 Response in log

We first compute the average of the *log* prices for the three regions:

```
Mwine <- wine %>% group_by(reg) %>%
    summarise(mprice = mean(price),
              lprice = mean(log(price)),
              mlprice = exp(lprice))
Mwine
```

```
# A tibble: 3 x 4
  reg          mprice lprice mlprice
  <fct>         <dbl>  <dbl>   <dbl>
1 Other          23.9   3.04    20.9
2 napavalley     41.3   3.56    35.2
3 sonoma         28.7   3.24    25.6
```

By definition, taking the exponential of the mean of the logs, we get the geometric mean of prices, which is always lower than the arithmetic mean. Estimating the model without intercept:

```
log_3 <- lm(log(price) ~ reg - 1, wine)
coef(log_3)
##       regOther regnapavalley      regsonoma
##          3.040         3.562          3.243
```

we get a model with a coefficient for the three regions. Taking the exponential of these coefficients, we get the geometric means of price previously computed for the three regions. Adding an intercept, only two dummies are introduced in the regression:

```
log_2 <- lm(log(price) ~ reg, wine)
coef(log_2)
##   (Intercept) regnapavalley     regsonoma
##        3.0399        0.5221        0.2033
```

The intercept is the mean of the logarithms of the prices for the reference region. The coefficient for the Napa Valley region is the difference of mean log prices for the Napa Valley and "other regions".

$$\beta^s = \ln y_n^s - \ln y_n^o = \ln \frac{y_n^s}{y_n^o} = \ln(1 + \tau_s) \approx \tau_s$$

where τ_s is the relative price difference between Sonoma and "other regions". β^s is an approximation of τ_s. The coefficient is 0.52 (or 52%) as the exact formula for τ gives 0.69 (or 69%). The approximation in this example is quite bad, because the value of $\hat{\beta}^s$ is high.

4.3 Several covariates

Most of the time, the model of interest includes several numerical and/or categorical covariates. A first step is to introduce these covariates separately, so that their linear effect is analyzed. But it is also sometimes interesting to introduce the product of two covariates in the regression. This enables to analyze the interaction effect of two covariates, i.e., the fact that the effect of x_1 on y depends on the value of x_2.

4.3.1 Separate effects

Consider the log-log model with the numerical `cases` covariate and add the `reg` categorical variable. We use the `+` operator in the formula to separate these two covariates:

```
di_us <- lm(log(price) ~ log(cases) + reg, wine)
di_us %>% coef
##   (Intercept)    log(cases) regnapavalley     regsonoma
##        4.0966       -0.1456        0.5186        0.2071
```

From previous subsections, we are able to interpret the results:

- a 1% increase in production leads to a decrease of −0.15% in price,
- wine produced in Napa Valley and Sonoma are more expensive than wine produced in other regions (by an amount approximately equal to 51.9% and 20.7%).

This model is represented in Figure 4.1. The slope is the same for the three regions, but the intercept is different. Therefore, the fitted model is represented by three parallel lines. Note that the sample mean for each region, depicted by a large point, is part of the corresponding regression line.

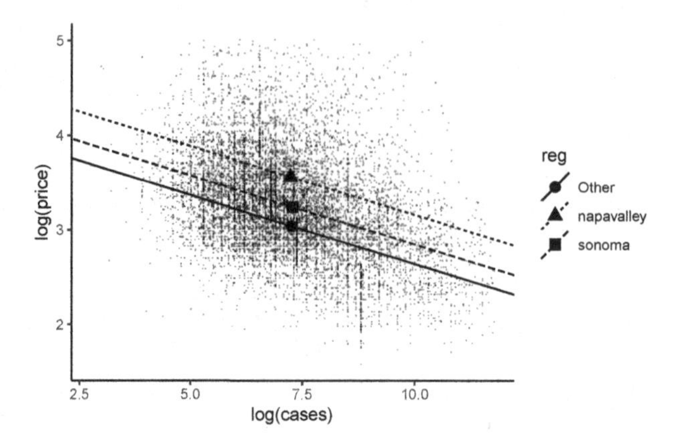

Figure 4.1: Additive effects of production and region

4.3.2 Multiplicative effects

A multiplicative effect is obtained by introducing in the regression the product of two covariates. Consider first the case where the two covariates are numeric; we'll consider as an example the production in m^3 (`prodcubm`) and `score`. The product can be computed before the estimation:

```
wine <- wine %>% mutate(prodcubm = cases * 0.009,
                        prod_score = prodcubm * score)
```

and introduced in the regression:

```
lm(price ~ prod_score, wine) %>% coef
## (Intercept)  prod_score
##    31.66271    -0.04194
```

The same result can be obtained by computing "on the fly" the product in the formula:

```
lm(price ~ I(prodcubm * score), wine) %>% coef
```

Note the use of the `I` function in the formula; arithmetic operators (`+`, `-`, `*`, `:` and `/`) have a special meaning in formulas and therefore they should be "protected" in order to perform the usual arithmetic operation. The last (and best solution) is to use the formula syntax: `x_1:x_2` introduces the interaction effect between `x_1` and `x_2`, i.e., their product:

```
lm(price ~ prodcubm:score, wine) %>% coef
```

This model is of the form: $y_n = \alpha + \beta x_1 x_2 + \epsilon_n$. Therefore, the marginal effect of x_1 is βx_2 and the one of x_2 is βx_1, which is difficult to justify. When one wants to introduce interaction effects, it's recommended to introduce them with the additive effects. This can be performed using any of the following commands:

```
lm(price ~ prodcubm + score + I(prodcubm * score), wine)
lm(price ~ prodcubm + score + prod_score, wine)
lm(price ~ prodcubm + score + prodcubm : score, wine)
```

the last expression uses the formula syntax, i.e., use `+` and `:`. However, introducing the additive and the interacting effect is a very common task, so that the `*` operator is devoted to this task:

```
z <- lm(price ~ prodcubm * score, wine)
z %>% coef
##      (Intercept)         prodcubm           score prodcubm:score
##        -191.9906           0.1487        259.2007          -0.2066
```

This model is now of the form: $y_n = \alpha + \beta_1 x_{n1} + \beta_2 x_{n2} + \beta_{12} x_{n1} x_{n2} \epsilon_n$ and the marginal effect for x_1 is now $\beta_1 + \beta_{12} x_{n2}$. This model has as a special case the additive model ($\beta_{12} = 0$) and this hypothesis can be tested, which was not the case for the model with only the interaction between the two covariates. The marginal effect of one more m^3 of production is (denoting x_2 the score): $0.149 - 0.207 x_2$ and is therefore a decreasing function of the score. Note that the coefficient of `prodcubm` (0.149) can be misleading. It indicates that, for a wine with a score equal to 0, the price increases with the production. But actually, such low scores are not observed, the lowest in the sample being 0.68 and for this value, the marginal effect of production is $0.149 - 0.207 \times 0.68 = 0.01$. The marginal effect of production has the expected negative sign for $x_2 > 0.149/0.207 = 0.72$, which is the case for 9560 wines out of 9580 in the sample. Therefore, the model predicts a negative marginal effect of the production on the price of a bottle for almost the whole sample, and this negative effect increases with the score obtained by the wine. We can easily test the hypothesis that the additive model is relevant using the value of the Student statistic for the interaction term:

```
coef(summary(z))[4, ]
##   Estimate Std. Error    t value   Pr(>|t|)
## -2.066e-01  4.972e-02 -4.155e+00  3.283e-05
```

which indicates that the interaction term is highly significant. For a model with the production and the age as covariates, we have:

```
z <- lm(price ~ prodcubm * age, wine)
coef(summary(z))[4, ]
##   Estimate Std. Error    t value   Pr(>|t|)
##  4.489e-05  2.837e-03  1.582e-02  9.874e-01
```

and this time, the interaction term is not significant and the additive model should be favored.

4.3.3 Polynomials

Until now, we have introduced the numerical covariates either in level or in log. In both cases, a unique coefficient $\hat{\beta}$ is estimated and the marginal effect of the covariate on the response is necessarily monotonic, which is not in practice always a relevant feature. For example, in a famous article, Kuznets (1955) analyzed the relation between economic development and inequality and his analysis has been formalized by Robinson (1976). A dual economy is considered, with a rural and an urban sector. In the rural sector, income is low and exhibits little variations as, in the urban sector, income is high and exhibits high variations. Denoting μ_r and μ_u the mean log-income in the rural and the urban sector and σ_r^2 and σ_u^2 the variance of log-income, it is therefore assumed that $\mu_r < \mu_u$ and $\sigma_r^2 < \sigma_u^2$. x is the share of the urban sector and, during the process of development, x increases. The mean log-income for the whole economy is:

$$\mu = (1 - x)\mu_r + x\mu_u = \mu_r + x(\mu_u - \mu_r) \tag{4.1}$$

As $\mu_u > \mu_r$, mean log-income increases during the process of development. To get the variance of log-income for the whole economy, which is a measure of inequality, we apply the variance decomposition formula

$$\sigma^2 = (1 - x)\sigma_r^2 + x\sigma_u^2 + (1 - x)(\mu_r - \mu)^2 + x(\mu_u - \mu)^2 \tag{4.2}$$

Replacing μ by its expression in Equation 4.1 and denoting $\Delta_\mu = \mu_u - \mu_r$ and $\Delta_{\sigma^2} = \sigma_u^2 - \sigma_r^2$, Equation 4.2 can be rewritten as:

$$\sigma^2 = \sigma_r^2 + (\Delta_{\sigma^2} + \Delta_\mu^2)x - \Delta_\mu^2 x^2$$

Therefore, the overall variance of log-income is the equation of a parabola with a positive first order term and a negative second order term. Graphically, the relation between the share of the modern sector and the overall variance of log-income is inverted U-shaped. From Equation 4.1, the mean of log income is positively correlated with the share of the urban sector. The relation between the variance and the mean of log-income is therefore also inverted U-shaped, and is called in the economic literature the **Kuznets curve**. We use the `kuznets` data set, which is used by Sudhir and Kanbur (1993):

```
kuznets
```

```
# A tibble: 60 x 5
  country  group        gnp  gini varlog
  <chr>    <fct>      <dbl> <dbl>  <dbl>
1 Chad     developing  79.5 0.369  0.368
2 Malawi   developing  80   0.470  0.576
3 Dahomey  developing  91.3 0.468  0.594
4 Pakistan developing  93.7 0.386  0.468
5 Tanzania developing 104.  0.503  0.661
# i 55 more rows
```

It contains data for 60 countries in the sixties; `group` categorizes developing, developed and socialist countries, `gnp` is the per capita gross national product in US$ of 1970, `gini` is the Gini index of inequality and `varlog` is the variance of the log-income. Figure 4.2 presents the relationship between log-income and inequality, the fitting line being a second-order polynomial. To perform the regression, `log(gnp)` and its square can be computed before the estimation or "on the fly" in the formula:

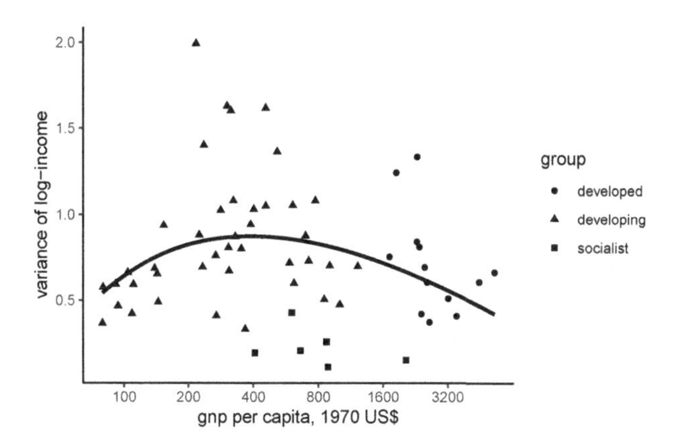

Figure 4.2: The Kuznets curve

```
kuznets <- kuznets %>%
  mutate(lgnp = log(gnp),
         lgnp2 = lgnp ^ 2)
lm(varlog ~ I(log(gnp)) + I(log(gnp) ^ 2), kuznets)
lm(varlog ~ lgnp + I(lgnp ^ 2), kuznets)
lm(varlog ~ lgnp + lgnp2, kuznets)
```

or more simply (especially if the order of the polynomial is high), using the `poly` function, which returns a matrix with a number of columns equal to the order of the polynomial:

```
lgnp3 <- poly(kuznets$lgnp, 3)
lgnp3 %>% head(3)
```

```
           1      2       3
[1,] -0.2200 0.2790 -0.2601
[2,] -0.2193 0.2764 -0.2547
[3,] -0.2040 0.2253 -0.1512
```

```
crossprod(lgnp3)
```

```
           1          2          3
1  1.000e+00 -3.062e-16 -7.980e-17
2 -3.062e-16  1.000e+00 -1.028e-16
3 -7.980e-17 -1.028e-16  1.000e+00
```

```
apply(lgnp3, 2, mean)
```

```
        1          2          3 
1.007e-17 -3.084e-18  1.382e-19 
```

By default, orthogonal polynomials are computed, i.e., polynomials for which all the inner products of the different terms are 0. Moreover, all the terms have a zero mean and a unit variance. "Raw" polynomials can be computed by setting the `raw` argument to `TRUE`. Using raw or orthogonal polynomials gives of course different coefficients, but the same fitted values and residuals. The `poly` function can be used inside a formula, so that the Kuznets model can be estimated using:

```
kc <- lm(varlog ~ poly(lgnp, 2, raw = TRUE), kuznets)
kc %>% coef %>% unname
## [1] -1.99919  0.92973 -0.07577
```

The results validate the Kuznets hypothesis, i.e., per capital income have first a positive, and then a negative effect on inequality. Denoting β_1 and β_2 the coefficients of log-income and its square, the summit of the curve is given by $\ln \text{gnp} = -\beta_1/(2\beta_2) = 6.14$, that corresponds to a per capita GNP equal to 462, which is very close to the median per capita income in the sample.

Using orthogonal polynomials, the covariance matrix of the estimators is now diagonal, but we get the same residuals and the same fitted values as previously:

```
kcp <- lm(varlog ~ poly(lgnp, 2), kuznets)
resid(kc) %>% head
##        1        2        3        4        5        6
## -0.25068 -0.04445 -0.05983 -0.19245 -0.02260 -0.27027
resid(kcp) %>% head
##        1        2        3        4        5        6
## -0.25068 -0.04445 -0.05983 -0.19245 -0.02260 -0.27027
```

```
vcov(kcp)
```

```
                (Intercept) poly(lgnp, 2)1 poly(lgnp, 2)2
(Intercept)       2.432e-03      3.138e-18      1.742e-33
poly(lgnp, 2)1    3.138e-18      1.459e-01      8.101e-17
poly(lgnp, 2)2    1.742e-33      8.101e-17      1.459e-01
```

4.4 Marginal effects

Until now, we have interpreted the coefficients in a linear model. However, except in the case where the covariate is in level, the coefficients are different from the marginal effects. For example, in a model with:

- the covariate in log: $y_n = \alpha + \beta \ln x + \epsilon_n$, the marginal effect is $dy_n/dx_n = \beta/x_n$,

- the covariate introduced as a second degree polynomial: $y_n = \alpha + \beta_1 x + \beta_2 x^2$, the marginal effect is $dy_n/dx_n = \beta_1 + 2\beta_2 x$.

Therefore, the marginal effect of x on y depends on the value of x and is therefore different from one observation to another. Moreover, the standard error of the marginal effect will also depend on the value of x. There are several **R** packages which enable the automatic computation of marginal effects. We'll present in this section the **marginaleffects** package (see Arel-Bundock 2023) which provides a very rich set of analytical and graphical tools to compute and analyze marginal effects.

4.4.1 Computation of the marginal effects with one covariate

Denote $m(\gamma, x_n) = dy/dx(\gamma, x_n)$. Until now, we have considered models which are linear in the parameters and are therefore of the form:

$$y_n = \alpha + \sum_{k=1}^{K} \beta_k f_k(x_n)$$

The marginal effect of a given covariate l is then: $m_l(\beta, x_n) = \sum_k \beta_k \frac{\partial f_k}{\partial x_{nl}}(x_n)$ and the fitted marginal effect is $\sum_k \hat{\beta}_k \frac{\partial f_k}{\partial x_{nl}}(x_n)$. Its variance is then a function of the matrix of covariance of the coefficients and of the partial derivatives of f_k with respect to x_{nl}. Consider as an example the Kuznets curve: $y_n = \alpha + \beta_1 \ln x_n + \beta_2 \ln^2 x_n$. The estimated marginal effect of the unique covariate is $\hat{\beta}_1/x_n + 2\hat{\beta}_2 \ln x_n / x_n$. It is therefore of the form $a\hat{\beta}_1 + b\hat{\beta}_2$, for which the variance is: $a^2\sigma^2_{\hat{\beta}_1} + b^2\sigma^2_{\hat{\beta}_1} + 2ab\sigma_{\hat{\beta}_1\hat{\beta}_2}$. Therefore, for the Kuznets model, the variance of the marginal effect of per capita GNP is:

$$(1/x_n)^2 \sigma^2_{\beta_1} + (2\ln(x_n)/x_n)^2 \sigma^2_{\beta_2} + 2 \times (1/x_n)(2\ln(x_n)/x_n)\sigma_{\beta_1\beta_2}$$

We first fit the Kuznets curve equation, and note that we use `log(gnp)` and not `lgnp` in the formula:

```
kc <- lm(varlog ~ poly(log(gnp), 2, raw = TRUE), kuznets)
kc %>% coef %>% unname
## [1] -1.99919  0.92973 -0.07577
```

Applying the preceding formulas, we compute the marginal effect and its standard error for each observation:

```
mekc <- kuznets %>%
  transmute(me = coef(kc)[2] / gnp + 2 * coef(kc)[3] * log(gnp) / gnp,
            sd = sqrt((1 / gnp) ^ 2 * vcov(kc)[2, 2] +
                        (2 * log(gnp) / gnp) ^ 2 * vcov(kc)[3, 3] +
                        2 * (1 / gnp) * (2 * log(gnp) / gnp) *
                        vcov(kc)[2, 3]))
mekc %>% head(3)
```

```
# A tibble: 3 x 2
       me      sd
    <dbl>   <dbl>
1 0.00335 0.00201
2 0.00332 0.00199
3 0.00269 0.00164
```

The **marginaleffects** package provides the `slopes` function to compute the marginal effects and their standard error for all the observations in the sample. It has a `variables` argument that is a character vector containing the variables for which the marginal effect has to be computed. It is unnecessary here, as we have only one covariate:[1]

```
library(marginaleffects)
kc_slps <- kc %>% slopes
kc_slps %>% as_tibble
```

```
# A tibble: 60 x 14
  rowid term  estimate std.error statistic p.value s.value  conf.low
  <int> <chr>    <dbl>     <dbl>     <dbl>   <dbl>   <dbl>     <dbl>
1     1 gnp    0.00335   0.00201      1.67  0.0949    3.40 -0.000582
2     2 gnp    0.00332   0.00199      1.67  0.0951    3.39 -0.000579
3     3 gnp    0.00269   0.00164      1.64  0.100     3.32 -0.000519
4     4 gnp    0.00258   0.00158      1.64  0.102     3.30 -0.000509
5     5 gnp    0.00218   0.00135      1.61  0.107     3.23 -0.000468
# i 55 more rows
# i 6 more variables: conf.high <dbl>, predicted_lo <dbl>,
#   predicted_hi <dbl>, predicted <dbl>, varlog <dbl>, gnp <dbl>
```

We can check that we get "almost" the same results as previously. The tiny difference is due to the fact that we used analytical derivatives, as `slopes` use numerical derivatives. We have seen previously that the same regression could have been performed by computing before the estimation the log of the per capita GNP:

```
kc2 <- lm(varlog ~ poly(lgnp, 2, raw = TRUE), kuznets)
```

although the coefficients are the same as previously, using `slopes` with this model returns completely different results:

```
kc2 %>% slopes %>% as_tibble
```

```
# A tibble: 60 x 14
  rowid term  estimate std.error statistic p.value s.value conf.low
  <int> <chr>    <dbl>     <dbl>     <dbl>   <dbl>   <dbl>    <dbl>
1     1 lgnp     0.267     0.160      1.67  0.0950    3.40  -0.0463
2     2 lgnp     0.266     0.159      1.67  0.0952    3.39  -0.0464
3     3 lgnp     0.246     0.150      1.64  0.100     3.32  -0.0474
```

[1]Note the use of `as_tibble`: objects returned by `slopes` are data frames with further attributes and don't print nicely in a book.

```
4      4 lgnp     0.242     0.148      1.64  0.101     3.30  -0.0475
5      5 lgnp     0.226     0.140      1.61  0.107     3.23  -0.0486
# i 55 more rows
# i 6 more variables: conf.high <dbl>, predicted_lo <dbl>,
#   predicted_hi <dbl>, predicted <dbl>, varlog <dbl>, lgnp <dbl>
```

The covariate is now `lgnp` and the marginal effect is no longer the effect of an increase of $1 of per capita on inequality, but the effect of a doubling of income.

The computation of the individual marginal effects results in a very large output, that is as such difficult to interpret. It is therefore customary to provide a unique value that summarizes the marginal effect of the covariate on the response for the whole sample. The preferred statistic is the **average marginal effects** (**AME**). It is simply the arithmetic mean of the individual effects. It can be obtained either using the `avg_slopes` function or applying the `summary` method to the object returned by `slope`:

```
z <- kc %>% slopes %>% summary
kc %>% avg_slopes %>% as_tibble
```

```
# A tibble: 1 x 12
  term  contrast estimate std.error statistic p.value s.value conf.low
  <chr> <chr>       <dbl>     <dbl>     <dbl>   <dbl>   <dbl>    <dbl>
1 gnp   mean(dY~ 0.000409  0.000304      1.35   0.178    2.49 -1.86e-4
# i 4 more variables: conf.high <dbl>, predicted_lo <dbl>,
#   predicted_hi <dbl>, predicted <dbl>
```

Note that the effect of **gnp** on inequality is not significant on average. This is hardly surprising as the effect is positive for about half of the sample and negative for the other half. Another popular choice to summarize marginal effects is the **marginal effect at the mean**. It consists of computing the marginal effect for a virtual observation for which all the covariates are set to their average in the sample. We get for our Kuznets curve:

```
tibble(gnp = mean(kuznets$gnp)) %>%
  transmute(me = coef(kc)[2] / gnp + 2 * coef(kc)[3] * log(gnp) / gnp,
            sd = sqrt((1 / gnp) ^ 2 * vcov(kc)[2, 2] +
                        (2 * log(gnp) / gnp) ^ 2 * vcov(kc)[3, 3] +
                        2 * (1 / gnp) * (2 * log(gnp) / gnp) *
                        vcov(kc)[2, 3]),
            z = me / sd)
## # A tibble: 1 x 3
##          me        sd     z
##       <dbl>     <dbl> <dbl>
## 1 -0.000116 0.0000597 -1.94
```

The effect is now negative and almost significant at the 5% level. This can be explained by the fact that the distribution of per capita income is highly asymmetric in our sample, so that the mean of **gnp** ($971) is much higher than the median ($456).

4.4.2 General computation of marginal effects

Consider now a model with several covariates, some being numerical, other categorical and with interactions. Using the `wine` data set, we fit the following model:

```
wine <- wine %>% mutate(lprodcubm = log(prodcubm))
u <- lm(log(price) ~ reg + log(score) * lprodcubm +
          reg : log(score) + vyd, wine)
u %>% summary %>% coef
```

```
                           Estimate Std. Error t value  Pr(>|t|)
(Intercept)                 3.84186    0.03345 114.869 0.000e+00
regnapavalley               0.69442    0.03290  21.109 1.013e-96
regsonoma                   0.19544    0.03833   5.099 3.476e-07
log(score)                  3.13420    0.21026  14.907 1.074e-49
lprodcubm                  -0.03561    0.01003  -3.550 3.878e-04
vyd                         0.09840    0.01066   9.232 3.216e-20
log(score):lprodcubm        0.51801    0.06182   8.379 6.068e-17
regnapavalley:log(score)    2.28783    0.21854  10.469 1.649e-25
regsonoma:log(score)        0.45321    0.24372   1.860 6.299e-02
```

The response is in log, so that every marginal effect measures a relative variation of the response. Two numerical variables are introduced in logs, but note that for the production, the log is computed before computing the regression, so that the covariate is `lprodcubm` and not `prodcubm`. On the contrary, for the score variable, the log is computed inside the formula. Therefore, while computing marginal effects, one has to keep in mind that what is measured is an increase of 100% of the production and of 1 point of the score. Note the use of the `*` operator between `log(score)` and `lprodcubm` so that the multiplicative effect of these two covariates is added to their separate additive effects. We also use the categorical variable `reg` (the region with three modalities) and a dummy variable for vineyard indication `vyd`. Moreover, the term `reg:log(score)` indicates that the effect of score on price will be different from one region to another. With such a complex linear regression, the coefficients are difficult to interpret and the computation of marginal effects is particularly useful. Using slopes without any further arguments, we get:

```
u_slps <- u %>% slopes
u %>% nobs
## [1] 9600
u_slps %>% nrow
## [1] 48000
```

The model is fitted on a sample of 9600 observations, but `slopes` return a data frame of $9600 \times 5 = 48000$ because there is one line for each observation and each covariate. The lines are sorted by covariate first and then by observation, the index of the observation being stored in a column called `rowid`. To get the marginal effects for the second observation, we can use:

```
u_slps %>% filter(rowid == 2) %>% as_tibble %>% select(1:6)
```

```
# A tibble: 5 x 6
  rowid term       contrast             estimate std.error statistic
  <int> <chr>      <chr>                   <dbl>     <dbl>     <dbl>
1     2 lprodcubm dY/dX                  -0.0902   0.00423     -21.3
2     2 reg        napavalley - Other    0.453     0.0131       34.7
3     2 reg        sonoma - Other        0.148     0.0152        9.69
4     2 score      dY/dX                 4.35      0.165        26.3
5     2 vyd        1 - 0                 0.0984    0.0107        9.23
```

The term of marginal effects is actually relevant only for numerical variables. The `contrast` column indicates the kind of effect that is computed. It is `dY/dX` (a derivative) for the two numerical covariates `lprobcubm` and `score`. For factors, the term is a contrast, i.e., the difference between one level (`sonoma` and `napavaley` for `reg`) and the reference level (`Other`). This is also the case for the dummy variable `vid`, for which the effect is the difference of prediction for `vid` equal to 1 and 0.[2] The `variables` argument can be used to compute the marginal effects only for a subset of covariate:

```
u %>% slopes(variables = c("score", "reg")) %>% as_tibble %>%
  filter(rowid == 2) %>% select(1:6)
```

As previously, we can compute the average marginal effect using `avg_slopes`:

```
u %>% avg_slopes %>% as_tibble
```

```
# A tibble: 5 x 12
  term      contrast     estimate std.error statistic   p.value s.value
  <chr>     <chr>           <dbl>     <dbl>     <dbl>     <dbl>   <dbl>
1 lprodcubm mean(dY/dX)  -0.113    0.00290     -39.1  0           Inf
2 reg       mean(napav~   0.351    0.0102       34.3  1.02e-257   854.
3 reg       mean(sonom~   0.127    0.0103       12.4  3.20e- 35   115.
4 score     mean(dY/dX)   6.07     0.110        55.2  0           Inf
5 vyd       mean(1) - ~   0.0984   0.0107        9.23 2.65e- 20    65.0
# i 5 more variables: conf.low <dbl>, conf.high <dbl>,
#   predicted_lo <dbl>, predicted_hi <dbl>, predicted <dbl>
```

Moreover, the **AME** can be computed for the different modalities of a factor, using the `by` argument:

```
u %>% avg_slopes(variables = c("score", "lprodcubm"),
                 by = "reg") %>% as_tibble %>% select(1:6)
```

```
# A tibble: 6 x 6
  term      contrast    reg        estimate std.error statistic
  <chr>     <chr>       <fct>         <dbl>     <dbl>     <dbl>
1 lprodcubm mean(dY/dX) Other        -0.120   0.00292     -41.0
2 lprodcubm mean(dY/dX) napavalley   -0.105   0.00319     -32.8
3 lprodcubm mean(dY/dX) sonoma       -0.112   0.00293     -38.3
```

[2] Actually in a linear model, this difference is equal to the derivative, see Section 4.1.5.

```
4 score      mean(dY/dX) Other          5.25     0.153           34.3
5 score      mean(dY/dX) napavalley     7.70     0.202           38.2
# i 1 more row
```

The marginal effects can also be computed for hypothetical observations, using the **newdata** argument, which should be a data frame containing one or several hypothetical observations. Such data frames can be easily computed using the **datagrid** function; it has a **model** argument and by default returns a one-line data frame containing the mean for the numerical variables and the mode for categorical variables:

```
datagrid(model = u)
##       reg  score lprodcubm vyd rowid
## 1 Other 0.8614       2.547   0     1
```

and can be used to compute the marginal effects at the mean:

```
u %>% slopes(newdata = datagrid()) %>% as_tibble %>% select(1:6)
```

```
# A tibble: 5 x 6
  rowid term      contrast             estimate std.error statistic
  <int> <chr>     <chr>                   <dbl>     <dbl>     <dbl>
1     1 lprodcubm dY/dX                 -0.113    0.00291    -38.8
2     1 reg       napavalley - Other     0.353    0.0102      34.6
3     1 reg       sonoma - Other         0.128    0.0103      12.4
4     1 score     dY/dX                  5.17     0.151       34.2
5     1 vyd       1 - 0                  0.0984   0.0107       9.23
```

datagrid can also include arguments whose names are the names of some covariates and their values are some values of the covariate. If, for example, two values are provided for the first covariate and three values for the second covariates, six observations are generated, obtained by taking all the combinations of the two covariates:

```
datagrid(model = u, score = c(20, 50, 100), vyd = c(0, 1))
```

```
    reg lprodcubm score vyd rowid
1 Other     2.547    20   0     1
2 Other     2.547    20   1     2
3 Other     2.547    50   0     3
4 Other     2.547    50   1     4
5 Other     2.547   100   0     5
6 Other     2.547   100   1     6
```

Note that for unspecified covariates the mean / mode are introduced for numerical / categorical variables. We've just presented a few features of the **marginaleffects** package. Detailed documentation can be found at https://marginaleffects.com/.

Part II

Beyond the OLS estimator

In the first part of the book, we've conducted regression analysis of the following form:

$$y_n = \alpha + \beta_1 x_{n1} + \beta_2 x_{n2} + ... + \beta_K x_{nK} + \epsilon_n = \alpha + \beta^\top x_n + \epsilon_n$$

The conditional expectation of y was then:

$$\mathrm{E}(y_n \mid x_n) = \alpha + \beta^\top x_n + \mathrm{E}(\epsilon_n \mid x_n)$$

The fundamental hypothesis is that $\mathrm{E}(y_n \mid x_n) = \alpha + \beta^\top x_n$ or equivalently that $\mathrm{E}(\epsilon_n \mid x_n) = 0$, which means that the population covariance between the covariates and the errors of the model is 0. We then supposed that the errors were spherical, which means that they are homoskedastic $\mathrm{V}(\epsilon_n \mid x_n) = \sigma_\epsilon^2$ and uncorrelated $\mathrm{cov}(\epsilon_n \epsilon_m \mid x_n, x_m) = 0 \;\forall n \neq m$. Finally, we supposed that the conditional expectation of the response is a linear function of the covariate.

The first three chapters of this part present estimators that enables to get rid of these three hypotheses. Chapter 5 presents the maximum likelihood estimator. This estimator is particularly useful when the desired conditional expectation function is non-linear, for example: $\mathrm{E}(y_n \mid x_n) = e^{\alpha + \beta^\top x_n}$. General results about this estimator will be presented in this chapter, and this estimator will be very intensively used in all the chapters of the last part of the book. Chapter 6 presents relevant tools to detect and deal with the common case where the errors are not spherical. This includes tests to detect heteroskedasticity and correlated errors, a more efficient estimator than OLS (the generalized least squares estimator) and robust estimates of the variance of the OLS estimators. Chapter 7 tackles the problem of endogeneity, i.e., situations where the errors are correlated with some covariates or, equivalently, where $\alpha + \beta^\top x_n$ is not the conditional expectation of y. Tests of endogeneity are presented, along with relevant estimators, namely the instrumental variable and the fixed effects estimators.

The last two chapters present two very dynamic fields in which the problems treated in these three chapters are particularly important. Chapter 8 presents relevant tools to analyze the effects of a treatment on an outcome of interest, dealing with the fundamental problem of the endogeneity of the treatment. Chapter 9 is devoted to spatial econometrics. These models consider samples with geolocalized units and neighbor units may be linked via the values of their response or of their errors. The first situation leads to a problem of endogeneity and the second one to a problem of non-spherical errors.

5

Maximum likelihood estimator

The **maximum likelihood** (**ML**) estimator is suitable in situations where the GDP of the response is perfectly known (or supposed to be), up to some unknown parameters. It is particularly useful for models where the response is not continuous but, for example, it is a count, a binomial or a multinomial variable.[1] Section 5.1 presents the ML estimator for the simplest case where there is a unique parameter to estimate. Section 5.2 extends this basic model to the case where there is a vector of parameters to estimate. Finally Section 5.3 present different tests that are particularly useful in the context of ML estimation.

5.1 ML estimation of a unique parameter

For a first informal view of this estimator, consider two very simple one-parameter distribution functions:

- the Poisson distribution, which is suitable for the distribution of count responses, i.e., responses which can take only non-negative integer values,
- the exponential distribution, which is often used for responses that represent a time span (for example, an unemployment spell).

5.1.1 Computation of the ML estimator for the Poisson distribution

As an example of count data, we consider the `cartel` data set of Miller (2009) who analyzed the role of commitment to the lenient prosecution of early confessors on cartel enforcement. The response `ncaught` is a count, the number of cartel discoveries per a 6-month period, for the 1985-2005 period. A Poisson variable is defined by the following mass probability function:

$$P(y;\theta) = \frac{e^{-\theta}\theta^y}{y!} \tag{5.1}$$

where θ is the unique parameter of the distribution. An important characteristic of this distribution is that $\mathrm{E}(y) = \mathrm{V}(y) = \theta$, which means that, for a Poisson variable, the variance equals the expectation. To have a first look at the relevance of this hypothesis for the response of our example, we compute the sample mean and variance:

[1] These models will be presented in details in the third part of the book.

```
cartels %>% summarise(m = mean(ncaught), v = var(ncaught))
##       m     v
## 1 5.175 7.789
```

The two moments are of the same order of magnitude so that we can confidently rely on the Poisson distribution hypothesis. The question is now how to use the information from the sample to estimate the unique parameter of the distribution. Figure 5.1 represents the empirical distribution of the response and adds several lines which represent the Poisson distribution for different values of θ.

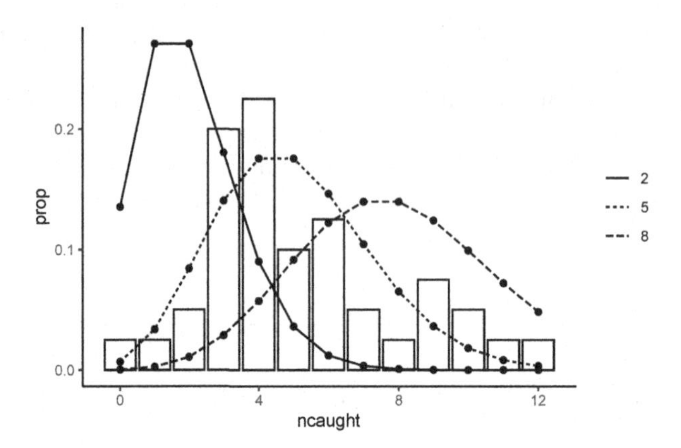

Figure 5.1: Empirical distribution and theoretical Poisson probabilities

The Poisson distribution is highly asymmetric and has a large mode for $\theta = 2$, it gets more and more flat and less asymmetric as θ increases. From Figure 5.1, we can see that the empirical distribution is very different from the Poisson one for a value of θ as small as 2 or as large as 8. For $\theta = 5$, there is a reasonable correspondence between the empirical and the theoretical distribution. To get unambiguously a value for our estimator, we need an objective function that is a function of θ which is, in the context of this chapter, the **likelihood function**. To construct it, it is much simpler to consider the sample as a random sample. For the first observation 1, given a value of θ, the probability of observing y_1 is $P(y_1; \theta)$. For the second one, with the random sample hypothesis, the probability of observing y_2 is $P(y_2; \theta)$ and is independent of y_1. Therefore, the joint probability of observing (y_1, y_2) is $P(y_1; \theta) \times P(y_2; \theta)$ and, more generally, the probability of observing the values of y for the whole sample is the likelihood function for a Poisson variable in a random sample which writes:

$$L(\theta, y) = \Pi_{n=1}^{N} P(y_n; \theta)$$

and the maximum likelihood estimator of θ is the value of θ which maximizes the likelihood function. Note the change of notations: $P(y_n; \theta)$ is the probability mass function, it is a function of $y^\top = (y_1, y_2, \dots, y_N)$ and computes a probability for a given (known) value of the parameter of the distribution. The likelihood function $L(\theta, y)$ is written as a function of the unknown value of the parameter of the distribution and its value depends on the realization of the response vector y on the sample.

There are several good reasons to consider the logarithm of the likelihood function instead of the likelihood itself. In particular, being a product of N terms, the likelihood will typically take very high or very low values for large samples. Moreover, the logarithm transforms a product in a sum, which is much more convenient. Note finally that, as the logarithm is a strictly increasing function, maximizing the likelihood or its logarithm leads to the same value of θ. Taking the log and replacing $P(y_n;\theta)$ by the expression given in Equation 5.1, we get:

$$\ln L(\theta, y) = \sum_{n=1}^{N} \ln \frac{e^{-\theta}\theta^{y}}{y!}$$

or, regrouping terms:

$$\ln L(\theta, y) = -N\theta + \ln\theta \sum_{n=1}^{N} y_n - \sum_{n=1}^{N} \ln y_n!$$

The log-likelihood is therefore the sum of three terms, and note that the last one doesn't depend on θ. For our sample, we have:

```
y <- cartels$ncaught
N <- length(y)
sum_y <- sum(y)
sum_log_fact_y <- sum(lfactorial(y))
lnl_poisson <- function(theta) - N * theta + log(theta) * sum_y -
    sum_log_fact_y
```

The last line of code define `lnl_poisson` as a function of `theta` which returns the value of the log likelihood function for a value of θ given the value of the responses in the `cartel` data set.

Taking some values of θ, we compute in Table 5.1 the corresponding values of the log likelihood function. The log likelihood function is inverse U-shaped, and the maximum value for the integer values we've provided is reached for $\theta = 5$. Figure 5.2 presents the log likelihood function as a smooth line in the neighborhood of $\theta = 5$, which indicates that the maximum of the log likelihood function occurs for a value of θ between 5.0 and 5.5. The first-order condition for a maximum is that the first derivative of $\ln L$ with respect to θ is 0.

Table 5.1: Poisson probabilities for different values of the parameter

θ	$\ln L$
1	-270.13
2	-166.65
3	-122.72
4	-103.16
5	-96.97
6	-99.23
7	-107.32
8	-119.68

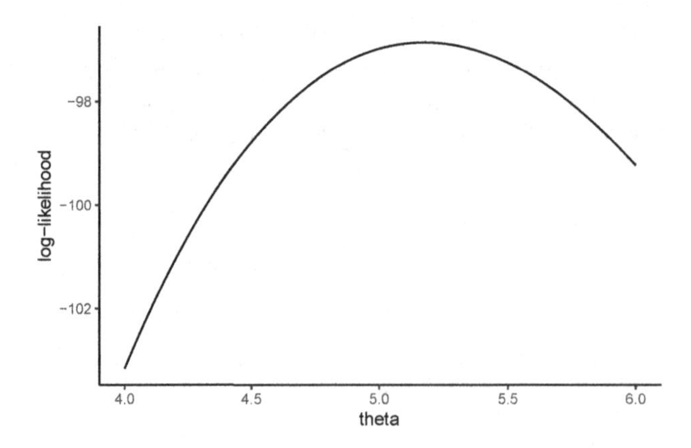

Figure 5.2: Log-likelihood curve for a Poisson variable

$$\frac{\partial \ln L}{\partial \theta} = -N + \frac{\sum_{n=1}^N y_n}{\theta} = 0$$

This leads to $\hat{\theta} = \frac{\sum_{n=1}^N y_n}{N}$. Therefore, in this simple case, we can explicitly obtain the ML estimator by solving the first-order condition, and, moreover, the ML estimator is the sample mean of the response, which is hardly surprising as θ is the expected value of y for a Poisson variable. The second derivative is:

$$\frac{\partial^2 \ln L}{\partial \theta^2} = -\frac{\sum_{n=1}^N y_n}{\theta^2} < 0$$

and is negative for all values of θ, which indicates that the log-likelihood is globally concave and therefore that optimum we previously obtained is the global maximum. For our sample, the ML estimator is:

```
hat_theta <- sum_y / N
hat_theta
## [1] 5.175
```

5.1.2 Computation of the ML estimator for the exponential distribution

The second example illustrates the computation of the ML estimator for a continuous variable. The density of a variable which follows an exponential distribution of parameter θ is:

$$f(y;\theta) = \theta e^{-\theta y}$$

The expected value and the variance of y are $\mathrm{E}(y) = 1/\theta$ and $\mathrm{V}(y) = 1/\theta^2$ so that, for a variable which follows an exponential distribution, the mean should be equal to the standard deviation. To illustrate the use of the exponential distribution, we use the `oil` data set of Favero, Pesaran, and Sharma (1994) for which the response `dur` is the time span between the discovery of an oil field and the date of the British government's approval to exploit this field.

```
oil %>% summarise(m = mean(dur), s = sd(dur))
## # A tibble: 1 x 2
##       m     s
##   <dbl> <dbl>
## 1  63.0  55.2
```

The sample mean and the standard deviation are close, in conformity with the features of the exponential distribution. Figure 5.3 presents the empirical distribution of `dur` by a histogram, and we add several exponential density lines for different values of θ (0.005, 0.01 and 0.03).

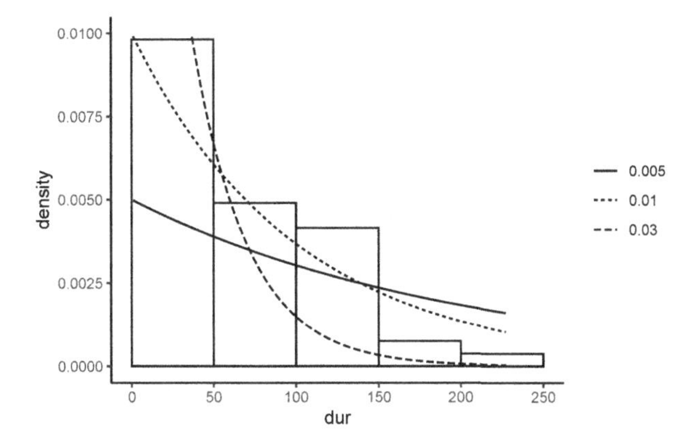

Figure 5.3: Empirical distribution and density of the exponential distribution

The adjustment between the empirical and the theoretical distribution is quite good with the dotted line which corresponds to $\theta = 0.01$. The reasoning to construct the log-likelihood function is exactly the same as for the discrete Poisson distribution, except that the mass probability function is replaced by the density function. The log-likelihood function therefore writes:

$$\ln L(\theta, y) = \sum_{n=1}^{N} \ln\left(\theta e^{-\theta y_n}\right)$$

or, regrouping terms:

$$\ln L(\theta, y) = N \ln \theta - \theta \sum_{n=1}^{N} y_n$$

The shape of the log-likelihood function is represented in Figure 5.4. As in the Poisson case, the log-likelihood seems to be globally concave, with a global maximum corresponding to a value of θ approximately equal to 0.015. The first-order condition for a maximum is:

$$\frac{\partial \ln L}{\partial \theta} = \frac{N}{\theta} - \sum_{n=1}^{N} y_n$$

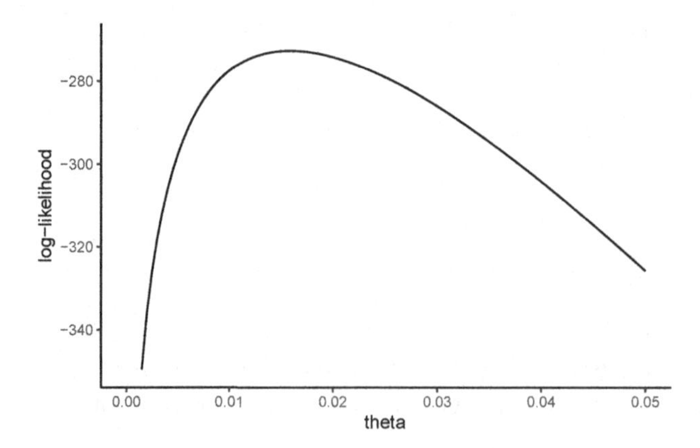

Figure 5.4: Log-likelihood curve for an exponential variable

which leads to the ML estimator: $\hat{\theta} = \frac{N}{\sum_{n=1}^{N} y_n} = \frac{1}{\bar{y}}$. The second derivative is:

$$\frac{\partial \ln^2 L}{\partial \theta^2} = -\frac{N}{\theta^2}$$

which is negative for all values of θ, so that $\hat{\theta}$ is the global maximum of the log-likelihood function. In our example, we get:

```
hat_theta <- N / sum_y
hat_theta
## [1] 0.01587
```

5.1.3 Properties of the ML estimator

We consider a variable y for which we observe N realizations in a random sample. We assume that y follows a distribution with a unique parameter θ. The density (if y is continuous) or the mass probability function (if y is discrete) is denoted $\phi(y;\theta)$. We also denote $\lambda(y;\theta) = \ln\phi(y;\theta)$ and $\gamma(y;\theta)$ and $\psi(y;\theta)$ the first two derivatives of λ with θ. The log likelihood function is then:

$$\ln L(\theta, y) = \sum_n \lambda(y_n;\theta) = \sum_n \ln \phi(y_n;\theta)$$

with $y^\top = (y_1, y_2, \dots, y_N)$ the vector containing all the values of the response in the sample. The "true value" of θ is denoted by θ_0 and the maximum likelihood estimator $\hat{\theta}$. The proof of the consistency of the ML estimator is based on Jensen's inequality, which states that, for a random variable X and a concave function f:

$$\mathrm{E}(f(X)) \leq f(\mathrm{E}(X))$$

As the logarithm is a concave function:

$$\mathrm{E}\ln\frac{L(\theta)}{L(\theta_0)} < \ln\mathrm{E}\frac{L(\theta)}{L(\theta_0)} \tag{5.2}$$

The expectation on the right side of the equation is obtained by integrating out $L(\theta)/L(\theta_0)$ using the density of y, which is $L(\theta_0)$. Therefore:

$$\mathrm{E}\frac{L(\theta)}{L(\theta_0)} = \int \frac{L(\theta)}{L(\theta_0)} L(\theta_0)dy = \int L(\theta)dy = 1$$

as any density sums to 1 for the whole support of the variable. Therefore Equation 5.2 implies that:

$$\mathrm{E}\ln L(\theta) \leq \mathrm{E}\ln L(\theta_0)$$

Dividing by N and using the law of large numbers, we also have:

$$\mathrm{plim}\frac{1}{N}\ln L(\theta) \leq \mathrm{plim}\frac{1}{N}\ln L(\theta_0) \tag{5.3}$$

As $\hat{\theta}$ maximizes $\ln L(\theta)$, it is also the case that $\ln L(\hat{\theta}) \geq \ln L(\theta)$. Once again, dividing by N and computing the limit, we have:

$$\mathrm{plim}\frac{1}{N}\ln L(\hat{\theta}) \geq \mathrm{plim}\frac{1}{N}\ln L(\theta) \tag{5.4}$$

The only solution for Equation 5.3 and Equation 5.4 to hold is that:

$$\mathrm{plim}\frac{1}{N}\ln L(\hat{\theta}) = \mathrm{plim}\frac{1}{N}\ln L(\theta_0) \tag{5.5}$$

Equation 5.5 indicates that, as N tends to infinity, the average likelihood for $\hat{\theta}$ converges to the average likelihood for θ_0. Using a regularity condition, this implies that the estimator is consistent, i.e., that $\mathrm{plim}\,\hat{\theta} = \theta_0$.

The first two derivatives of the log-likelihood functions, which are respectively called the **gradient** (or the **score**) and the **hessian** of the log-likelihood are:

$$g(\theta, y) = \frac{\partial \ln L}{\partial \theta}(\theta, y) = \sum_{n=1}^{N} \frac{\partial \lambda}{\partial \theta}(y_n; \theta) = \sum_{n=1}^{N} \gamma(y_n; \theta) \tag{5.6}$$

and

$$h(\theta, y) = \frac{\partial^2 \ln L}{\partial \theta^2}(\theta, y) = \sum_{n=1}^{N} \frac{\partial^2 \lambda}{\partial \theta^2}(y_n; \theta) = \sum_{n=1}^{N} \psi(y_n; \theta) \tag{5.7}$$

As there is only one parameter to estimate, both functions are scalar functions. For a discrete distribution, the probabilities for every possible K values of y (denoted by y_k) sum to unity:

$$\sum_{k=1}^{K} \phi(y_k;\theta) = 1$$

The same result applies with the continuous sum for a continuous random variable:

$$\int \phi(y;\theta)dy = 1$$

As $\phi = e^{\lambda}$, taking the derivative with respect to θ, we get:

$$\left\{ \begin{array}{rcl} \sum_{k=1}^{K} \gamma(y_k;\theta)\phi(y_k;\theta) &=& 0 \\ \int \gamma(y;\theta)\phi(y;\theta)dy &=& 0 \end{array} \right. \tag{5.8}$$

Evaluated for the true value θ_0, Equation 5.8 is the expectation of an individual contribution to the gradient. Therefore, evaluated at the true value of θ, $\gamma(y;\theta)$ is a random variable with 0 expectation. Taking now the second derivative with respect to θ, we get:

$$\left\{ \begin{array}{rcl} \sum_{k=1}^{K} \gamma(y_k;\theta)^2\phi(y_k;\theta) + \sum_{k=1}^{K} \psi(y_k;\theta)\phi(y_k;\theta) &=& 0 \\ \int \gamma(y;\theta)^2\phi(y;\theta)dy + \int \psi(y;\theta)\phi(y;\theta)dy &=& 0 \end{array} \right. \tag{5.9}$$

We denote the first term by $v_\gamma(\theta)$: evaluated for the true value of the parameter, this is the variance of $\gamma(y;\theta)$:, $v_\gamma(\theta_0) = \sigma^2_\gamma$. We denote the second term by $m_\psi(\theta)$, which is, evaluated for the true value of the parameter, the expectation of $\psi(y;\theta)$: $m_\psi(\theta_0) = \mu_\psi$. Therefore, Equation 5.9 indicates that $\sigma^2_\gamma = -\mu_\psi$.

The gradient $g(\theta, y)$ (Equation 5.6) is the sum of N contributions $\gamma(y_n;\theta)$ which have 0 expectation. Therefore, its expectation is 0. With the random sample hypothesis, $\gamma(y_n;\theta)$ and $\gamma(y_m;\theta)$ are independent for all $m \neq n$ and the variance of the gradient is therefore the sum of the variances of its N contributions, which are all equal to σ^2_γ. Therefore, $\mathrm{V}(g(\theta_0, y)) = N\sigma^2_\gamma$. The variance of the gradient is called the **information matrix** in the general case, but it is actually in our one parameter case a scalar that we'll denote $\iota(\theta_0)$. The hessian being a sum of N contributions $\psi(y_n;\theta)$, which have an expectation equal to μ_ψ, its expected value is $\mathrm{E}(h(\theta_0, y)) = N\mu_\psi$. The result we previously established ($\sigma^2_\gamma = -\mu_\psi$) implies that the variance of the gradient (the information) equals the opposite of the expectation of the hessian:

$$\iota(\theta_0) = \mathrm{V}(g(\theta_0, y)) = -\mathrm{E}(h(\theta_0, y)) = N\sigma^2_\gamma = -N\mu_\psi \tag{5.10}$$

This important result is called the **information equality**. Denoting θ_0 as the true (unknown) value of the parameter, and omitting for convenience the y vector, a first-order Taylor expansion of $g(\hat{\theta})$ around θ_0 is:

$$g(\hat{\theta}) \approx g(\theta_0) + h(\theta_0)(\hat{\theta} - \theta_0)$$

If we use instead an exact first-order Taylor expansion:

$$g(\hat{\theta}) = g(\theta_0) + h(\bar{\theta})(\hat{\theta} - \theta_0) \tag{5.11}$$

where $\bar{\theta}$ lies between $\hat{\theta}$ and θ_0. As $g(\hat{\theta}) = 0$, solving for $\hat{\theta} - \theta_0$, we get:

$$(\hat{\theta} - \theta_0) = -h(\bar{\theta})^{-1} g(\theta_0)$$

or, multiplying by $\sqrt{N}$:

$$\sqrt{N}(\hat{\theta} - \theta_0) = -\left(\frac{h(\bar{\theta})}{N}\right)^{-1} \frac{g(\theta_0)}{\sqrt{N}}$$

Assuming that the estimator is consistent, as N grows, $\hat{\theta}$ converges to θ_0 and so does $\bar{\theta}$. As $\mathrm{E}(h(\theta_0)) = N\mu_\psi$, the expectation of the first term is μ_ψ, and it is also its probability limit. The second term has a 0 expected value and a variance equal to σ_γ^2. Therefore, $\sqrt{N}(\hat{\theta} - \theta_0)$ has a zero expectation and an asymptotic covariance equal to $\mu_\psi^{-2}\sigma_\gamma^2$. Moreover, applying the central-limit theorem, its asymptotic distribution is normal. Therefore:

$$\sqrt{N}(\hat{\theta} - \theta_0) \xrightarrow{p} N(0, \mu_\psi^{-2}\sigma_\gamma^2) \text{ and } \hat{\theta} \overset{a}{\sim} \mathcal{N}(\theta_0, \mu_\psi^{-2}\sigma_\gamma^2/N) \tag{5.12}$$

Applying the information equality, $\iota(\theta_0) = N\sigma_\gamma^2(\theta_0) = -N\mu_\psi(\theta_0)$ and therefore:

$$\hat{\theta} \overset{a}{\sim} \mathcal{N}(\theta_0, \iota(\theta_0)^{-1}) \tag{5.13}$$

We have seen that the variance of the ML estimator is the inverse of the information, which can be either obtained using the variance of the gradient or the opposite of the expectation of the hessian. In terms of the average information ($\iota(\theta)/N$), we have:

$$\left\{\begin{array}{rcl} \frac{\iota(\theta)}{N} & = & \frac{1}{N}\mathrm{V}\left(\frac{g(\theta,y)}{\sqrt{N}}\right) = \frac{1}{N}\sum_{n=1}^N \mathrm{E}(\gamma(y;\theta)^2) = v_\gamma(\theta) \\ \frac{\iota(\theta)}{N} & = & \frac{1}{N}\mathrm{E}\left(-\frac{h(\theta,y)}{N}\right) = -\frac{1}{N}\sum_{n=1}^N \mathrm{E}(\psi(y_n;\theta)) = -m_\psi(\theta) \end{array}\right. \tag{5.14}$$

If this variance/expectation can be computed, then a natural estimator of $\iota(\theta_0)$ is $\iota(\hat{\theta})$. This **information-based estimator** of the variance is obtained by inverting the information evaluated for the maximum likelihood value of θ:

$$\hat{\sigma}^2_{\hat{\theta}i} = \iota(\hat{\theta})^{-1}$$

On the contrary, if it is impossible to compute the expectations, two natural estimators of the information are based on the gradient and the hessian and are obtained by evaluating one of the two expressions in Equation 5.14 without the expectation. Denoting ι_g and ι_h, these two estimations of the information, we have from Equation 5.14:

$$\left\{\begin{array}{rcl} \frac{\iota_g(\theta)}{N} & = & \frac{1}{N}\sum_{n=1}^N \gamma(y_n;\theta)^2 = \hat{v}_\gamma(\theta) \\ \frac{\iota_h(\theta)}{N} & = & -\frac{h(\theta,y)}{N} = -\frac{1}{N}\sum_{n=1}^N \psi(\theta;y_n) = -\hat{m}_\psi(\theta) \end{array}\right.$$

Evaluated for the maximum likelihood estimator value of θ, we then get the **gradient-based estimator**:

$$\hat{\sigma}^2_{\hat{\theta}g} = \iota_g(\hat{\theta})^{-1} = \left(\sum_{n=1}^{N} \gamma(y;\hat{\theta})^2\right)^{-1} = \frac{1}{N\hat{v}_\gamma(\hat{\theta})} \tag{5.15}$$

and the **hessian-based estimator** of the variance of θ:

$$\hat{\sigma}^2_{\hat{\theta}h} = \iota_h(\hat{\theta})^{-1} = -h(\hat{\theta}, y)^{-1} = -\frac{1}{N\hat{m}_\psi(\hat{\theta})} \tag{5.16}$$

A fourth estimator is based on Equation 5.12, which states that, before applying the information equality,

$$\sigma^2_{\hat{\theta}} = \frac{1}{N}\frac{v_\gamma(\theta_0)}{m_\psi(\theta_0)^2} \tag{5.17}$$

Removing the expectation from Equation 5.17 and evaluating for the maximum likelihood estimator of θ, we get the **sandwich estimator** of the variance of $\hat{\theta}$.

$$\hat{\sigma}^2_{\hat{\theta}s} = \frac{1}{N}\left(\sum_{n=1}^{N} \gamma(y_n;\hat{\theta})^2/N\right) / \left(\sum_{n=1}^{N} \psi(y_n;\hat{\theta})/N\right)^2 = \frac{1}{N}\frac{\hat{v}_\gamma(\hat{\theta})}{\hat{m}_\psi(\hat{\theta})} \tag{5.18}$$

Equation 5.18 is called a sandwich estimator for a reason that will be clear when we'll compute it in the general case where more than one parameter are estimated. It is a more general estimator than the previous three, as its consistency doesn't rely on the information equality property, which is only valid if the distribution of y is correctly specified.

5.1.4 Computation of the variance for the Poisson and the exponential distribution

For the Poisson model, we have:

$$g(\theta) = -N + \frac{\sum_{n=1}^N y_n}{\theta} \text{ and } h(\theta) = -\frac{\sum_{n=1}^N y_n}{\theta^2}$$

The variance of the gradient is:

$$\mathrm{V}\left(g(\theta)\right) = \frac{1}{\theta^2}\mathrm{V}\left(\sum_{n=1}^N y_n\right) = \frac{1}{\theta^2}\sum_{n=1}^N \mathrm{V}\left(y_n\right) = \frac{1}{\theta^2}N\theta = \frac{N}{\theta}$$

The first equality holds because of the random sample hypothesis and the second one because $\mathrm{V}(y) = \theta$. The expected value of the hessian is:

$$\mathrm{E}\left(h(\theta)\right) = -\frac{\sum_{n=1}^{N}\mathrm{E}(y_n)}{\theta^2} = -\frac{N\theta}{\theta^2} = -\frac{N}{\theta}$$

because $\mathrm{E}(y) = \theta$. Therefore, we are in the case where the information can be computed and the result illustrates the information equality. The information based estimator of $\hat{\theta}$ is:

$$\iota(\hat{\theta}) = \frac{N}{\hat{\theta}} = \frac{N}{\bar{y}}$$

The individual contributions to the gradient are: $\gamma(y_n; \theta) = -1 + y_n/\theta$, so that the gradient-based estimator of the information is:

$$\iota_g(\theta) = \sum_{n=1}^{N}\gamma(y_n;\theta)^2 = \sum_{n=1}^{N}\frac{(y_n - \theta)^2}{\theta^2}$$

Evaluating ι_g for the maximum likelihood estimator, we finally get:

$$\iota_g(\hat{\theta}) = \sum_{n=1}^{N}\frac{(y_n - \bar{y})^2}{\bar{y}^2} = N\frac{\hat{\sigma}_y^2}{\bar{y}^2}$$

For the hessian-based estimator of the information, we consider the opposite of the hessian evaluated for the ML estimator:

$$\iota_h(\hat{\theta}) = \frac{\sum_n y_n}{\hat{\theta}^2} = \frac{N}{\bar{y}}$$

Finally, from Equation 5.18, the sandwich estimator of the variance is:

$$\hat{\sigma}_{\hat{\theta}s} = \frac{1}{N}\frac{\hat{\sigma}_y^2/\bar{y}^2}{(1/\bar{y})^2} = \frac{\hat{\sigma}_y^2}{N}$$

To summarize, the four estimators of the variance of $\hat{\theta}$ are:

$$\left\{\begin{array}{ccccc}
\hat{\sigma}^2_{\hat{\theta}i} & = & \iota(\hat{\theta})^{-1} & = & \bar{y}/N \\
\hat{\sigma}^2_{\hat{\theta}g} & = & \iota_g(\hat{\theta})^{-1} & = & \frac{\bar{y}^2}{\hat{\sigma}^2_y}/N \\
\hat{\sigma}^2_{\hat{\theta}h} & = & \iota_h(\hat{\theta})^{-1} & = & \bar{y}/N \\
\hat{\sigma}^2_{\hat{\theta}s} & & & = & \hat{\sigma}^2_y/N
\end{array}\right.$$

Note that in this case, $\iota(\hat{\theta}) = \iota_h(\hat{\theta})$ and that, if the Poisson distribution hypothesis is correct, $\mathrm{plim}\,\bar{y} = \mathrm{plim}\,\hat{\sigma}_y^2 = \theta_0$ so that the four estimators are consistent as $\bar{y}$ and $\hat{\sigma}_y^2$ both converge to θ_0. Computing these four estimations of the variance of $\hat{\theta}$ for the **ncaught** variable, we get:

```
N <- nrow(cartels) ; y <- cartels$ncaught
mean_y <- mean(y) ; var_y <- mean( (y - mean_y) ^ 2)
c(info = mean_y / N, gradient = mean_y ^ 2 / var_y / N,
  hessian = mean_y / N, sandwich = var_y / N) %>%
```

```
  sqrt %>% round(3)
##     info gradient  hessian sandwich
##    0.360    0.297    0.360    0.436
```

For the exponential distribution, remember that $\lambda(y;\theta) = \ln\theta - \theta y$, $\gamma(y;\theta) = 1/\theta - y$ and $\psi(y;\theta) = -1/\theta^2$. As $h(\theta, y) = \sum_n \psi(y;\theta)$ the hessian is obviously $h(\theta, y) = -N/\theta^2$ and equals its expected value, as it doesn't depend on y. Therefore, $\iota(\theta) = N/\theta^2$. Computing the variance of the gradient, we get:

$$
\begin{array}{rcl}
\mathrm{V}\left(g(\theta)\right) & = & \mathrm{E}\left(\sum_n \gamma(y_n;\theta)^2\right) = \sum_n \mathrm{E}(\gamma(y_n;\theta)^2) \\
 & = & \sum_n \mathrm{E}\left((y - 1/\theta)^2\right) = N\mathrm{V}(y) = N/\theta^2
\end{array}
$$

because the expected value and the variance of y are respectively equal to $1/\theta$ and $1/\theta^2$ for an exponential distribution. Therefore $\iota(\theta) = N/\theta^2$ and the information-based estimator of the information for $\theta = \hat{\theta}$ is, as $\hat{\theta} = 1/\bar{y}$, $\iota(\hat{\theta}) = N\bar{y}^2$. The same result obviously applies to the hessian-based approximation of the information, which is: $\iota_h(\hat{\theta}) = \frac{N}{\hat{\theta}^2} = N\bar{y}^2$. Considering now the gradient-based estimate of the information, we have:

$$
\iota_g(\theta, y) = \sum_{n=1}^{N}(1/\theta - y_n)^2 = \sum_{n=1}^{N}(y_n - 1/\theta)^2
$$

as $\hat{\theta} = 1/\bar{y}$, evaluated for the ML estimator, we have $\iota_g(\hat{\theta}, y) = N\hat{\sigma}_y^2$. Finally, the sandwich estimator of the variance of $\hat{\theta}$ is:

$$
\hat{\sigma}^2_{\hat{\theta}s} = \frac{1}{N}\frac{\hat{\sigma}_y^2}{(\bar{y}^2)^2} = \frac{1}{N}\frac{\hat{\sigma}_y^2}{\bar{y}^4}
$$

The four estimators of the variance for the exponential distribution are then:

$$
\left\{
\begin{array}{rclcl}
\hat{\sigma}^2_{\hat{\theta}i} & = & \iota(\hat{\theta})^{-1} & = & 1/(N\bar{y}^2) \\
\hat{\sigma}^2_{\hat{\theta}g} & = & \iota_g(\hat{\theta})^{-1} & = & 1/(N\hat{\sigma}_y^2) \\
\hat{\sigma}^2_{\hat{\theta}h} & = & \iota_h(\hat{\theta})^{-1} & = & 1/(N\bar{y}^2) \\
\hat{\sigma}^2_{\hat{\theta}s} & & & = & \hat{\sigma}_y^2/(N\bar{y}^4)
\end{array}
\right.
$$

Computing this four estimations of the variance of $\hat{\theta}$, we get for the `oil` variable:

```
y <- oil$dur ; N <- length(y)
mean_y <- mean(y) ; var_y <- mean( (y - mean_y) ^ 2)
c(info = 1 / (N * mean_y ^ 2), gradient = 1 / (N * var_y),
  hessian = 1 / (N * mean_y ^ 2),
  sandwich = var_y / (N * mean_y ^ 4)) %>%
    sqrt %>% round(4)
##     info gradient  hessian sandwich
##   0.0022   0.0025   0.0022   0.0019
```

5.2 ML estimation in the general case

Compared to the simple case analyzed in the previous section, we consider in this section two extensions:

- θ is now a vector of unknown parameters that we seek to estimate, which means that the gradient is a vector and the hessian is a matrix,
- the density for observation n not only depends on the value of the response y_n, but also on the value of a vector of covariates x_n.

5.2.1 Computation and properties of the ML estimator

The density (or the probability mass) for observation n is now: $\phi(y_n; \theta, x_n) = \phi_n(y_n; \theta)$; therefore, written as a function of y and θ only, the density is now indexed by n, as it is a function of x_n. Denoting as previously $\lambda_n(y_n; \theta) = \ln \phi_n(y_n; \theta)$, $\gamma_n = \frac{\partial \lambda_n}{\partial \theta}$ and $\Psi_n = \frac{\partial^2 \lambda_n}{\partial \theta \partial \theta^\top}$[2], the log-likelihood is:

$$\ln L(\theta, y, X) = \sum_{n=1}^{N} \ln \phi_n(y_n; \theta) = \sum_{n=1}^{N} \lambda_n(y_n; \theta)$$

The gradient and the hessian are:

$$\left\{ \begin{array}{rcl} g(\theta, y, X) & = & \sum_{n=1}^{N} \gamma_n(y_n; \theta) \\ H(\theta, y, X) & = & \sum_{n=1}^{N} \Psi_n(y_n; \theta) \end{array} \right.$$

The variance of the score is the **information matrix**, denoted by $I(\theta, X)$ and, by virtue of the **information matrix equality** demonstrated previously in the scalar case, it is equal to the opposite of the expected value of the hessian:

$$I(\theta, X) = \mathrm{V}(g(\theta, y, X)) = -\mathrm{E}(H(\theta, y, X))$$

Note that now, each individual contribution to the gradient and to the hessian depends on x_n; therefore, their variance (for the gradient) and their expectation (for the hessian) are not constant as previously. In terms of the individual observations, the information matrix equality states that:

$$\mathrm{I}(\theta; X) = \sum_{n=1}^{N} \mathrm{E}\left(\gamma_n(y_n; \theta) \gamma_n(y_n; \theta)^\top\right) = -\sum_{n=1}^{N} \mathrm{E}\left(\Psi_n(y_n; \theta)\right)$$

Define the asymptotic information and the asymptotic hessian as:

[2]We now have a matrix of second derivatives, denoted by Ψ, which replaces the scalar second derivative ψ in the previous section.

$$\left\{ \begin{array}{lcl} \mathcal{I} & = & \frac{1}{N} \lim_{n\rightarrow+\infty} \sum_{n=1}^{N} \gamma_n(y_n;\theta)\gamma_n(y_n;\theta)^\top \\ \mathcal{H} & = & \frac{1}{N} \lim_{n\rightarrow+\infty} \sum_{n=1}^{N} \Psi_n(y_n;\theta) \end{array} \right.$$

The information matrix equality implies that: $\mathcal{I} = -\mathcal{H}$. At the ML estimate, the gradient is 0: $g(\hat{\theta}, y, X) = 0$. Using a first-order Taylor expansion around the true value θ_0, we have:

$$g(\hat{\theta}, y, X) = g(\theta_0, y, X) + H(\bar{\theta}, y, X)(\hat{\theta} - \theta_0) = 0$$

The equivalent of $\bar{\theta}$ lying in the $\theta_0 - \hat{\theta}$ interval for the scalar case (see Equation 5.11) is that $\|\bar{\theta} - \theta_0\| \leq \|\hat{\theta} - \theta_0\|$. Solving this equation for $\hat{\theta} - \theta_0$, we get, multiplying by $\sqrt{N}$:

$$\sqrt{N}(\hat{\theta} - \theta_0) = \left(-\frac{H(\bar{\theta}, y, X)}{N}\right)^{-1} \frac{g(\theta_0, y, X)}{\sqrt{N}}$$

The probability limit of the term in brackets is $-\mathcal{H}$ (as $\bar{\theta}$ converges to θ_0) and therefore:

$$\sqrt{N}(\hat{\theta} - \theta_0) \overset{a}{=} (-\mathcal{H})^{-1} \frac{g(\theta_0, y, X)}{\sqrt{N}} \quad (5.19)$$

where $w \overset{a}{=} z$ means that w is asymptotically equal to z, i.e., it tends to the same limit in probability (see Davidson and MacKinnon 2004, 205).

The asymptotic variance of the second term is:

$$\mathrm{V}\left(\lim_{n\rightarrow\infty} \frac{g(\theta_0; y, X)}{\sqrt{N}}\right) \overset{a}{=} \lim_{n\rightarrow\infty} \frac{1}{N} \sum_{n=1}^{N} \gamma_n(y_n;\theta_0)\gamma_n(y_n;\theta_0)^\top = \mathcal{I}$$

Therefore, the asymptotic variance of $\sqrt{N}(\hat{\theta} - \theta_0)$ is $\mathcal{H}^{-1}\mathcal{I}\mathcal{H}^{-1}$, which reduces to, applying the information matrix equality result, $\mathcal{I}^{-1}$. Applying the central-limit theorem, we finally get:

$$\sqrt{N}(\hat{\theta} - \theta_0) \overset{p}{\rightarrow} \mathcal{N}(0, \mathcal{I}^{-1})$$

$$\hat{\theta} \overset{a}{\sim} \mathcal{N}(\theta_0, \mathcal{I}^{-1}/N)$$

The asymptotic variance can be estimated using the information evaluated at $\hat{\theta}$ if the variance of the gradient or the expectation of the hessian can be computed:

$$\hat{\mathrm{V}}_I(\hat{\theta}) = \left(\sum_{n=1}^{N} \mathrm{E}\left(\gamma_n(y_n;\hat{\theta})\gamma_n(y_n;\hat{\theta})^\top\right)\right)^{-1} = \left(-\sum_{n=1}^{N} \mathrm{E}\left(\Psi_n(y_n;\hat{\theta})\right)\right)^{-1}$$

Two other possible estimators are obtained by evaluating the two previous expressions without the expectation. The gradient-based estimator, also called the **outer product of the gradient** or the **BHHH**[3] estimator is:

[3]The initials of the authors of Berndt et al. (1974) who first proposed this estimator.

$$\hat{\mathrm{V}}_g(\hat{\theta}) = \left(\sum_{n=1}^{N}\left(\gamma_n(y_n;\hat{\theta})\gamma_n(y_n;\hat{\theta})^\top\right)\right)^{-1}$$

and the hessian-based estimator is:

$$\hat{\mathrm{V}}_H(\hat{\theta}) = \left(-\sum_{n=1}^{N}\Psi_n(y_n;\hat{\theta})\right)^{-1} = \left(-H(\hat{\theta},y,X)\right)^{-1}$$

Finally, the sandwich estimator is based on the expression of the asymptotic covariance of $\hat{\theta}$ before applying the information matrix equality theorem. Then:

$$\hat{\mathrm{V}}_s(\hat{\theta}) = \left(-H(\hat{\theta},y,X)\right)^{-1}\left(\sum_{n=1}^{N}\left(\gamma_n(y_n;\hat{\theta})\gamma_n(y_n;\hat{\theta})^\top\right)\right)\left(-H(\hat{\theta},y,X)\right)^{-1}$$

This estimator actually looks like a sandwich, the "meat" (the estimation of the variance of the gradient) being surrounded by two slices of "bread" (the inverse of the opposite of the hessian).

5.2.2 Computation of the estimators for the exponential distribution

We have seen in Section 5.1.2 that we can get an explicit solution for the maximization of the likelihood of the one parameter exponential distribution. Once covariates are introduced, there are $K+1$ parameters to estimate and there is no longer an explicit solution. Then, a numerical optimization algorithm should be used. We'll present in this section the simplest algorithm, called the **Newton-Raphson** algorithm, using the `oil` data set with two covariates:

- `p98` is the adaptive expectations for the real after-tax oil prices formed at the time of the approval,
- `varp98` is the volatility of the adaptive expectations for the real after-tax oil prices.

For the exponential model, remember that $\mathrm{E}(y) = 1/\theta$; θ should therefore be positive. The way the linear combination of the covariates $\gamma^\top z_n$ is related to the parameter of the distribution θ_n is called the **link**. It is customary to define $\theta_n = e^{-\gamma^\top z_n}$ (so that $\ln \mathrm{E}(y \mid x_n) = -\ln\theta_n = \gamma^\top z_n$).

Then:

$$\left\{\begin{array}{rcl}\ln L &=& -\sum_{n=1}^{N}\left(\gamma^\top z_n + e^{-\gamma^\top z_n}y_n\right)\\ \frac{\partial \ln L}{\partial \gamma} &=& -\sum_{n=1}^{N}\left(1-e^{-\gamma^\top z_n}y_n\right)z_n\\ \frac{\partial^2 \ln L}{\partial\gamma\partial\gamma^\top} &=& -\sum_{n=1}^{N}e^{-\gamma^\top z_n}y_n z_n z_n^\top\end{array}\right. \quad (5.20)$$

Starting from an initial vector of parameters γ_i, we use a first-order Taylor expansion of the gradient for γ_{i+1} "close to" γ_i:

$$g(\gamma_{i+1}) \approx g(\gamma_i) + H(\gamma_i)(\gamma_{i+1}-\gamma_i)$$

Solving the first order-conditions for a maximum, we should have $g(\gamma_i)+H(\gamma_i)(\gamma_{i+1}-\gamma_i) = 0$, which leads to:

$$\gamma_{i+1} = \gamma_i - H(\gamma_i)^{-1}g(\gamma_i) \tag{5.21}$$

Except if the gradient is a linear function of γ, γ_{i+1} is not the maximum, but it is closer to the maximum than γ_i and successive iterations enable to reach a value of γ as close as desired to the maximum. We first begin by describing the model we want to estimate using a formula and extracting the relevant components of the model, namely the vector of response and the matrix of covariates.

```
form <- dur ~ p98 + varp98
mf <- model.frame(form, oil)
Z <- model.matrix(form, mf)
y <- model.response(mf)
N <- length(y)
```

We then define functions for the log-likelihood, the gradient and the hessian, as a function of the vector of parameters γ.

```
theta <- function(gamma) theta <- exp(- drop(Z %*% gamma))
f <- function(gamma) - sum(log(theta(gamma)) + theta(gamma) * y)
G <- function(gamma) - (1 - y * theta(gamma)) * Z
g <- function(gamma) apply(G(gamma), 2, sum)
H <- function(gamma) - crossprod(sqrt(theta(gamma) * y) * Z)
```

Starting from an initial vector of coefficients, we use the preceding formula to update the vector of coefficients. The choice of good starting values is crucial when the log-likelihood function is not concave. This is not the case here, but, anyway, the choice of good starting values limits the number of iterations. In our example, a good candidate is the ordinary least squares estimator, with $\ln y$ as the response:

```
gamma_0 <- coef(lm(log(dur) ~ p98 + varp98, oil))
gamma_0
## (Intercept)         p98      varp98
##      1.2007      0.9480      0.2262
g(gamma_0)
## (Intercept)         p98      varp98
##       15.40       28.62       56.42
```

We then update γ using Equation 5.21:

```
gamma_1 <- gamma_0 - solve(H(gamma_0), g(gamma_0))
gamma_1
## (Intercept)         p98      varp98
##      1.4077      0.8615      0.2785
g(gamma_1)
## (Intercept)         p98      varp98
##       1.687       3.198       6.646
```

We can see that we obtain an updated vector of coefficients which seems closer to the maximum than the initial one, as the elements of the gradient are much smaller than previously. The gradient being still quite different from 0, it is worth iterating again. We'll stop the iterations when a scalar value obtained from the gradient is less than an arbitrary small real value. As a simple criterion, we consider the mean of the squares of the elements of the score, and we iterate as long as this scalar is greater than 10^{-07}:

```
gamma_0 <- coef(lm(log(dur) ~ p98 + varp98, oil))
i <- 0
gamma <- gamma_0
crit <- 1
while (crit > 1E-07){
    i <- i + 1
    gamma <- gamma - solve(H(gamma), g(gamma))
    crit <- mean(g(gamma) ^ 2)
    cat(paste("iteration", i, "crit = ", crit, "\n"))
}
```

```
iteration 1 crit =  19.0801422479605
iteration 2 crit =  0.00824133223009916
iteration 3 crit =  2.88870943821704e-09
```

Only three iterations were necessary to reach the maximum (as defined by our criteria).

```
gamma
## (Intercept)          p98       varp98
##      1.4405       0.8303       0.2952
g(gamma)
## (Intercept)          p98       varp98
##   1.809e-05    3.445e-05    8.457e-05
```

We then use Equation 5.20 and the functions H and G defined previously to compute the three estimations of the information matrix and the four estimators of the covariance matrix of the estimator:

```
Info_g <- crossprod(G(gamma))
Info_H <- - H(gamma)
Info <- crossprod(Z)
V_g <- solve(Info_g)
V_i <- solve(Info)
V_h <- solve(Info_H)
V_sand <- solve(Info_H) %*% Info_g %*% solve(Info_H)
```

We then compare the resulting estimated standard errors of the estimator. To obtain a compact output, we use the `sapply` function, which is a specialized version of `lapply`. `lapply` takes as argument a list and a function, and the outcome is a list containing the result of applying the function to each element of the list. `sapply`, when possible, returns a matrix:

```
sapply(list(V_i, V_h, V_g, V_sand), function(x) stder(x))
```

```
               [,1]   [,2]   [,3]    [,4]
(Intercept) 0.5541 0.5509 0.9621 0.32300
p98         0.4858 0.4935 0.8793 0.29246
varp98      0.1584 0.1568 0.2739 0.09506
```

Not that the gradient-based estimate gives fairly different results, compared to the other estimators. This is often the case, this estimator being known to perform poorly in small samples.

5.2.3 Linear gaussian model

The first part of the book was devoted to the OLS estimator, suitable in situations where $\mathrm{E}(y_n \mid x_n) = \alpha + \beta^\top x_n$, with $\mathrm{V}(\epsilon_n \mid x_n) = \sigma_\epsilon^2$ and $\mathrm{cov}(\epsilon_n, \epsilon_m) = 0$. This model can also be estimated by maximum likelihood by specifying the conditional density of y, and not only its conditional expectation. Assuming that y follows a normal distribution, we have:

$$y_n \mid x_n \sim \mathcal{N}(\alpha + \beta^\top x_n, \sigma)$$

and the conditional density of y is:

$$f(y_n \mid x_n) = \frac{1}{\sqrt{2\pi}\sigma} e^{-\frac{1}{2}\left(\frac{y_n - \gamma^\top z_n}{\sigma}\right)^2} = \frac{1}{\sigma}\phi\left(\frac{y_n - \gamma^\top z_n}{\sigma}\right)$$

The log-likelihood is then:

$$\ln L = -\frac{N}{2}\ln 2\pi - N\ln\sigma - \frac{1}{2\sigma^2}\sum_{n=1}^{N}(y_n - \gamma^\top z_n)^2 = -\frac{N}{2}\ln 2\pi - \frac{N}{2}\ln\sigma^2 - \frac{1}{2\sigma^2}\epsilon^\top\epsilon \quad (5.22)$$

The gradient is:

$$g = \begin{pmatrix} \frac{1}{\sigma^2}\sum_n (y_n - \gamma^\top z_n) z_n \\ -\frac{N}{\sigma} + \frac{1}{\sigma^3}\sum_n (y_n - \gamma^\top z_n)^2 \end{pmatrix} = \begin{pmatrix} \frac{1}{\sigma^2} Z^\top \epsilon \\ -\frac{N}{\sigma} + \frac{1}{\sigma^3}\epsilon^\top\epsilon \end{pmatrix} \quad (5.23)$$

and the ML estimator is obtained by computing the vector of parameters $(\gamma^\top, \sigma)$ that set this system of $K + 2$ equations to 0. The hessian is:

$$H = \begin{pmatrix} -\frac{1}{\sigma^2} Z^\top Z & 0 \\ 0 & \frac{N}{\sigma^2} - \frac{3}{2\sigma^4}\epsilon^\top\epsilon \end{pmatrix} \quad (5.24)$$

Note that it is block diagonal, as the cross-derivatives are $-2/\sigma^3 Z^\top\epsilon$, which is 0 for the ML estimator. Solving the last line in Equation 5.23 for σ, we get:

$$\sigma^2 = \frac{\epsilon^\top\epsilon}{N} \quad (5.25)$$

Replacing σ^2 by this expression in Equation 5.22, we get the **concentrated** log-likelihood:

$$\ln L = -\frac{N}{2}(1 - \ln N + \ln 2\pi) - \frac{N}{2}\ln \epsilon^\top \epsilon = C - \frac{N}{2}\ln \epsilon^\top \epsilon \tag{5.26}$$

which makes clear that maximizing the log-likelihood is equivalent to minimizing the residual sum of squares, and, therefore, that the ML and the OLS estimators of γ are the same. Once $\hat{\gamma}$ has been computed, $\hat{\sigma}^2$ can be estimated using Equation 5.25. Note that the residual sum of squares divided by the number of observations and not by the number of degrees of freedom, contrary to the OLS estimator.

5.2.4 Transformation of the response

The ML estimator is based on the density (or probability mass) function of the response, given a set of covariates. Sometimes, it is more interesting to consider the density of a parametric transformation of the response. To illustrate this kind of model, we'll consider the estimation of production functions, using the `apples` data set, already described in Section 3.7. We consider models of the form:

$$w_n = v(y_n, \lambda) \sim \mathcal{N}(\mu_n, \sigma) \text{ with } \mu_n = \alpha + \beta^\top x_n$$

The parametric transformation of y_n may depends on a set of unknown parameters λ, and it follows a normal distribution with an expectation that is a linear function of some covariates and a constant variance. As the density of $v(y_n, \gamma)$ is normal, the one for y_n can be obtained using the following formula:

$$f(y_n) = \phi(v(y_n, \gamma)) \times \left|\frac{dv}{dy}\right|$$

where the last term is called the Jacobian of the transformation of w on y. Consider the simple case where $v(y_n) = \ln y_n$. Then, v doesn't contain any unknown parameter, the Jacobian of the transformation is $1/y$ and the density of y_n:

$$f(y_n) = \frac{1}{y}\frac{1}{\sigma}\phi\left(\frac{\ln y - \mu_n}{\sigma}\right)$$

is simply the log-normal density, which implies the following log-likelihood function:

$$\ln L(\gamma) = -\frac{N}{2}\ln 2\pi - N\ln\sigma - \sum_{n=1}^{N}\ln y_n + \frac{1}{2\sigma^2}\sum_{n=1}^{N}(\ln y_n - \gamma^\top z_n)^2 \tag{5.27}$$

Maximizing Equation 5.27 is obviously equivalent to minimizing $\sum_{n=1}^{N}(\ln y_n - \gamma^\top z_n)^2$ which is the residual sum of squares of a regression with $\ln y_n$ as the response. Therefore the ML estimation of γ can be performed using OLS.

We consider only one year (1986) of production for the `apples` data set, we construct a unique output variable (`y`), we rename for convenience the three factors as `k`, `l` and `m` (respectively for capital, labor and materials) and, for a reason that will be clear later, we divide all the variables by their sample mean:

```
aps <- apples %>%
    filter(year == 1986) %>%
    transmute(y = apples + otherprod, y = y / mean(y),
              k = capital / mean(capital),
              l = labor / mean(labor),
              m = materials / mean(materials))
```

We first consider the Cobb-Douglas production function which is linear in logs. Denoting $j = 1 \dots J$ the inputs, we get:

$$\ln y_n = \alpha + \sum_{j=1}^{J} \beta_j \ln q_j + \epsilon_n \tag{5.28}$$

We then fit Equation 5.28 using `lm`:

```
cobb_douglas <- lm(log(y) ~ log(k) + log(l) + log(m), aps)
cobb_douglas %>% coef
## (Intercept)      log(k)      log(l)      log(m)
##     -0.1331      0.2596      0.3537      0.5111
```

The problem of using `lm` is that, although the estimates are the ML estimates, the response is $\ln y$ and not y. Therefore, the log likelihood reported by `lm` is computed using the density of $\ln y$:

```
logLik(cobb_douglas)
## 'log Lik.' -120.2 (df=5)
```

and is incorrect as the term $-\sum_{n=1}^{N} \ln y_n$ in Equation 5.27 is missing. Adding this term to the log-likelihood returned by the `logLik` function, we get:

```
as.numeric(logLik(cobb_douglas)) - sum(log(aps$y))
## [1] -62
```

The `micsr::loglm` function estimates models for which the response is in logarithm. The estimation of the parameters is performed using `lm`, but the result is a `micsr` object, from which the correct log-likelihood, gradient and hessian can be extracted:

```
cd_loglm <- loglm(y ~ log(k) + log(l) + log(m), aps)
logLik(cd_loglm)
## 'log Lik.' -62 (df=5)
```

The scale elasticity, which measures the relative growth of output for a proportional increase of all the inputs is $\sum_{j=1}^{J} \beta_j$ and therefore constant returns to scale imply that $\sum_{j=1}^{J} \beta_j = 1$, or $\beta_J = 1 - \sum_{j=1}^{J-1} \beta_j$. Replacing in Equation 5.28, we get:

$$\ln y_n = \alpha + \sum_{j=1}^{J-1} \beta_j \ln q_j / q_J + \left(\sum_{j=1}^{J} \beta_j - 1 \right) \ln q_J + \ln q_J + \epsilon_n \tag{5.29}$$

Therefore, regressing $\ln y_n$ on $\ln q_j/q_J, \forall j = 1 ... J-1$ and $\ln q_J$, the hypothesis of constant returns to scale is $\beta^*_J = \sum_{j=1}^J \beta_j - 1 = 0$ and the constant returns to scale are imposed if $\ln q_J$ is removed from the regression. Note the presence of a constant term ($\ln q_J$) on the right side of the equation. This is called an offset and can be introduced in a formula with the `offset(w)` syntax:

```
cd_repar <- loglm(y ~ log(k / m) + log(l / m) + log(m) +
                    offset(log(m)), aps)
coef(cd_repar)
## (Intercept)      log(k/m)       log(l/m)        log(m)        sigma
##     -0.1331       0.2596         0.3537        0.1244       0.5896
logLik(cd_repar)
## 'log Lik.' -62 (df=5)
```

The log-likelihood is obviously the same as previously and $\beta^*_J = 0.12$, which implies a scale elasticity equal to 1.12.

Zellner and Revankar (1969) proposed a generalization of the Cobb-Douglas production function of the form:

$$\ln y + \lambda y \sim \mathcal{N}(\mu, \sigma)$$

where $\mu = \alpha + \sum_{j=1}^J \beta_j \ln q_j$. The scale elasticity is:

$$\psi = \frac{\sum_{j=1}^J \beta_j}{1 + \lambda y} \tag{5.30}$$

If $\lambda = 0$, this function reduces to the Cobb-Douglas production function with normal errors. For $\lambda > 0$, the scale elasticity is equal to $\sum_{j=1}^J \beta_j$ for $y = 0$ and tends to 0 as y tends to $+\infty$. Therefore, if $\sum_{j=1}^J \beta_j > 1$, returns to scale are increasing for a low level of production, get constant for a level of production that is equal to $(\sum_{j=1}^J \beta_j - 1)/\lambda$ and decreasing above this level of production. Denoting ϵ_n the difference between $\ln y + \lambda y$ and its conditional expectation, the model can be rewritten as:

$$\ln y_n + \lambda y_n = \alpha + \sum_{j=1}^J \beta_j \ln q_{nj} + \epsilon_n$$

The hypothesis of constant scale elasticity is simply $\lambda = 0$. The hypothesis of constant return to scale adds the condition: $\sum_j \beta_j = 1$. It can be easily tested using the same reparametrization of the model as in Equation 5.29:

$$\ln y_n + \lambda y_n = \alpha + \sum_{j=1}^{J-1} \beta_j \ln q^*_{nj} + \beta^*_J \ln q_{nJ} + \ln q_J + \epsilon_n$$

where $q^*_{nj} = q_{nj}/q_{nJ}$ and $\beta^*_J = \sum_{j=1}^J \beta_j - 1$. The hypothesis of constant returns to scale is the joint hypothesis that $\lambda = 0$ and that the coefficient of $\ln q_{nJ}$ in this reparametrized version of the model is also 0. The Jacobian is $\frac{1}{y} + \lambda = \frac{1+\lambda y}{y}$, which leads to the following density:

$$f(y_n;\theta,\beta) = \frac{1}{\sigma}\frac{1+\lambda y_n}{y_n}\phi\left(\frac{\ln y_n + \lambda y_n - \mu_n}{\sigma}\right)$$

where μ_n is $\alpha + \sum_{j=1}^{J}\beta_j \ln q^*_{nj} + \beta^*_J \ln q_{nJ} + \ln q_J$. Taking the logarithm of this density, we get the individual contribution of an observation to the log-likelihood function:

$$l_n = -\ln\sigma - \frac{1}{2}\ln 2\pi + \ln(1+\lambda y_n) - \ln y_n - \frac{1}{2\sigma^2}\left(\ln y_n + \lambda y_n - \mu_n\right)^2$$

Denoting $\gamma^\top = (\alpha, \beta_1, ...\beta^*_J)$, $z_n = (1, q^*_{n1}, q^*_{n2}, ... q^*_{nJ-1}, q_{nJ})$ and $\epsilon_n = \ln y_n + \lambda y_n - \mu_n$, the derivatives with respect to the unknown parameters $(\gamma, \lambda, \sigma)$ are:

$$\left\{\begin{array}{rcl}
\frac{\partial l_n}{\partial \gamma} &=& \frac{1}{\sigma^2}\epsilon_n z_n \\
\frac{\partial l_n}{\partial \lambda} &=& \frac{y_n}{1+\lambda y_n} - \frac{1}{\sigma^2}\epsilon_n y_n \\
\frac{\partial l_n}{\partial \sigma} &=& -\frac{1}{\sigma} + \frac{1}{\sigma^3}\epsilon_n^2
\end{array}\right.$$

and the second derivatives give the following individual contributions to the hessian:

$$\left(\begin{array}{ccc}
-\frac{1}{\sigma^2}z_n z_n^\top & \frac{1}{\sigma^2}y_n z_n & -\frac{2}{\sigma^3}\epsilon_n z_n \\
\frac{1}{\sigma^2}y_n z_n^\top & -\frac{y_n^2}{(1+\lambda y_n)^2} - \frac{y_n^2}{\sigma^2} & \frac{2}{\sigma^3}\epsilon_n y_n \\
-\frac{2}{\sigma^3}\epsilon_n z_n^\top & \frac{2}{\sigma^3}\epsilon_n y_n & \frac{1}{\sigma^2} - \frac{3}{\sigma^4}\epsilon_n^2
\end{array}\right)$$

To estimate the generalized production function, we use the **maxLik** package (see Henningsen and Toomet 2011), which is dedicated to ML estimation. This package provides different algorithms to compute the maximum of the likelihood, the Newton-Raphson method that we used previously being the default. A `maxLik` object is returned, and specific methods as `coef` and `summary` are provided. To use **maxLik**, we need to define a function of the unknown parameters which returns the value of the log-likelihood function. This function can return either a scalar (the log-likelihood) or a N-length vector that contains the individual contribution to the log-likelihood. Moreover, it is advisable to provide a function that returns the gradient: it can either return a $K+1$-length vector or a $N \times (K+1)$ matrix on which each line is the contribution of an observation to the gradient. The analytical hessian matrix can also be provided (otherwise, a numerical approximation is computed, which is more time-consuming and less precise). Moreover, **maxLik** allows to provide a function that returns the maximum-likelihood and the gradient and the hessian as attributes. It is a good idea to do so, as there are often common code while writing the likelihood, the gradient and the hessian.

The `micsr::zellner_revankar` function returns the log-likelihood function for the generalized production function. Its mandatory arguments are `theta`, a vector of starting values, `y`, the vector of response and `Z`, a matrix of covariates. By default, the function computes the log-likelihood (a vector of contributions), the gradient (a matrix of contributions) and the hessian. The `gradient` and `hessian` arguments (by default `TRUE`) are booleans which enable to return only the log-likelihood if set to `FALSE`. If the `sum` argument is `FALSE` (the default), the log-likelihood and the gradient are returned as a vector and a matrix of individual contributions. If set to `TRUE`, a scalar and a vector are returned. Finally, the `repar` argument enables to write the likelihood in its "raw" form (`repar = FALSE`) or in its reparametrized form (the default).

We first extract the response and the covariates matrix:

```
form <- y ~ log(k) + log(l) + log(m) ;
mf <- model.frame(form, aps)
y <- model.response(mf) ;
Z <- model.matrix(form, mf)
```

We then use as starting values the coefficients of the Cobb-Douglas previously estimated, and we set the starting value of λ to 0.

```
st_val <- c(coef(cd_repar)[1:4], lambda = 0, coef(cd_repar)[5])
```

< We now proceed to the estimation. The two mandatory arguments of `maxLik` are the first argument `logLik` which should be a function returning the log-likelihood and `start` which is a vector of starting values. We use also here `y` and `Z` arguments that are passed to `zellner_revankar`:

```
gpf <- maxLik::maxLik(zellner_revankar, start = st_val, y = y, Z = Z)
gpf %>% summary %>% coef
```

```
            Estimate Std. error t value   Pr(> t)
(Intercept)   0.0478    0.12869  0.3714 7.103e-01
log(k/m)      0.2574    0.11950  2.1543 3.122e-02
log(l/m)      0.4712    0.18287  2.5769 9.969e-03
log(m)        0.3245    0.16921  1.9179 5.512e-02
lambda        0.1362    0.08506  1.6011 1.093e-01
sigma         0.6565    0.05961 11.0145 3.254e-28
```

The probability values for the hypothesis that $\sum_j \beta_j = 1$ and that $\lambda = 0$ are respectively about 5 and 10%. The hypothesis of constant returns to scale is the joint hypothesis that $\sum_j \beta_j = 1$ and $\lambda = 0$ and will be tested in the next section. Applying Equation 5.30, the level of output for which the scale elasticity is unity is 2.38, reminding that the average production is 1. We then compute the elasticity for different values of the production that are presented in Table 5.2. Returns to scale are increasing for more than three quarters of the sample.

Table 5.2: Scale elasticity

production	y	elast
min	0.05	1.315
Q1	0.40	1.257
median	0.60	1.225
Q3	1.04	1.161
max	6.55	0.700

The "raw" version of the model can also be estimated by setting the **repar** argument to **FALSE**. In this case, we use the coefficients of **cd_loglm** as starting values for all the coefficients except λ (for which the starting value is set to 0).

```
st_val2 <- c(coef(cd_loglm)[1:4], lambda = 0, coef(cd_loglm)[5])
gpf2 <- maxLik::maxLik(zellner_revankar, y = y, Z = Z,
                       start = st_val2, repar = FALSE)
```

5.3 Tests

Three kinds of tests will be considered:

- tests for nested models: in this context, there is a "large" model that reduces to a "small" model when the hypotheses are imposed. Depending on whether H_0 is true or false, the small or the large model are assumed to be the "true" model,
- conditional moment tests for which only one model is considered and moment conditions are constructed from the fitted model that should be zero if the tested hypothesis (for example, normality or homoskedasticity) are true,
- tests for non-nested models and especially the test proposed by Vuong (1989): in this case, whatever the values of the parameters, there is no way for one model to reduce to the other model. Moreover, the test have the interesting feature that the two models are compared without hypothesizing that one of them is the true model.

5.3.1 Tests for nested models: the three classical tests

We have seen in Section 3.6.3 that a set of hypotheses defines constraints on the parameters and therefore leads to two models. The unconstrained or large model doesn't take these constraints into account, so that the log-likelihood is maximized without constraints. The constrained or small model is obtained by maximizing the log-likelihood under the constraints corresponding to the set of hypotheses. This leads to three test principles: the likelihood ratio test (based on the comparison of both models), the Wald test (based only on the unconstrained model) and the Lagrange multiplier test (based only on the constrained model). Remember that, when applying these test principles on a model fitted by OLS, the three statistics are exactly the same.[4] On the, contrary the three tests give different results for non-linear models but, if the hypothesis are true, the three test statistics follow a χ^2 distribution with a number of degrees of freedom equal to the number of hypotheses and converge in probability to the same value.

We'll consider in this section a set of J linear hypotheses, such that $R\theta - q = 0$, and we'll denote $\hat{\theta}_{nc}$ and $\hat{\theta}_c$ the unconstrained and constrained maximum likelihood estimators.

[4] Actually, the Lagrange multiplier test is numerically different, only because the estimation of σ_ϵ^2 is based on the constrained model.

5.3.1.1 Three tests for the linear gaussian model

We first compute the three tests for the linear gaussian model developed in Section 5.2.3. We denote $\hat{\epsilon}_{nc}$ and $\hat{\epsilon}_c$ the vectors of residuals for the unconstrained and the constrained model. The F statistic for the hypothesis that $R\gamma = q$ is, denoting $A = R(Z^\top Z)^{-1}R^\top$:

$$F = \frac{(R\hat{\gamma}_{nc} - q)^\top A^{-1}(R\hat{\gamma}_{nc} - q)}{\hat{\epsilon}_{nc}^\top \hat{\epsilon}_{nc}} \frac{N-K-1}{J} = \frac{(R\hat{\gamma}_{nc} - q)^\top A^{-1}(R\hat{\gamma}_{nc} - q)}{\dot{\sigma}_{nc}^2} / J$$

where $\dot{\sigma}_{nc}^2 = \hat{\epsilon}_{nc}^\top \hat{\epsilon}_{nc}/(N-K-1)$ is the unbiased estimator of σ^2. With the hypothesis of iid normal errors, this statistic follows a Fisher-Snedecor distribution with J and $N-K-1$ degrees of freedom and the asymptotic distribution of $J \times F$ is a χ^2 with J degrees of freedom. Remember from Equation 3.27 that $\hat{\gamma}_c = \hat{\gamma}_{nc} - (Z^\top Z)^{-1}R^\top A^{-1}(R\hat{\gamma}_{nc} - q)$, then:

$$\hat{\epsilon}_c = \hat{\epsilon}_{nc} + Z(Z^\top Z)^{-1}R^\top A^{-1}(R\hat{\gamma}_{nc} - q) \tag{5.31}$$

and, because $Z^\top \hat{\epsilon}_{nc} = 0$, the relation between the two residual sum of squares is:

$$\hat{\epsilon}_c^\top \hat{\epsilon}_c = \hat{\epsilon}_{nc}^\top \hat{\epsilon}_{nc} + (R\hat{\gamma}_{nc} - q)^\top A^{-1}(R\hat{\gamma}_{nc} - q) \tag{5.32}$$

The likelihood ratio statistic is, using Equation 5.26:

$$LR = 2\left(\ln L(\hat{\gamma}_{nc}) - \ln L(\hat{\gamma}_c)\right) = N \ln \frac{\hat{\epsilon}_c^\top \hat{\epsilon}_c}{\hat{\epsilon}_{nc}^\top \hat{\epsilon}_{nc}} \tag{5.33}$$

and the Wald statistic is:

$$W = N\frac{(R\hat{\gamma}_{nc} - q)^\top A^{-1}(R\hat{\gamma}_{nc} - q)}{\hat{\epsilon}_{nc}^\top \hat{\epsilon}_{nc}} = \frac{(R\hat{\gamma}_{nc} - q)^\top A^{-1}(R\hat{\gamma}_{nc} - q)}{\hat{\sigma}_{nc}^2} \tag{5.34}$$

For the Lagrange multiplier test, as the hessian is block diagonal, we can consider only the part of the gradient and the hessian that concerns γ. For the constrained model, the subset of the gradient and of the information matrix are $Z^\top \epsilon_c/\hat{\sigma}_c^2$ and $Z^\top Z/\hat{\sigma}_c^2$, where $\hat{\sigma}_c^2 = \hat{\epsilon}_c^\top \hat{\epsilon}_c/N$ Then the statistic is $\hat{\epsilon}_c^\top Z(Z^\top Z)^{-1}Z^\top \hat{\epsilon}_c/\hat{\sigma}_c^2$. Using Equation 5.31, we get:

$$LM = N\frac{(R\hat{\gamma}_{nc} - q)^\top A^{-1}(R\hat{\gamma}_{nc} - q)}{\hat{\epsilon}_c^\top \hat{\epsilon}_c} = \frac{(R\hat{\gamma}_{nc} - q)^\top A^{-1}(R\hat{\gamma}_{nc} - q)}{\hat{\sigma}_c^2} \tag{5.35}$$

which is the same expression as Equation 5.34 except that the estimation of σ^2 is based on the constrained model. From Equation 5.32, $e^{LR/N} = 1 + W/N$. Therefore:

$$LR/N = \ln(1 + W/N)$$

Defining, $f(x) = \ln(1+x) - x$ for $x \geq 0$, $f'(x) = -x/(1+x) < 0$. As $f(0) = 0$ and f is a strictly decreasing function, $f(x) < 0$ for $x > 0$. Therefore $\ln(1 + W/N) - W/N = LR/N - W/N < 0$ and therefore $W > LR$.

Using Equation 5.34 and Equation 5.35, we get: $W/LM = \hat{\epsilon}_c^\top \hat{\epsilon}_c/\hat{\epsilon}_{nc}^\top \hat{\epsilon}_{nc}$. Finally, using Equation 5.32, we get: $W/LM = 1 + W/N$ or, rearranging terms:

$$LM/N = \frac{W/N}{1+W/N}$$

Denoting $f(x) = \ln(1+x) - x/(1+x)$, for $x \geq 0$, $f'(x) = x/(1+x)^2$. Therefore, $f(x)$ is strictly increasing for $x > 0$ and, as $f(0) = 0$, $f(x) > 0$ $\forall x > 0$. With $x = W/N$, $\ln(1+W/N) - W/N/(1+W/N) = LR/N - LM/N > 0$ and therefore $LR > LM$. Therefore, we have proved that:[5]

$$LM < LR < W$$

5.3.1.2 Pseudo-R^2 for models estimated by maximum likelihood

Consider now the case where we test the set of $J = K$ hypothesis that $\beta = 0$. Then, TSS $= \hat{\epsilon}_c^\top \hat{\epsilon}_c$ and RSS $= \hat{\epsilon}_{nc}^\top \hat{\epsilon}_{nc}$ and, from Equation 5.32:

$$\frac{\text{TSS}}{\text{RSS}} = 1 + W/N = 1 + \frac{\text{TSS}}{\text{RSS}} \times LM/N = e^{LR/N}$$

Therefore, the three statistics are:

$$\left\{ \begin{array}{rcl} W &=& N(TSS - RSS)/RSS \\ LM &=& N(TSS - RSS)/TSS \\ LR &=& N \ln TSS/RSS \end{array} \right.$$

The R^2 being equal to $1 - RSS/TSS$, it can be easily expressed, for the linear gaussian model, as a function of the three statistics:

$$R^2 = \frac{W}{N+W} = 1 - e^{LR/N} = LM/N \tag{5.36}$$

These pseudo-R^2 can be used for any model computed by maximum likelihood and will be denoted respectively by R^2_W, R^2_{LR} and R^2_{LM}. They have been proposed by Magee (1990). R^2_{LR} is known as the R^2 of D. R. Cox and Snell (1989), but it has been previously proposed by Maddala (1983). A variant, proposed by Aldrich and Nelson (1984), is obtained using the formula of R^2_W, but using the LR statistic: $LR/(N + LR)$. A problem with the pseudo-R^2 computed using the LR statistic is that, for a perfect or **saturated** model, it can be lower than 1. Denoting $\ln L_0$ and $\ln L_*$ the values of the log-likelihood for the null (intercept only) and the saturated model, the LR statistic for the saturated model is $LR_* = 2(\ln L_* - \ln L_0)$. In case of discrete response (as for the Poisson model developed in the beginning of this chapter), the individual contributions to the log-likelihood are logs of probabilities (all equal to 1 for a saturated model), so that $\ln L_* = 0$ and $LR_* = -2\ln L_0$. Therefore, R^2_{LR} equals $1 - e^{LR_*/N} \neq 1$. Scaled versions of R^2_{LR} and Aldrich and Nelson (1984)'s R^2 have been proposed respectively by Nagelkerke (1991):[6]

$$R^2 = \frac{1 - e^{-LR/N}}{1 - e^{-LR_*/N}}$$

[5]See Vandaele (1981).

[6]And previously used by Cragg and Uhler (1970).

and by Veall and Zimmermann (1996):

$$R^2 = \frac{LR/(LR+N)}{LR_*/(LR_*+N)}$$

5.3.1.3 General formula and application to the generalized production function

Although the three tests are not limited to linear constraints, we'll consider only linear hypothesis in this section. Moreover, as seen previously, a linear model can always be reperametrized so that a set of linear hypothesis reduces to the test that a subset of the parameters (θ_2) is zero.

In our generalized production function example, the original set of parameters are $\theta^\top = (\alpha, \beta_k, \beta_l, \beta_s, \lambda, \sigma)$ and the constant returns to scale hypothesis is $\beta_k + \beta_l + \beta_s = 1$ and $\lambda = 0$.

$$R\theta - q = \begin{pmatrix} 0 & 1 & 1 & 1 & 0 & 0 \\ 0 & 0 & 0 & 0 & 1 & 0 \end{pmatrix} \begin{pmatrix} \alpha \\ \beta_k \\ \beta_l \\ \beta_s \\ \lambda \\ \sigma \end{pmatrix} - \begin{pmatrix} 1 \\ 0 \end{pmatrix} = 0$$

After the reparametrization, the set of parameters is $\theta^{*\top} = (\alpha, \beta_k, \beta_l, \beta_s^*, \lambda, \sigma)$ and the constant returns to scale hypothesis is $\lambda = \beta_s^* = 0$:

$$R^*\theta^* - q^* = \begin{pmatrix} 0 & 0 & 0 & 1 & 0 & 0 \\ 0 & 0 & 0 & 0 & 1 & 0 \end{pmatrix} \begin{pmatrix} \alpha \\ \beta_k \\ \beta_l \\ \beta_s^* \\ \lambda \\ \sigma \end{pmatrix} - \begin{pmatrix} 0 \\ 0 \end{pmatrix} = 0$$

R^* is then a matrix which selects a subset of coefficients, that we'll call θ_2.

The Wald test is based on the unconstrained model and more precisely on $R\hat{\theta}_{nc} - q$. If H_0 is true, this vector of length J (here $J = 2$) should be close to 0. Using the central-limit theorem, the asymptotic distribution of $\hat{\theta}$ is normal, so that, under H_0: $R\hat{\theta}_{nc} - q \overset{a}{\sim} \mathcal{N}(0, R\hat{V}_{\hat{\theta}}R^\top)$. The Wald statistic is obtained by computing the quadratic form of this vector using the inverse of its covariance matrix:

$$W = (R\hat{\theta}_{nc} - q)^\top (R\hat{V}_{\hat{\theta}}R^\top)^{-1}(R\hat{\theta}_{nc} - q)$$

With the reparametrized model, denoting $\hat{V}_{\hat{\theta}_2}$ the subset of $\hat{V}_{\hat{\theta}}$ that concerns θ_2, the Wald statistic simplifies to $W = \hat{\theta}_2^\top \hat{V}_{\hat{\theta}_2}^{-1}\hat{\theta}_2$. We first extract the vector and the matrix used in the quadratic form:

```
R <- matrix(0, 2, 6)
R[1, 2:4] <- R[2, 5] <- 1
```

```
q <- c(1, 0)
d <- drop(R %*% coef(gpf2) - q)
d <- R %*% coef(gpf2) - q
V <- R %*% vcov(gpf2) %*% t(R)
```

The quadratic form of the vector `d` with the inverse of the matrix `V` can be obtained using the basic tools provided by **R** for matrix algebra:

```
drop(crossprod(d, solve(V, d)))
## [1] 3.753
```

Note the use of `drop` to get a scalar and not a 1×1 matrix. The `micsr::quad_form` function enables to compute more simply the quadratic form:

```
wald_test <- quad_form(d, V)
```

A subset of the elements of the vector and the matrix can be used with the `subset` argument:

```
wald_test2 <- quad_form(gpf, subset = c("log(m)", "lambda"))
c(wald_test, wald_test2)
## [1] 3.753 3.753
```

which illustrates the equivalence of the two formulas for the Wald test using either the “raw” or the reparametrized model.

We can also use the approximate formula that indicates that the Wald statistic is close to the sum of the squares of the corresponding t statistics if the correlation between the two parameters is not too high:

```
csgpf <- coef(summary(gpf))
t1 <- csgpf["log(m)", 3] ; t2 <- csgpf["lambda", 3] ; t1 ^ 2 + t2 ^ 2
## [1] 6.242
```

The approximation is very bad, which should be explained by a high correlation between the two coefficients:

```
s1 <- csgpf["log(m)", 2] ; s2 <- csgpf["lambda", 2] ;
v12 <- vcov(gpf)["log(m)", "lambda"] ; r12 <- v12 / (s1 * s2)
r12
## [1] 0.7387
```

“Correcting” this correlation using Equation 3.23, we get the correct value of the statistic:

```
(t1 ^ 2 + t2 ^ 2 - 2 * r12 * t1 * t2) / (1 - r12 ^ 2)
## [1] 3.753
```

The statistic can also easily obtained using the `car::linearHypothesis` function described in Section 3.6.4.1:

```
car::linearHypothesis(gpf, c("log(m) = 0", "lambda = 0")) %>% gaze
## Chisq = 3.753, df: 2, pval = 0.153
car::linearHypothesis(gpf2, c("log(l) + log(k) + log(m) = 1",
                              "lambda = 0")) %>% gaze
## Chisq = 3.753, df: 2, pval = 0.153
```

The likelihood ratio statistic is very easy to compute if the two models have been estimated as it is simply twice the difference of the log-likelihood of the two models. The coefficients of the constrained model can be obtained by least squares using the reparametrized version of the Cobb-Douglas (Equation 5.29) and imposing $\beta_m = 0$. Using `micsr::loglm`, we get:

```
crs <- loglm(y ~ log(k / m) + log(l / m) + offset(log(m)), aps)
lr_test <- 2 * as.numeric(logLik(gpf) - logLik(crs))
lr_test
## [1] 4.922
```

The Lagrange multiplier is based on the gradient evaluated with the estimates of the constrained model: $g(\hat{\theta}_{nc})$, which should be closed to 0. Applying the central-limit theorem, this vector is normally distributed, with, under H_0, a zero expectation and a variance equal to the information matrix, which can be estimated by minus the hessian: $g(\hat{\theta}_{nc}) \overset{a}{\sim} \mathcal{N}(0, -H_{nc})$. The statistic is then:

$$LM = g(\hat{\theta}_{nc})^\top(-H_{nc})^{-1}g(\hat{\theta}_{nc}) \tag{5.37}$$

For the reparametrized model, denoting g_2 and H_2 the parts of the gradient and of the hessian that concern θ_2, we get $g_2^\top(-H_2)^{-1}g_2$. We consider here the case where the model is parametrized in a way that the hypothesis simply states that a subset of the coefficient $\theta_2^\top = (\beta_m, \lambda)$ is zero. For the constrained model, $\theta_1^\top = (\alpha, \beta_l, \beta_k, \sigma)$ is estimated, so that elements of the gradient that are the derivatives of the log-likelihood with respect to θ_1 should be 0. The other elements of the gradient are not 0, but should be close to 0 if the hypotheses are true. Using the `zellner_revankar` function, we compute all the necessary information (the log-likelihood, the gradient and the hessian) for the constrained model, which corresponds to the coefficients of the `crs` object fitted by least squares using `loglm`:

```
const_model <- zellner_revankar(c(coef(crs)[1:3], "log(m)" = 0,
                                  lambda = 0, coef(crs)[4]),
                                y = y, Z = Z, sum = FALSE)
G <- attr(const_model, "gradient")
g <- apply(G, 2, sum)
H <- attr(const_model, "hessian")
g
## (Intercept)      log(k)      log(l)      log(m)      lambda
##   6.883e-15  -4.708e-14  -1.901e-14   1.173e+01   1.364e+01
##       sigma
##  -3.297e-14
```

Applying Equation 5.37, or using only the part of the gradient and of the inverse of the hessian that concerns θ_2, we get:[7]

```
lm_test_H <- quad_form(g, - H)
g2 <- g[c("log(m)", "lambda")]
H2m1 <- solve(H)[c("log(m)", "lambda"), c("log(m)", "lambda")]
lm_test_H2 <- quad_form(g2, - H2m1, inv = FALSE)
c(lm_test_H, lm_test_H2)
## [1] 3.923 3.923
```

Instead of using the opposite of the hessian, we could also have used the outer product of the gradient to estimate the information matrix:

```
Ig <- crossprod(G)
lm_test_G <- quad_form(g, Ig)
lm_test_G
## [1] 13.58
```

But in this case, the statistic can be more simply computed using the results of a regression, because the statistic is:

$$g^\top (G^\top G)^{-1} g$$

The gradient g being the columnwise sum of G, it can be written as $g = G^\top j$, where j is a vector of 1s of length N. Therefore, the test statistic is also:

$$j^\top G(G^\top G)^{-1} G^\top j = j^\top P_G j$$

where P_G is the projection matrix on the subspace defined by the columns of G. Therefore, regressing j on G, the fitted values are $P_G j$ and the sum of squares of the fitted values are $j^\top P_G j$, P_G being idempotent. In a regression without intercept, this is the explained sum of squares. The total sum of squares being: $j^\top j = N$, the (uncentered) R-squared is equal to $\frac{j^\top P_G j}{N}$ and the test statistic is therefore N times the R-squared of a regression on a vector of 1 on the column of the individual contributions to the gradient.

```
N <- length(y)
areg <- lm(rep(1, N) ~ G - 1)
lm_test_G <- sum(fitted(areg) ^ 2)
lm_test_G2 <- rsq(areg) * N
c(lm_test_G, lm_test_G2)
## [1] 13.58 13.58
```

To summarize the results of this section, the Wald test statistic is 3.75, the likelihood-ratio test statistic is 4.92 and the score test statistic, when computed using the hessian based estimation of the information is 3.92. The values are quite similar and the hypotheses are not rejected at the 5% level (the critical value is 5.99) and are rejected at the 10% level (the critical value is equal to 4.61) only for the likelihood ratio test. On the contrary, the

[7]Note the use of the `inv` argument set to `FALSE` in the second call to `quad_form` as the matrix provided (`H2m1`) is a subset of the inverse of the covariance matrix of the estimator.

score test computed using the gradient-based estimate of the information matrix has a much higher value (13.58) and leads to a rejection of the hypothesis, even at the 1% level (the critical value is 9.21).

To compute the pseudo R^2, we should define the "null" or "intercept-only" model. In the generalized production function, it is a model with only two parameters, α and σ. Therefore, the R matrix selects all the coefficients except these two and, using this matrix, we can compute the Wald statistic:

```
R <- diag(6)[- c(1, 6), ]
W_0 <- drop(t(R %*% coef(gpf2)) %*% solve(R %*% vcov(gpf2) %*% t(R)) %*%
              (R %*% coef(gpf2)))
```

For the other two tests, the constrained model should be estimated. Actually, this is a log linear model with only an intercept, and the LM estimates of α and σ^2 are simply the mean and the variance of $\ln y$.

```
alpha_c <- mean(log(y))
sigma_c <- sqrt(mean((log(y) - alpha_c) ^ 2))
null_model <- zellner_revankar(c(alpha_c, 0, 0, 0, 0, sigma_c),
                                 y = y, Z = Z, repar = FALSE)
lnL_c <- sum(null_model)
G_c <- attr(null_model, "gradient")
g_c <- apply(G_c, 2, sum)
```

The likelihood ratio statistic is just twice the difference between the two values of the log-likelihood:

```
LR_0 <- as.numeric(2 * (logLik(gpf2) - lnL_c))
```

Finally, the Lagrange multiplier statistic can be obtained by considering N times the R^2 of the regression of a vector of 1 on the matrix of the individual contributions to the gradient:

```
LM_0 <- N * rsq(lm(rep(1, N) ~ G_c - 1))
```

We can then compute the three pseudo-R^2:

```
R2_w <- W_0 / (N + W_0)
R2_lr <- 1 - exp(- LR_0 / N)
R2_lm <- LM_0 / N
c(wald = R2_w, lik_ratio = R2_lr, lagr_mult = R2_lm) %>% print
##       wald lik_ratio lagr_mult
##     0.5464    0.6054    0.7134
```

5.3.2 Conditional moment test

Compared to the three classical tests, conditional moment tests don't define two nested models (a "large" one and a "small" one). These tests, first presented by Tauchen (1985) and Newey (1985) are based on moment conditions that should be 0 under H_0. They are particularly useful for models fitted by maximum likelihood, although they can be used for models fitted by other estimation methods. Consider the example where the distribution of the response is related to the normal distribution. This is the case for the generalized production function estimated in the previous section, and it is also the case for the probit and the tobit model that will be developed in the last part of the book. With the OLS estimator, the most important properties of the estimator, especially unbiasedness and consistency only rely on the hypothesis that the conditional expectation of the response is correctly specified. This is not the case for models fitted by ML. Therefore, if the conditional distribution of the response is not normal with a constant conditional variance, the estimator may be inconsistent. Testing the hypothesis of normality and of homoskedasticity is therefore crucial in this context.

Denote $\mu_n = \mu(\theta, w_n)$ a vector of length J for observation n, that depends on a vector of parameters (θ) and on a vector of variables (w_n, which typically contains the response and a vector of covariate). The hypothesis is that $\mathrm{E}(\mu_n) = 0$. For example, to test normality, the hypothesis will be that the third moment of the errors is zero and that the fourth (standardized) moment is three and therefore: $\mu_n^\top = (\epsilon_n^3, \epsilon_n^4 - 3\sigma_\epsilon^4)$. Denote $m(\theta, W) = \sum_{n=1}^N \mu(\theta, w_n)$. The test is based on the sample equivalent of the moment conditions, which is:

$$\hat{\tau} = m(\hat{\theta}, W)/N = \frac{1}{N}\sum_{n=1}^N \mu(\hat{\theta}, w_n) = \frac{1}{N}\sum_{n=1}^N \hat{\mu}_n$$

and the hypotheses won't be rejected if $\hat{\tau}$ is sufficiently close to a vector of 0. The derivation of its variance is quite complicated because there are two sources of stochastic variations, as both $\hat{\theta}$ and $\hat{\mu}_n$ are random.[8] Using a first-order Taylor expansion around the true value θ_0, we have:

$$\hat{\tau} = \frac{1}{N}m(\theta_0, Z) + \frac{1}{N}\frac{\partial m}{\partial \theta^\top}(\bar{\theta}, Z)(\hat{\theta} - \theta_0)$$

Denote: $\mathcal{W} = \lim\limits_{N\rightarrow\infty} \dfrac{\partial m(\theta, Z)}{\partial \theta^\top} = \lim\limits_{N\rightarrow\infty} \dfrac{1}{N}\sum_{n=1}^N \dfrac{\partial \mu(\theta, z_n)}{\partial \theta^\top}$. As the estimator is consistent, we have:

$$\hat{\tau} \overset{a}{=} \frac{1}{N}m(\theta_0, Z) + \mathcal{W}(\hat{\theta} - \theta_0)$$

Using Equation 5.19:

$$\hat{\tau} \overset{a}{=} \frac{m(\theta_0, Z)}{N} - \mathcal{W}\mathcal{H}^{-1}\frac{g(\theta_0, Z)}{N} = \frac{1}{N}\sum_{n=1}^N \mu(\theta_0, z_n) - \mathcal{W}\mathcal{H}^{-1}\frac{1}{N}\sum_{n=1}^N \gamma(\theta_0, Z)$$

Or:

$$\sqrt{N}\hat{\tau} \overset{a}{=} [I, -\mathcal{W}\mathcal{H}^{-1}] \begin{pmatrix} \frac{\sum_n \mu(\theta_0, z_n)}{\sqrt{N}} \\ \frac{\sum_n \gamma(\theta_0, z_n)}{\sqrt{N}} \end{pmatrix} \tag{5.38}$$

[8] Cameron and Trivedi (2005), page 260.

Define V as the variance of the vector in parentheses in Equation 5.38, obtained by concatenating $g(\theta_0, Z)$ (the gradient for the true value of the parameters) and $m(\theta_0, Z)$, both divided $\sqrt{N}$. The probability limit of V is:

$$\mathcal{V} = \lim_{N\to\infty} \frac{1}{N} \begin{pmatrix} \sum_n \mu_n \mu_n^\top & \sum_n \mu_n \gamma_n^\top \\ \sum_n \gamma_n \mu_n^\top & \sum_n \gamma_n \gamma_n^\top \end{pmatrix}$$

The probability limit of the variance of $\sqrt{N}\hat{\tau}$ is therefore:

$$\mathrm{V}(\sqrt{N}\hat{\tau}) \xrightarrow{p} [I, -\mathcal{W}\mathcal{H}^{-1}]\, \mathcal{V}\, [I, -\mathcal{W}\mathcal{H}^{-1}]^\top \tag{5.39}$$

and $\mathcal{V}$ can be consistently estimated by:

$$\frac{1}{N} \begin{pmatrix} \hat{M}^\top \hat{M} & \hat{M}^\top \hat{G} \\ \hat{G}^\top \hat{M} & \hat{G}^\top \hat{G} \end{pmatrix}$$

with $\hat{M}$ the $N \times J$ matrix containing the individual contributions to $\hat{m}$ (with a nth row equal to $\hat{\mu}_n^\top$) and $\hat{G}$ the $N \times (K+1)$ matrix containing the individual contributions to the gradient (with a nth row equal to $\hat{\gamma}_n^\top$). Replacing $\mathcal{V}$ in Equation 5.39, developing the quadratic form and regrouping terms, we finally get:[9]

$$\mathrm{V}(\sqrt{N}\hat{\tau}) \xrightarrow{p} \frac{1}{N} \left(\hat{M} - \hat{G}\mathcal{H}^{-1}\mathcal{W}^\top\right)^\top \left(\hat{M} - \hat{G}\mathcal{H}^{-1}\mathcal{W}^\top\right)$$

and the statistic is:

$$\hat{\tau}^\top \left[\frac{1}{N^2} \left(\hat{M} - \hat{G}\mathcal{H}^{-1}\mathcal{W}^\top\right)^\top \left(\hat{M} - \hat{G}\mathcal{H}^{-1}\mathcal{W}^\top\right)\right]^{-1} \hat{\tau}$$

or, in terms of $\hat{m} = N\hat{\tau}$:

$$\hat{m}^\top \left[\left(\hat{M} - \hat{G}\mathcal{H}^{-1}\mathcal{W}^\top\right)^\top \left(\hat{M} - \hat{G}\mathcal{H}^{-1}\mathcal{W}^\top\right)\right]^{-1} \hat{m} \tag{5.40}$$

which is, under H_0, a chi-squared with J degrees of freedom. Different flavors of the test are obtained using different estimators of $\mathcal{W}$ and $\mathcal{H}$:

- the first uses the expected value of the estimators of $\mathcal{H}$ and $\mathcal{W}$ which are respectively: $\mathrm{E}\frac{\partial \ln L}{\partial\theta\partial\theta^\top}(\hat{\theta}, Z)/N$ and $\mathrm{E}\frac{\partial m(\hat{\theta}, Z)}{\partial\theta}/N$,
- the second uses the same expressions without the expectation: $\frac{\partial \ln L}{\partial\theta\partial\theta^\top}(\hat{\theta}, Z)/N$ and $\frac{\partial m(\hat{\theta}, Z)}{\partial\theta}/N$,
- the third uses the information equality to estimate $\mathcal{H}$ by $-\hat{G}^\top\hat{G}/N$ and the generalized information equality to estimate $\mathcal{W}$ by $-\hat{G}^\top\hat{M}/N$.

[9] Skeels and Vella (1999), eq. 2.13.

The last one is particularly convenient, as it only requires the $\hat{G}$ matrix of the contributions to the gradient and the $\hat{M}$ matrix containing the contributions to the empirical moment vector. Rearranging terms and using the fact that $\hat{m} = \hat{M}j$ with j a vector of 1s, Equation 5.40 becomes:

$$j^\top \hat{M}\left[M^\top\left(I - \hat{G}(\hat{G}^\top\hat{G})^{-1}\hat{G}^\top\right)\hat{M}\right]^{-1}\hat{M}^\top j \tag{5.41}$$

which is just the explained sum of squares of a regression of a vector of 1s on $\hat{G}$ and $\hat{M}$. To see that, start with the expression of the explained sum of squares which is, denoting y the response and X the matrix of covariates: $y^\top X(X^\top X)^{-1}X^\top y$. In our case, the response is ι and the matrix of covariates $(\hat{G}\,\hat{M})$. Therefore, the explained sum of squares is:

$$j^\top(\hat{G}\,\hat{M})\begin{pmatrix}\hat{G}^\top\hat{G} & \hat{G}^\top\hat{M}\\ \hat{M}^\top\hat{G} & \hat{M}^\top\hat{M}\end{pmatrix}^{-1}\begin{pmatrix}\hat{G}^\top\\ \hat{M}^\top\end{pmatrix}j \tag{5.42}$$

But all the columns of $\hat{G}$ sum to 0 ($\hat{G}^\top j = 0$), so that Equation 5.42 reduces $j^\top\hat{M}C\hat{M}^\top j$, where C is the lower right square matrix in the formula of the partitioned inverse of the matrix in Equation 5.42 and is the matrix in bracket in Equation 5.41 (see for example Greene (2003), equation A-74, page 824).

The conditional moment test can be used to test the hypothesis of normality, homoskedasticity and omitted variables. For the normality hypothesis, the theoretical moments are $\mbox{E}(\epsilon_n^3 \mid x_n) = 0$ and $\mbox{E}(\epsilon_n^4 - 3\sigma^2 \mid x_n) = 0$ and the empirical counterparts are:

$$\left\{\begin{array}{l}\frac{1}{N}\sum_n \hat{\epsilon}_n^3\\ \frac{1}{N}\sum_n \hat{\epsilon}_n^4 - 3\hat{\sigma}^2\end{array}\right. \tag{5.43}$$

For the homoskedasticity hypothesis, the theoretical moments are $\mbox{E}\left(w_n(\epsilon_n^2 - \sigma^2)|x_n\right) = 0$, where, under the alternative hypothesis, w are variables that enter the skedastic function and the empirical counterparts are:

$$\frac{1}{N}\sum_n w_n(\hat{\epsilon}_n^2 - \hat{\sigma}^2) \tag{5.44}$$

For the omitted variables test, the theoretical moments are $\mbox{E}(w_n\hat{\epsilon}_n|x_n) = 0$ and the empirical counterparts are:

$$\frac{1}{N}\sum_n w_n\hat{\epsilon}_n \tag{5.45}$$

As an example, we test the normality hypothesis in the context of the generalized production function previously estimated:

$$\epsilon = \ln y + \theta \ln y - \left(\alpha + \sum_{j=1}^{J-1}\beta_j \ln z_j^* + \beta_J^* \ln z_J + \ln z_J\right) \sim \mathcal{N}(0, \sigma^2)$$

We first extract the fitted coefficients of the model `htheta`, we matrix of the individual contribution to the gradient `G` and the hessian `H`:

```
htheta <- coef(gpf)
lnlest <- zellner_revankar(htheta, y = y, Z = Z)
G <- attr(lnlest, "gradient")
H <- attr(lnlest, "hessian")
```

Working on the reparametrized version of the model, we then transform the matrix of covariates, compute ϵ the matrix of the individual contribution to the moments M and the moment conditions m which is a vector containing the sum of the columns of M.

```
Zm <- Z
Zm[, 2:3] <- Z[, 2:3] - Z[, 4]
mu <- drop(Zm %*% htheta[1:4])
eps <- (log(y) - Z[, 4] + htheta["lambda"] * y - mu)
M <- cbind(eps ^ 3, eps ^ 4 - 3 * htheta["sigma"] ^ 4)
m <- M %>% apply(2, sum)
```

Then, the derivatives of m with respect to the parameters of the model are computed to obtain the W matrix.

```
W <- matrix(c(-3 * apply(eps ^ 2 * Z, 2, sum),
              3 * sum(eps ^ 2 * y), 0,
              -4 * apply(eps ^ 3 * Z, 2, sum),
              4 * sum(eps ^ 3 * y),
              - 12 * htheta["sigma"] ^ 3),
            ncol = 2)
```

We first compute the test using minus the hessian to estimate the information and the matrix of the analytical derivatives of m just computed:

```
N <- length(y)
Q1 <- crossprod(M - G %*% solve(H) %*% W)
quad_form(m, Q1)
## [1] 1.284
```

The statistic is 1.28, which is far less than the critical value for a χ^2 with 2 degrees of freedom, even at the 10% level. The hypothesis of normality is therefore not rejected. We then compute the version of the test based on the OPG to estimate the information matrix and on $G^\top M$ to estimate W. The statistic can be computed by using matrix algebra or by taking the explained sum of squares in a regression of a column of 1 on G and M:

```
Q2 <- crossprod(M - G %*% solve(crossprod(G)) %*%
                crossprod(G, M))
quad_form(m, Q2)
## [1] 11.78
j <- rep(1, N)
rsq(lm(j ~ G + M - 1)) * N
## [1] 11.78
```

Note that this second version of the test leads to a much higher value of the statistic and a probability value of 0.003.

5.3.3 Tests for non-nested models

5.3.3.1 Vuong test

Vuong (1989) proposed a test for non-nested models. He considered two competing models characterized by densities $f(y|z;\beta)$ and $g(y|z;\gamma)$. Denoting $h(y|z)$ the true conditional density, the distance of the first model to the true model is measured by the minimum Kullback-Leibler information criterion (**KLIC**):

$$D_f = \mathrm{E}^0\left[\ln h(y \mid z)\right] - \mathrm{E}^0\left[\ln f(y \mid z;\beta_*)\right]$$

where E^0 is the expected value using the true joint distribution of (y, X) and β_* is the pseudo-true value of β.[10] As the true model is unobserved, denoting $\theta^\top = (\beta^\top, \gamma^\top)$, we consider the difference of the KLIC distance to the true model of model G_γ and model F_β:

$$\Lambda(\theta) = D_g - D_f = \mathrm{E}^0\left[\ln f(y \mid z;\beta_*)\right] - \mathrm{E}^0\left[\ln g(y \mid z;\gamma_*)\right] = \mathrm{E}^0\left[\ln \frac{f(y \mid z;\beta_*)}{g(y \mid z;\gamma_*)}\right]$$

The null hypothesis is that the distances of the two models to the true models are equal or, equivalently, that: $\Lambda = 0$. The alternative hypothesis is either $\Lambda > 0$, which means that the first model is better than the second or $\Lambda < 0$, which means that the second model is better than the first. Denoting, for a given random sample of size N, $\hat{\beta}$ and $\hat{\gamma}$ the maximum likelihood estimators of the two models and $\ln L_f(\hat{\beta})$ and $\ln L_g(\hat{\gamma})$ the corresponding values of the log-likelihood functions, Λ can be consistently estimated by:

$$\hat{\Lambda}_N = \frac{1}{N}\sum_{n=1}^N \left(\ln f(y_n \mid x_n, \hat{\beta}) - \ln g(y_n \mid x_n, \hat{\gamma})\right) = \frac{1}{N}\left(\ln L_f(\hat{\beta}) - \ln L_g(\hat{\gamma})\right)$$

which is the likelihood ratio divided by the sample size. Note that the statistic of the standard likelihood ratio test, suitable for nested models is $2\left(\ln L^f(\hat{\beta}) - \ln L^g(\hat{\gamma})\right)$, which is $2N\hat{\Lambda}_N$. The variance of Λ is:

$$\omega_*^2 = \mathrm{V}^o\left[\ln \frac{f(y \mid x;\beta_*)}{g(y \mid x;\gamma_*)}\right]$$

which can be consistently estimated by:

$$\hat{\omega}_N^2 = \frac{1}{N}\sum_{n=1}^N \left(\ln f(y_n \mid x_n, \hat{\beta}) - \ln g(y_n \mid x_{,n} \hat{\gamma})\right)^2 - \hat{\Lambda}_N^2$$

Three different cases should be considered when the two models are:

- nested, ω_*^2 is necessarily 0,
- overlapping (which means than the models coincide for some values of the parameters), ω_*^2 *may be* equal to 0 or not,
- strictly non-nested, ω_*^2 is necessarily strictly positive.

[10] β_* is called the pseudo-true value because f may be an incorrect model.

The distribution of the statistic depends on whether ω_*^2 is zero or positive. If ω_*^2 is positive, the statistic is $\hat{T}_N = \sqrt{N}\frac{\hat{\Lambda}_N}{\hat{\omega}_N}$ and, under the null hypothesis that the two models are equivalent, follows a standard normal distribution. This is the case for strictly non-nested models.

On the contrary, if $\omega_*^2 = 0$, the distribution is much more complicated. We need to define two matrices: A contains the expected values of the second derivatives of Λ:

$$A(\theta_*) = \mathrm{E}^0 \left[\frac{\partial^2 \Lambda}{\partial\theta\partial\theta^\top}\right] = \mathrm{E}^0 \left[\begin{array}{cc} \frac{\partial^2 \ln f}{\partial\beta\partial\beta^\top} & 0 \\ 0 & -\frac{\partial^2 \ln g}{\partial\beta\partial\beta^\top} \end{array}\right] = \left[\begin{array}{cc} A_f(\beta_*) & 0 \\ 0 & -A_g(\gamma_*) \end{array}\right]$$

and B the variance of its first derivatives:

$$\begin{array}{rcl} B(\theta_*) = \mathrm{E}^0 \left[\frac{\partial\Lambda}{\partial\theta}\frac{\partial\Lambda}{\partial\theta^\top}\right] & = & \mathrm{E}^0 \left[\left(\frac{\partial \ln f}{\partial\beta}, -\frac{\partial \ln g}{\partial\gamma}\right)\left(\frac{\partial \ln f}{\partial\beta^\top}, -\frac{\partial \ln g}{\partial\gamma^\top}\right)\right] \\ & = & \mathrm{E}^0 \left[\begin{array}{cc} \frac{\partial \ln f}{\partial\beta}\frac{\partial \ln f}{\partial\beta^\top} & -\frac{\partial \ln f}{\partial\beta}\frac{\partial \ln g}{\partial\gamma^\top} \\ -\frac{\partial \ln g}{\partial\gamma}\frac{\partial \ln f}{\partial\beta^\top} & \frac{\partial \ln g}{\partial\gamma}\frac{\partial \ln g}{\partial\gamma^\top} \end{array}\right] \end{array}$$

or:

$$B(\theta_*) = \left[\begin{array}{cc} B_f(\beta_*) & -B_{fg}(\beta_*, \gamma_*) \\ -B_{gf}(\beta_*, \gamma_*) & B_g(\gamma_*) \end{array}\right]$$

Then:

$$W(\theta_*) = B(\theta_*)\left[-A(\theta_*)\right]^{-1} = \left[\begin{array}{cc} -B_f(\beta_*)A_f^{-1}(\beta_*) & -B_{fg}(\beta_*, \gamma_*)A_g^{-1}(\gamma_*) \\ B_{gf}(\gamma_*, \beta_*)A_f^{-1}(\beta_*) & B_g(\gamma_*)A_g^{-1}(\gamma_*) \end{array}\right]$$

Denote λ_* the eigenvalues of W. When $\omega_*^2 = 0$ (which is always the case for nested models), the statistic is the one used in the standard likelihood ratio test: $2(\ln L_f - \ln L_g) = 2N\hat{\Lambda}_N$ which, under the null, follows a weighted χ^2 distribution with weights equal to λ_*. The Vuong test can be seen in this context as a more robust version of the standard likelihood ratio test, because it doesn't assume, under the null, that the larger model is correctly specified.

Note that, if the larger model is correctly specified, the information matrix equality implies that $B_f(\theta_*) = -A_f(\theta_*)$. In this case, the two matrices on the diagonal of W reduces to $-I_{K_f}$ and I_{K_g}, the trace of W to $K_g - K_f$ and the distribution of the statistic under the null reduce to a χ^2 with $K_g - K_f$ degrees of freedom.

The W matrix can be consistently estimated by computing the first and the second derivatives of the likelihood functions of the two models for $\hat{\theta}$. For example,

$$\hat{A}_f(\hat{\beta}) = \frac{1}{N}\sum_{n=1}^{N}\frac{\partial^2 \ln f}{\partial\beta\partial\beta^\top}(\hat{\beta}, x_n, y_n)$$

$$\hat{B}_{fg}(\hat{\theta}) = \frac{1}{N}\sum_{n=1}^{N}\frac{\partial \ln f}{\partial\beta}(\hat{\beta}, x_n, y_n)\frac{\partial \ln g}{\partial\gamma^\top}(\hat{\gamma}, x_n, y_n)$$

For the overlapping case, the test should be performed in two steps:

- the first step consists of testing whether ω_*^* is 0 or not. This hypothesis is based on the statistic $N\hat{\omega}^2$ which, under the null ($\omega_*^2 = 0$) follows a weighted χ^2 distributions with weights equal to λ_*^2. If the null hypothesis is not rejected, the test stops at this step, and the conclusion is that the two models are equivalent;
- if the null hypothesis is rejected, the second step consists of applying the test for non-nested models previously described.

Shi (2015) provides an example of simulations of non-nested linear models that shows that the distribution of the Vuong statistic can be very different from a standard normal. The data generating process used for the simulations is:

$$y = 1 + \sum_{k=1}^{K_f} z_k^f + \sum_{k=1}^{K_g} z_k^g + \epsilon$$

where z^f is the set of K_f covariates that are used in the first model and z^g the set of K_g covariates used in the second model and $\epsilon \sim N(0, 1 - a^2)$. $z_k^f \sim N(0, a/\sqrt{K_f})$ and $z_k^g \sim N(0, a/\sqrt{K_g})$, so that the variance explained by the two competing models is the same (equal to a^2) and the null hypothesis of the Vuong test is true. The `micsr::vuong_sim` enables to simulate values of the Vuong test. As in Shi (2015), we use a very different degree of parametrization for the two models, with $K_f = 15$ and $K_G = 1$.

```
Vuong <- vuong_sim(N = 100, R = 1000,
                   Kf = 15, Kg = 1, a = 0.5)
head(Vuong)
## [1]  0.4757  0.9187  0.9234  2.5316 -0.3940  1.4318
mean(Vuong)
## [1] 1.073
mean(abs(Vuong) > 1.96)
## [1] 0.183
```

We can see that the mean of the statistic for the 1000 replications is far away from 0, which means that the numerator of the Vuong statistic is seriously biased. A total of 18.3% of the values of the statistic are greater than the critical value so that the Vuong test will lead in such context to a noticeable over-rejection. The empirical density function is shown in Figure 5.5, along with the normal density.

Shi (2015) proposed a non-degenerate Vuong test which corrects the small sample bias of the numerator of the Vuong statistic and inflates the denominator by adding a constant.

5.3.3.2 An example: generalized production function vs. translog function

A popular alternative to the generalized production function is the translog function, which is:

$$\ln y = \alpha + \sum_{j=1}^{J} \beta_j \ln q_j + \frac{1}{2} \sum_{j=1}^{J} \sum_{k=1}^{J} \beta_{jk} \ln q_j \ln q_k \tag{5.46}$$

The elasticity of the production with a factor is:

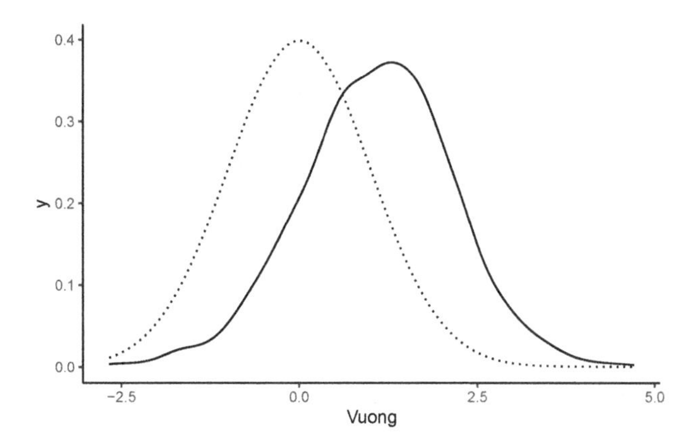

Figure 5.5: Distribution of the Vuong statistic

$$\frac{\partial \ln y}{\partial \ln x_j} = \beta_j + \sum_{k=1}^{J} \beta_{jk} \ln q_k$$

and the scale elasticity is just the sum of these J elasticities:

$$\epsilon = \sum_{j=1}^{J} \beta_j + \sum_{j=1}^{J} \sum_{k=1}^{J} \beta_{jk} \ln q_k = \sum_{j=1}^{J} \beta_j + \sum_{k=1}^{J} \ln q_k \sum_{j=1}^{J} \beta_{jk}$$

The constant returns to scale hypothesis therefore implies that $\sum_j \beta_j = 1$ and $\sum_{j=1}^{J} \beta_{jl} = 0 \; \forall \; k$. We fit Equation 5.46 using `loglm`:

```
trsl <- loglm(y ~ log(k) + log(l) + log(m) + log(k):log(l) +
                log(k):log(m) + log(l):log(m) + I(log(k) ^ 2) +
                I(log(l) ^ 2) + I(log(m) ^ 2), aps)
```

All that is required to compute the Vuong test for non-nested models is the contributions to the log-likelihood for both models:[11]

```
est_gpf <- zellner_revankar(coef(gpf), y = y, Z = Z)
lnl_gpf <- as.numeric(est_gpf)
lnl_trsl <- trsl$value
```

We can then compute the average likelihood ratio statistic (`L`), its variance (`w2`) and the statistic:

```
N <- length(lnl_gpf)
L <- mean(lnl_gpf) - mean(lnl_trsl)
w2 <- mean((lnl_gpf - lnl_trsl) ^ 2) - L ^ 2
vuong_stat <- sqrt(N) * L / sqrt(w2)
```

[11]The object returned by `loglm` contains an element called `values` that is a vector of individual contributions to the log-likelihood.

```
vuong_stat
## [1] -1.147
```

The probability value is 0.126 so that the test concludes that the two models are indistinguishable, although the difference of their log-likelihood is quite high:

```
sum(lnl_gpf)
## [1] -60.28
sum(lnl_trsl)
## [1] -55.86
```

6

Non-spherical disturbances

In the first part of the book, we considered a model of the form: $y_n = \alpha + \beta x_n + \epsilon_n$ with the hypothesis that the errors were homoskedastic: $\mbox{V}(\epsilon_n) = \sigma_\epsilon^2$ and uncorrelated: $\mbox{E}(\epsilon_n \epsilon_m) = 0 \; \forall n \neq m$. In this case, the errors (or disturbances) are **spherical** and the covariance matrix of the errors Ω is, up to a multiplicative constant σ_ϵ^2, the identity matrix:

$$\Omega = \mbox{V}(\epsilon) = \mbox{E}(\epsilon\epsilon^\top) = \sigma_\epsilon^2 I$$

In this chapter, we analyze cases where these hypothesis are violated. This has the following consequences concerning the results established in the first part of the book:

- the OLS estimator is still consistent: this means that $\hat{\beta}$ estimates consistently β and $\hat{\epsilon}$ estimates also consistently ϵ; this is an important result, as the residuals of the OLS estimator can therefore be used to test whether the errors are spherical or not,
- the OLS estimator is no longer BLUE, i.e., it is no longer the best linear unbiased estimator,
- there is another linear unbiased estimator (the **generalized least squares**, or **GLS**) which is more efficient (which means that it has a smaller variance) than the OLS estimator and which is the BLUE estimator when the errors are non-spherical,
- the simple formula for the variance of the OLS estimator, $\mbox{V}(\hat{\beta}) = \sigma_\epsilon^2 \left(\tilde{X}^\top \tilde{X}\right)^{-1}$ is no longer an unbiased estimator of the true covariance matrix of the OLS estimator. However, as we'll see in this chapter, **sandwich** estimators, which are consistent, can be used.

Section 6.1 reviews some important cases where the errors are non-spherical. Section 6.2 presents tests that enable to detect whether the errors are spherical or not. Section 6.3 presents robust estimators of the variance of the OLS estimators. Finally, Section 6.4 is devoted to the GLS estimator.

6.1 Situations where the errors are non-spherical

As stated previously, the hypothesis of spherical disturbances implies that the errors are homoskedastic and uncorrelated. We'll describe in the next three subsections important situations where this hypothesis is violated. In each case, we'll establish the expression of the matrix of covariance of the errors Ω and, for a reason that will be clear in the subsequent sections, we'll also compute the inverse of this matrix.

6.1.1 Heteroskedasticity

In a linear model: $y_n = \alpha + \beta x_n + \epsilon_n$, heteroskedasticity occurs when the conditional variance of the response y_n (the variance of ϵ_n) is not a constant. As an example, Houthakker (1951) analyzed electric consumption in the United Kingdom, and his data set (called `uk_elec`) is a sample of 42 British cities.[1] The response is the per capita consumption in kilowatt hours. Denoting c_{ni} the consumption of an individual i in city n, the response is then $y_n = \frac{1}{I_n}\sum_i^{I_n} c_{ni}$ where I_n is the total number of consumers in city n. Then, if the standard deviation of the individual consumption is σ_c (the same in every city), the variance of y_n is σ_c^2/I_n and therefore depends on the number of consumer units of the city. Even if covariates are taken into account, it is doubtful that the conditional variance of y will be the same for every city, and a more reasonable hypothesis is that, as the unconditional variance, it is inversely proportional to the number of consumer units. With heteroskedastic but uncorrelated errors, the matrix of covariance of the errors is a diagonal matrix with non-constant diagonal terms:

$$\Omega = \begin{pmatrix} \sigma_1^2 & 0 & \dots & 0 \\ 0 & \sigma_2^2 & \dots & 0 \\ \vdots & \vdots & \ddots & \vdots \\ 0 & 0 & \dots & \sigma_N^2 \end{pmatrix} \tag{6.1}$$

The inverse of Ω is easily obtained:

$$\Omega^{-1} = \begin{pmatrix} 1/\sigma_1^2 & 0 & \dots & 0 \\ 0 & 1/\sigma_2^2 & \dots & 0 \\ \vdots & \vdots & \ddots & \vdots \\ 0 & 0 & \dots & 1/\sigma_N^2 \end{pmatrix}$$

and, more generally, Ω^r (r being any integer or rational number) is a diagonal matrix with typical element $(\sigma_n^2)^r$.

6.1.2 Correlation of the errors

Consider now the case where we have several observations from the same **entity**. An example is the case where the unit of observation is the individual, but siblings are observed. In this case, each observation is doubly indexed, the first index being the family and the second one the rank of the sibling's birth. Another very important case is when the same individual (in a wide sense, it can be a household, firm, country, etc.) is observed several times, for example for different periods like years or months. Such a data set is called a **panel data**. In the subsequent sections, we'll use two data sets. The first data set, called `twins`, is from Bonjour et al. (2003) who studied the return to education with a sample of twins. The sample contains 428 observations (214 pairs of twins), and it is reasonable to assume that, for a given pair of twins, the two errors are correlated as they partly contain unobserved characteristics that are common to both twins. The second data set, called `tobinq`, is from Schaller (1990) who tests the relevance of Tobins' Q theory of investment by regressing the investment rate (the ratio of the investment and the stock of capital) to Tobin's Q, which

[1] This data set is used extensively by Berndt (1991), chapter 7.

is the ratio of the value of the firm and the stock of capital. The data set is a panel of 188 firms observed for 35 years (from 1951 to 1985).

For such data, we'll denote $n = 1, 2, \dots N$ the entity / individual index and $t = 1, 2, \dots T$ the index of the observation in the entity / the time period. Then, the simple linear model can be written: $y_{nt} = \alpha + \beta x_{nt} + \epsilon_{nt}$ and it is useful to write the error term as the sum of two components:

- an entity / individual effect η_n,
- an idiosyncratic effect ν_{nt}.

We therefore have $\epsilon_{nt} = \eta_n + \nu_{nt}$. This leads to the so-called **error-component** model, which can be easily analyzed with the following hypothesis:

- the two components are homoskedastic and uncorrelated: $\mathrm{V}(\eta_n) = \sigma^2_\eta, \forall n$, $\mathrm{V}(\nu_{nt}) = \sigma^2_\nu, \forall n, t$ and $\mathrm{cov}(\eta_n, \nu_{nt}) = 0, \forall n, t$,
- the idiosyncratic terms for the same entity are uncorrelated: $\mathrm{E}(\nu_{nt}\nu_{ns}) = 0 \; \forall \; n, t \neq s$,
- the two components of the errors are uncorrelated for two observations of different entities $\mathrm{cov}(\nu_{nt}, \nu_{ms}) = \mathrm{cov}(\eta_n, \eta_m) = 0 \; \forall \; n \neq m, t, s$.

With these hypotheses, we have:

$$\left\{\begin{array}{rcl} \mathrm{E}(\epsilon^2_{nt}) &=& \sigma^2_\eta + \sigma^2_\nu \\ \mathrm{E}(\epsilon_{nt}\epsilon_{mt}) &=& \sigma^2_\eta \\ \mathrm{E}(\epsilon_{nt}\epsilon_{ms}) &=& 0; \; \forall \; n \neq m \end{array}\right.$$

and Ω is a block-diagonal matrix with identical blocks. For example, with $N = 2$ and $T = 3$:

$$\Omega = \left(\begin{array}{cccccc} \sigma^2_\eta + \sigma^2_\nu & \sigma^2_\eta & \sigma^2_\eta & 0 & 0 & 0 \\ \sigma^2_\eta & \sigma^2_\eta + \sigma^2_\nu & \sigma^2_\eta & 0 & 0 & 0 \\ \sigma^2_\eta & \sigma^2_\eta & \sigma^2_\eta + \sigma^2_\nu & 0 & 0 & 0 \\ 0 & 0 & 0 & \sigma^2_\eta + \sigma^2_\nu & \sigma^2_\eta & \sigma^2_\eta \\ 0 & 0 & 0 & \sigma^2_\eta & \sigma^2_\eta + \sigma^2_\nu & \sigma^2_\eta \\ 0 & 0 & 0 & \sigma^2_\eta & \sigma^2_\eta & \sigma^2_\eta + \sigma^2_\nu \end{array}\right)$$

The blocks of this matrix can be written as, denoting j a vector of ones and $J = jj^\top$ a square matrix of 1:

$$\sigma^2_\nu I_T + \sigma^2_\eta J_T$$

and, using Kronecker product, Ω is:

$$\Omega = I_N \otimes \left(\sigma^2_\nu I_T + \sigma^2_\eta J_T\right) = \sigma^2_\nu I_{NT} + \sigma^2_\eta I_N \otimes J_T$$

Another equivalent expression which will prove to be particularly useful is:

$$\Omega = \sigma^2_\nu \left(I_N - I_N \otimes J_T/T\right) + \left(T\sigma^2_\eta + \sigma^2_\nu\right)\left(I_N \otimes J_T/T\right) = \sigma^2_\nu W + \sigma^2_\iota B \tag{6.2}$$

where $\sigma_\iota^2 = T\sigma_\eta^2 + \sigma_\nu^2$. With our $N = 2$ and $T = 3$ simple case, the two matrices are:

$$W = \begin{pmatrix} 2/3 & -1/3 & -1/3 & 0 & 0 & 0 \\ -1/3 & 2/3 & -1/3 & 0 & 0 & 0 \\ -1/3 & -1/3 & 2/3 & 0 & 0 & 0 \\ 0 & 0 & 0 & 2/3 & -1/3 & -1/3 \\ 0 & 0 & 0 & -1/3 & 2/3 & -1/3 \\ 0 & 0 & 0 & -1/3 & -1/3 & 2/3 \end{pmatrix}$$

and:

$$B = \begin{pmatrix} -1/3 & -1/3 & -1/3 & 0 & 0 & 0 \\ -1/3 & -1/3 & -1/3 & 0 & 0 & 0 \\ -1/3 & -1/3 & -1/3 & 0 & 0 & 0 \\ 0 & 0 & 0 & -1/3 & -1/3 & -1/3 \\ 0 & 0 & 0 & -1/3 & -1/3 & -1/3 \\ 0 & 0 & 0 & -1/3 & -1/3 & -1/3 \end{pmatrix}$$

The first matrix is called the **within** matrix. Premultiplying a vector by W transforms it as deviations from the individual means. The second matrix is called the **between** matrix, and premultiplying a vector by B transforms it as a vector of individual means. These two symmetric matrices have interesting properties:

- they are **idempotent**, which means that $BB = B$ and $WW = W$. For example $W(Wz) = Wz$, as taking the deviations from the individual means of a vector of deviations from the individual means leaves this vector unchanged,
- they are **orthogonal**, which means that $WB = BW = 0$. For example $W(Bz) = 0$ because the deviations from the individual means of a vector of individual means are zero,
- they sum to the **identity matrix**, $W + B = I$. $Wz + Bz = z$, because the sum of the deviations from the individual means of a vector and its individual means is the vector itself.

W and B therefore perform an **orthogonal decomposition** of a vector. One advantage of this decomposition is that it is very easy to obtain powers of Ω. For example, the inverse of Ω is:

$$\Omega^{-1} = \frac{1}{\sigma_\nu^2} W + \frac{1}{\sigma_\iota^2} B$$

and, more generally, for any power r (either an integer or a rational):

$$\Omega^r = \sigma_\nu^{2^r} W + \sigma_\iota^{2^r} B \tag{6.3}$$

6.1.3 System of equations

We have seen in Section 3.7 that, for example in fields such as consumption or production analysis, it is more relevant to consider the estimation of system of equations instead of the estimation of a single equation. In matrix form, the model corresponding to the whole system was presented in Equation 3.26. We've seen in Section 3.7 that a first advantage of considering the whole system of equations, and not an equation in isolation, is that restrictions on coefficients that concern different equations can be taken into account using the constrained least squares estimator.

A second advantage is that, if the errors of the different equations for the same observation are correlated, these correlations can be taken into account if the whole system of equations is considered. Denoting ϵ_l the vector of length N containing the errors for the $l^{\mbox{th}}$ equation and $\Xi = (\epsilon_1, \epsilon_2, ... \epsilon_L)$ the $N \times L$ matrix containing errors for the whole system, the covariance matrix of the errors for the system is:

$$\Omega = \mbox{E}(\Xi\Xi^\top) = \mbox{E}\left(\begin{array}{cccc} \epsilon_1\epsilon_1^\top & \epsilon_1\epsilon_2^\top & ... & \epsilon_1\epsilon_L^\top \\ \epsilon_2\epsilon_1^\top & \epsilon_2\epsilon_2^\top & ... & \epsilon_2\epsilon_L^\top \\ \vdots & \vdots & \ddots & \vdots \\ \epsilon_L\epsilon_1^\top & \epsilon_L\epsilon_2^\top & ... & \epsilon_L\epsilon_L^\top \end{array}\right)$$

Assume that the errors of two equations l and m for the same observation are correlated and that the covariance, denoted σ_{lm}, is constant. The variance of errors for each equation l is denoted σ_{ll} and may be different from one equation to another. Moreover, we assume that errors for different individuals are uncorrelated. With these hypotheses, the covariance matrix is, denoting I the identity matrix of dimension N:

$$\Omega = \left(\begin{array}{cccc} \sigma_1^2 I & \sigma_{12} I & ... & \sigma_{1L} I \\ \sigma_{12} I & \sigma_2^2 I & ... & \sigma_{2L} I \\ \vdots & \vdots & \ddots & \vdots \\ \sigma_{1L} I & \sigma_{2L} I & ... & \sigma_L^2 I \end{array}\right)$$

Denoting Σ the $L \times L$ matrix of inter-equation variances and covariances, we have:

$$\Sigma = \left(\begin{array}{cccc} \sigma_1^2 & \sigma_{12} & ... & \sigma_{1L} \\ \sigma_{12} & \sigma_2^2 & ... & \sigma_{2L} \\ \vdots & \vdots & \ddots & \vdots \\ \sigma_{1L} & \sigma_{2L} & ... & \sigma_L^2 \end{array}\right)$$

and $\Omega = \Sigma \otimes I$. The inverse of the covariance matrix of the errors is easily obtained, as it requires only to compute the inverse of Σ: $\Omega^{-1} = \Sigma^{-1} \otimes I$.

6.2 Testing for non-spherical disturbances

Numerous tests have been proposed to investigate whether, in different contexts, the disturbances are spherical or not. Among them, we'll present a family of tests that are based on OLS residuals. Even if the disturbances are non-spherical, OLS is a consistent estimator and therefore OLS's residuals are a consistent estimate of the errors of the model. Therefore, one can use these residuals to analyze the unknown features of the errors.

6.2.1 Testing for heteroskedasticity

Breusch and Pagan (1979) consider the following heteroskedastic model: $y_n = \gamma^\top z_n + \epsilon_n$ with $\epsilon_n \sim \mathcal{N}(0, \sigma_n^2)$. Assume that σ_n^2 is a function of a set of J covariates denoted by w_n:

$$\sigma_n^2 = h(\delta^\top w_n)$$

The first element of w is 1, so that the homoskedasticity hypothesis is that all the elements of δ except the first one are 0: $\delta_0^\top = (\alpha, 0, \dots, 0)$, so that $\sigma_n^2 = h(\delta_0^\top w_n) = h_0$. The log-likelihood function is:

$$\ln L = -\frac{N}{2}\ln 2\pi - \frac{1}{2}\sum_n \ln \sigma_n^2 - \frac{1}{2}\sum_n \frac{(y_n - \gamma z_n)^2}{\sigma_n^2}$$

The derivative of $\ln L$ with δ is, denoting $h'_n = \frac{\partial h}{\partial \delta}(w_n)$:

$$\frac{\partial \ln L}{\partial \delta} = \frac{1}{2}\sum_n \left(\frac{\epsilon^2}{\sigma_n^4} - \frac{1}{\sigma_n^2}\right) h'_n w_n \tag{6.4}$$

With the homoskedasticity hypothesis, $\sigma_n = \sigma$ and $h'_n = h'_n(\delta_0) = h'_0$ and Equation 6.4 simplifies to:

$$d = \frac{\partial \ln L}{\partial \delta}(\delta_0) = \frac{h'_0}{2\sigma^2}\sum_n \left(\frac{\epsilon_n^2}{\sigma^2} - 1\right) w_n$$

The second derivatives are:

$$\frac{\partial \ln^2 L}{\partial \delta \partial \delta^\top} = \frac{1}{2}\sum_n \left[h''_n \left(\frac{\epsilon_n^2}{\sigma_n^4} - \frac{1}{\sigma_n^2}\right) - h'^2_n \left(\frac{2\epsilon_n^2}{\sigma_n^6} - \frac{1}{\sigma_n^4}\right)\right] w_n w'_n$$

To get the information matrix, we take the expectation of the opposite of this matrix. If the errors are homoskedastic, the first term disappears and $h'_n = h'_0$ so that:

$$I_0 = \mathrm{E}\left(-\frac{\partial \ln^2 L}{\partial \delta \partial \delta^\top}(\delta_0)\right) = \frac{h'^2_0}{2\sigma^4}\sum_n w_n w'_n \tag{6.5}$$

Denoting $\hat{\epsilon}$ the vector of OLS residuals and $\hat{\sigma}^2 = \hat{\epsilon}^\top \hat{\epsilon}/N$ the estimate of σ^2, the estimated score is:

$$\hat{d} = \frac{h'_0}{2\hat{\sigma}^2}\sum_n \left(\frac{\hat{\epsilon}_n^2}{\hat{\sigma}^2} - 1\right) w_n$$

and the test statistic is the quadratic form of $\hat{d}$ with the inverse of its variance given by Equation 6.5:

$$LM = \frac{1}{2}\left[\sum_n \left(\frac{\hat{\epsilon}_n^2}{\hat{\sigma}^2} - 1\right) w_n^\top\right] \left(\sum_n w_n w_n^\top\right)^{-1} \left[\sum_n \left(\frac{\hat{\epsilon}_n^2}{\hat{\sigma}^2} - 1\right) w_n\right]$$

or, in matrix form, denoting f the N-length vector with typical element $\left(\frac{\hat{\epsilon}_n^2}{\hat{\sigma}^2} - 1\right)$ and W the matrix of covariates:

$$LM = \frac{1}{2} f^\top W (W^\top W)^{-1} W f = \frac{1}{2} f^\top P_W f \tag{6.6}$$

which is half the explained sum of squares of a regression of f_n on w_n and is a χ^2 with J degrees of freedom in case of homoskedasticity. Note also that $f^\top f / N = \sum_n \left(\frac{\hat{\epsilon}_n^2}{\hat{\sigma}^2} - 1\right)^2 / N = \sum_n \left(\frac{\hat{\epsilon}_n^4}{\hat{\sigma}^4} + 1 - 2\frac{\hat{\epsilon}_n^2}{\hat{\sigma}^2}\right) / N$ is the total sum of squares divided by N. It converges to 2 as the first term is the fourth center moment of a normal variable, which is 3. Therefore, a second version of the statistic can be computed as N times the R^2 of a regression of the first-step residuals on w:

$$NR^2 = \frac{f^\top P_W f}{f' f / N}$$

White (1980) proposed a test that is directly linked to its proposition of the heteroskedasticity-robust matrix of covariance of the OLS estimates.[2] This matrix depends on the squares and on the cross-products of the covariates. Therefore, he proposed to run a regression of the squares of the first-step residuals on the covariates, their squares and their cross-product. NR^2 of this regression is asymptotically distributed as a χ^2 with $K(K+1)/2$ degrees of freedom. Therefore, White's test can be viewed as a special case of Breusch and Pagan's test.

In his electric consumption regression, Houthakker (1951) used as covariates `inc` (average yearly income in pounds), the inverse of `mc6` (the marginal cost of electricity), `gas6` (the marginal price of gas) and `cap` (the average holdings of heavy electric equipment). The OLS estimation is:

```
lm_elec <- lm(kwh ~ inc + I(1 / mc6) +  gas6 + cap, uk_elec)
```

The f vector in Equation 6.6 is:

```
f <- 1 - resid(lm_elec) ^ 2 / mean(resid(lm_elec) ^ 2)
```

We then regress f on W (which is here a constant and the inverse of `cust`) and get half the explained sum of squares:

```
lm_elec_resid <- lm(f ~ I(1 / cust), uk_elec)
bp1_elec <- sum(fitted(lm_elec_resid) ^ 2) / 2
```

For the second version of the test, we compute N times the R^2:

[2]See Section 6.3.

```
bp2_elec <- nobs(lm_elec_resid) * rsq(lm_elec_resid)
```

The values and the probability values for the two versions of the Breusch-Pagan test are:

```
c(bp1_elec, bp2_elec)
## [1] 16.49 11.38
pchisq(c(bp1_elec, bp2_elec), lower.tail = FALSE, df = 1)
## [1] 4.887e-05 7.441e-04
```

The homoskedasticity hypothesis is therefore highly rejected. The `lmtest::bptest` computes automatically the Breusch-Pagan test for heteroskedasticity, with two formulas: the first being the formula of the model and the second being a one-side formula for the skedasticity equation. An alternative syntax is to provide a `lm` model as the first argument:

```
lmtest::bptest(kwh ~ inc + I(1 / mc6) + gas6 + cap, ~ I(1 / cust),
               data = uk_elec, studentize = FALSE)
lmtest::bptest(lm_elec, ~ I(1 / cust),
               data = uk_elec, studentize = FALSE) %>% gaze
```

Note that we set the `studentize` argument to `FALSE`. The default value is `TRUE` and in this case, a modified version of the test due to Koenker (1981) is used.

6.2.2 Testing for individual effects

Breusch and Pagan (1980) extend their Lagrange multiplier test to the problem of individual (or entity) effects in a panel (or in a pseudo-panel) setting. Assuming a normal distribution, the joint density for the whole sample is:

$$f(y \mid X) = \frac{1}{(2\pi)^{NT/2} \mid \Omega \mid} e^{-\frac{1}{2}\epsilon^\top \Omega^{-1} \epsilon} \tag{6.7}$$

We have seen (Equation 6.2) that $\Omega = \sigma_\nu^2 W + \sigma_\iota^2 B$, with $\sigma_\iota^2 = \sigma_\nu + T\sigma_\eta$. Then,

$$\epsilon^\top \Omega^{-1} \epsilon = \frac{1}{\sigma_\nu^2} \epsilon^\top W \epsilon + \frac{1}{\sigma_\iota^2} \epsilon^\top B \epsilon$$

The determinant of Ω is the product of its eigenvalues, which are σ_ν^2 with periodicity $N(T-1)$ and σ_ι^2 with periodicity N. Then, taking the logarithm of Equation 6.7 and denoting $\theta^\top = (\sigma_\nu^2, \sigma_\eta^2)$, we get the following log-likelihood function:

$$\ln L(\theta) = \frac{NT}{2} \ln 2\pi - \frac{N(T-1)}{2} \ln \sigma_\nu^2 - \frac{N}{2} \ln(T\sigma_\eta^2 + \sigma_\nu^2) - \frac{\epsilon^\top W \epsilon}{2\sigma_\nu^2} - \frac{\epsilon^\top W \epsilon}{2(\sigma_\nu^2 + T\sigma_\eta^2)}$$

The gradient and the hessian are respectively:

$$g(\theta) = \begin{pmatrix} \frac{\partial \ln L}{\partial \sigma_\nu^2} \\ \frac{\partial \ln L}{\partial \sigma_\eta^2} \end{pmatrix} = \begin{pmatrix} -\frac{N(T-1)}{2\sigma_\nu^2} - \frac{N}{2\sigma_\iota^2} + \frac{\epsilon^\top W \epsilon}{2\sigma_\nu^4} + \frac{\epsilon^\top B_\eta \epsilon}{2\sigma_\iota^2} \\ -\frac{NT}{2\sigma_\iota^2} + \frac{T\epsilon^\top B \epsilon}{2\sigma_\iota^2} \end{pmatrix}$$

$$H(\theta) = \begin{pmatrix} -\frac{N(T-1)}{2\sigma_\nu^4} + \frac{N}{2\sigma_\iota^4} - \frac{\epsilon^\top W\epsilon}{\sigma_\nu^6} - \frac{\epsilon^\top B\epsilon}{\sigma_\iota^6} & \frac{NT}{2\sigma_\iota^4} - \frac{T\epsilon^\top B\epsilon}{\sigma_\iota^6} \\ \frac{NT}{2\sigma_\iota^4} - \frac{T\epsilon^\top B\epsilon}{\sigma_\iota^6} & \frac{NT^2}{2\sigma_\iota^4} - \frac{T^2\epsilon^\top B\epsilon}{\sigma_\iota^6} \end{pmatrix}$$

To compute the expectation of this matrix, we note that $\mathrm{E}(\epsilon^\top W_\eta \epsilon) = N(T-1)\sigma_\nu^2$ and $\mathrm{E}(\epsilon^\top B_\eta \epsilon) = N\sigma_\iota^2$:

$$\mathrm{E}(H(\theta)) = \begin{pmatrix} -\frac{N(T-1)}{2\sigma_\nu^4} - \frac{N}{2\sigma_\iota^4} & -\frac{NT}{2\sigma_\iota^4} \\ -\frac{NT}{2\sigma_\iota^4} & -\frac{NT^2}{2\sigma_\iota^4} \end{pmatrix}$$

To compute the test statistic, we impose the null hypothesis: $H_0 : \sigma_\eta^2 = 0$ (no individual effects), so that $\sigma_\iota^2 = \sigma_\nu^2$. In this case, the OLS estimator is BLUE and $\hat{\sigma}_\nu^2$ is $\hat{\epsilon}^\top\hat{\epsilon}/NT$. The estimated score and the information matrix are, with $\hat{\theta}^\top = (\hat{\sigma}_\nu^2, 0)$

$$\hat{g}(\hat{\theta}) = \begin{pmatrix} 0 \\ -\frac{NT}{2\hat{\sigma}_\nu^2}\left(\frac{\hat{\epsilon}^\top B\hat{\epsilon}}{N\hat{\sigma}_\nu^2} - 1\right) \end{pmatrix}$$

$$\hat{I}(\hat{\theta}) = \mathrm{E}\left(-H(\hat{\theta})\right) = \frac{NT}{2\hat{\sigma}_\nu^4}\begin{pmatrix} 1 & 1 \\ 1 & T \end{pmatrix}$$

and the inverse of the estimated information matrix is:

$$\hat{I}(\hat{\theta})^{-1} = \frac{2\hat{\sigma}_\nu^4}{NT(T-1)}\begin{pmatrix} T & -1 \\ -1 & 1 \end{pmatrix}$$

Finally, the test statistic is computed as the quadratic form: $\hat{g}(\hat{\theta})^\top \hat{I}(\hat{\theta})^{-1}\hat{g}(\hat{\theta})$ which simplifies to:

$$LM = \left(-\frac{NT}{2\hat{\sigma}_\nu^2}\left(\frac{\hat{\epsilon}^\top B\hat{\epsilon}}{N\hat{\sigma}_\nu^2} - 1\right)\right)^2 \times \frac{2\hat{\sigma}_\nu^4}{NT(T-1)} = \frac{NT}{2(T-1)}\left(\frac{\hat{\epsilon}^\top B\hat{\epsilon}}{N\hat{\sigma}_\nu^2} - 1\right)^2$$

Or, replacing $\hat{\sigma}_\nu^2$ by $\hat{\epsilon}^\top\hat{\epsilon}/NT$:

$$LM = \frac{NT}{2(T-1)}\left(T\frac{\hat{\epsilon}^\top B\hat{\epsilon}}{\hat{\epsilon}^\top\hat{\epsilon}} - 1\right)^2 \tag{6.8}$$

which is asymptotically distributed as a χ^2 with 1 degree of freedom.

Bonjour et al. (2003) estimated a Mincer equation with a sample of twins, the entity index being `family`. The response is the log of wage and the covariates are education (`educ`) and potential experience and its square (approximated by the age `age`):

```
lm_twins <- lm(log(earning) ~ poly(age, 2) + educ, twins)
lm_twins %>% coef
##   (Intercept) poly(age, 2)1 poly(age, 2)2          educ
##       1.03397       0.06237      -1.93282       0.07675
```

We then add the residuals to the data and we compute the individual mean of the residuals by grouping by entities and then using `mutate` and not `summarise` to compute the mean, $B\epsilon$ being a vector of length $N \times T$ where each value is returned T times:

```
twins <- twins %>% add_column(e = resid(lm_twins)) %>%
  group_by(family) %>%
  mutate(Be = mean(e)) %>% ungroup
```

We finally compute the statistic using Equation 6.8:

```
N_tw <- twins %>% pull(family) %>% unique %>% length
T_tw <- 2
twins %>%
  summarise(bp = 0.5 * T_tw * N_tw / (T_tw - 1)   *
              (T_tw * sum(Be ^ 2) / sum(e ^ 2) - 1) ^ 2) %>% pull
## [1] 4.222
```

The **plm** package (Croissant and Millo 2008, 2018) provides different tools to deal with panel or pseudo-panel data. In particular, Breusch and Pagan's test can easily be obtained using `plm::plmtest`. We set the `type` argument to `"bp"` to get the statistic of the original Breusch-Pagan test:

```
library(plm)
plmtest(log(earning) ~ poly(age, 2) + educ, twins, type = "bp") %>% gaze
## chisq = 4.222, df: 1, pval = 0.040
```

The absence of individual effects is rejected at the 5% level, but not at the 1% level.

The model fitted by Schaller (1990) is a simple linear model, the response being the rate of investment (`ikn`) and the unique covariate Tobin's Q (`qn`):

```
tobinq %>% print(n = 3)
```

```
# A tibble: 6,580 x 15
  cusip  year  isic    ikb   ikn    qb    qn kstock     ikicb  ikicn
  <int> <dbl> <int>  <dbl> <dbl> <dbl> <dbl>  <dbl>     <dbl>  <dbl>
1  2824  1951  2835 0.230  0.205  5.61 10.9    27.3 NA        0.0120
2  2824  1952  2835 0.0403 0.200  6.01 12.2    30.5  0.193    0.0245
3  2824  1953  2835 0.0404 0.110  4.19  7.41   31.7  0.00292 0.0976
# i 6,577 more rows
# i 5 more variables: omphi <dbl>, qicb <dbl>, qicn <dbl>, sb <dbl>,
#   sn <dbl>
```

The first two columns contain the firm and the time index. The Breusch-Pagan statistic is:

```
plmtest(ikn ~ qn, tobinq, type = "bp") %>% gaze
## chisq = 8349.686, df: 1, pval = 0.000
```

The statistic is huge, the hypothesis of no individual effects is therefore very strongly rejected, which is a quite customary result for panel data, especially when the time dimension is high, which is the case for the `tobinq` data (35 years).

6.2.3 System of equations

Breusch and Pagan (1980) also proposed a test of correlation between equations in a system of equation. Remember that in this case, the covariance matrix of the errors for the whole system is $\Omega = \Sigma \otimes I$, where Σ contains the variances (on the diagonal) and the covariances (off diagonal) of the errors of the L equations. This symmetric matrix contains $L \times (L+1)/2$ distinct elements, $L \times (L+1)/2 - L = L \times (L-1)/2$ being covariances. The Breusch-Pagan test is based on the estimation of the covariance matrix using OLS residuals. Denoting $\hat{\Xi} = (\hat{\epsilon}_1, \hat{\epsilon}_2, ... \hat{\epsilon}_L)$ the matrix where each column contains the vector of residuals for one equation:

$$\hat{\Omega} = \hat{\Xi}^\top \hat{\Xi} = \begin{pmatrix} \hat{\epsilon}_1^\top \hat{\epsilon}_1 & \hat{\epsilon}_1^\top \hat{\epsilon}_2 & \dots & \hat{\epsilon}_1^\top \hat{\epsilon}_L \\ \hat{\epsilon}_2^\top \hat{\epsilon}_1 & \hat{\epsilon}_2^\top \hat{\epsilon}_2 & \dots & \hat{\epsilon}_2^\top \hat{\epsilon}_L \\ \vdots & \vdots & \ddots & \vdots \\ \hat{\epsilon}_L^\top \hat{\epsilon}_1 & \hat{\epsilon}_L^\top \hat{\epsilon}_2 & \dots & \hat{\epsilon}_L^\top \hat{\epsilon}_L \end{pmatrix}$$

The coefficients of correlations are then estimated, denoting $\hat{\sigma}_l^2 = \hat{\epsilon}_l^\top \hat{\epsilon}_l / N$ the estimated variances:

$$\hat{\rho}_{lm} = \frac{\hat{\epsilon}_l^\top \hat{\epsilon}_m / N}{\hat{\sigma}_l \hat{\sigma}_m}$$

The statistic is then $N \sum_{l=1}^{L-1} \sum_{m=l+1}^{L} \hat{\rho}_{lm}^2$ and is a χ^2 with $L(L-1)/2$ degrees of freedom if the hypothesis of no correlation is true.

We use the apple production example estimated in Section 3.7. The estimation for the whole system by OLS taking constraints into account was stored in an object called `ols_const`. We extract the residuals and arrange them in a $N \times L$ matrix. Taking the cross-product of this matrix and dividing by N, we get $\hat{\Sigma}$:

```
N_ap <- nobs(ols_const) / 3
EPS <- ols_const %>% resid %>% matrix(ncol = 3)
Sigma <- crossprod(EPS) / N_ap
Sigma
```

```
          [,1]      [,2]      [,3]
[1,]  0.088417 -0.002245  0.001650
[2,] -0.002245  0.008088 -0.006070
[3,]  0.001650 -0.006070  0.007718
```

We then compute the estimated standard deviation of the errors for every equation $\hat{\sigma}_l$ and we divide $\hat{\Sigma}$ by a matrix containing the products of the standard deviations, using the `outer` function:

```
sig <- Sigma %>% stder
d <- Sigma / outer(sig, sig)
d
```

```
          [,1]     [,2]     [,3]
[1,]  1.00000 -0.08394  0.06315
[2,] -0.08394  1.00000 -0.76827
[3,]  0.06315 -0.76827  1.00000
```

so that we get a matrix with ones on the diagonal and coefficients of correlations off-diagonal. Then, we extract the off-diagonal elements using `upper.tri`, which returns a logical matrix with values of `TRUE` above the diagonal:

```
upper.tri(d)
```

```
      [,1]  [,2]  [,3]
[1,] FALSE  TRUE  TRUE
[2,] FALSE FALSE  TRUE
[3,] FALSE FALSE FALSE
```

Then, indexing `d` by `upper.tri(d)` returns a vector containing the three elements of the matrix that are above the diagonal:[3]

```
d[upper.tri(d)]
## [1] -0.08394  0.06315 -0.76827
```

Finally, we sum the squares of the elements of this vector and multiply by the sample size to get the statistic:

```
bp_stat <- sum( d[upper.tri(d)] ^ 2) * N_ap
pval_pb <- pchisq(bp_stat, df = 3, lower.tail = FALSE)
c(bp_stat, pval_pb)
## [1] 8.237e+01 9.497e-18
```

The hypothesis of no correlation is clearly rejected.

6.3 Robust inference

If the errors are not spherical, the simple estimator of the covariance of the OLS estimates is biased. More general estimators can then be used instead. These estimators use the residuals of the OLS estimator which, in the context of this chapter, is an inefficient but consistent estimator. We'll present the robust estimator of the covariance matrix of the OLS estimates first in the context of the simple linear model and then for the multiple linear model.

[3] As the matrix is symmetric, `lower.tri` could also have been used.

6.3.1 Simple linear model

The variance of the slope estimated by OLS is:

$$\sigma^2_{\hat{\beta}} = \mathrm{V}(\hat{\beta} \mid x) = \frac{\mathrm{E}\left(\left[\sum_n (x_n - \bar{x})\epsilon_n\right]^2 \mid x\right)}{\left(\sum_n (x_n - \bar{x})^2\right)^2}$$

The numerator is the sum of the expectations of N^2 terms. For $N = 4$, replacing the errors ϵ_n by the OLS residuals $\hat{\epsilon}_n$ and dropping the expectation operator, these 16 terms can be presented conveniently in the following matrix:

$$\left(\begin{array}{cccc}
(x_1 - \bar{x})^2\hat{\epsilon}_1^2 & (x_1 - \bar{x})(x_2 - \bar{x})\hat{\epsilon}_1\hat{\epsilon}_2 & (x_1 - \bar{x})(x_3 - \bar{x})\hat{\epsilon}_1\hat{\epsilon}_3 & (x_1 - \bar{x})(x_4 - \bar{x})\hat{\epsilon}_1\hat{\epsilon}_4 \\
(x_2 - \bar{x})(x_1 - \bar{x})\hat{\epsilon}_2\hat{\epsilon}_1 & (x_2 - \bar{x})^2\hat{\epsilon}_2^2 & (x_2 - \bar{x})(x_3 - \bar{x})\hat{\epsilon}_2\hat{\epsilon}_3 & (x_2 - \bar{x})(x_4 - \bar{x})\hat{\epsilon}_2\hat{\epsilon}_4 \\
(x_3 - \bar{x})(x_1 - \bar{x})\hat{\epsilon}_3\hat{\epsilon}_1 & (x_3 - \bar{x})(x_2 - \bar{x})\hat{\epsilon}_3\hat{\epsilon}_2 & (x_3 - \bar{x})^2\hat{\epsilon}_3^2 & (x_3 - \bar{x})(x_4 - \bar{x})\hat{\epsilon}_3\hat{\epsilon}_4 \\
(x_4 - \bar{x})(x_1 - \bar{x})\hat{\epsilon}_4\hat{\epsilon}_1 & (x_4 - \bar{x})(x_2 - \bar{x})\hat{\epsilon}_4\hat{\epsilon}_2 & (x_4 - \bar{x})(x_3 - \bar{x})\hat{\epsilon}_4\hat{\epsilon}_3 & (x_4 - \bar{x})^2\hat{\epsilon}_4^2
\end{array}\right)$$

The robust estimator is obtained by taking the sum of *some* of this terms. Note first that the sum of all these terms is $\left(\sum_{n=1}^N (x_n - \bar{x})\hat{\epsilon}_n\right)^2$, which is equal to zero as: $\sum_{n=1}^N (x_n - \bar{x})\hat{\epsilon}_n = 0$. Therefore, it is not relevant to sum *all* the terms of this matrix to get an estimator of the variance of $\hat{\beta}$. The first possibility is to take only the diagonal terms of this matrix, which is relevant if we maintain the hypotheses that the errors are uncorrelated. In this case, we get the so-called **heteroskedastic-consistent (HC)** estimator of $\sigma_{\hat{\beta}}$ proposed by White (1980):

$$\hat{\sigma}^2_{\mathrm{HC}\hat{\beta}} = \frac{1}{S^2_{xx}} \sum_{n=1}^N (x_n - \bar{x})^2 \hat{\epsilon}_n^2 \tag{6.9}$$

Consider now the case where some errors are correlated. This often happens when some observations share some common unobserved characteristics which are included in their (therefore correlated) errors. For example, if observations belong to different regions, their errors may share some common unobserved features of the regions. In our four observations case, suppose that the first two observations belong to one group, and the two others to another group. Then, a consistent estimator is obtained by summing the following subset of elements of the preceding matrix:

$$\left(\begin{array}{cccc}
(x_1 - \bar{x})^2\hat{\epsilon}_1^2 & (x_1 - \bar{x})(x_2 - \bar{x})\hat{\epsilon}_1\hat{\epsilon}_2 & - & - \\
(x_2 - \bar{x})(x_1 - \bar{x})\hat{\epsilon}_2\hat{\epsilon}_1 & (x_2 - \bar{x})^2\hat{\epsilon}_2^2 & - & - \\
- & - & (x_3 - \bar{x})^2\hat{\epsilon}_3^2 & (x_3 - \bar{x})(x_4 - \bar{x})\hat{\epsilon}_3\hat{\epsilon}_4 \\
- & - & (x_4 - \bar{x})(x_3 - \bar{x})\hat{\epsilon}_4\hat{\epsilon}_3 & (x_4 - \bar{x})^2\hat{\epsilon}_4^2
\end{array}\right)$$

which leads to the clustered estimated variance. More generally, for N observations belonging to G groups, this estimator is:

$$\hat{\sigma}^2_{\mathrm{CL}\hat{\beta}} = \frac{1}{S^2_{xx}} \sum_{g=1}^G \left(\sum_{n \in g} (x_n - \bar{x})\hat{\epsilon}_n\right)^2 \tag{6.10}$$

which is consistent with the hypothesis that errors are correlated *within* a group, but uncorrelated *between* groups. To illustrate the computation of robust covariance estimators, we use the data set `urban_gradient` of Duranton and Puga (2020). It contains the population, the area and the distance to the central business district for 2315 block groups[4] in Alabama.[5]

```
urban_gradient %>% print(n = 5)
```

```
# A tibble: 2,315 x 7
  msa        county  tract  blkg  area population distance
  <chr>      <chr>   <dbl> <dbl> <dbl>      <dbl>    <dbl>
1 Montgomery Autauga 20100     1  4.24        530     19.6
2 Montgomery Autauga 20100     2  5.57       1282     20.8
3 Montgomery Autauga 20200     1  2.06       1274     19.3
4 Montgomery Autauga 20200     2  1.28        944     18.0
5 Montgomery Autauga 20300     1  3.87       2538     18.2
# i 2,310 more rows
```

A classic model in urban economics states that urban density is a negative exponential function of the distance to the central business district: $y = Ae^{\beta x}$ where y is measured in inhabitants per square kilometers, x is measured in kilometers and $\beta < 0$ is called the urban gradient. Taking logs, this leads to a semi-log linear regression model:

$$\ln y_n = \alpha + \beta x_n + \epsilon_n$$

We first compute the `density` variable and then estimate the urban gradient model.

```
urban_gradient <- urban_gradient %>% mutate(density = population / area)
ols_ug <- lm(log(density) ~ distance, urban_gradient)
ols_ug %>% gaze
##           Estimate Std. Error t value Pr(>|t|)
## distance -0.07057    0.00182   -38.9   <2e-16
```

The estimated standard deviation of the slope is 0.0018, but it may be seriously biased if the errors are heteroskedastic and/or correlated. We first plot the data and the regression line in Figure 6.1, the shape of the points depending on the metropolitan statistical area (MSA, there are 12 of them in the data set).

```
urban_gradient %>% ggplot(aes(distance, log(density))) +
  geom_point(aes(shape = msa), size = .3) +
  geom_smooth(method = "lm", se = FALSE, color = "black") +
  scale_shape_manual(values = c(3, 16, 17, 8, 5, 2, 1, 4, 6, 9, 0, 12))
```

[4]A block group is a geographical unit which is between the census tract and the census block.

[5]Actually, the whole data set covers the whole United States, but we use here the small subsample that concerns the state of Alabama.

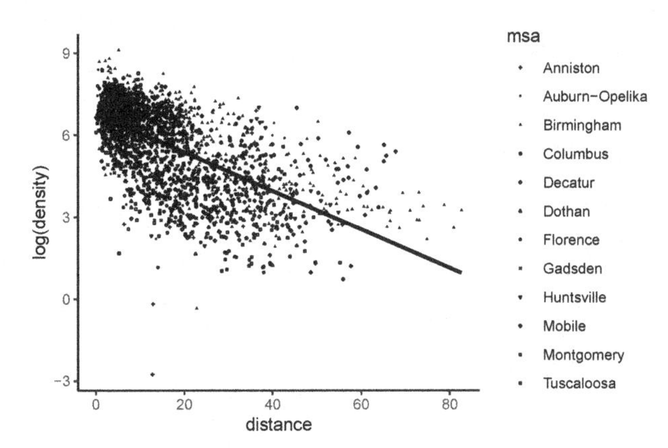

Figure 6.1: Data and regression line for the **urban_gradient** data set

Heteroskedasticity seems to be present in this data set, as the size of the residuals seems to be an increasing function of the unique covariate. We first compute the HC standard deviation of the slope, computing the mean of the covariate and S_{xx} and then applying Equation 6.9:

```
dist_mean <- urban_gradient %>% pull(distance) %>% mean
Sxx <- sum( (pull(urban_gradient, distance) - dist_mean) ^ 2)
sd_hc <- urban_gradient %>%
  summarise( sqrt(sum( (distance - dist_mean) ^ 2 *
                         resid(ols_ug) ^ 2) / Sxx ^ 2)) %>%
  pull
sd_hc
## [1] 0.002007
```

In this example, the heteroskedastic-robust standard error is just slightly higher than the one computed using the simple formula. However, we also have to investigate the potential correlation between the errors of some observations. There are 12 MSA and 22 counties in Alabama. It is possible that errors for block groups of the same county or of the same MSA are correlated (because of some unobserved common features of block groups in the same county or MSA). We compute the estimation of the clustered standard deviation of the slope (Equation 6.10) at the MSA level:

```
G <- urban_gradient %>% pull(msa) %>% unique %>% length
sd_CL <- urban_gradient %>%
  add_column(eps = resid(ols_ug)) %>%
  group_by(msa) %>%
  summarise(z = sum( (distance - dist_mean) * eps) ^ 2) %>%
  summarise(sd = sqrt(sum(z)) / Sxx) %>%
  pull
sd_CL
## [1] 0.006035
```

This time, we get a much higher estimate of the standard deviation of the slope (about three times larger than the one obtained with the simple formula).

6.3.2 Multiple linear model

Consider now the multiple regression model. The vector of slopes can be written as a linear combination of the response and the error vectors:

$$\begin{aligned}\hat{\beta} &= (\tilde{X}^\top\tilde{X})^{-1}\tilde{X}^\top\tilde{y} \\ &= (\tilde{X}^\top\tilde{X})^{-1}\tilde{X}^\top(\tilde{X}\beta+\epsilon) \\ &= \beta + (\tilde{X}^\top\tilde{X})^{-1}\tilde{X}^\top\epsilon\end{aligned}$$

$\tilde{X}^\top\epsilon$ is a K-length vector containing the product of the covariates (the column of X) in deviation from their sample mean and the vector of errors:

$$\tilde{X}^\top\epsilon = \begin{pmatrix} \sum_{n=1}^N(x_{1n}-\bar{x}_1)\epsilon_n \\ \sum_{n=1}^N(x_{2n}-\bar{x}_2)\epsilon_n \\ \vdots \\ \sum_{n=1}^N(x_{1n}-\bar{x}_K)\epsilon_n \end{pmatrix} = \sum_{n=1}^N \psi_n$$

ψ_n is called the vector of score, as it is proportional to the vector of the first derivatives of the sum of square residuals and is therefore equal to 0 when evaluated for $\hat{\beta}$, the OLS estimator. The general form of the covariance of the OLS estimates was given in Equation 3.16:

$$\hat{V}(\hat{\beta}) = \frac{1}{N}\left(\frac{1}{N}\tilde{X}^\top\tilde{X}\right)^{-1}\frac{1}{N}\mathrm{E}(\tilde{X}^\top\epsilon\epsilon^\top\tilde{X} \mid X)\left(\frac{1}{N}\tilde{X}^\top\tilde{X}\right)^{-1}$$

This is a sandwich formula, the meat (the variance of the score) being surrounded by two slices of bread (the inverse of the covariance matrix of the covariates). Remember from Section 3.5.2 that the meat can be written, for $K = 2$, as the expected value of:

$$\frac{1}{N}\begin{pmatrix} \left(\sum_{n=1}^N(x_{1n}-\bar{x}_1)\epsilon_n\right)^2 & \left(\sum_{n=1}^N(x_{1n}-\bar{x}_1)\epsilon_n\right)\left(\sum_{n=1}^N(x_{2n}-\bar{x}_2)\epsilon_n\right) \\ \left(\sum_{n=1}^N(x_{1n}-\bar{x}_1)\epsilon_n\right)\left(\sum_{n=1}^N(x_{2n}-\bar{x}_2)\epsilon_n\right) & \left(\sum_{n=1}^N(x_{2n}-\bar{x}_2)\epsilon_n\right)^2 \end{pmatrix}$$

If the errors are uncorrelated, but potentially heteroskedastic, we use the following estimation:

$$\frac{1}{N}\sum_{n=1}^N \hat{\epsilon}_n^2 \begin{pmatrix} (x_{1n}-\bar{x}_1)^2 & (x_{1n}-\bar{x}_1)(x_{2n}-\bar{x}_2) \\ (x_{1n}-\bar{x}_1)(x_{2n}-\bar{x}_2) & (x_{2n}-\bar{x}_2)^2 \end{pmatrix}$$

which generalizes the scalar case of the heteroskedastic covariance matrix computed for the single regression case. This estimator of the variance of the score can easily be obtained by defining the **estimating function** F which is obtained by multiplying (elements by elements) the columns of the matrix of covariates (in deviation from the sample means) by the vector of residuals:

$$F = \begin{pmatrix} \hat{\epsilon}_1 x_{11} & \dots & \hat{\epsilon}_1\tilde{x}_{1K} \\ \hat{\epsilon}_2 x_{21} & \dots & \hat{\epsilon}_2\tilde{x}_{2K} \\ \vdots & \ddots & \vdots \\ \hat{\epsilon}_N x_{N1} & \dots & \hat{\epsilon}_N\tilde{x}_{NK} \end{pmatrix}$$

Then $F^\top F = \sum_{n=1}^N \hat{\epsilon}_n^2 \tilde{x}_n \tilde{x}_n^\top$ is N times the HC estimator of the meat and the HC estimator is then:

$$\hat{V}(\hat{\beta}) = \frac{1}{N}\left(\frac{1}{N}\tilde{X}^\top \tilde{X}\right)^{-1}\left(\frac{1}{N}\sum_{n=1}^N \hat{\epsilon}_n^2 \tilde{x}_n \tilde{x}_n^\top\right)\left(\frac{1}{N}\tilde{X}^\top \tilde{X}\right)^{-1} \tag{6.11}$$

To get the clustered estimator of the variance, we define for each cluster $\psi_g = \sum_{n\in g} \psi_n$, and the clustered expression of the meat is obtained as the sum of the outer products of ψ_g divided by N:

$$\frac{1}{N}\sum_{g=1}^G \hat{\psi}_g \hat{\psi}_g^\top \tag{6.12}$$

Going back to the electricity consumption estimation, `lm_elec` is the model fitted by OLS. We first compute the estimator manually, computing the meat (the cross-products of the model matrix multiplied by the vector of residuals divided by the sample size) and the bread (the inverse of the cross-products of the model matrix divided by the sample size). The covariance matrix is then computed using Equation 6.11, and we extract the standard errors of the estimates:

```
N_el <- nobs(lm_elec)
Z <- model.matrix(lm_elec)
M <- crossprod(resid(lm_elec) * Z) / N_el
B <- solve(crossprod(Z) / N_el)
(B %*% M %*% B / N_el) %>% stder
## (Intercept)          inc    I(1/mc6)        gas6        cap
##     458.5529       0.2012    157.5829     31.5546    91.6099
```

The **sandwich** package (Zeileis 2004, 2006; Zeileis, Köll, and Graham 2020.), provides specialized functions to construct the different pieces of the estimator, namely `estfun`, `meat` and `bread`:

```
library(sandwich)
F <- estfun(lm_elec)
M <- meat(lm_elec)
B <- bread(lm_elec)
(B %*% M %*% B / N_el) %>% stder
## (Intercept)          inc    I(1/mc6)        gas6        cap
##     458.5529       0.2012    157.5829     31.5546    91.6099
```

`sandwich::vcovHC` uses these functions to compute the HC estimator; its first argument is a fitted model and a `type` argument can also be supplied. Setting `type` to `"HC0"` gives the simplest version of the estimator, the one we have previously computed "by hand". Other flavors of this heteroskedasticity-robust estimator can be obtained by setting the `type` to `"HC1"`, ..., `"HC5"` to perform different flavors of degrees of freedom correction. For example, when `type = "HC1"` (which is the default), the covariance matrix is multiplied by $N/(N-K)$:[6]

[6]See Zeileis (2006) for more details concerning the other values of the `type` argument.

```
vcovHC(lm_elec, type = "HC0") %>% stder
## (Intercept)          inc     I(1/mc6)        gas6         cap
##    458.5529       0.2012     157.5829     31.5546     91.6099
```

Comparing the standard and the robust estimation of the standard deviations of the estimates, we get:

```
robust_sd <- vcovHC(lm_elec, type = "HC0") %>% stder
standard_sd <- vcov(lm_elec) %>% stder
rbind(standard = standard_sd, robust = robust_sd,
      ratio = robust_sd / standard_sd)
```

```
         (Intercept)    inc I(1/mc6)    gas6     cap
standard    498.0150 0.1819 164.8722 34.2927 98.6943
robust      458.5529 0.2012 157.5829 31.5546 91.6099
ratio         0.9208 1.1060   0.9558  0.9202  0.9282
```

In this example, the robust standard errors are very close to the ones obtained using the simple formula, although the Breusch-Pagan test rejected the hypothesis of homoskedasticity. This is because the source of heteroskedasticity (the number of customers) is not a covariate of the model.

To illustrate the use of the clustered sandwich estimator, we use the `twins` data set. `lm_twins` is the OLS estimation of the Mincer equation. In this context, the meat can easily be computed using the `apply` and `tapply` functions:

- `apply` performs an operation on one of the margins (1 for rows, 2 for columns) of a matrix,
- `tapply` performs an operation (the third argument) on a vector (the first argument) conditional on the values of another vector (the second argument).

```
Z <- model.matrix(lm_twins)
e <- resid(lm_twins)
id <- twins$family
Ze <- Z * e
M <- crossprod(apply(Ze, 2, tapply, id, sum)) / N_tw
B <- solve(crossprod(Z) / N_tw)
V_1 <- (B %*% M %*% B / N_tw)
```

We can more simply use the `sandwich::vcovCL` function (Zeileis, Köll, and Graham 2020) to compute the clustered estimator. The clustering variable is defined using the `cluster` argument that can be set to a one-sided formula. The `type` argument is similar to the one of `vcovHC` and there is also a `cadjust` argument which, if `TRUE`, multiplies the covariance matrix by $G/(G-1)$, G being the number of clusters. The default behavior of `vcovCL` is to set `cadjust` to `TRUE` and `type` to `"HC1"`, so that the adjustment is done for the number of observations and for the number of groups. For the `twins` data set, the clustering variable is `family`:

```
V_2 <- vcovCL(lm_twins, ~ family, type = "HC0", cadjust = FALSE)
```

Finally, `twins` being a pseudo-panel, the **plm** package can also be used. The OLS estimate can be fitted with `plm::plm` by setting the `model` argument to `"pooling"`.

```
plm_twins <- plm(log(earning) ~ poly(age, 2) + educ, twins,
                 model = "pooling")
```

A `plm` object is returned and the `vcovHC` method for a `plm` object returns by default the same clustered covariance matrix as previously:

```
V_3 <- vcovHC(plm_twins)
```

The `vcovHC` and `vcovCL` functions are particularly useful while using the testing functions of the **lmtest** package. For example, `lmtest::coeftest` computes the usual table of coefficients (the same as the one obtained using `summary`), but a matrix or a function that computes a covariance matrix can be passed as a supplementary argument. Therefore, `lmtest::coeftest(lm_twins)` returns exactly the same table of coefficients as `summary(lm_twins)`, but other standard errors are obtained by filling the second argument. For example, to get the clustered standard errors with our preferred specification without degrees of freedom correction (`type = "HC0", cadjust = FALSE`), we can use any of the following three equivalent syntaxes:

```
cl <- vcovCL(lm_twins, ~ family, cadjust = FALSE, type = "HC0")
library(lmtest)
coeftest(lm_twins, cl)
coeftest(lm_twins, function(x)
  vcovCL(x, ~ family, cadjust = FALSE, type = "HC0"))
coeftest(lm_twins, vcovCL, cluster = ~ family,
         cadjust = FALSE, type = "HC0")
```

In the first expression, we provide a matrix that was previously computed. In the second expression, we use an anonymous function with our preferred options. In the last one, the two arguments of `vcovCL` are indicated as arguments of `coeftest` and are passed internally to `vcovCL`. We finally compare the ordinary and the robust estimates of the standard errors of the OLS estimate:

```
ord_se <- vcov(lm_twins) %>% stder
cl_se <- V_1 %>% stder
rbind(ordinary = ord_se, cluster = cl_se,
ratio = cl_se / ord_se)
```

```
         (Intercept) poly(age, 2)1 poly(age, 2)2    educ
ordinary      0.1516        0.5451        0.5309 0.01059
cluster       0.1620        0.5744        0.6000 0.01103
ratio         1.0685        1.0539        1.1300 1.04106
```

As for the electricity consumption example, we can see that the robust standard errors are almost the same as those computed using the simple formula.

6.4 Generalized least squares estimator

For a linear model $y_n = \alpha + \beta^\top x_n + \epsilon_n = \gamma^\top z_n + \epsilon_n$ with non-spherical disturbances ($\Omega \neq \sigma_\epsilon^2 I$), the OLS estimator is no longer BLUE. A more efficient estimator called the **generalized least squares (GLS)** can then be used instead.

6.4.1 General formulation of the GLS estimator

The GLS estimator is:

$$\hat{\gamma} = (Z^\top \Omega^{-1} X)^{-1} Z^\top \Omega^{-1} y \tag{6.13}$$

Replacing y by $Z\gamma + \epsilon$, we get:

$$\hat{\gamma} = (Z^\top \Omega^{-1} Z)^{-1} Z^\top \Omega^{-1} (Z\gamma + \epsilon) = \gamma + (Z^\top \Omega^{-1} Z)^{-1} Z^\top \Omega^{-1} \epsilon$$

As for the OLS estimator, the estimator is unbiased if $\mathrm{E}(\epsilon \mid Z) = 0$. The variance is:

$$\begin{array}{rcl} \mathrm{V}(\hat{\gamma}) &=& \mathrm{E}\left[(Z^\top \Omega^{-1} Z)^{-1} Z^\top \Omega^{-1} \epsilon\epsilon^\top \Omega^{-1} Z (Z^\top \Omega^{-1} Z)^{-1}\right] \\ &=& (Z^\top \Omega^{-1} Z)^{-1} Z^\top \Omega^{-1} \mathrm{E}(\epsilon\epsilon^\top) \Omega Z (Z^\top \Omega^{-1} Z)^{-1} \\ &=& (Z^\top \Omega^{-1} Z)^{-1} \end{array}$$

Written this way, the GLS estimator is unfeasible for two reasons:

- the Ω matrix is a square matrix of dimension $N \times N$, and it is computationally difficult (or impossible) to store and to invert for large samples,
- it uses a matrix Ω which contains $N(N+1)/2$ unknown parameters.

A **feasible GLS** estimator is obtained by imposing some structure on Ω so that the number of unknown parameters becomes much less than $N(N+1)/2$ and by estimating these unknown parameters using residuals of a first step consistent estimation. Actually, in practice, the GLS estimator is obtained by performing OLS on transformed data. More precisely, consider the matrix C such that $C^\top C = \Omega^{-1}$. Then, Equation 6.13 can be rewritten, denoting $w^* = Cw$:

$$\hat{\gamma} = (Z^\top C^\top C Z)^{-1} Z^\top C^\top C y = \left(Z^{*\top} Z^*\right)^{-1} Z^{*\top} y^* \tag{6.14}$$

which is the OLS estimator of the linear model: $y^* = Z^* \gamma + \epsilon^*$, with $\epsilon^* = C\epsilon$. Replacing y^* by $Z^*\gamma + \epsilon^*$ in Equation 6.14, we get:

$$\hat{\gamma} = \gamma + \left(Z^{*\top} Z^*\right)^{-1} Z^{*\top} C\epsilon$$

And the variance of the estimator is:

$$\mathrm{V}(\hat{\gamma}) = \left(Z^{*\top} Z^*\right)^{-1} Z^{*\top} C \Omega^{-1} C^\top Z^* \left(Z^{*\top} Z^*\right)^{-1} \tag{6.15}$$

But $C\Omega^{-1}C^\top = C(C^\top C)^{-1}C^\top = CC^{-1}C^{\top-1}C^\top = I$[7], and therefore, Equation 6.15 simplifies to:

$$\mathrm{V}(\hat{\gamma}) = \left(Z^{*\top}Z^*\right)^{-1}$$

which is very similar to the formula used for the OLS estimator. Note that σ_ϵ^2 doesn't appear in this formula because the variance of the transformed errors is 1.

6.4.2 Weighted least squares

With heteroskedastic, but uncorrelated errors, Ω is diagonal and each element is the specific variance of one observation. From Equation 6.1, it is obvious that the transformation matrix C can be written as:

$$C = \begin{pmatrix} 1/\sigma_1 & 0 & 0 & \ldots & 0 \\ 0 & 1/\sigma_2 & 0 & \ldots & 0 \\ \vdots & \vdots & \vdots & \ddots & \vdots \\ 0 & 0 & 0 & \ldots & 1/\sigma_N \end{pmatrix}$$

and therefore, premultiplying any vector by C leads to a transformed vector where each value is divided by the standard deviation of the corresponding error: $z^{*\top} = (z_1/\sigma_1, z_2/\sigma_2, \ldots, z_N/\sigma_N)$. Performing OLS on the transformed data, we get the **weighted-least square (WLS)** estimator. The name of this estimator comes from the fact that the estimator can be obtained by minimizing $\sum_n \epsilon_n^2/\sigma_n^2$, i.e., by minimizing not the sum of the squares of the residuals, but a linear combination of the squares of the residuals, the weight of each observation being $1/\sigma_n^2$. Therefore, an observation n for which σ_n^2 is high will receive a smaller weight in WLS compared to OLS. The weights are unknown and therefore need to be estimated. The simplest solution is to assume that the variance (or the standard deviation) of the errors is proportional to an observed variable (which may or may not be a covariate of the regression). We then get either $\sigma_n^2 = \sigma^2 w_n$ or $\sigma_n^2 = \sigma^2 w_n^2$ and the weights are then respectively $1/w_n$ or $1/w_n^2$. The WLS estimator can then be obtained by OLS with all the variables divided either by $\sqrt{w_n}$ or w_n. A more general solution is to assume a functional form for the skedastic function: $\sigma_n^2 = h(\delta^\top w_n)$ where h is a monotonous increasing function that returns only positive values, w is a set of covariates and δ a vector of parameters. If, for example, h is the exponential function (which is a very common choice), the skedastic function is: $\ln \sigma_n^2 = \delta^\top w_n$ and δ can be consistently estimated by performing the following regression:

$$\ln \hat{\epsilon}_n^2 = \delta^\top w_n + \nu_n \tag{6.16}$$

where $\hat{\epsilon}$ are the OLS residuals which are consistent estimates of the errors. The WLS estimator is then performed in three steps:

- estimate the model by OLS and retrieve the vector of residuals $\hat{\epsilon}$,
- estimate $\hat{\gamma}$ by using OLS on Equation 6.16 and compute $\hat{\sigma}_n^2 = e^{\hat{\gamma} w_n}$,
- divide every variable (the response and the covariates) by $\hat{\sigma}_n$ and perform OLS on the transformed variables.

[7] $(AB)^{-1} = B^{-1}A^{-1}$ if the inverse of the two matrices exists, see Greene (2018), online appendix p 1074.

Note that there is no intercept in the third estimation as the "covariate" associated with the intercept (a vector of 1) becomes a vector with typical element $1/\hat{\sigma}_n$.

For the electricity consumption example, we know that the unconditional variance of y in city n is $\sigma^2_{yn} = \sigma^2_c/I_n$, I_n being the number of consumption units in city n and σ^2_c the variance of the individual consumption. Assuming that the same relation applies for the conditional variance, then $\sigma^2_{\epsilon n} = \sigma^2/I_n$ and the WLS estimator can then be obtained by computing OLS on series multiplied by $\sqrt{I_n}$:

```
wls_elec <- lm(I(kwh * sqrt(cust)) ~ sqrt(cust) + I(inc * sqrt(cust)) +
                 I(1 / mc6 * sqrt(cust)) +  I(gas6 * sqrt(cust)) +
                 I(cap * sqrt(cust)) - 1, uk_elec)
```

Or more simply by setting the `weights` argument of `lm` to `cust`:

```
wls_elec2 <- lm(kwh ~ inc + I(1 / mc6) +  gas6 + cap, uk_elec,
                weights = cust)
```

Comparing the robust standard errors of the OLS estimator and those of the WLS estimator, we get:

```
std_ols <- vcovHC(lm_elec) %>% stder
std_wls <- vcov(wls_elec2) %>% stder
tibble(ols = std_ols, wls = std_wls, ratio = wls / ols)
```

```
# A tibble: 5 x 3
      ols     wls ratio
    <dbl>   <dbl> <dbl>
1 535.    310.    0.581
2   0.256   0.201 0.784
3 195.    125.    0.639
4  37.9    21.2   0.559
5 120.     61.9   0.515
```

The efficiency gain of using WLS is substantial, as the standard errors reduce by about 25$-$50% depending on the coefficient.

6.4.3 Error component model

The error component model is suitable for panel data or pseudo panel data. In the remainder of this section, we'll mention individual means, which is the proper term for panel data, but should be replaced by entity means for pseudo-panel data. Remember that for the error component model, Equation 6.3 is a general formula that can be used to compute any power of Ω, and C is in this context obtained by taking $r = -0.5$:

$$C = \Omega^{-0.5} = \frac{1}{\sigma_\nu}W + \frac{1}{\sigma_\iota}B = \frac{1}{\sigma_\nu}\left(W + \frac{\sigma_\nu}{\sigma_\iota}B\right) \tag{6.17}$$

As $W = I - B$, C can also be rewritten as:

$$C = \Omega^{-0.5} = \frac{1}{\sigma_\nu}\left(I - \left[1 - \frac{\sigma_\nu}{\sigma_\iota}\right]B\right) = \frac{1}{\sigma_\nu}(I - \theta B) \tag{6.18}$$

with $\theta = 1 - \frac{\sigma_\nu}{\sigma_\iota}$. θ can be further written as:

$$\theta = 1 - \frac{\sigma_\nu}{\sqrt{T\sigma_\eta^2 + \sigma_\nu^2}} = 1 - \frac{1}{\sqrt{T\sigma_\eta^2/\sigma_\nu^2 + 1}}$$

Therefore $0 \leq \theta \leq 1$, so that the $\sigma_\nu C$ matrix performs, in this context, a quasi-difference from the individual mean:

$$z_n^* = z_{nt} - \theta \bar{z}_{n.}$$

The share of the individual mean that is subtracted depends on:

- the relative weights of the two variances: $\theta \rightarrow 0$ when $\sigma_\eta^2/\sigma_\nu^2 \rightarrow 0$, which means that there are no individual effects. As $\sigma_\eta^2/\sigma_\nu^2 \rightarrow +\infty$, $\theta \rightarrow 1$ and the transformation is the difference from the individual mean;
- the number of observations for each individual, $\theta \rightarrow 1$ when $T \rightarrow +\infty$; therefore the transformation is close to a difference from individual mean for large T.

σ_ν and σ_η are unknown parameters and have to be estimated. Consider the errors of the model ϵ_{nt}, their individual mean $\bar{\epsilon}_{n.}$ and the deviations from these individual means $\epsilon_{nt} - \bar{\epsilon}_{n.}$. By hypothesis, we have: $\mathrm{V}(\epsilon_{nt}) = \sigma_\nu^2 + \sigma_\eta^2$. For the individual means, we get:

$$\bar{\epsilon}_{n.} = \frac{1}{T}\sum_{t=1}^T \epsilon_{nt} = \eta_n + \frac{1}{T}\sum_{t=1}^T \nu_{nt}$$

for which the variance is:

$$\mathrm{V}(\bar{\epsilon}_{n.}) = \sigma_\eta^2 + \frac{1}{T}\sigma_\nu^2 = \sigma_\iota^2/T$$

The variance of the deviation from the individual means is easily obtained by isolating terms in ϵ_{nt}:

$$\epsilon_{nt} - \bar{\epsilon}_{n.} = \epsilon_{nt} - \frac{1}{T}\sum_{t=1}^T \epsilon_{nt} = \left(1 - \frac{1}{T}\right)\epsilon_{nt} - \frac{1}{T}\sum_{s \neq t}\epsilon_{ns}$$

The variance is, noting that the sum now contains $T-1$ terms:

$$\mathrm{V}(\epsilon_{nt} - \bar{\epsilon}_{n.}) = \left(1 - \frac{1}{T}\right)^2 \sigma_\nu^2 + \frac{1}{T^2}(T-1)\sigma_\nu^2 = \frac{T-1}{T}\sigma_\nu^2$$

If ϵ were known, natural estimators of these two variances σ_ι^2 and σ_ν^2 would be:

$$\hat{\sigma}_\iota^2 = T\frac{\sum_{n=1}^N \bar{\epsilon}_{n.}^2}{N} = T\frac{\sum_{n=1}^N \sum_{t=1}^T \bar{\epsilon}_{n.}^2}{NT} = T\frac{\epsilon^\top B \epsilon}{NT} = \frac{\epsilon^\top B \epsilon}{N} \tag{6.19}$$

$$\hat{\sigma}_\nu^2 = \frac{T}{T-1}\frac{\sum_{n=1}^N \sum_{t=1}^T (\epsilon_{nt} - \bar{\epsilon}_{n.})^2}{NT} = \frac{\sum_{n=1}^N \sum_{t=1}^T (\epsilon_{nt} - \bar{\epsilon}_{n.})^2}{N(T-1)} = \frac{\epsilon^\top W \epsilon}{N(T-1)} \quad (6.20)$$

Several estimators of the two components of the variance have been proposed in the literature. They all consist of replacing ϵ_{nt} in the previous two equations by consistent estimates (and for some of them by applying some degrees of freedom correction). The estimator proposed by Wallace and Hussain (1969) is particularly simple because it uses the residuals of the OLS estimation.[8]

The residuals of the OLS regression were already added to the `twins` data set and $B\hat{\epsilon}$ was computed. We then compute $W\hat{\epsilon} = \hat{\epsilon} - B\hat{\epsilon}$ and we use Equation 6.19 and Equation 6.20 to compute $\hat{\sigma}_\iota^2$, $\hat{\sigma}_\nu^2$, $\theta = 1 - \hat{\sigma}_\nu / \hat{\sigma}_\iota$ and then $\hat{\sigma}_\eta^2 = (\hat{\sigma}_\iota^2 - \hat{\sigma}_\nu^2)/T$.

```
twins <- twins %>%
  mutate(We = e - Be)
sigs <- twins %>%
  summarise(s2iota = sum(Be ^ 2) / N_tw,
            s2nu = sum(We ^ 2) / (N_tw * (T_tw - 1))) %>%
  mutate(s2eta = (s2iota - s2nu) / T_tw,
         theta = 1 - sqrt(s2nu / s2iota))
sigs
## # A tibble: 1 x 4
##   s2iota  s2nu  s2eta theta
##    <dbl> <dbl>  <dbl> <dbl>
## 1  0.316 0.238 0.0389 0.132
```

The `plm` function performs the GLS estimation, i.e., an OLS regression on data transformed using quasi-differences ($w^*_{nt} = w_{nt} - \theta\bar{w}_{n.}$) if the `model` argument is set to `"random"`:

```
gls_twins <- plm(log(earning) ~ poly(age, 2) + educ, twins,
                 random.method = "walhus", model  = "random")
summary(gls_twins)
```

```
Oneway (individual) effect Random Effect Model
   (Wallace-Hussain's transformation)

Call:
plm(formula = log(earning) ~ poly(age, 2) + educ, data = twins,
    model = "random", random.method = "walhus")

Balanced Panel: n = 214, T = 2, N = 428

Effects:
                 var std.dev share
idiosyncratic 0.2380  0.4878  0.86
individual    0.0389  0.1972  0.14
theta: 0.132
```

[8]See also Swamy and Arora (1972), Amemiya (1971) and Nerlove (1971).

```
Residuals:
   Min. 1st Qu.  Median 3rd Qu.    Max.
-2.9921 -0.2492 -0.0325  0.1922  2.3954

Coefficients:
                Estimate Std. Error z-value Pr(>|z|)
(Intercept)       1.0642     0.1573    6.76  1.3e-11 ***
poly(age, 2)1     0.0355     0.5811    0.06  0.95129
poly(age, 2)2    -1.9428     0.5668   -3.43  0.00061 ***
educ              0.0746     0.0110    6.79  1.1e-11 ***
---
Signif. codes:  0 '***' 0.001 '**' 0.01 '*' 0.05 '.' 0.1 ' ' 1

Total Sum of Squares:    117
Residual Sum of Squares: 102
R-Squared:      0.133
Adj. R-Squared: 0.127
Chisq: 65.1443 on 3 DF, p-value: 4.67e-14
```

Note that we set the `random.method` argument to `"walhus"` to select the Wallace and Hussain estimator. The output is quite similar to the one for `lm` objects, except the two parts that appear at the beginning. The dimensions of the panel are indicated (number of individuals / entities and number of time series / observations in each entity) and whether the data set is balanced or not. A panel data is balanced if all the individuals are observed for the same set of time periods. In a pseudo-panel (which is the case here), the data set is balanced if there is the same number of observations for each entity (which is obviously the case for our sample of twins). This information can be obtained using `pdim`. The first argument is a data frame, the second one, called `index`, is a vector of two characters indicating the name of the individual and of the time index. It can be omitted if the first two columns contains these indexes, which is the case for the `twins` data set:

```
pdim(twins, index = c("family", "twin"))
## Balanced Panel: n = 214, T = 2, N = 428
```

The second specific part of the output gives information about the variances of the two components of the error. We can see here that the individual effects (in this example, a family effect) account for only 15% of the total variance of the error. Therefore, only a small part of the individual mean is removed while performing GLS (14.3%). This information can be obtained using the `ercomp` function:

```
ercomp(log(earning) ~ poly(age, 2) + educ, twins, method = "walhus")
```

The model can also be estimated by maximum likelihood, using `pglm::pglm`. This function adapts the behavior of the `stats::glm` function which fits generalized linear model for panel data. In particular, it has a `family` argument that is set to `"gaussian"`:

```
ml_twins <- pglm::pglm(log(earning) ~ poly(age, 2) + educ,
                       twins, family = gaussian)
ml_twins %>% gaze
```

```
               Estimate Std. Error z-value Pr(>|z|)
poly(age, 2)1    0.0352     0.5789    0.06  0.95149
poly(age, 2)2   -1.9429     0.5646   -3.44  0.00058
educ             0.0746     0.0110    6.79  1.1e-11
sd.id            0.1981     0.0471    4.21  2.6e-05
sd.idios         0.4875     0.0217   22.49  < 2e-16
```

σ_η and σ_ν are now two parameters that are directly estimated. We can see that the estimated values are very close to the ones obtained using GLS and that σ_η is statistically significant. Comparing OLS and GLS standard deviations, we get:

```
ols_se <- vcovCL(lm_twins, ~ family, type = "HC0", cadjust = FALSE) %>%
  stder
gls_se <- vcov(gls_twins) %>% stder
rbind(ols = ord_se, gls = gls_se, ratio = gls_se / ols_se)
```

```
      (Intercept) poly(age, 2)1 poly(age, 2)2    educ
ols        0.1516        0.5451        0.5309 0.01059
gls        0.1573        0.5811        0.5668 0.01098
ratio      0.9712        1.0115        0.9448 0.99593
```

and therefore, there seems to be no gain of efficiency while using OLS instead of GLS. With Tobin's Q example, we get:

```
gls_q <- plm(ikn ~ qn, tobinq, model = "random")
ercomp(gls_q)
```

```
                  var std.dev share
idiosyncratic 0.00533 0.07303  0.73
individual    0.00202 0.04493  0.27
theta: 0.735
```

The share of the individual effect is now 27.5%, and the GLS is now OLS on series for which 73.5% of the individual mean has been removed, mostly because the time dimension of the panel is high (35 years). Comparing the robust standard errors of OLS and those of GLS, we get:

```
ols_q <- lm(ikn ~ qn, tobinq)
sd_ols_q <- vcovCL(ols_q, ~ cusip) %>% stder
sd_gls_q <- vcov(gls_q) %>% stder
rbind(ols = sd_ols_q, gls = sd_gls_q, ratio = sd_gls_q / sd_ols_q)
```

```
      (Intercept)        qn
ols      0.003186 0.0006679
gls      0.003425 0.0001683
ratio    1.074968 0.2519458
```

GLS is much more efficient than OLS as the standard error of the slope is about four times smaller.

6.4.4 Seemingly unrelated regression

Because of the inter-equation correlations, the efficient estimator is the GLS estimator: $\hat{\gamma} = (Z^\top \Omega^{-1} Z)^{-1} Z^\top \Omega^{-1} y$. This estimator, first proposed by Zellner (1962), is known as a **seemingly unrelated regression** (**SUR**). It can be obtained by applying OLS on transformed data, each variable being premultiplied by $\Omega^{-0.5}$. This matrix is simply $\Omega^{-0.5} = \Sigma^{-0.5} \otimes I$. Denoting δ_{lm} the elements of $\Sigma^{-0.5}$, the transformed response and covariates are:

$$y^* = \left(\begin{array}{c} \delta_{11} y_1 + \delta_{12} y_2 + ... + \delta_{1L} y_L \\ \delta_{21} y_1 + \delta_{22} y_2 + ... + \delta_{2L} y_L \\ \vdots \\ \delta_{L1} y_1 + \delta_{L2} y_2 + ... + \delta_{LL} y_L \end{array}\right), Z^* = \left(\begin{array}{cccc} \delta_{11} Z_1 & \delta_{12} Z_2 & ... & \delta_{1L} Z_L \\ \delta_{21} Z_1 & \delta_{22} Z_2 & ... & \delta_{2L} Z_L \\ \vdots & \vdots & \ddots & \vdots \\ \delta_{L1} Z_1 & \delta_{L2} Z_2 & ... & \delta_{LL} Z_L \end{array}\right) \quad (6.21)$$

Σ is a matrix that contains unknown parameters, which can be estimated using residuals of a consistent but inefficient preliminary estimator, like OLS. The efficient estimator is then obtained the following way:

- first, estimate each equation separately by OLS and denote $\hat{\Xi} = (\hat{\epsilon}_1, \hat{\epsilon}_2, ..., \hat{\epsilon}_L)$ the $N \times L$ matrix for which every column is the residual vector of one of the equations in the system,
- then, estimate the covariance matrix of the errors: $\hat{\Sigma} = \hat{\Xi}^\top \hat{\Xi} / N$,
- compute the matrix $\hat{\Sigma}^{-0.5}$ and use it to transform the response and the covariates of the model,
- finally, estimate the model by applying OLS on transformed data.

$\Sigma^{-0.5}$ can conveniently be computed using the Cholesky decomposition, i.e., the upper-triangular matrix C which is such that $C^\top C = \Sigma^{-1}$.

To illustrate the use of the **SUR** estimator, we go back to the estimation of the system of three equations (one cost function and two factor share equations) for the production of apples started in Section 3.7. In this section, we computed a tibble containing the three responses `Y` and the model matrices for the three equations `Z_c`, `Z_l` and `Z_m` (respectively for the cost, the labor and the materials equations). In Section 6.2.3, we estimated the covariance matrix of the errors of the three equations `Sigma`. To implement the **SUR** estimator, we compute the Cholesky decomposition of the inverse of the estimated covariance matrix of the errors of the three equations:

```
V <- chol(solve(Sigma))
V
```

```
      [,1]    [,2]     [,3]
[1,] 3.375  0.9645  0.03717
[2,] 0.000 17.3707 13.66170
[3,] 0.000  0.0000 11.38294
```

We then transform the response and the covariates using Equation 6.21:

```
Zs <- rbind(cbind(V[1, 1] * Z_c, V[1, 2] * Z_l, V[1, 3] * Z_m),
            cbind(V[2, 1] * Z_c, V[2, 2] * Z_l, V[2, 3] * Z_m),
```

```
              cbind(V[3, 1] * Z_c, V[3, 2] * Z_l, V[3, 3] * Z_m))
ys <- as.matrix(Y) %*% t(V) %>% as.numeric
```

Then the SUR estimator is computed, using `lm` on the transformed data and then using `clm` in order to impose the linear restrictions.

```
sur <- lm(ys ~ Zs - 1) %>% clm(R = R)
sur %>% coef
```

```
Zs(Intercept)             Zsy            Zsy2            Zspl            Zspm
      0.01478         0.43854         0.11661         0.49207         0.32453
        Zspl2          Zsplm           Zspm2 Zs(Intercept)            Zspl
      0.11775        -0.09496         0.10277         0.49207         0.11775
         Zspm Zs(Intercept)            Zspl            Zspm
     -0.09496         0.32453        -0.09496         0.10277
```

More simply, the `systemfit::systemfit` function can be used, [9] with the `method` argument set to `"SUR"`:[10]

```
library(systemfit)
sur <- systemfit(list(cost = eq_ct, labor = eq_sl, materials = eq_sm),
                 data = ap, restrict.matrix = R, method = "SUR",
                 methodResidCov = "noDfCor")
```

The coefficients of the fitted model can be used to compute the Allen elasticities of substitution and the price elasticities. The former are defined as:

$$\sigma_{ij} = \frac{\beta_{ij}}{s_i s_j} - 1 \;\; \forall i \neq j \text{ and } \sigma_{ii} = \frac{\beta_{ij} - s_i(1 - s_i)}{s_i^2}$$

Denote as B the matrix containing the coefficients β_{ij}. Remember that, by imposing the homogeneity of degree one of the cost function, we imposed that $\beta_{iI} = -\sum_{i=1}^{I-1} \beta_{ij}$. Therefore β_{iI} was not estimated and we must add it to the B matrix using this formula:

```
B <- matrix(coef(sur)[- (1:8)], ncol = 2)[-1, ]
add <- - apply(B, 1, sum)
B <- cbind(rbind(B, add), c(add, - sum(add)))
shares <- ap %>% summarise(sl = mean(sl),
                           sm = mean(sm), sk = 1 - sl - sm) %>%
    as.numeric
elast <- B /outer(shares, shares) + 1
diag(elast) <- diag(elast) - 1 / shares
dimnames(elast) <- list(c("l", "m", "k"), c("l", "m", "k"))
elast
```

[9] The **systemfit** package was presented in Section 3.7.2.

[10] Note the use of the `methodResidCov` argument: setting it to `"noDfCor"`, the cross-product of the vectors of residuals is divided by the number of observations to get the estimation of the covariance matrix. Other values of this argument enables to perform different kinds of degrees of freedom correction.

```
        l       m       k
l -0.5214  0.4064  0.7469
m  0.4064 -1.1295  0.8624
k  0.7469  0.8624 -3.6371
```

The three factors are substitutes, all the Allen elasticities of substitution being positive. The price elasticities are given by: $\epsilon_{ij} = s_j \sigma_{ij}$.

```
elast * rbind(shares, shares, shares)
```

```
        l       m       k
l -0.2626  0.1291  0.1335
m  0.2047 -0.3588  0.1542
k  0.3761  0.2740 -0.6501
```

Note that this matrix is not symmetric: for example, 0.1675 is the elasticity of the demand of materials with the price of labor whereas 0.1039 is the elasticity of the demand of labor with the price of materials. The price elasticities indicate that the demand for the three inputs is inelastic, and it is particularly the case for labor and materials.

7

Endogeneity

The unbiasedness and the consistency of the OLS estimator rest on the hypothesis that the conditional expectation of the error is constant (and can safely be set to zero if the model contains an intercept). Namely, starting with the simple linear model: $y_n = \alpha + \beta x_n + \epsilon_n$, $\mathrm{E}(\epsilon \mid x) = 0$, or equivalently $\mathrm{E}(y \mid x) = \alpha + \beta x_n$. The same property can also be described using the covariance between the covariate and the error that can be written, using the rule of repeated expectation:

$$\mathrm{cov}(x, \epsilon) = \mathrm{E}\left((x - \mu_x)\epsilon\right) = \mathrm{E}_x\left[\mathrm{E}_\epsilon\left((x - \mu_x)\epsilon \mid x\right)\right] = \mathrm{E}_x\left[(x - \mu_x)\mathrm{E}_\epsilon\left(\epsilon \mid x\right)\right]$$

If the conditional expectation of ϵ is a constant $\mathrm{E}_\epsilon(\epsilon \mid x) = \mu_\epsilon$ (not necessarily 0), the covariance is $\mathrm{cov}(x, \epsilon) = \mu_\epsilon \mathrm{E}_x(x - \mu_x) = 0$. Stated differently, x is supposed to be exogenous, or x is assumed to be uncorrelated with ϵ. This is a reasonable assumption in an experimental setting, when the values of x in the sample are set by the researcher. Unfortunately, most of the data used in microeconometrics are not experimental, so the problem of endogeneity is severe.[1]

Section 7.1 presents the circumstances where some covariates are endogenous. Section 7.2 presents the simple instrumental variable estimator (one endogenous covariate and one instrument), Section 7.3 extends this model to the case where there may be more than one endogenous covariate and several instruments and Section 7.4 to the estimation of systems of equations. Section 7.5 presents the fixed effects model. Finally, several testing procedures are presented in Section 7.6.

7.1 Sources of endogeneity

In economics, the problem of endogenous covariates happens mainly in three circumstances: errors in variables, omitted variables and simultaneity.

7.1.1 Errors in variables

Data used in economics, especially micro-data, are prone to errors of measurement. This problem can affect either the response or some of the covariates. Suppose that the model

[1]The terminology used to distinguish variables that are uncorrelated (exogenous) or correlated (endogenous) with the error of the model comes from the literature on system of equations estimation where variables determined within and outside the model are respectively called endogenous and exogenous variables (see Stock and Watson 2015, 423).

that we seek to estimate is: $y_n^* = \alpha + \beta x_n^* + \epsilon_n^*$, where the covariate is exogenous, which implies that $\mbox{cov}(x^*, \epsilon^*) = 0$. Suppose that the response is observed with error, namely that the observed value of the response is $y_n = y_n^* + \nu_n$, where ν_n is the measurement error of the response. In terms of the observed response, the model is now:

$$y_n = \alpha + \beta x_n^* + (\epsilon_n^* + \nu_n)$$

The error of the estimable model is then $\epsilon_n = \epsilon_n^* + \nu_n$ which is still uncorrelated with x if ν is uncorrelated with x, which means that the error of measurement of the response is uncorrelated with the covariate. In this case, the measurement error only increases the size of the error, which implies that the coefficients are estimated less precisely and that the R^2 is lower compared to a model with a correctly measured response.

Now consider that the covariate is measured with error and that the observable values of the covariate is $x_n = x_n^* + \nu_n$. If the measurement error is uncorrelated with the value of the covariate, the variance of the observed covariate is therefore $\sigma_x^2 = \sigma_x^{*2} + \sigma_\nu^2$. Moreover, the covariance between the observed covariate and the measurement error is equal to the variance of the measurement error: $\sigma_{x\nu} = \mbox{E}\left((x^* + \nu - \mu_x)\nu\right) = \sigma_\nu^2$, because the measurement error is uncorrelated with the covariate. Rewriting the model in terms of x, we get: $y_n = \alpha + \beta x_n + \epsilon_n$, with $\epsilon_n = \epsilon_n^* - \beta\nu_n$. The error of this model is now correlated with x, as $\mbox{cov}(x, \epsilon_n) = \mbox{cov}(x^* + \nu, \epsilon^* - \beta\nu) = -\beta\sigma_\nu^2$. The OLS estimator can be written as usual as:

$$\hat{\beta} = \frac{\sum_n (x_n - \bar{x})(y_n - \bar{y})}{\sum_n (x_n - \bar{x})^2} = \beta + \frac{\sum_n (x_n - \bar{x})\epsilon_n}{\sum_n (x_n - \bar{x})^2}$$

Taking the expectations, we have $\mbox{E}\left[(x - \bar{x})\epsilon\right] = -\beta\sigma_\nu^2$ and the expected value of the estimator is then:

$$\mbox{E}(\hat{\beta}) = \beta\left(1 - \frac{\sigma_\nu^2}{\sum(x_n - \bar{x})^2/N}\right) = \beta\left(1 - \frac{\sigma_\nu^2}{\hat{\sigma}_x^2}\right)$$

Therefore, the OLS estimator is biased and the term in brackets is the minus the share of the variance of x that is due to measurement errors.Therefore $\mid \hat{\beta} \mid < \beta$. This kind of bias is called an **attenuation bias** (the absolute value of the estimator is lower than the true value), which can be either a lower or an upper bias depending on the sign of β. This bias clearly doesn't attenuate in large samples. As N grows, the empirical variances/covariances converge to the population ones, and the estimator therefore converges to: $\mbox{plim}\, \hat{\beta} = \beta\left(1 - \sigma_\nu^2/\sigma_x^2\right)$. For example, if the measurement error accounts for 20% of the total variance of x, $\hat{\beta}$ converges to 80% of the true parameter.

7.1.2 Omitted variable bias

Suppose that the true model is: $y_n = \alpha + \beta_x x_n + \beta_z z_n + \epsilon_n$, where the conditional expectation of ϵ with respect to x and z is 0. Therefore, this model can be consistently estimated by least squares. Consider now that z is unobserved. Therefore, the model to be estimated is $y_n = \alpha + \beta_x x_n + \eta_n$, with $\eta_n = \beta_z z_n + \epsilon_n$. The omission of a relevant ($\beta_z \neq 0$) covariate has two consequences:

- the variance of the error is now $\sigma_\eta^2 = \beta_z^2\sigma_z^2 + \sigma_\epsilon^2$, and is therefore greater than the one of the initial model for which z is observed and used as a covariate,
- the covariance between the error and x is $\mbox{cov}(x, \eta) = \beta_z \mbox{cov}(x, z)$; therefore, if the covariate is correlated with the omitted variable, the covariate and the error of the model are correlated.

As the variance of the OLS estimator is proportional to the variance of the errors, omission of a relevant covariate will always induce a less precise estimation of the slopes and a lower $\mbox{R}^2$. Moreover, if the omitted covariate is correlated with the covariate used in the regression, the estimation will be biased and inconsistent. This **omitted variable bias** can be computed as follows:

$$\hat{\beta}_x = \beta_x + \frac{\sum_n (x_n - \bar{x})(\beta_z z_n + \epsilon_n)}{\sum_n (x_n - \bar{x})^2} = \beta_x + \beta_z \frac{\sum_n (x_n - \bar{x})(z_n - \bar{z})}{\sum_n (x_n - \bar{x})^2} + \frac{\sum_n (x_n - \bar{x})\epsilon_n}{\sum_n (x_n - \bar{x})^2}$$

Taking the conditional expectation, the last term disappears, so that:

$$\mbox{E}(\hat{\beta}_x \mid x, z) = \beta_x + \beta_z \frac{\hat{\sigma}_{xz}}{\hat{\sigma}_x^2}$$

There is an upper bias if the signs of the covariance between x and z and β_z are the same, and a lower bias if they have opposite signs. As N tends to infinity, the OLS estimator converges to:

$$\mbox{plim}\, \hat{\beta}_x = \beta_x + \beta_z \frac{\sigma_{xz}}{\sigma_x^2} = \beta_x + \beta_z \beta_{z/x}$$

where $\beta_{z/x}$ is the true value of the slope of the regression of z on x. This formula makes clear what $\hat{\beta}_x$ really estimates in a linear regression:

- the direct effect of x on y which is β_x,
- the indirect effect of x on y which is the product of the effect of x on z ($\beta_{z/x}$) times the effect of z on η and therefore on y (β_z).

A classic example of omitted variable bias occurs in the estimation of a Mincer earning function which relates wage (w), education (e in years) and experience (s in weeks):

$$\ln w = \beta_o + \beta_e e + \beta_s s + \beta_{ss} s^2 + \epsilon$$

$\beta_e = \frac{d \ln w}{de} = \frac{dw/w}{de}$ is the percentage increase of the wage for one more year of education. To illustrate the estimation of a Mincer function, we use the data of Koop, Poirier, and Tobias (2005), which is a sample of 303 white males taken from the National Longitudinal Survey of Youth and is available as `sibling_educ`.

```
sibling_educ <- sibling_educ %>%
    mutate(experience = experience / 52)
lm(log(wage) ~ educ + poly(experience, 2), sibling_educ) %>% gaze
```

```
                      Estimate Std. Error t value Pr(>|t|)
educ                    0.1000     0.0118    8.48  1.0e-15
poly(experience, 2)1    2.3913     0.4554    5.25  2.9e-07
poly(experience, 2)2    0.4822     0.4542    1.06     0.29
```

Results indicate that one more year of education increases on average the wage by 10%. One concern about this kind of estimation is that individuals have different abilities (a), and that more abilities have a positive effect on wage, but may also have a positive effect on education. If this is the case, $\beta_a > 0$, $\beta_{a/e} > 0$ and therefore plim $\hat{\beta}_e = \beta_e + \beta_a\beta_{a/e} > \beta_e$ and the OLS estimator is upward biased. This is the case because more education:

- increases, for a given level of ability, the expected wage by β_e,
- means that, on average, the level of ability is higher, this effect being $\beta_{a/e}$ (which in this case doesn't imply a causality of e on a, but simply a correlation), so the wage will also be higher ($\beta_a > 0$).

Numerous studies of the Mincer function deal with this problem of endogeneity of the education level. But in the data set we used, there is a measure of the ability, which is the standardized AFQT test score. If we introduce ability in the regression, education is no more endogenous and least squares will give a consistent estimation of the effect of education on wage.[2] We first check that education and ability are positively correlated:

```
sibling_educ %>% summarise(cor(educ, ability)) %>% pull
## [1] 0.6056
```

Therefore, adding ability as a covariate in the previous regression should decrease the coefficient on education:

```
lm(log(wage) ~ educ + poly(experience, 2) + ability, sibling_educ) %>%
  gaze(coef = "educ")
##        Estimate Std. Error t value Pr(>|t|)
## educ     0.0887     0.0150    5.92  8.7e-09
```

The effect of one more year of education is now an increase of 8.7% of the wage (compared to 10% previously).

7.1.3 Simultaneity bias

Often in economics, the phenomenon of interest is not described by a single equation, but by a system of equations. Consider for example a market equilibrium. The two equations relate the quantity demanded / supplied (q^d and q^o) to the unit price and to some specific covariates to the demand and to the supply side of the market. Maddala and Lahiri (2009), pp. 376-377 studied the market for commercial loans using monthly US data for 1979-1984. The data set is available as `loan_market`. Total commercial loans (`loans`) are in billions of dollars; the price is the average prime rate charged by banks (`prime_rate`). `aaa_rate` is the AAA corporate bond rate, which is the price of an alternative financing to firms. Therefore, it enters only the demand equation, with an expected positive sign. `treas_rate`

[2]Needless to say, the hypothesis that abilities in all their dimensions can be measured by the AFQT test is very strong.

is the treasure bill rate. As it is a substitute to commercial loans from a bank, it should enter only the supply equation, with an expected negative sign. For the sake of simplicity, we denote q and p as the quantity (in logarithm) and the price for this loan market model and, d and s as the two rates that enter only, respectively, the demand and the supply equation:

```
loan <- loan_market %>%
    transmute(q = log(loans),  p = prime_rate,
              d = aaa_rate, s = treas_rate)
```

The equilibrium on the loan market is then defined by a system of three equations:

$$\left\{\begin{array}{lcl} q_d &=& \alpha_d + \beta_d p + \gamma_d d + \epsilon_d \\ q_s &=& \alpha_s + \beta_s p + \gamma_s s + \epsilon_s \\ q_d &=& q_s \end{array}\right.$$

The last equation is non-stochastic and states that the market should be in equilibrium. The demand curve should be decreasing ($\beta_d < 0$) and the supply curve increasing ($\beta_s > 0$). The equilibrium is depicted in Figure 7.1.

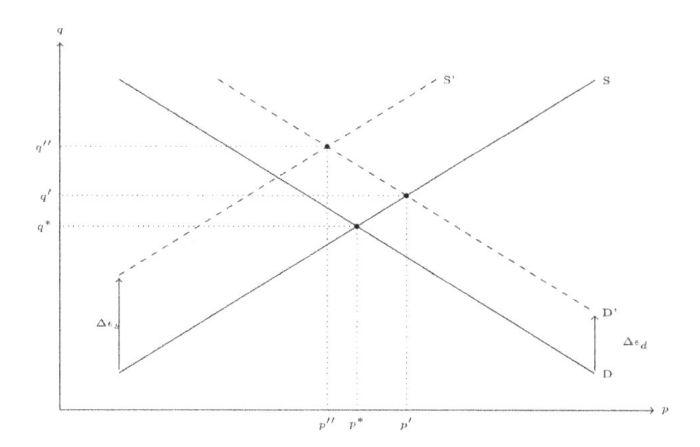

Figure 7.1: Market equilibrium

The OLS estimation of the demand and the supply equations is given below:

```
ols_d <- lm(q ~ p + d, data = loan)
ols_s <- lm(q ~ p + s, data = loan)
ols_d %>% gaze
##    Estimate Std. Error t value Pr(>|t|)
## p -0.04337    0.00429   -10.1  3.1e-15
## d  0.10346    0.00817    12.7  < 2e-16
ols_s %>% gaze
##    Estimate Std. Error t value Pr(>|t|)
## p  0.00136    0.01478    0.09     0.93
## s -0.01838    0.01911   -0.96     0.34
```

The two coefficients of the demand equation have the predicted sign and are highly significant: a 1 point of percentage increase of the prime rate decreases loans by 4.3%, and a one point of percentage increase of the corporate bond rate increases loans by 10.3%. The fit of the supply equation is very bad and, even if the two coefficients have the predicted sign, the values are very low and insignificant.

What is actually observed for each observation in the sample is a price-quantity combination at an equilibrium. A positive shock on the demand equation will move upward the demand curve and will lead to a new equilibrium with a higher equilibrium quantity q' and also a higher equilibrium price p' (except in the special case where the supply curve is vertical, which means that the price elasticity of supply is infinite). This means that p is correlated with ϵ_d, which leads to a bias in the estimation of β_d by OLS. The same reasoning applies of course to the supply curve.

One solution would be to use the equilibrium condition and then to solve for p, and then for q. This **reduced-form** system of two equations:

$$\left\{\begin{array}{ccccccccc} p & = & \frac{\alpha_d-\alpha_s}{\beta_s-\beta_d} & + & \frac{\gamma_d}{\beta_s-\beta_d}d & - & \frac{\gamma_s}{\beta_s-\beta_d}s & + & \frac{\epsilon_d-\epsilon_s}{\beta_s-\beta_d} \\ & = & \pi_o & + & \pi_d d & + & \pi_s s & + & \nu_p \\ q & = & \frac{\alpha_d\beta_s-\beta_d\alpha_s}{\beta_s-\beta_d} & + & \frac{\beta_s\gamma_d}{\beta_s-\beta_d}d & - & \frac{\beta_d\gamma_s}{\beta_s-\beta_d}s & + & \frac{-\beta_d\epsilon_s+\beta_s\epsilon_d}{\beta_s-\beta_d} \\ & = & \delta_o & + & \delta_d d & + & \delta_s s & + & \nu_q \end{array}\right.$$

makes clear that both p and q depend on ϵ_s and ϵ_d. More precisely, as $\beta_s - \beta_d > 0$, ϵ_d and ϵ_s are respectively positively and negatively correlated with the price. $\Delta\epsilon_d > 0$ will shift the demand curve upward and therefore will increase the equilibrium price. Therefore, the OLS estimator of the slope of the demand curve is biased upward which means, as it is negative, that it is biased downward in absolute value. A positive shock on the supply side ($\Delta\epsilon_s > 0$) will move the supply curve upward and therefore will decrease the equilibrium price. The OLS estimator of the slope of the supply curve is then downward biased. The parameters of these two equations can be consistently estimated by least squares, as d and s are uncorrelated with the two error terms.

```
ols_q <- lm(q ~ d + s, data = loan)
ols_p <- lm(p ~ d + s, data = loan)
ols_q %>% gaze
##    Estimate Std. Error t value Pr(>|t|)
## d  0.09437    0.00837    11.3   < 2e-16
## s -0.05061    0.00569    -8.9   4.6e-13
ols_p %>% gaze
##    Estimate Std. Error t value Pr(>|t|)
## d   0.2876     0.1096    2.62     0.011
## s   1.0667     0.0745   14.32    <2e-16
```

The reduced form coefficients are not meaningful by themselves, but only if they enable to retrieve the structural parameters. This is actually the case here as $\frac{\delta_s}{\pi_s} = \beta_d$ and $\frac{\delta_d}{\pi_d} = \beta_s$.

```
price_coefs <- unname(coef(ols_q) / coef(ols_p))[2:3]
price_coefs
## [1]  0.32810 -0.04744
```

The price coefficients are as expected higher in absolute values than those obtained using OLS on the demand and on the supply equation. It is particularly the case for β_s

(0.328 vs. 0.001). On the contrary, the absolute value of β_d increases very slightly (−0.047 vs. −0.043).

Actually, the case we have considered is special because there is one and only one extra covariate in both equations. Consider as an example the following system of equations:

$$\left\{ \begin{array}{rcl} y_d & = & \alpha_d + \beta_d p + \gamma_1 d_1 + \gamma_2 d_2 + \epsilon_d \\ y_s & = & \alpha_s + \beta_s p + \epsilon_s \\ y_d & = & y_s \end{array} \right.$$

where there are two extra covariates in the demand equation and no covariates (except the price) in the supply equation. Solving for the two endogenous variables p and q, we get in this case:

$$\left\{ \begin{array}{rcccccccc} p & = & \frac{\alpha_d - \alpha_s}{\beta_s - \beta_d} & + & \frac{\gamma_1}{\beta_s - \beta_d} d_1 & + & \frac{\gamma_2}{\beta_s - \beta_d} d_2 & + & \frac{\epsilon_d - \epsilon_s}{\beta_s - \beta_d} \\ & = & \pi_s & + & \pi_1 d_1 & + & \pi_2 d_2 & + & \nu_p \\ q & = & \frac{\beta_s \alpha_d - \beta_d \alpha_s}{\beta_s - \beta_d} & + & \frac{\beta_s \gamma_1}{\beta_s - \beta d} d_1 & + & \frac{\beta_s \gamma_2}{\beta_s - \beta_d} d_2 & + & \frac{\beta_s \epsilon_d - \beta_d \epsilon_s}{\beta_s - \beta_d} \\ & = & \delta_s & + & \delta_1 d_1 & + & \delta_2 d_2 & + & \nu_q \end{array} \right.$$

we now have $\frac{\delta_1}{\pi_1} = \frac{\delta_2}{\pi_2} = \beta_s$; there are two ratios of the reduced parameters that give the value of the slope of the supply curve, but it is very implausible that these two values will be equal. On the contrary, there is no way to retrieve the slope of the demand curve from the reduced form parameters. Therefore, this **indirect least squares** approach is of limited interest. In the general case, as we have seen, some coefficients like β_s in our example may be **over-identified** and some other like β_d are **under-identified**.

7.2 Simple instrumental variable estimator

The general idea of the **instrumental variable (IV)** estimator is to find variables which are correlated with the endogenous covariates and uncorrelated with the error of the structural equation. This means that the instruments don't have a direct effect on the response, but only an indirect effect because of their correlation with the endogenous covariates. These instruments allow to get an exogenous source of variation of the covariate, i.e., a source of variation that has nothing to do with the process of interest. We'll start with the simple case where the number of instruments equals the number of endogenous covariates, i.e., the **just-identified** case.

7.2.1 Computation of the simple instrumental variable estimator

Consider a simple linear regression: $y_n = \alpha + \beta x_n + \epsilon_n$, with $\mathrm{E}(\epsilon \mid x) \neq 0$. From Equation 1.9, the first-order conditions for the minimization of the sum of square residuals can be written as:

$$\frac{1}{N}\sum_n \left[(y_n - \bar{y}) - \hat{\beta}(x_n - \bar{x})\right](x_n - \bar{x}) = \frac{1}{N}\sum_n \hat{\epsilon}_n(x_n - \bar{x}) = \hat{\sigma}_{\hat{\epsilon}x} = 0 \tag{7.1}$$

Equation 7.1 states that the OLS estimator is obtained by imposing the sample equivalent to the moment condition $\mathrm{E}(\epsilon \mid x) = 0$, i.e., that the covariance between the residuals and the covariate is exactly 0 in the sample. This of course leads to a biased and inconsistent estimator if $\mathrm{E}(\epsilon \mid x) \neq 0$. Now suppose that a variable w exists, which is correlated with the covariate, but not with the error of the model (this latter hypothesis is called the **exclusion restriction**). Such a variable has no direct effect on the response, but only an indirect effect due to its correlation with the covariate. As by hypothesis, $\mathrm{E}(\epsilon \mid w) = 0$, a consistent estimator can be obtained by imposing the sample equivalent to this moment condition, i.e., by setting the estimator to a value such that the covariance between the residuals and the instrument is 0:

$$\frac{1}{N}\sum_n \left[(y_n - \bar{y}) - \hat{\beta}(x_n - \bar{x})\right](w_n - \bar{w}) = \frac{1}{N}\sum_n \hat{\epsilon}_n(w_n - \bar{w}) = \hat{\sigma}_{\hat{\epsilon}w} = 0 \tag{7.2}$$

Solving Equation 7.2 for $\hat{\beta}$, we get the instrumental variable estimator:

$$\hat{\beta} = \frac{\sum_n (w_n - \bar{w})(y_n - \bar{y})}{\sum_n (w_n - \bar{w})(x_n - \bar{x})} = \frac{\hat{\sigma}_{wy}}{\hat{\sigma}_{wx}} \tag{7.3}$$

which is the ratio of the empirical covariances of w with y and with x. Dividing both sides of the ratio by the empirical variance of w ($\hat{\sigma}_w^2$), Equation 7.3 can also be written as:

$$\hat{\beta} = \frac{\hat{\sigma}_{wy}/\hat{\sigma}_w^2}{\hat{\sigma}_{wx}/\hat{\sigma}_w^2} = \frac{\hat{\beta}_{y/w}}{\hat{\beta}_{x/w}} = \frac{dy/dw}{dx/dw} \tag{7.4}$$

where $\hat{\beta}_{y/w}$ and $\hat{\beta}_{x/w}$ are the slopes of the regressions of y and x on w. Therefore, the IV estimator is also the ratio of the marginal effects of w on y and on x.[3] Replacing $y_n - \bar{y}$ by $\beta(x_n - \bar{x}) + \epsilon_n$ in Equation 7.3, we get:

$$\hat{\beta} = \beta + \frac{\sum_n (w_n - \bar{w})\epsilon_n}{\sum_n (w_n - \bar{w})(x_n - \bar{x})} = \beta + \frac{\hat{\sigma}_{w\epsilon}}{\hat{\sigma}_{wx}}$$

As N grows, the two empirical covariances converge to the population covariances and, as by hypothesis, the instrument is uncorrelated with the error ($\sigma_{w\epsilon} = 0$) and is correlated with the covariate ($\sigma_{wx} \neq 0$), the instrumental variable estimator is consistent ($\mathrm{plim}\, \hat{\beta} = \beta + \frac{\sigma_{w\epsilon}}{\sigma_{wx}} = \beta$). Assuming spherical disturbances, the variance of the IV estimator is:

$$\mathrm{V}(\hat{\beta}) = \mathrm{E}\left[(\hat{\beta} - \beta)^2\right] = \frac{\sigma_\epsilon^2 \sum_n (w_n - \bar{w})^2}{\left[\sum_n (w_n - \bar{w})(x_n - \bar{x})\right]^2} = \frac{\sigma_\epsilon^2 \hat{\sigma}_w^2}{N\hat{\sigma}_{wx}^2} = \frac{\sigma_\epsilon^2}{N\hat{\sigma}_x^2 \hat{\rho}_{wx}^2} = \left(\frac{\sigma_\epsilon}{\sqrt{N}\hat{\sigma}_x \mid \hat{\rho}_{wx} \mid}\right)^2$$

The last expression, which introduces the coefficient of correlation between x and z, is particularly appealing as it shows that, compared to the standard error of the OLS estimator,

[3] See Heckman (2000), page 58 and Cameron and Trivedi (2005), page 98.

the standard error of the IV estimator is inflated by $1/\mid \hat{\rho}_{wx}\mid$. This term is close to 1 if the correlation between x and w is high and, in this case, the loss of precision implied by the use of the IV estimator is low. On the contrary, if the correlation between x and w is low, the IV estimator will be very imprecisely estimated.

The IV estimator can also be obtained by first regressing x on w and then by regressing y on the fitted values of the previous regression. Consider the linear model that relates the instrument to the covariate: $x_n = \gamma + \delta w_n + \nu_n$. We estimate this model by OLS and then we express the fitted values ($\hat{x}$) as a function of $\hat{\delta}$: $\hat{x}_n - \bar{x} = \hat{\delta}(w_n - \bar{w})$. We can therefore rewrite the numerator and the denominator of the IV estimator given by Equation 7.3 as:

- $\sum_n (w_n - \bar{w})(y_n - \bar{y}) = \sum_n (\hat{x}_n - \bar{x})(y_n - \bar{y})/\hat{\delta}$,
- $\sum_n (w_n - \bar{w})(x_n - \bar{x}) = \sum_n (w_n - \bar{w})(\hat{x}_n + \hat{\epsilon}_n - \bar{x}) = \sum_n (\hat{x}_n - \bar{x})^2/\hat{\delta}$ (as $\sum_n (w_n - \bar{w})\hat{\epsilon}_n = 0$),

so that:

$$\hat{\beta} = \frac{\sum_n (\hat{x}_n - \bar{x})(y_n - \bar{y})}{\sum_n (\hat{x}_n - \bar{x})^2}$$

which is the OLS estimator of y on $\hat{x}$, $\hat{x}$ being the fitted values of the regression of x on w. Therefore, the IV estimator can be obtained by running two OLS regressions and is for this reason also called the **two-stage least square (2SLS)** estimator. As x is correlated with ϵ, but w is not, the idea of the 2SLS estimator is to replace x by a linear transformation of w which is as close as possible to x, and this is simply the fitted values of the regression of x on w.

The extension of the case when there are $K > 1$ covariates and K instruments is straightforward. The two sets of variables x and w can overlap because, if some of the covariates are exogenous, they should also be used as instruments. Denoting w as the vector of instruments, the moment conditions are $\mbox{E}(\epsilon \mid w) = 0$ and the sample equivalent is:

$$\frac{1}{N} W^\top (y - Z\gamma) = 0$$

where Z and W are $N \times (K+1)$ matrix containing respectively the covariates and the instruments (with a vector of 1). Solving for γ, we get the instrumental variable estimator:

$$\hat{\gamma} = (W^\top X)^{-1} W^\top y$$

7.2.2 Small sample properties of the IV estimator

Although it is consistent, the instrumental variable estimator isn't unbiased. Actually, it doesn't even have an expected value in the just-identified case. This result can be easily shown starting with the following system of equations:[4]

$$\left\{ \begin{array}{rcl} y &=& \alpha + \beta x + \sigma_\epsilon \epsilon \\ x &=& \gamma + \delta w + \sigma_\nu \nu \end{array} \right.$$

[4]See Davidson and MacKinnon (2004), pages 326-327.

where, for convenience, the two error terms are written as standard normal deviates. Moreover, we can write $\epsilon = \rho\nu + \iota$, so ρ is the coefficient of correlation between the two error terms and ι is by construction uncorrelated with ν. The IV estimator is then:

$$\hat{\beta} = \frac{\sum_n (w_n - \bar{w})(y_n - \bar{y})}{\sum_n (w_n - \bar{w})(x_n - \bar{x})} = \beta + \frac{\sigma_\epsilon \sum_n (w_n - \bar{w})(\rho\nu_n + \iota_n)}{\sum_n (w_n - \bar{w})(x_n - \bar{x})} \tag{7.5}$$

The denominator is closely linked to the OLS estimator of γ, which is:

$$\hat{\delta} = \frac{\sum_n (w_n - \bar{w})(x_n - \bar{x})}{\sum_n (w_n - \bar{w})^2} = \delta + \frac{\sigma_\nu \sum_n (w_n - \bar{w})\nu_n}{\sum_n (w_n - \bar{w})^2} \tag{7.6}$$

From Equation 7.5 and Equation 7.6, we get:

$$\hat{\beta} = \beta + \frac{\sigma_\epsilon \rho \sum_n (w_n - \bar{w})\nu_n + \sigma_\epsilon \sum_n (w_n - \bar{w})\iota_n}{\delta \sum_n (w_n - \bar{w})^2 + \sigma_\nu \sum_n (w_n - \bar{w})\nu_n}$$

Denoting $c_n = \frac{w_n - \bar{w}}{\sqrt{\sum_n (w_n - \bar{w})^2}}$, with $\sum_n c_n^2 = 1$, we get:

$$\hat{\beta} = \beta + \frac{\sigma_\epsilon \rho \sum_n c_n \nu_n + \sigma_\epsilon \sum_n c_n \iota_n}{\gamma\sqrt{\sum_n (w_n - \bar{w})^2} + \sigma_\nu \sum_n c_n \nu_n}$$

As, by construction, $\mathrm{E}(\iota \mid \nu) = 0$, denoting $\omega = \sum_n c_n \nu_n$, which is a standard normal deviate and $a = \delta\sqrt{\sum_n (w_n - \bar{w})^2} / \sigma_\nu$ we finally get:

$$\hat{\beta} = \beta + \frac{\sigma_\epsilon \rho}{\sigma_\nu} \frac{\omega}{\omega + a}$$

Then the expected value of $\hat{\beta}$ is obtained by integrating out this expression with respect to ω:

$$\mathrm{E}(\hat{\beta}) = \beta + \frac{\sigma_\epsilon \rho}{\sigma_\nu} \int_{-\infty}^{\infty} \frac{\omega}{\omega + a} \phi(\omega) d\omega$$

but this integral is divergent as $\omega/(a + \omega)$ tends to infinity as ω approach $-a$. Therefore, $\hat{\beta}$ has no expected value. The 2SLS derivation of the IV estimator also gives an intuition of the reason why the IV estimator is not unbiased. It would be if, for the second OLS estimation, $\mathrm{E}(x_n \mid w_n) = \gamma + \delta w_n$ were used as the regressor. But actually, $\hat{x}_n = \hat{\gamma} + \hat{\delta} w_n$ is used and, as the OLS estimator over-fits, the fitted values of x will be partly correlated with ϵ. Of course, when the sample size grows, as the OLS estimator is consistent, $\hat{x}_n$ converges to $\gamma + \delta z_n$ and the asymptotic bias vanishes. This can be usefully illustrated by simulation. The `iv_data` function draws a sample of y, x and one instrument w:

```
iv_data <- function(N = 5E01, R = 1E03,
                    r_xe = 0.5, r_xw = 0.2, r_we = 0,
                    alpha = 1, beta = 1,
                    sds = c(x = 1, e = 1, w = 1),
                    mns = c(x = 0, w = 0)){
```

```
        nms <- c("x", "e", "w")
        names(sds) <- nms ;  names(mns) <- c("x", "w")
        b_wx <- r_xw * sds["x"] / sds["w"]
        a_wx <- mns["x"] - b_wx * mns["w"]
        cors <- matrix(c(1, r_xe, r_xw, r_xe, 1, r_we, r_xw, 0, 1), nrow = 3)
        XEW <- matrix(rnorm(N * R * 3), nrow = N * R) %*%
            chol(cors)
        colnames(XEW) <- nms
        XEW %>%
            as_tibble %>%
            mutate(x = x * sds["x"] + mns["x"],
                   w = w * sds["w"] + mns["w"],
                   e = e * sds["e"],
                   Exw = a_wx + b_wx * w,
                   y = alpha + beta * x + e) %>%
            add_column(id = factor(rep(1:R, each = N)), .before = 1)
}
```

The arguments of the function are the sample size (`N`), the number of samples (`R`), the correlations between x and ϵ (the default is 0.5), between x and w (0.2 by default) and between w and ϵ. The default value for this last correlation is 0, which is a necessary condition for the IV estimator to be consistent. x, ϵ and w are assumed by default to be standard normal deviates, but the means of x and w and the standard deviations of x, w and ϵ can be customized using the `mns` and `sds` argument. Finally, the coefficients of the linear relation between y and x are `alpha` and `beta` and these two values are set by default to 1. First, a matrix of normal standard deviates `XEW` is constructed. This matrix is post-multiplied by the Cholesky decomposition of the matrix of correlation, which introduces the desired correlation between x, ϵ and w, which are then adjusted for non-zero means and non-unity standard deviations if necessary. Finally the vector of response is computed, along with the conditional expectation of x: $\mbox{E}(x \mid w) = \gamma + \delta w$. The `iv_coefs` function computes the IV estimator using the 2SLS approach.

```
iv_coefs <- function(i){
    xh <- lm.fit(cbind(1, i$w), i$x)$fitted.values
    ols <- coef(lm.fit(cbind(1, i$x), i$y))[2] %>% unname
    ivo <- coef(lm.fit(cbind(1, i$Exw), i$y))[2] %>% unname
    iv <- coef(lm.fit(cbind(1, xh), i$y))[2] %>% unname
    tibble(ols = ols, ivo = ivo, iv = iv)
}
```

It uses `lm.fit` to regress x on w.[5] We compute `xh` which is $\hat{x} = \hat{\gamma} + \hat{\delta} w_n$, the fitted values of the regression of x on w. `iv_coefs` computes three estimators:

- `ols`, which is the OLS estimator of y on x,
- `iv`, which is the 2SLS estimator of y on x using w as instruments,
- `ivo`, which is an estimator that can only be computed in the context of simulations and uses $\mbox{E}(x_n \mid w_n)$ instead of $\hat{x}_n$ as the regressor in the second OLS estimation.

[5]`lm.fit` is used internally by `lm`, and its two first arguments are a matrix of covariates and the response.

We first generate a unique sample of 100 observations and we compute the three estimators.

```
set.seed(1)
d <- iv_data(R = 1, N = 1E02)
iv_coefs(d)
## # A tibble: 1 x 3
##     ols   ivo    iv
##   <dbl> <dbl> <dbl>
## 1  1.50 0.623 0.752
```

To empirically analyze the distribution of these three estimators, we then generate several samples by setting the R argument, compute the estimators for every sample and analyze the empirical distribution of the estimators. This can be easily done using `tidyr::nests` and `tidyr::unnests` functions, as in Section 1.6.2.

```
set.seed(1)
d <- iv_data(R = 1E04, N = 50) %>%
    nest(.by = id) %>%
    transmute(model = map(data, iv_coefs)) %>%
    unnest(cols = model)
d %>% print(n = 3)
```

```
# A tibble: 10,000 x 3
    ols    ivo     iv
  <dbl>  <dbl>  <dbl>
1  1.56 0.437  0.769
2  1.42 0.605  0.415
3  1.64 0.0185 0.0786
# i 9,997 more rows
```

We then compute the mean, the median and the standard deviation for the three estimators:

```
d %>% summarise(across(everything(),
                       list(mean = mean, median = median,
                            sd = sd))) %>%
    pivot_longer(1:6) %>%
    separate(name, into = c("model", "stat")) %>%
    pivot_wider(names_from = stat, values_from = value)
```

```
# A tibble: 2 x 7
  iv_mean iv_median iv_sd model  mean median    sd
    <dbl>     <dbl> <dbl> <chr> <dbl>  <dbl> <dbl>
1   -6.39      1.10  747. ols    1.50  1.50  0.126
2   -6.39      1.10  747. ivo    1.00  0.996 1.26
```

The distribution of the three estimators is presented in Figure 7.2.

```
dsties <- d %>% as_tibble %>% pivot_longer(1:3) %>% ggplot(aes(value)) +
    geom_density(aes(linetype = name)) +
    scale_x_continuous(limits = c(-4, 3))
dsties
```

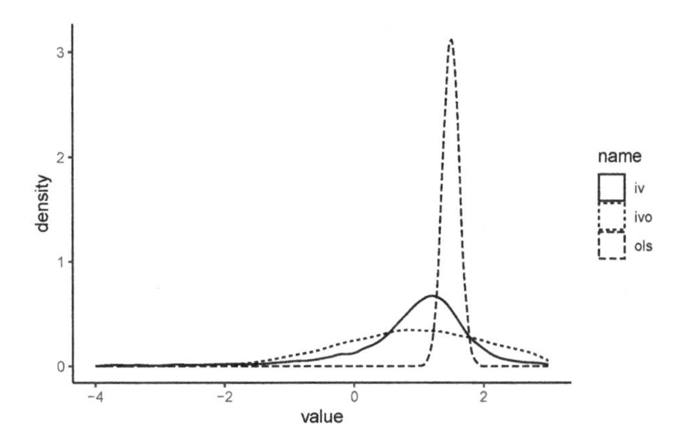

Figure 7.2: Distribution of the IV estimator

The OLS estimator is severely biased, as the central value of its distribution is about 1.5 and it has a small variance. The "pseudo-IV" estimator seems unbiased, as its mean (and median) is very close to one. The standard deviation is about 10 times larger than that of the OLS estimator, so the density curve is extremely flat. The mode of the density curve of the IV estimator is slightly larger than 1. Moreover, it has extremely fat tails (much more than the ones of the pseudo-IV estimator), which explains why the expected value and the variance don't exist. This feature becomes obvious if we zoom in extreme values of the estimator, for example on the $(-4, -3)$ range (see Figure 7.3).

```
dsties + coord_cartesian(xlim = c(-4, -3), ylim = c(0, 0.015))
```

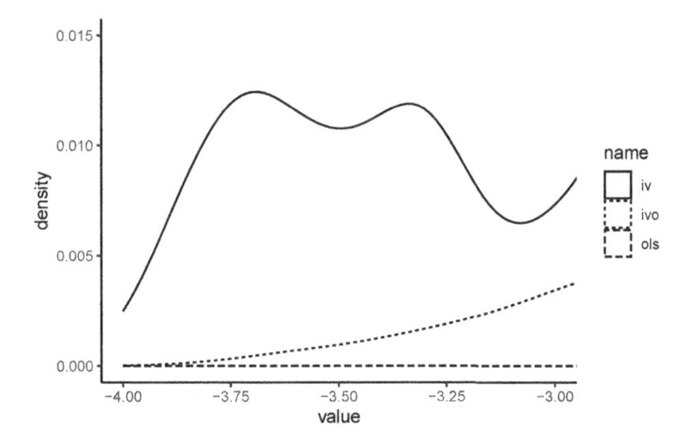

Figure 7.3: Distribution of the IV estimator (zoom)

7.2.3 An example: Segregation effects on urban poverty and inequality

Ananat (2011) investigates the causal effect of segregation on urban poverty and inequality. Several responses are used, especially the Gini index and the poverty rate for the black populations of 121 American cities. The data set is called `tracks_side`. The level of segregation is measured by the following dissimilarity index:

$$\frac{1}{2}\sum_{n=1}^{N}|b_n - w_n|$$

where b_n and w_n are respectively the share of the black and white population of the whole city that lives in census track n of the city. This index ranges from 0 (no segregation) to 1 (perfect segregation). In the sample of 121 cities used, the segregation index ranges from 0.33 to 0.87, with a median value of 0.57. We'll focus on the effect of segregation on the poverty rate of black people:

```
lm_yx <- lm(povb ~ segregation, tracks_side)
lm_yx %>% gaze
##             Estimate Std. Error t value Pr(>|t|)
## segregation   0.1818     0.0514    3.54  0.00058
```

The coefficient of segregation is positive and highly significant. It indicates that a 1 point increase of the segregation index raises the poverty rate of black people by about 0.18 point. The correlation between segregation and bad economic outcome for black people is well established but, according to the author, the OLS estimator cannot easily be considered as a measure of the causal relationship of segregation on income, as there are some other variables that both influence segregation and outcome for black people. As an example, the situation of Detroit is described, which is a highly segregated city with poor economic outcomes, but other characteristics of the city (political corruption, legacy of a manufacturing economy) can be the cause of these two phenomena (Ananat 2011, 35). Therefore, the OLS estimator is suspected to be biased and inconsistent because of the omitted variable bias. The instrumental variable estimator can be used in this context, but it requires the use of a good instrumental variable, i.e., a variable which is correlated with the endogenous covariate (segregation), but not directly with the response (the poverty rate). The author suggests that the way cities were subdivided by railroads into a large number of neighborhoods can be used as an instrument. Moreover, the tracks were mostly built during the nineteenth century, prior to the great migration (between 1915 to 1950) of African Americans from the south. More precisely, the index is defined as follow:

$$1-\sum_n\left(\frac{a_n}{A}\right)^2$$

where a_n is the area of neighborhood n defined by the rail tracks and A is the total area of the city. The index is 0 if the city is completely undivided and tends to 1 if the number of neighborhoods tends to infinity. This index ranges from 0.24 to 0.99 in the sample, with a median value of 0.74. The regression of x (`segregation`) on w (`raildiv`) gives:

```
lm_xw <- lm(segregation ~ raildiv, tracks_side)
lm_xw %>% gaze
```

```
##            Estimate Std. Error t value Pr(>|t|)
## raildiv     0.3995      0.0796    5.02  1.8e-06
```

In the 2SLS interpretation of the IV estimator, this is the **first stage regression**. The coefficient of `raildiv` is, as expected, positive and highly significant. It is important to check that the correlation between the covariate and the instrument is strong enough to get a precise IV estimator. This can be performed by computing their coefficient of correlation, or using the R^2 or the F statistic of the first stage regression:

```
tracks_side %>% summarise(cor(segregation, raildiv)) %>% pull
## [1] 0.418
lm_xw %>% ftest %>% gaze
## F = 25.190, df: 1-119, pval = 0.000
lm_xw %>% rsq
## [1] 0.1747
```

The IV estimator can be obtained by regressing the response on the fitted values of the first stage regression:

```
lm_yhx <- lm(povb ~ fitted(lm_xw), tracks_side)
lm_yhx %>% gaze
##                 Estimate Std. Error t value Pr(>|t|)
## fitted(lm_xw)      0.231      0.128    1.81    0.072
```

As expected, the IV estimator is larger than the OLS estimator (0.23 vs. 0.18). It can also be obtained by dividing the OLS coefficients of the regressions of y on w and of x on w. The latter has already been computed, the former is:

```
lm_yw <- lm(povb ~ raildiv, tracks_side)
coef(lm_yw)[2]
## raildiv
## 0.09233
coef(lm_yw)[2] / coef(lm_xw)[2]
## raildiv
##  0.2311
```

A 1 point increase of `raildiv` is associated with a 0.4 point increase of the discrimination index and with a 0.09 point of the poverty rate. Therefore, the 0.4 point increase of `segregation` increases `povb` by 0.09, which means that an increase of 1 point of `segregation` would increase `povb` by $0.09/0.4 = 0.23$, which is the value of the IV estimator.

7.2.4 Wald estimator

The Wald estimator is the special case of the instrumental variable estimator where the instrument w is a binary variable and therefore defines two groups ($w = 0$ and $w = 1$). In this case, the slope of the regression of y on z is $\hat{\beta}_{y/w} = \bar{y}_1 - \bar{y}_0$, where $\bar{y}_i$ is the sample mean of y in group $i = 0, 1$ and, similarly, the regression of x on w is $\hat{\beta}_{x/w} = \bar{x}_1 - \bar{x}_0$. The Wald estimator is then:

$$\hat{\beta} = \frac{\bar{y}_1 - \bar{y}_0}{\bar{x}_1 - \bar{x}_0} = \frac{\hat{\beta}_{yw}}{\hat{\beta}_{xw}}$$

As $(\bar{x}_1 - \bar{x}_0)$ converges to a constant, the asymptotic standard deviation of $\hat{\beta}$ is:[6]

$$\hat{\sigma}_{\hat{\beta}} = \frac{\hat{\sigma}_{\bar{y}_1 - \bar{y}_0}}{\bar{x}_1 - \bar{x}_0} \tag{7.7}$$

Angrist (1990) studied whether veterans (in his article the Vietnam war) experience a long-term loss of income, and therefore should legitimately receive a benefit to compensate this loss. An obvious way to detect a loss would be to consider a sample with two groups (veterans and non-veterans) and to compare the mean income in these two groups. However, enrollment in the army is not random, and therefore it is probable that "certain types of men are more likely to serve in the armed forces than others".[7] In this situation, income difference is likely to be a biased estimator of the effect of military enrollment. To overcome this difficulty, Angrist (1990) used the fact that a draft lottery was used during the Vietnam war to select the young men who were enrolled in the army. More precisely, the 365 possible days of birth were randomly drawn and ordered and, later on, the army announced the number of days of birth that would lead to an enrollment (depending on the year, this number was between 95 and 195). All the lottery-eligible men didn't go to the war and some that were not eligible fought, but the lottery produces an exogenous variation of the probability to be enrolled. Two data sources are used and contained in the `vietnam` data set. The first one is a 1% sample of all the social security numbers and indicates the yearly income. A subset of the `vietnam` data set (defined by `variable == income`) contains the average and the standard deviation of income (the FICA definition) for every combination of birth year (`birth`, from 1950 to 1953), draft eligibility (`eligible` a factor with levels `yes` or `no`), race (`white` or `nonwhite`) and year (from 1966 to 1984 for men born in 1950 and starting in 1967, 1968 and 1969 for men born respectively in 1951, 1952 and 1953).

```
vietnam %>% print(n = 2)
```

```
# A tibble: 234 x 8
  variable birth eligible race   year  mean    sd   cpi
  <chr>    <dbl> <fct>    <chr> <dbl> <dbl> <dbl> <dbl>
1 income    1950 no       white  1966  479.  8.35  97.2
2 income    1950 no       white  1967  825. 10.8  100
# i 232 more rows
```

We divide the mean income and its standard deviation by the consumer price index (`cpi`), we then create two columns for eligible and non-eligible men and we compute the mean difference (`dmean`) and its standard deviation (`sd_dmean`):

```
dinc_elig <- vietnam %>%
    filter(variable == "income") %>%
    mutate(mean = mean / cpi * 100, sd = sd / cpi * 100) %>%
```

[6]See Angrist (1990), note 7, pp. 321-322.
[7]Angrist (1990), p. 313.

```
    select(- variable, - cpi) %>%
    pivot_wider(names_from = eligible, values_from = c(mean, sd)) %>%
    mutate(dmean = mean_yes - mean_no,
           sd_dmean = sqrt(sd_no ^ 2 + sd_yes ^ 2)) %>%
    select(-(mean_no:sd_yes))
dinc_elig %>% print(n = 2)
```

```
# A tibble: 108 x 5
  birth race   year  dmean sd_dmean
  <dbl> <chr> <dbl>  <dbl>    <dbl>
1  1950 white  1966 -22.4      12.1
2  1950 white  1967  -8.02     15.1
# i 106 more rows
```

The results match table 1 of Angrist (1990) page 318 (except that, in the table, the income is not divided by the cpi). The mean income difference is the numerator of the Wald estimator. The denominator is the difference of enrollment probability between draft eligible and ineligible young men. The author estimates this difference using the data of the 1984 Survey of Income and Program Participation (SIPP), which are available in the subset of the `vietnam` data set obtained with `variable == "veteran"` and contains the share of veterans and the standard deviation of this share for eligible and non-eligible young men by year of birth and race (`birth` and `race`):

```
veterans <- vietnam %>%
    filter(variable == "veteran") %>%
    select(- year, - cpi, - variable)
veterans %>% print(n = 3)
```

```
# A tibble: 18 x 5
  birth eligible race   mean     sd
  <dbl> <fct>    <chr> <dbl>  <dbl>
1  1950 no       white 0.193 0.0169
2  1950 yes      white 0.353 0.0223
3  1950 total    white 0.267 0.0140
# i 15 more rows
```

We then create one column for eligible and non-eligible young men, and we compute the difference of shares of enrollment and its standard deviation:

```
dshare_elig <- veterans %>%
    pivot_wider(names_from = eligible, values_from = c(mean, sd)) %>%
    mutate(dshare = mean_yes - mean_no,
           sd_dshare = sqrt(sd_yes ^ 2 + sd_no ^ 2)) %>%
    select(- (mean_no:sd_total))
dshare_elig %>% print(n = 2)
```

```
# A tibble: 6 x 4
  birth race  dshare sd_dshare
```

```
  <dbl> <chr>  <dbl>      <dbl>
1  1950 white  0.159     0.0280
2  1951 white  0.136     0.0277
# i 4 more rows
```

Finally, we join the two tables by race and year of birth; we compute the Wald estimator of the income difference, its standard deviation (using Equation 7.7) and the Student statistic, which is their ratio:

```
wald <- dinc_elig %>%
    left_join(dshare_elig, by = c("birth", "race")) %>%
    mutate(wald = dmean / dshare,
           sd = sd_dmean / dshare,
           z = wald / sd) %>%
    select(birth, race, year, wald, sd, z)
wald %>% print(n = 2)
```

```
# A tibble: 108 x 6
  birth race   year   wald    sd      z
  <dbl> <chr> <dbl>  <dbl> <dbl>  <dbl>
1  1950 white  1966 -141.   76.0 -1.85
2  1950 white  1967  -50.3  94.5 -0.532
# i 106 more rows
```

The income differentials are depicted in Figure 7.4; note the use of `geom_ribbon` to draw a confidence interval for the mean income difference.

```
wald %>%
    filter(race == "white") %>%
    ggplot(aes(year, wald)) +
    geom_ribbon(aes(ymin = wald - 1.96 * sd, ymax = wald + 1.96 * sd),
                fill = "lightgrey") +
    geom_hline(yintercept = 0, linetype = "dashed") +
    geom_smooth(se = FALSE, span = 0.2, color = "black") +
    facet_grid(~ birth)
```

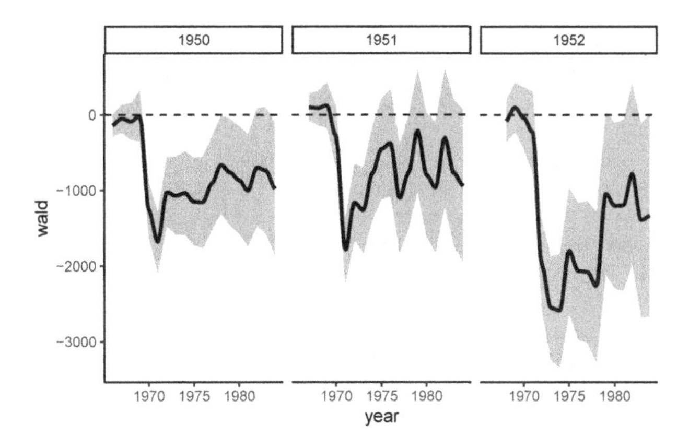

Figure 7.4: Income differentials between veterans and non-veterans

The income differential between veterans and non-veterans is substantial and persistent (about \$1000 of 1984) and is most of the time significant at the 5% level (except for men born in 1951).

7.3 General IV estimator

Consider now the general case. Among the covariates, some of them are endogenous and other are not and should be included in the instrument list. Moreover, there may be more instruments than endogenous covariates.

7.3.1 Computation of the estimator

There are K covariates, J endogenous covariates, $K-J$ exogenous covariates and G external instruments. We denote Z and W the matrix of covariates and instruments. The number of columns of these two matrices are respectively $K+1$ and $G+K-J+1$. For the model to be identified, we must have $G+K-J+1 \geq K+1$ or $G \geq J$. Therefore, there must be at least as many external instruments as there are endogenous covariates. We'll now denote $L+1 = G+K-J+1$ the number of columns of W. The $L+1$ moment conditions are $\mbox{E}(\epsilon|w) = 0$ which implies $\mbox{E}(\epsilon w) = 0$. The sample equivalent is the vector of $L+1$ empirical moments: $m = \frac{1}{N}W^\top\hat{\epsilon}$. As $\hat{\epsilon} = y - Z\hat{\gamma}$, this is a system of $L+1$ equation with $K+1$ unknown parameter. The system is over-identified if $L > K$ and in this case it is not possible to find a vector of estimates $\hat{\beta}$ for which all the empirical moments are 0. The instrumental variable estimator is, in this setting, the vector of parameters that makes the vector of empirical moments as close as possible to a vector of 0. If the errors are spherical, the variance of the vector of empirical moments is:

$$\mbox{V}(m) = \mbox{V}\left(\frac{1}{N}W^\top\epsilon\right) = \frac{1}{N^2}\mbox{E}\left(W^\top\epsilon\epsilon^\top W\right) = \frac{\sigma_\epsilon^2}{N^2}W^\top W \quad (7.8)$$

and the IV estimator minimizes the quadratic form of the vector of moments with the inverse of its covariance matrix:

$$\frac{N^2}{\sigma_\epsilon^2} m^\top (W^\top W)^{-1} m = \frac{1}{\sigma_\epsilon^2} \epsilon^\top W (W^\top W)^{-1} W^\top \epsilon = \frac{1}{\sigma_\epsilon^2} \epsilon^\top P_W \epsilon$$

where P_W is the projection matrix of W, i.e., $P_W z$ is the vector of fitted values of z obtained by regressing z on W. Therefore, the IV estimator minimizes:

$$\frac{1}{\sigma_\epsilon^2}(y - Z\gamma)^\top P_W (y - Z\gamma) = \frac{1}{\sigma_\epsilon^2}(P_W y - P_W Z\gamma)^\top (P_W y - P_W Z\gamma)$$

and therefore (because P_W is idempotent):

$$\hat{\gamma} = (Z^\top P_W Z)^{-1}(Z^\top P_W y) \tag{7.9}$$

The 2SLS interpretation of the IV estimator is clear, as it is the OLS estimator of a model with y or $(P_W y)$ as the response and $P_W Z$ the covariate. Therefore, it can be obtained in a first step by regressing all the covariates on the instruments and in a second steps by regressing the response on the fitted values of all the covariates obtained in the first step. Replacing y by $Z\gamma + \epsilon$ in Equation 7.9, we get:

$$\hat{\gamma} = \gamma + \left(\frac{1}{N} Z^\top P_W Z\right)^{-1} \left(\frac{1}{N} Z^\top P_W \epsilon\right)$$

and the IV estimator is consistent if $\mbox{plim} \frac{1}{N} Z^\top P_W \epsilon = 0$, or $\mbox{plim} \frac{1}{N} W^\top \epsilon = 0$. With spherical disturbances, the variance of the IV estimator is:

$$\mbox{V}(\hat{\gamma}) = \left(\frac{1}{N} Z^\top P_W Z\right)^{-1} \left(\frac{1}{N^2} Z^\top P_W \epsilon\epsilon^\top P_W Z\right) \left(\frac{1}{N} Z^\top P_W Z\right)^{-1} = \sigma_\epsilon^2 \left(Z^\top P_W Z\right)^{-1}$$

If the errors are heteroskedastic (or correlated), Equation 7.8 is a biased estimator of the variance of the moments, as $\mbox{E}(\epsilon\epsilon^\top) \neq \sigma_\epsilon^2 I$. In case of heteroskedasticity, $\mbox{E}(W^\top \epsilon\epsilon^\top W)$ can be consistently estimated by $\hat{S} = \sum_n \hat{\epsilon}_n^2 w_n w_n^\top$ where $\hat{\epsilon}$ are the residuals of a consistent estimation, for example the residuals of the IV estimator previously described. Then, the objective function is:

$$(y - Z\gamma)^\top W \hat{S}^{-1} W^\top (y - Z\gamma)$$

Minimizing this quadratic form leads to the **general method of moments (GMM)** estimator, also called the **two-stage IV** estimator.[8]

$$\hat{\gamma} = \left(Z^\top W \hat{S}^{-1} W^\top Z\right)^{-1} Z^\top W \hat{S}^{-1} W^\top y \tag{7.10}$$

Replacing y in Equation 7.10 by $Z\gamma + \epsilon$, we get:

[8]Cameron and Trivedi (2005), page 187.

$$\hat{\gamma} = \gamma + \left(Z^\top W \hat{S}^{-1} W^\top Z\right)^{-1} Z^\top W \hat{S}^{-1} W^\top \epsilon$$

which leads to the following covariance matrix:

$$\hat{\mathrm{V}}(\hat{\gamma}) = \left(Z^\top W \hat{S}^{-1} W^\top Z\right)^{-1}$$

To estimate this covariance matrix, one can use an estimation of S based on the residuals of the regression of the second step.

7.3.2 An example: long-term effects of slave trade

Africa experienced poor economic performances during the second half of the twentieth century, which can be explained by its experience of slave trade and colonialism. In particular, slave trade may induce long-term negative effects on the economic development of African countries because of induced corruption, ethnic fragmentation and weakening of established states. Africa experienced, between 1400 and 1900 four slave trades: the trans-Atlantic slave trade (the most important), but also the trans-Saharan, the Red Sea and the Indian Ocean slave trades. Not including those who died during the slave trade process, about 18 millions slaves were exported from Africa. Nunn (2008) conducted a quantitative analysis of the effects of slave trade on economic performances, by regressing the 2000 GDP per capita of 52 African countries on a measure of the level of slave extraction. The `slave_trade` data set is provided by the **necountries** package:

```
sltd <- necountries::slave_trade
```

The response is `gdp` and the main covariate is a measure of the level of slave extraction, which is the number of slaves normalized by the area of the country. In Figure 7.5, we first use a scatterplot, using log scales for both variables, which clearly indicates a negative relationship between slaves extraction and per capita GDP in 2000. Note the use of `ggrepel::geom_label_repel`: **ggplot2** provides two geoms to draw labels (`geom_text` and `geom_label`), but the labels may overlap. `geom_label_repel` computes a position for the labels that prevent this overlapping.

```
sltd %>% ggplot(aes(slavesarea, gdp)) + geom_point() +
  scale_x_continuous(trans = "log10", expand = expansion(mult = c(.1))) +
  scale_y_log10() + geom_smooth(method = "lm", se = FALSE, color = "black") +
  ggrepel::geom_label_repel(aes(label = country),
                            size = 2, max.overlaps = Inf)
```

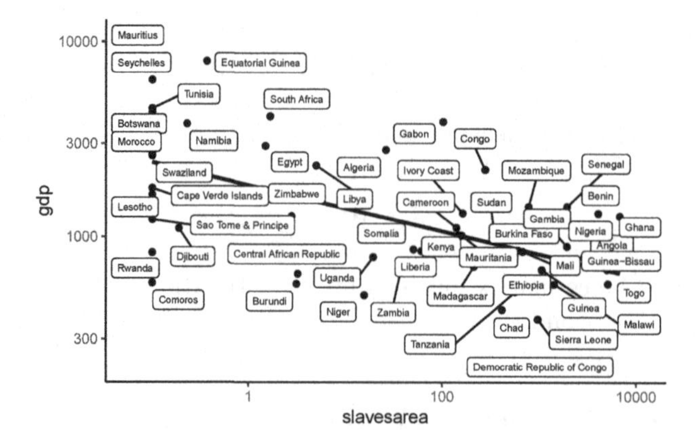

Figure 7.5: Per capita GDP and slave extraction

Nunn (2008) in table 2 presents a series of linear regressions, with different sets of controls. We just consider Nunn's first specification, which includes only dummies for the colonizer as supplementary covariates. `colony` is a factor with eight levels: we use `forcats::fct_lump_min` to merge the most infrequent levels: `"none"` (2 countries), `"spain"`, `"germany"` and `"italy"` (1 country).

```
sltd <- sltd %>% mutate(colony = fct_lump_min(colony, 3))
slaves_ols <- lm(log(gdp) ~ log(slavesarea) + colony, sltd)
slaves_ols %>% gaze(coef = "log(slavesarea)")
##                 Estimate Std. Error t value Pr(>|t|)
## log(slavesarea)  -0.1231     0.0234   -5.26  3.7e-06
```

The coefficient is negative and highly significant; it implies that a 10% increase of slave extraction induces a reduction of 1% of GDP per capita. As noticed by Nunn (2008), the estimation of the effect of slave trade on GDP can be inconsistent for two reasons:

- the level of slave extraction, which is based on information of the ethnicity of individual slaves and then aggregated at the current countries' level can be prone to error of measurement; moreover, for countries inside the continent (compared to coastal countries), a lot of slaves died during the journey to the coastal port of export, so the level of extraction may be underestimated for these countries,
- the average economic conditions may be different for countries that suffered a large extraction, compared to the others; in particular, if countries where the trade was particularly important were poor, their current poor economic conditions can be explained by their poor economic conditions 600 years ago and not by slave trades.

Measurement error induces an attenuation bias, which means that without measurement error, the negative effect of slave trades on GDP per capita would be stronger. The second effect would induce an upward-bias (in absolute value) of the coefficient on slave trades. But, actually, Nunn (2008) showed that areas of Africa that suffered the most slave trade were in general not the poorest areas, but the most developed ones. In this case, the OLS estimator would underestimate the effect of slave trades on GDP per capita. Nunn (2008) then performs instrumental variable regressions, using as instruments the distance between the centroid of the countries and the closest major market for the four slave trades (for example Mauritius and Oman for the Indian Ocean slave trade and Massawa, Suakin and Djibouti for the Red Sea slave trade). The IV regression can be performed by first regressing the endogenous covariate on the external instruments (**atlantic**, **indian**, **redsea** and **sahara**) and on the exogenous covariates (here **colony**, the factor indicating the previous colonizer). This is the so-called **first-stage regression**:

```
slaves_first <- lm(log(slavesarea) ~ colony + atlantic + indian +
                       redsea + sahara, sltd)
slaves_first %>% gaze
##                  Estimate Std. Error t value Pr(>|t|)
## colonyfrance       -1.873      1.323   -1.42  0.16391
## colonyportugal     -0.294      1.846   -0.16  0.87423
## colonybelgium      -3.260      2.236   -1.46  0.15213
## colonyOther        -1.360      1.946   -0.70  0.48840
## atlantic           -1.516      0.382   -3.97  0.00027
## indian             -1.108      0.410   -2.70  0.00982
## redsea             -0.384      0.765   -0.50  0.61845
## sahara             -2.569      0.881   -2.92  0.00561
slaves_first %>% rsq
## [1] 0.3315
```

Except for the **redsea** variable, the coefficients are highly significant, and the four instruments and the exogenous covariate explain more than one-fourth of the variance of slave extraction. The second stage is obtained by regressing the response on the fitted values of the first-step estimation:

```
sltd <- sltd %>% add_column(hlslarea = fitted(slaves_first))
slaves_second <- lm(log(gdp) ~ hlslarea, sltd)
slaves_second %>% gaze
##            Estimate Std. Error t value Pr(>|t|)
## hlslarea    -0.1736     0.0459   -3.78  0.00041
```

The coefficient has almost doubled, compared to the OLS estimator, which confirms that this latter estimator is biased, with an attenuation bias due to measurement error and a downward-bias caused by the fact that the most developed African regions were more affected by slave trade. The two-stage IV estimator is then computed. We use the **Formula** package which enables to write complex formulas with multiple set of variables, separated by the | operator:[9]

```
library(Formula)
form <- Formula(log(gdp) ~ log(slavesarea) + colony| colony +
                   redsea + atlantic + sahara + indian)
```

The first part contains the covariates and the second part the instruments. We then compute the model frame and we extract the covariates, the instruments and the response:

[9]We've already presented briefly this package in Section 3.7.1.

```
mf <- model.frame(form, sltd)
W <- model.matrix(form, sltd, rhs = 2)
Z <- model.matrix(form, sltd, rhs = 1)
y <- model.part(form, mf, lhs = 1) %>% pull
```

We then compute the cross-products of the instruments and the covariates ($Z^\top W$) and $\hat{S}$

```
ZPW <- crossprod(Z, W)
S <- crossprod(abs(resid(slaves_second)) * W)
vcov_1 <- solve(ZPW %*% solve(S) %*% t(ZPW))
iv2s <- vcov_1 %*% (ZPW %*% solve(S) %*% crossprod(W, y)) %>% drop
resid2 <- (y - Z %*% iv2s) %>% drop
S2 <- crossprod(abs(resid2) * W)
vcov_2 <- solve(ZPW %*% solve(S2) %*% t(ZPW))
cbind(coef = iv2s, sd1 = stder(vcov_1),
      sd2 = stder(vcov_2))[2, ] %>% print(digits = 2)
##    coef    sd1    sd2
## -0.213  0.044  0.044
```

The results are very similar to those of the one-step IV estimator. The IV estimator can also be computed using the **ivreg::ivreg** function. The main argument is a two-part formula, the first part containing the covariates and the second part the instruments:

```
ivreg::ivreg(log(gdp) ~ log(slavesarea) + colony| colony +
                 redsea + atlantic + sahara + indian, data = sltd)
```

The output of the **ivreg** function will be presented in Section 7.6.6.

7.4 Three-stage least squares

Consider now the case where the model is defined by a system of equations, some of the covariates entering these equations being endogenous. We consider therefore a system of L equations denoted by $y_l = Z_l\gamma_l + \epsilon_l$, with $l = 1 \ldots L$. This situation has already encountered in Section 3.7.1, Section 6.2.3 and Section 6.4.4. In this latter section, we considered that all the covariates were exogenous and we presented the seemingly unrelated regression estimator, which is a GLS estimator that takes into account the correlation between the errors of the different equations. Remember that, in matrix form, the system can be written as follows:

$$\begin{pmatrix} y_1 \\ y_2 \\ \vdots \\ y_L \end{pmatrix} = \begin{pmatrix} Z_1 & 0 & \ldots & 0 \\ 0 & Z_2 & \ldots & 0 \\ \vdots & \vdots & \ddots & \vdots \\ 0 & 0 & \ldots & Z_L \end{pmatrix} \begin{pmatrix} \gamma_1 \\ \gamma_2 \\ \vdots \\ \gamma_L \end{pmatrix} + \begin{pmatrix} \epsilon_1 \\ \epsilon_2 \\ \vdots \\ \epsilon_L \end{pmatrix}$$

And the covariance of the error vector for the whole system is assumed to be:

$$\Omega = \begin{pmatrix} \sigma_{11}I & \sigma_{12}I & \dots & \sigma_{1L}I \\ \sigma_{12}I & \sigma_{22}I & \dots & \sigma_{2L}I \\ \vdots & \vdots & \ddots & \vdots \\ \sigma_{1L}I & \sigma_{2L}I & \dots & \sigma_{LL}I \end{pmatrix} = \Sigma \otimes I$$

where $\otimes$ is the Kronecker product and Σ is a symmetric matrix of dimensions L for which the diagonal elements are the variance of the errors for a given equation and the off-diagonal elements the covariances between the pairs of errors of two different equations.

7.4.1 Computation of the three-stage least square estimator

If some covariates are endogenous, we should consider, for each equation, a matrix of instruments:

$$W = \begin{pmatrix} W_1 & 0 & \dots & 0 \\ 0 & W_2 & \dots & 0 \\ \vdots & \vdots & \ddots & \vdots \\ 0 & 0 & \dots & W_L \end{pmatrix}$$

The moment conditions for the whole system are then:

$$m = \frac{1}{N} W^\top \epsilon = \frac{1}{N} \begin{pmatrix} W_1^\top \epsilon_1 \\ W_2^\top \epsilon_2 \\ \vdots \\ W_L^\top \epsilon_L \end{pmatrix}$$

and the variance of the vector of moments is:

$$\mathrm{V}(m) = \frac{1}{N^2} \mathrm{E}\left(mm^\top\right) = \frac{1}{N^2} W^\top \mathrm{E}\left(\epsilon\epsilon^\top\right) W = \frac{1}{N^2} W^\top \Omega W = \frac{1}{N^2} W^\top (\Sigma \otimes I) W$$

The **three-stage least squares (3SLS)** estimator[10] minimizes the quadratic form of the moments with the inverse of this variance matrix:

$$m^\top (W^\top \Omega W)^{-1} m = (y - Z\gamma)^\top (W^\top \Omega W)^{-1} (y - Z\gamma)$$

which leads to the following estimator:

$$\hat{\gamma} = \left(Z^\top W (W^\top \Omega W)^{-1} W^\top Z\right)^{-1} \left(Z^\top W (W^\top \Omega W)^{-1} W^\top y\right) \tag{7.11}$$

This estimator can actually be computed using least squares on transformed data. Denote $\Psi = \Sigma^{-0.5} \otimes I$ the matrix such that $\Psi^\top \Psi = \Sigma^{-1} \otimes I = \Omega^{-1}$. Then, premultiply the covariates and the response by Ψ and the instruments by $(\Psi^{-1})^\top$. Then the projection matrix of $\tilde{W} = {\Psi^{-1}}^\top W$ is:

$$P_{\tilde{W}} = (\Psi^{-1})^\top W \left(W^\top \Psi^{-1} (\Psi^{-1})^\top W\right)^{-1} W^\top \Psi^{-1}$$

[10] Zellner and Theil (1962).

but $\Psi^{-1}(\Psi^{-1})^\top = \Psi^{-1}(\Psi^\top)^{-1} = (\Psi^\top\Psi)^{-1} = \Omega$. Therefore:

$$P_{\tilde{W}} = (\Psi^{-1})^\top W \left(W^\top \Omega W\right)^{-1} W^\top \Psi^{-1}$$

The transformed covariates and response are $\tilde{Z} = \Psi Z$ and $\tilde{y} = \Psi y$, so that performing the instrumental variable estimator on the transformed data, we get:

$$\begin{array}{rcl}
\hat{\gamma} &=& \left(\tilde{Z}^\top P_{\tilde{W}} \tilde{Z}\right)^{-1} \left(\tilde{Z}^\top P_{\tilde{W}} \tilde{y}\right) \\
&=& \left(Z^\top \Psi^\top (\Psi^{-1})^\top W \left(W^\top \Omega W\right)^{-1} W^\top \Psi^{-1} \Psi Z\right)^{-1} \\
&\times& \left(Z^\top \Psi^\top (\Psi^{-1})^\top W \left(W^\top \Omega W\right)^{-1} W^\top \Psi^{-1} \Psi y\right) \\
&=& \left(Z^\top W (W^\top \Omega W)^{-1} W^\top Z\right)^{-1} \left(Z^\top W (W^\top \Omega W)^{-1} W^\top y\right)
\end{array}$$

which is Equation 7.11. Therefore, the 3SLS estimator can be computed in the following way:

1. First compute the 2SLS estimator and retrieve the vectors of residuals ($\hat{\Xi} = (\hat{\epsilon}_1, \dots, \hat{\epsilon}_L)$).
2. Estimate Σ using the cross-products of the vectors of residuals: $\hat{\Sigma} = \hat{\Xi}^\top \hat{\Xi} / N$,
3. Use the Cholesky decomposition of $\hat{\Sigma}^{-1}$ to get V such that $V^\top V = \Sigma^{-1}$,
4. Premultiply the covariates and the response by $\hat{\Psi} = V \otimes I$ and the instruments by $\left(\hat{\Psi}^{-1}\right)^\top = (V^{-1})^\top \otimes I$,
5. Regress the transformed covariates on the transformed instruments and retrieve the fitted values,
6. Regress the transformed response on the fitted values of the previous regression.

7.4.2 An example: the watermelon market

Suits (1955) built an econometric model of the watermelon market using a time series for the United States, and his study was complemented by Wold (1958) who rebuilt the data set. This data set (called `watermelon`) is a good example of the use of system estimation with endogeneity, and its use for teaching purposes is advocated by Stewart (2019):

```
watermelon %>% print(n = 2)
```

```
# A tibble: 22 x 12
   year     q     h     y     n     p    pc    pv     w    pf    d1
  <dbl> <dbl> <dbl> <dbl> <dbl> <dbl> <dbl> <dbl> <dbl> <dbl> <dbl>
1  1930  1.93  1.90  4.87  2.09  2.07 0.976 0.367  1.46  1.10     0
2  1931  1.89  1.88  4.80  2.09  2.00 0.753 1.18   1.36  1.11     0
# i 20 more rows
# i 1 more variable: d2 <dbl>
```

On the supply side, two quantities of watermelons are distinguished: `q` are crop of watermelons available for harvest (millions) and `h` are watermelons actually harvested (millions).

`q` depends on planting decisions made on information of the previous season; more specifically, `q` depends on lag values of the average farm price of watermelon `p` (in dollars per thousand), the average annual net farm price per pound of cottons `pc` (in dollars), the average farm price of vegetables `pv` (an index) and on two dummy variables for government cotton acreage allotment program `d1` (one for the 1934-1951 period) and for World War 2 `d2` (one for 1943-1946 period).

The amount of watermelons actually harvested `h` depends on current price farm of watermelons `p`, wages `w` (the major cost of harvesting) and is of course bounded by `q`, the amount of watermelon available for harvest. More specifically, the relative price of watermelon and wage is considered. On the demand side, farm price depends on per capita harvest, per capita income (`yn`) and transportation cost (`pf`). An inverse demand function is estimated, of the form:

$$p = \alpha + \beta_q q + \beta_r r + ...$$

and, all the variables being in logarithm, the price and income elasticities are therefore respectively $1/\beta_q$ and $-\beta_r/\beta_q$. We compute the relative price of watermelons in terms of wage (`pw`), the income per capita (`yn`) and harvest (`hn`) and the first lags for `p`, `pc` and `pv`. We also remove the first and the last observation (because of the lag for the first one and because of the missing value for `pv` and `w` for the last one).

```
wm <- watermelon %>%
    mutate(yn = y - n, hn = h - n,
           pw = p - w, lp = lag(p),
           lpc = lag(pc), lpv = lag(pv)) %>%
    filter(! year %in% c(1930, 1951))
```

We can now define the set of three equations:

```
eq_c <- q ~ lp + lpc + lpv + d1 + d2
eq_s <- h ~ pw + q
eq_d <- p ~ hn + yn + pf
```

The exogenous variables are `w`, `n`, `yn`, `pf`, `d1`, `d2` and the lagged values of the price of watermelons (`lp`), cotton (`lpc`) and vegetables (`lpv`). We form a one-sided formula for this set of instruments:

```
eq_inst <- ~ w + n + yn + lp + pf + d1 + d2 + lpc + lpv
```

We then extract the three matrices of covariates, the matrix of instruments and the matrix of responses:

```
W <- model.matrix(eq_inst, wm)
X1 <- model.matrix(eq_c, wm)
X2 <- model.matrix(eq_s, wm)
X3 <- model.matrix(eq_d, wm)
N <- nrow(W)
Y <- select(wm, q, h, p) %>% as.matrix
```

We first compute the 2SLS estimator, using the **systemfit** package:

```
library(systemfit)
twosls <- systemfit(list(crop = eq_c, supply = eq_s, demand = eq_d),
                    inst = eq_inst, method = "2SLS", data = wm)
```

From this consistent, but inefficient estimator, we extract the data frame of residuals (one column per equation, one line per observation), we coerce it to a matrix and we estimate the matrix of covariance for the system of equation (Σ):

```
Sigma <- crossprod(as.matrix(resid(twosls))) / N
```

We then compute `V` using the Cholesky decomposition of the inverse of Σ:

```
V <- Sigma %>% solve %>% chol
```

Using V, we apply the relevant transformation for the response and for the covariate:

```
Xt <- rbind(cbind(V[1, 1] * X1, V[1, 2] * X2, V[1, 3] * X3),
            cbind(V[2, 1] * X1, V[2, 2] * X2, V[2, 3] * X3),
            cbind(V[3, 1] * X1, V[3, 2] * X2, V[3, 3] * X3))
yt <- Y %*% t(V) %>% as.numeric
```

We then apply the transformation for the instruments, using $\left(V^{-1}\right)^{\top}$. The matrix of instruments being the same for all the equations, the transformation can be obtained more simply using a Kronecker product:

```
Wt <- t(solve(V)) %x% W
```

Then, 2SLS is performed by first regressing `Zt` on `Xt`:

```
first <- lm(Xt ~ Wt - 1)
```

and then by regressing the fitted values of this first step regression on the transformed response:

```
second <- lm(yt ~ fitted(first) - 1)
```

Identical results are obtained using `systemfit` and setting `method` to `"3SLS"`:[11]

```
threesls <- systemfit(list(crop = eq_c, supply = eq_s,
                           demand = eq_d), inst = eq_inst,
                      method = "3SLS", data = wm,
                      methodResidCov=  "noDfCor")
coef(threesls)
```

[11]As for the SUR model estimated in Section 6.4.4, we set the `methodResidCov` argument to `"noDfCor"`, so that no degrees of freedom correction is used for the estimation of the covariance matrix of the errors.

```
  crop_(Intercept)             crop_lp           crop_lpc
           1.05294             0.57926           -0.32113
          crop_lpv             crop_d1            crop_d2
          -0.12373             0.03173           -0.15660
supply_(Intercept)           supply_pw           supply_q
          -0.21490             0.12145            1.06224
demand_(Intercept)           demand_hn          demand_yn
          -1.43808            -0.89257            1.55942
         demand_pf
          -0.85591
```

The relative price of watermelon is significantly positive in the supply equation, with a value of 0.12 which is the price elasticity of supply. In the inverse demand function, the coefficients of per capita quantity of watermelons and of per capita income have the expected sign (respectively negative and positive) and are highly significant. The estimated price and income elasticities are $1/-0.89 = -1.12$ and $-1.56/-0.89 = 1.75$.

7.5 Fixed effects model

In Section 6.1.2, we developed the error component model for panel or pseudo-panel data sets. Remember that when we have several observations (t) for the same entities (n), the simple linear model can be written as:

$$y_{nt} = \alpha + \beta x_{nt} + \epsilon_{nt} = \alpha + \beta x_{nt} + \eta_n + \nu_{nt}$$

the error term being the sum of two components: an entity / individual effect η_n and an idiosyncratic effect ν_{nt}. In Section 6.1.2, we assumed that x was uncorrelated with ϵ_{nt}. OLS was still a consistent estimator, but the BLUE estimator was GLS which takes into account the correlation between errors of the observations of the same entity / individual caused by the presence of a common individual effect η_n. Consider now that x is endogenous; the OLS and GLS estimators are then biased and inconsistent. If x is correlated with η: $\mathrm{E}(\eta \mid x) \neq 0$ but not with ν: $\mathrm{E}(\nu \mid x) = 0$, unbiased and consistent OLS estimators can be obtained when the individual effect η is either estimated or if the estimation is performed on a transformation of the covariate and the response that removes the individual effect. This is called the **fixed effects** estimator.

7.5.1 Computation of the fixed effects estimator

The linear model: $y_{nt} = \beta^\top x_{nt} + \eta_n + \nu_{nt}$ can be written in matrix form as:

$$\begin{pmatrix} y_{11} \\ y_{12} \\ \vdots \\ y_{1T} \\ y_{21} \\ y_{22} \\ \vdots \\ y_{2T} \\ \vdots \\ y_{N1} \\ y_{N2} \\ \vdots \\ y_{NT} \end{pmatrix} = \begin{pmatrix} x_{11}^\top \\ x_{12}^\top \\ \vdots \\ x_{1T}^\top \\ x_{21}^\top \\ x_{22}^\top \\ \vdots \\ x_{2T}^\top \\ \vdots \\ x_{N1}^\top \\ x_{N2}^\top \\ \vdots \\ x_{NT}^\top \end{pmatrix} \beta + \begin{pmatrix} 1 & 0 & \dots & 0 \\ 1 & 0 & \dots & 0 \\ \vdots & \vdots & \ddots & \vdots \\ 1 & 0 & \dots & 0 \\ 0 & 1 & \dots & 0 \\ 0 & 1 & \dots & 0 \\ \vdots & \vdots & \ddots & \vdots \\ 0 & 1 & \dots & 0 \\ \vdots & \vdots & \ddots & \vdots \\ 0 & 0 & \dots & 1 \\ 0 & 0 & \dots & 1 \\ \vdots & \vdots & \ddots & \vdots \\ 0 & 0 & \dots & 1 \end{pmatrix} \begin{pmatrix} \eta_1 \\ \eta_2 \\ \vdots \\ \eta_N \end{pmatrix} + \begin{pmatrix} \nu_{11} \\ \nu_{12} \\ \vdots \\ \nu_{1T} \\ \nu_{21} \\ \nu_{22} \\ \vdots \\ \nu_{2T} \\ \vdots \\ \nu_{N1} \\ \nu_{N2} \\ \vdots \\ \nu_{NT} \end{pmatrix}$$

or:

$$y = X\beta + D\eta + \nu$$

Note that this model doesn't contain an intercept: as η are considered as parameters to be estimated, with an intercept α, only N parameters like $\alpha + \eta_n$ can be estimated and therefore, α can safely be set to 0. $D = j_T \otimes I_N$ is a $NT \times N$ matrix where each column contains a dummy variable for one individual. The **least squares dummy variable** (**LSDV**) estimator consists of estimating by least squares β and η. Instead of estimating the N η parameters, it is simpler and more efficient to use the Frisch-Waugh theorem: first X and y are regressed on D, then the residuals of the first regression of y are regressed on those of X.

The OLS estimate of a variable z on D is: $\hat{\delta} = (D^\top D)^{-1} D^\top z$. But $D^\top D = TI$, so that $(D^\top D)^{-1} = I/T$. Moreover:

$$D^\top z = \begin{pmatrix} \sum_t z_{1t} \\ \sum_t z_{2t} \\ \vdots \\ \sum_t z_{Nt} \end{pmatrix}$$

and therefore $\hat{\delta}$ is a vector of individual mean of z, with typical element $\hat{\delta}_n = \bar{z}_{n.} = \sum_t z_{nt}/T$. Therefore, applying the Frisch-Waugh theorem implies that regressing y on X and D is equivalent to regressing y on X with both the response and the covariates measured in deviations from their individual means. For one observation, we have: $y_{nt} = \beta^\top x_{nt} + \eta_n + \nu_{nt}$. Taking the individual mean of this equation, we get: $\bar{y}_{n.} = \beta^\top \bar{x}_{n.} + \eta_n + \bar{\nu}_{n.}$. Taking deviations from the individual means, the model to estimate is then:

$$y_{nt} - \bar{y}_{n.} = \beta^\top (x_{nt} - \bar{x}_{n.}) + (\nu_{nt} - \bar{\nu}_{n.})$$

Therefore, if x is correlated with η, but not with ν, the OLS estimator of this model is consistent because the error doesn't contain η anymore and is therefore uncorrelated with x. This model is called the fixed effects estimator and also the **within** estimator in the panel data literature. With a single regressor, the estimator of the unique slope is:

$$\hat{\beta} = \frac{\sum_{n=1}^N \sum_{t=1}^T (y_{nt} - \bar{y}_{n.})(x_{nt} - \bar{x}_{n.})}{\sum_{n=1}^N \sum_{t=1}^T (x_{nt} - \bar{x}_{n.})^2} \tag{7.12}$$

The deviation from individual means is obviously not the only transformation that enables to get rid of the entity effects. Consider the special case where $T = 2$ for all individuals. An interesting example of this particular case is samples of twins. In this case:

$$\left\{ \begin{array}{l} y_{n1} = \beta x_{n1} + \eta_n + \nu_{n1} \\ y_{n2} = \beta x_{n2} + \eta_n + \nu_{n2} \end{array} \right.$$

And the difference between the two equations enables to get rid of the entity effect:

$$y_{n1} - y_{n2} = \beta(x_{n1} - x_{n2}) + \nu_{n1} - \nu_{n2}$$

The least squares estimator of the slope is the following **first-difference** estimator:

$$\hat{\beta} = \frac{\sum_n (y_{n1} - y_{n2})(x_{n1} - x_{n2})}{\sum_n (x_{n1} - x_{n2})}$$

which is in this case identical to the within estimator. This result can be established by remarking that the numerator of Equation 7.12 is a sum of N terms of the form:

$$(y_{n1} - \bar{y}_{n.})(x_{n1} - \bar{x}_{n.}) + (y_{n2} - \bar{y}_{n.})(x_{n2} - \bar{x}_{n.})$$

But $z_{n1} - \bar{z}_{n.} = z_{n1} - \frac{1}{2}(z_{n1} + z_{n2}) = \frac{1}{2}(z_{n1} - z_{n2})$. Similarly, $z_{n2} - \bar{z}_{n.} = -\frac{1}{2}(z_{n1} - z_{n2})$. Therefore, each term of the numerator of Equation 7.12 reduces to $\frac{1}{2}(y_{n1} - y_{n2})(x_{n1} - x_{n2})$ and similarly, each term of the denominator reduces to $\frac{1}{2}(x_{n1} - x_{n2})^2$ which proves the equivalence between the first-difference and the within estimators. When $T > 2$ the within and the first-difference estimators differ:

- the within estimator is more efficient, as it uses all the observations; on the contrary, while performing the first difference, the first observation for every individual is lost,
- the error of the within estimator is $\nu_{nt} - \bar{\nu}_{n.}$ and therefore contains the whole series of ν_n; on the contrary, the error of the first difference is $\nu_{nt} - \nu_{n(t-1)}$ and therefore contains the values of ν_n only for the current and the previous period.

7.5.2 Application: Mincer earning function using a sample of twins

In Section 6.4.3, we've estimated a Mincer earning function using the `twins` data set.

```
twins <- twins %>% mutate(age2 = age ^ 2 / 100)
lm(log(earning) ~ educ + age + age2, twins) %>% gaze
```

```
     Estimate Std. Error t value Pr(>|t|)
educ   0.0768     0.0106    7.25    2e-12
age    0.0778     0.0214    3.64  0.00031
age2  -0.0968     0.0266   -3.64  0.00031
```

The estimated return of education is about 7.7%, but it may be biased if the error is correlated with education. In particular, "abilities" are unobserved and may be correlated with education. If this correlation is positive, then the OLS estimator is upward-biased. The solution here is to consider that abilities should be similar for the pair of twins, as they are genetically identical and had the same familial environment. In this case, the fixed effects model can be performed. It can be obtained either by: estimating coefficients for all family dummies, using OLS on the within transformed variables or using OLS on the differences. Note that identical twins have obviously the same age, so this covariate disappears in the fixed effects model. Let's start with the LSDV estimator:

```
lsdv <- lm(log(earning) ~ educ + factor(family), twins)
lsdv %>% coef %>% head(4)
##      (Intercept)             educ factor(family)2 factor(family)3
##          1.57897          0.03935        -0.20232        -0.43463
lsdv %>% coef %>% length
## [1] 215
```

The estimated return of education is much lower (3.9%) than the OLS estimator. Note that numerous parameters are estimated (215), and the computation can be unfeasible if the number of entities is very large. Whatever the size of the sample, it is simpler and more efficient to use OLS on the transformed data, either using the within or the first-difference estimator. This can easily be done using:

```
twins %>% group_by(family) %>%
    mutate(learning = log(earning) - mean(log(earning)),
           educ = educ - mean(educ)) %>%
    lm(formula = learning ~ educ - 1) %>% coef
##    educ
## 0.03935
```

or:

```
twins %>% group_by(family) %>%
    summarise(learning = diff(log(earning)),
              educ = diff(educ)) %>%
    lm(formula = learning ~ educ - 1) %>% coef
##    educ
## 0.03935
```

In both cases, we grouped the rows by family, i.e., there are 214 groups of two lines, one for each pair of twins. In the first case, we use `mutate` to remove the individual mean; for example, `mean(educ)` is the mean of education computed for every family and is repeated two times, and we therefore have 428 observations (one for each individual). In the second case, we use `summarise` and we therefore get 214 observations (one for each family), the response and the covariate being the twin difference of earning and education.

7.5.3 Application: Testing Tobin's Q theory of investment using panel data

Schaller (1990) tested the relevance of Tobin's Q theory of investment by regressing the investment rate (the ratio of the investment and the stock of capital) to Tobin's Q, which is the ratio of the value of the firm and the stock of capital. This data set has already been presented in Section 6.4.3, where we've estimated by GLS an investment equation using the Tobin's Q theory of investment. We consider now the estimation of a fixed effects model. It is obtained using `plm` (of the **plm** package) by setting the `model` argument to `"within"` (which is actually the default value).

```
library(plm)
qw <- plm(ikn ~ qn, tobinq)
qw %>% summary
```

```
Oneway (individual) effect Within Model

Call:
plm(formula = ikn ~ qn, data = tobinq)

Balanced Panel: n = 188, T = 35, N = 6580

Residuals:
    Min.  1st Qu.   Median  3rd Qu.     Max.
-0.21631 -0.04525 -0.00849  0.03365  0.61844

Coefficients:
   Estimate Std. Error t-value Pr(>|t|)
qn 0.003792   0.000173      22   <2e-16 ***
---
Signif. codes:  0 '***' 0.001 '**' 0.01 '*' 0.05 '.' 0.1 ' ' 1

Total Sum of Squares:    36.7
Residual Sum of Squares: 34.1
R-Squared:        0.0702
Adj. R-Squared: 0.0428
F-statistic: 482.412 on 1 and 6391 DF, p-value: <2e-16
```

The individual effects are not estimated because the estimator uses the within transformation and then performs OLS on transformed data. However, the individual effects can be computed easily because: $\hat{\eta}_n = \bar{y}_{n.} - \hat{\beta}^\top \bar{x}_{n.}$. The `fixef` method for `plm` objects retrieves the fixed effects and a `type` argument can be set to:

- `"level"`: the effects are then the same as those obtained by LSDV without intercept,
- `"dfirst"`: only $N-1$ effects are estimated, the first one being set to 0; these are the effects obtained by LSDV with an intercept,
- `"dmean"`: N effects and an overall intercept are estimated, but the N effects have a 0 mean.

```
qw %>% fixef(type = "level") %>% head
##    2824   6284   9158  13716  17372  19411
## 0.1453 0.1281 0.2581 0.1100 0.1267 0.1695
qw %>% fixef(type = "dfirst") %>% head
##       6284      9158      13716      17372      19411      19519
```

```
## -0.01723  0.11279 -0.03528 -0.01856  0.02420 -0.01038
qw %>% fixef(type = "dmean") %>% head
##      2824      6284      9158     13716     17372     19411
## -0.014213 -0.031448  0.098581 -0.049492 -0.032778  0.009986
```

There is a `summary` method that reports the usual table of coefficients for the effects:

```
qw %>% fixef(type = "dmean") %>% summary %>% head(3)
```

```
     Estimate Std. Error t-value  Pr(>|t|)
2824 -0.01421    0.01240  -1.146 2.519e-01
6284 -0.03145    0.01234  -2.548 1.087e-02
9158  0.09858    0.01235   7.984 1.673e-15
```

7.6 Specification tests

7.6.1 Hausman test

Models estimated in this chapter, either using the IV or the fixed effects estimator, treat the endogeneity of some covariates, either by using instruments or by removing individual effects. These estimates are consistent if some covariates are actually endogenous, although other estimators like OLS or GLS are not. However, if endogeneity is actually not a problem, these later estimators are also consistent and are moreover more efficient. The Hausman test is based on the comparison of these two estimators and on the null hypothesis of the absence of endogeneity:

- $\hat{\beta}_0$, with variance $\hat{V}_0$ is only consistent if the hypothesis is true and is in this case efficient,
- $\hat{\beta}_1$, with variance $\hat{V}_1$ is always consistent, but is less efficient than $\hat{\beta}_0$ if the hypothesis is true.

Hausman's test (Hausman 1978) test is based on $\hat{q} = \hat{\beta}_1 - \hat{\beta}_0$, for which the variance is:

$$\mathrm{V}(\hat{q}) = \mathrm{V}(\hat{\beta}_1) + \mathrm{V}(\hat{\beta}_0) - 2\mathrm{cov}(\hat{\beta}_1, \hat{\beta}_0) \tag{7.13}$$

Hausman showed that the covariance between $\hat{q}$ and $\hat{\beta}_0$ is 0. Therefore:

$$\mathrm{cov}(\hat{\beta}_1 - \hat{\beta}_0, \hat{\beta}_0) = \mathrm{cov}(\hat{\beta}_1, \hat{\beta}_0) + \mathrm{V}(\hat{\beta}_0) = 0$$

and therefore Equation 7.13 simplifies to: $\mathrm{V}(\hat{q}) = \mathrm{V}(\hat{\beta}_1) - \mathrm{V}(\hat{\beta}_0)$. The asymptotic distribution of the difference of the two vectors of estimates is, under H_0:

$$\hat{\beta}_1 - \hat{\beta}_0 \overset{a}{\sim} \mathcal{N}\left(0, \mathrm{V}(\hat{\beta}_1) - \mathrm{V}(\hat{\beta}_0)\right)$$

and therefore $(\hat{\beta}_1 - \hat{\beta}_0)^\top \left(\mathrm{V}(\hat{\beta}_1) - \mathrm{V}(\hat{\beta}_0)\right)^{-1} (\hat{\beta}_1 - \hat{\beta}_0)$ is a χ^2 with degrees of freedom equal to the number of estimated parameters. Note that the length of $\hat{\beta}_0$ and $\hat{\beta}_1$ may be different. This is the case for panel data with individual specific covariates which are estimated using OLS or GLS but which are not with the within estimator. In this case, the test should be performed on the subset of common parameters. For tests involving OLS and IV estimator, the test can also be performed only on the subset of parameters associated with covariates that are suspected to be endogenous.

7.6.2 Weak instruments

Instruments should be not only uncorrelated with the error of the model, but they should also be correlated with the covariates. We have seen in Section 7.2.2 that the expected value of the IV estimator doesn't exist when there is only one instrument. It exists if the number of instruments is at least 2 and in this case, it can be shown that the bias of the IV estimator is approximately inversely proportional to the F statistic of the regression of the endogenous variable on the instruments.[12] Therefore, if the correlation between the endogenous variable and the instruments is weak, i.e., if the IV is performed using **weak instruments**, the estimator will not only be highly imprecise, but it will also be seriously biased, in the direction of the OLS estimator. While performing IV estimation it is therefore important to check that the instruments are sufficiently correlated with the endogenous covariate. This can be performed using an F test for the first stage regression, comparing the fit for the regression of the endogenous covariates on the set of the exogenous covariates, on the set of the exogenous covariates, and on the external instruments. A rule of thumb often used is that the F statistic should be at least equal to 10 (a less strict rule is $F > 5$).

7.6.3 Sargan test

The IV can be obtained as a moment estimator, using moment conditions $m = W^\top \epsilon / N$. If the instruments are relevant, they are uncorrelated with the errors, so that $m \overset{a}{\sim} \mathcal{N}(0, V)$ and $S = m^\top V^{-1} m$ is a χ^2 with a degree of freedom equal to the number of instruments. In the just-identified case, $m = 0$, but in the over-identified case, this statistic is positive and the hypothesis of orthogonal instruments implies that m is close enough to a vector of 0 or that S is sufficiently small.

7.6.4 Individual effects

In the fixed effects model, the absence of individual effects can be tested using a standard F test that all the estimated individual effects are zero. More simply, it can be obtained by comparing the sum of squares residuals of the OLS and the fixed effects models.

$$\frac{SCR_{\text{OLS}} - SCR_{\text{FE}}}{SCR_{\text{FE}}} \times \frac{N(T-1)-K}{N-1} \sim F_{N-1, N(T-1)-K}$$

[12]See Bound, Jaeger, and Baker (1995) and Cameron and Trivedi (2005) pages 108-109 for a discussion and further references.

7.6.5 Panel application: Testing Tobin's Q theory of investment

The presence of individual effects was already tested using a score test based on the OLS residuals. We compute here an F test, using `plm::pFtest`:

```
pFtest(ikn ~ qn, tobinq) %>% gaze
## F = 14.322, df: 187-6391, pval = 0.000
```

and the hypothesis is strongly rejected. Once we have concluded that there are individual effects, the question is whether these effects are correlated with the covariates or not. If this is the case, the fixed effects model should be used because the GLS model is inconsistent. On the contrary, both models are consistent, and the GLS estimator, which is more efficient, should be used.

```
phtest(ikn ~ qn, tobinq) %>% gaze
## chisq = 3.304, df: 1, pval = 0.069
```

The hypothesis of uncorrelated individual effects is not rejected at the 5%, which leads to the choice of the GLS estimator.

7.6.6 Instrumental variable application: slave trade

The IV estimator for the `slave_trade` data set was presented in Section 7.6.6. We use this time the `ivreg::ivreg` functions, which computes all the relevant specification tests.

```
sltd_iv <- ivreg::ivreg(log(gdp) ~ log(slavesarea) + colony |
                             colony + redsea + atlantic + sahara + indian,
                           data = sltd)
summary(sltd_iv)
```

```
Call:
ivreg::ivreg(formula = log(gdp) ~ log(slavesarea) + colony |
    colony + redsea + atlantic + sahara + indian, data = sltd)

Residuals:
    Min      1Q  Median      3Q     Max
-1.9262 -0.4487  0.0698  0.4580  1.3327

Coefficients:
                Estimate Std. Error t value Pr(>|t|)
(Intercept)       7.8498     0.2238   35.08   <2e-16 ***
log(slavesarea)  -0.1960     0.0461   -4.25   0.0001 ***
colonyfrance     -0.0101     0.2281   -0.04   0.9649
colonyportugal   -0.1119     0.3581   -0.31   0.7561
colonybelgium    -1.3941     0.4479   -3.11   0.0032 **
colonyOther       0.1865     0.3615    0.52   0.6084

Diagnostic tests:
```

```
                 df1 df2 statistic p-value
Weak instruments   4  43      4.89  0.0024 **
Wu-Hausman         1  45      4.76  0.0344 *
Sargan             3  NA      3.63  0.3042
---
Signif. codes:  0 '***' 0.001 '**' 0.01 '*' 0.05 '.' 0.1 ' ' 1

Residual standard error: 0.707 on 46 degrees of freedom
Multiple R-Squared: 0.338,  Adjusted R-squared: 0.266
Wald test: 5.34 on 5 and 46 DF,  p-value: 0.000592
```

The F statistic for the first stage regression is only 4.89, so that the instruments can be considered as weak and we can suspect the IV estimator to be severely biased. Anyway, remind that the bias is in the direction of the OLS estimator so that the effect of slave trades on current GDP would be underestimated. The p-value for the Hausman test is 0.03 so that the exogeneity hypothesis of the covariate is rejected at the 5% level, but not at the 1% level. Finally, as there are 4 instruments, the Sargan statistic, which is the quadratic form of the 4 moment conditions with the inverse of its variance is 3.63 and is a χ^2 with 3 degrees of freedom if the instruments are valid. This hypothesis is not rejected, even at the 10% level.

8

Treatment effect

In treatment evaluation analysis, one is interested in the effect of a treatment on a relevant outcome. Popular examples are the effect of microcredit access on the economic outcomes of households in developing countries, effects of a reduction of class size on the academic outcomes of pupils, effects of a youth job training program on employment status. In the simplest case, the outcome y is a continuous variable and the treatment variable D is binomial (1 for treatment, 0 otherwise). The causal impact of D on y is measured by the difference between the value of y if $D = 0$, denoted by y_0 and if $D = 1$, denoted by y_1. To an individual for which $D = 0$, y_0 is the observed or factual situation. The value of y if the individual had received the treatment is y_1 and is called the **counterfactual**. The fundamental problem is that the counterfactual is unobserved. This framework leads to the **potential outcome** model, which became increasingly popular in econometrics. On a sample of size N, denoting y_{0n} and y_{1n} the two values of the outcome, the natural estimator of the effect of the treatment would be:

$$\frac{\sum_{n=1}^{N}(y_{1n} - y_{0n})}{N} \tag{8.1}$$

but is unfeasible for a missing data problem: either y_{0n} or y_{1n} is observed, but not both, as an individual cannot be at the same time treated and untreated. In real settings, we have a sample that contains:

- a subsample of individuals who received the treatment ($D = 1$) called the **treatment group** (denoted by T),
- a subsample of individuals who didn't receive the treatment ($D = 0$) called the **control group** (denoted by C).

and an alternative estimator is obtained using the difference between the mean values of the outcome in the treatment and in the control group:

$$\frac{\sum_{n\in T} y_n}{N_T} - \frac{\sum_{n\in C} y_n}{N_C} \tag{8.2}$$

Moreover the data can be either **experimental** or **observational**, the difference between the two being that in the first case, the treatment is randomly assigned to some individuals. With experimental data, Equation 8.2 should be a reliable estimator of the effect of the treatment. On the contrary, in case of observational data, the observed and unobserved characteristics of the individuals in the two groups may be different and therefore, the estimator given in Equation 8.2 may include partly these differences and therefore may be biased. To overcome these difficulties, different estimators have been suggested for observational data, which rely on different assumptions that are detailed, for example, in Cameron and Trivedi (2005), pp. 862-865.

Section 8.1 presents the estimation of the treatment effect with experimental data. Section 8.2 explains how the instrumental variable estimator can be used to estimate the effect of a treatment. Section 8.3 presents the regression discontinuity estimator, Section 8.4 the difference in difference estimator and Section 8.5 the matching estimator. Finally, Section 8.6 is devoted to the synthetic control estimator.

8.1 Randomized experiment

Randomized experiences are the ideal setting to analyze treatment effects. The treatment and the control groups are composed of individuals who are randomly drawn from the same population, and therefore the average observable and unobserved characteristics in both groups should be similar.

To illustrate the use of randomized experiments, we consider the study of Burde and Linden (2013) who study the effect of village-based schools on children's academic performance. The sample comprises villages from the Ghor province in northwestern Afghanistan. More specifically, 31 villages were selected and formed 11 village groups. Five of them received a village-based school in summer 2007. In fall 2007, a survey was conducted in the 31 villages. The two outcomes of interest are the enrollment status of children between the ages of 6 and 11 and the score obtained to a short test covering math and language skills.

The first step of the analysis consists of checking that the observable characteristics are similar in the treatment and in the control group. For numerical variables, this can be performed using a t test of equal means. Assume that x is drawn from a normal distribution, with potentially a different mean for the treatment and for the control group, but the same variance σ_x^2. Then, $\bar{x}_T = \sum_{n\in T} x_n/N_T$ and $\bar{x}_C = \sum_{n\in C} x_n/N_C$ are normal variables with means equal to μ_T and μ_C and a variance equal respectively to σ_x^2/N_T and σ_x^2/N_C. Moreover, the difference of the means is also a normal variable with mean $\mu_T - \mu_C$ and a variance equal to the sum of the two variances, the covariance being null for a random sample. Therefore:

$$\frac{(\bar{x}_T - \bar{x}_C) - (\mu_T - \mu_C)}{\sigma_x\sqrt{1/N_T + 1/N_C}} \sim \mathcal{N}(0,1) \tag{8.3}$$

Replacing the unknown σ_x^2 by the unbiased estimate:

$$\hat{\sigma}_x^2 = \frac{\sum_{n\in T}(x_n - \bar{x}_T)^2 + \sum_{n\in C}(x_n - \bar{x}_C)^2}{N_T + N_C - 2}$$

and, with the hypothesis of equal mean ($H_0 : \mu_T = \mu_C$), $\frac{\bar{x}_T - \bar{x}_C}{\hat{\sigma}_x\sqrt{1/N_T + 1/N_C}}$ is a Student t with $N_T + N_C - 2$ degrees of freedom. The Student t-test assumes that the variance of x is the same in both groups. Welch's t-test, or the unequal variance t-test, extends the simple t-test to the case where the variance of x is different in the two groups.

`stats::t.test` computes the two flavors of the test. It uses a formula-data interface and has a supplementary argument called `var.equal` which is `FALSE` by default. Therefore, Welch's version of the test is computed by default and the simple t-test is obtained by setting `var.equal` to `TRUE`. We apply the two flavors of the test to the `distance` variable, which is the distance to the nearest formal school, using the `group` variable which is a factor with levels `"control"` and `"treatment"`:

```
t.test(distance ~ group, afghan_girls) %>% gaze
## diff = 0.253 (0.058), t = 4.393, df: 1483, pval = 0.000
```

and the hypothesis of equal mean is highly rejected. Computing the simple version of the t-test:

```
t.test(distance ~ group, afghan_girls, var.equal = TRUE) %>% gaze
## diff = 0.253 (0.058), t = 4.384, df: 1488, pval = 0.000
```

we get almost the same result. This last version of the test is equivalent to the result of a regression of the outcome variable on the group dummy:

```
lm(distance ~ group, afghan_girls) %>% gaze
##                 Estimate Std. Error t value Pr(>|t|)
## grouptreatment  -0.2529     0.0577   -4.38  1.2e-05
```

The coefficient of the intercept is the mean for the control group and the coefficient of `grouptreatment` is the difference of the means for the two groups. The t-value is exactly the one obtained using `t.test` with `var.equal = TRUE`.

Balance between groups of factors can be tested using Pearson's χ^2 test. It is based on the comparison of the observed joint distribution of the two variables and the one obtained using the independence hypotheses (obtained by multiplying the marginal frequencies of the two variables). Denote o and e the observed and the hypothesized frequencies, i the index of the cell and N the size of the sample:

$$\sum_i \frac{(o_i - e_i)^2}{e_i} \tag{8.4}$$

the statistic is, under the hypothesis of independence, a χ^2 with a number of degrees of freedom equal to $(J_1 - 1) \times (J_2 - 1)$ where J_1 and J_2 are the number of modalities for the two factors.

In the `afghan_girls` data set, `ethny` is a factor with levels `"other"`, `"farsi"` and `"tajik"`. We first compute the frequency table using the `table` function, which is similar to `dplyr::count`, but the input is a series and not a data frame and the result is an object of class `table`, which is much alike an object of class `matrix`, and not a data frame. We then compute the relative frequencies using the `prop.table` function. Note that `prop.table` has a `margin` argument that we don't use here.[1]

```
obs_freq <- prop.table(table(afghan_girls$group, afghan_girls$ethny))
```

In case of independence, the frequencies are obtained using the outer product of the marginal frequencies:

[1]If `margin = 1`, the frequencies sum to one for each line, and if `margin = 2`, they sum to one for each column.

```
freq_group <- prop.table(table(afghan_girls$group))
freq_ethny <- prop.table(table(afghan_girls$ethny))
ind_freq <- outer(freq_group, freq_ethny)
```

Applying Equation 8.4, we obtain:

```
sum((obs_freq - ind_freq) ^ 2 / ind_freq) * nrow(afghan_girls)
## [1] 2.846
```

This test can be conveniently computed using the `stats:chisq.test` function:

```
chisq.test(table(afghan_girls$group, afghan_girls$ethny)) %>% gaze
## X-squared = 2.846, df: 2, pval = 0.241
```

and the hypothesis of independence is not rejected. The whole table of balance is often called "Table 1", as it is the first table that appears in most articles in medical reviews. It can be computed using the `gtsummary::tbl_summary` function, which has a `by` argument that indicates the grouping variable. `tbl_summary` has a lot of arguments, and auxiliary functions that enables to customize the results.[2] The balance table for the `afghan_girls` data set is presented in Table 8.1.[3]

```
library("gtsummary") ; library("kableExtra")
afghan_girls %>%
    select(group, head_child:ethny) %>%
    tbl_summary(by = group,
                missing = "no",
                statistic = all_continuous() ~ "{mean} ({sd})",
                type = list(age ~ "continuous"),
                digits = all_continuous() ~ 2) %>%
    add_p(all_continuous() ~ "t.test",
          pvalue_fun = ~ style_pvalue(., digits = 2)) %>%
    as_kable_extra(booktabs = TRUE) %>% kable_styling() %>%
    column_spec(1, width = "15em")
```

The set of tests doesn't detect any significant differences between the two groups for the covariates `head_child`, `sex`, `age`, `occup`, `age_head`, `jeribs`, `distance` and `ethny`. On the contrary, for the `duration`, `hsize`, `sheeps` and `distance` covariates, the equal mean hypothesis is strongly rejected. The next step is to apply the same tests to the outcomes which are, in this study, `enrollment` and `test`. For the `test` variable, a boxplot is presented in Figure 8.1, with a separate plot for boys and girls.

```
afghan_girls %>% ggplot(aes(group, test)) +
  geom_boxplot() + facet_wrap(~ sex)
```

[2]Detailed documentation of the package can be found on the website of the package https://www.danieldsjoberg.com/gtsummary/ and on Sjoberg et al. (2021).

[3]Note that we also used some functions of the **kableExtra** package for fine-tuning the table.

Table 8.1: Balance table for the Afghan girls data set

Characteristic	**control** N = 708	**treatment** N = 782	**p-value**
head_child	645 (91%)	731 (93%)	0.085
sex			0.45
boy	386 (55%)	411 (53%)	
girl	322 (45%)	371 (47%)	
Child's age, fall 2007	8.31 (1.64)	8.32 (1.66)	0.92
Length of time family has lived in the village, fall 2007	27.59 (15.66)	30.30 (15.51)	<0.001
occup			0.67
farmer	515 (73%)	561 (72%)	
other	193 (27%)	221 (28%)	
Age of head of the household, fall 2007	39.97 (11.40)	40.14 (11.19)	0.77
Years of education of the head of the household, fall 2007	3.08 (3.51)	3.31 (3.54)	0.19
Number of people in the household, fall 2007	7.82 (2.56)	8.40 (2.92)	<0.001
Number of jeribs of land owned by household, fall 2007	1.27 (1.63)	1.34 (1.56)	0.39
Number of sheeps and goats owned by the household, fall 2007	5.63 (6.99)	7.55 (8.09)	<0.001
Distance (miles) to the nearest non-community based school, fall 2007	3.16 (1.09)	2.91 (1.13)	<0.001
ethny			0.24
other	413 (58%)	429 (55%)	
farsi	148 (21%)	163 (21%)	
tajik	147 (21%)	190 (24%)	

n (%); Mean (SD)

Pearson's Chi-squared test; Welch Two Sample t-test

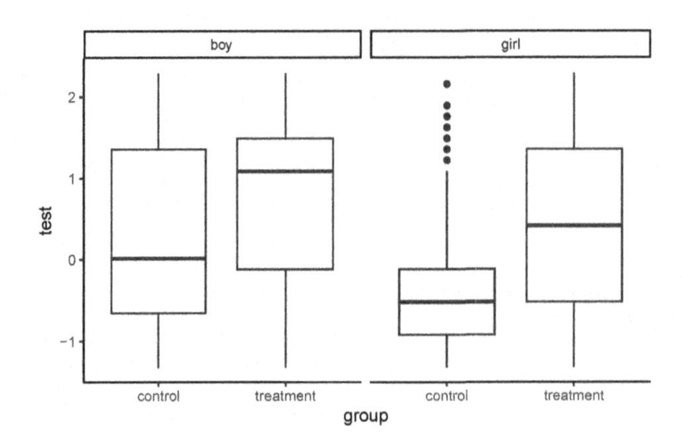

Figure 8.1: Boxplot for the test variable between the control and the treatment group

There seems to be a large positive effect of the treatment on the results of the test. Moreover, the scores for boys are much higher than for girls. For girls, we get:

```
t.test(test ~ group, afghan_girls, subset = sex == "girl") %>% gaze
## diff = -0.746 (0.068), t = -10.910, df: 653, pval = 0.000
```

The effect is strong (about 0.75, which means three-quarters of a standard deviation of the score, as this variable is standardized) and highly significant. As some covariates are unbalanced, this estimator may be biased. Therefore, it is recommended to measure the effect as the coefficient of `group` in a multiple regression with all the available controlling variables. Moreover, adding relevant variables will increase the precision of the estimation.

```
lm(test ~ group + chagcharan + head_child + age + duration + occup +
       age_head + educ + hsize + jeribs + sheeps + distance + ethny,
   afghan_girls, subset = sex == "girl") %>%
  gaze(coef = 2)
##                   Estimate Std. Error t value Pr(>|t|)
## grouptreatment     0.6542     0.0637    10.3   <2e-16
```

The coefficient is slightly lower but still very significant.

8.2 Instrumental variable estimator

Imbens and Angrist (1994) and Angrist, Imbens, and Rubin (1996) consider the use of instrumental variables in the context of the potential outcome model. We consider the simple case where the treatment d is binary and the instrument w is also binary.[4] The value of the outcome for individual n for a given combination of d and w is denoted by: $y_n(w_n, d_n)$; w can be used as an instrumental variable if w is related to y only because of its influence on d. Stated differently, for a given value of d, y is the same whatever

[4]This is the setting of the Wald estimator that was presented in Section 7.2.4.

the value of w: $y_n(0, d_n) = y_n(1, d_n)$. This is the **exclusion restriction**, standard in the instrumental variable literature. Therefore, potential outcome can be defined as a function of the treatment variable only: $y_n(w_n, d_n) = y_n(d_n)$ and d_n can be expressed as a function of d_n: $d_n(w_n)$. The next assumption is that on average, w has a causal effect on d: $\mathrm{E}(d(1) - d(0)) \neq 0$. The last assumption is **monotonicity** (Imbens and Angrist 1994) which states that $d_n(1) \geq d_n(0)\ \forall n$. With these assumptions in hand, the causal effect of w on y is:

$$\begin{array}{rcl} y_n(1, d_n(1)) - y_n(0, d_n(0)) & = & y_n(d_n(1)) - y_n(d_n(0)) \\ & = & [y_n(1)d_n(1) + y_n(0)(1 - d_n(1))] \\ & - & [y_n(1)d_n(0) - y_n(0)(1 - d_n(0))] \\ & = & (y_n(1) - y_n(0))(d_n(1) - d_n(0)) \end{array} \quad (8.5)$$

Therefore, the causal effect of w on y is the product of the causal effects of d on y and of w on d. Consider now the relation between w and d at the individual level. Four categories can be considered (represented on Figure 8.2):

- compliers: $d_n(1) = 1$ and $d_n(0) = 0$,
- always takers: $d_n(1) = d_n(0) = 1$,
- never takers: $d_n(1) = d_n(0) = 0$,
- deniers: $d_n(1) = 0$ and $d_n(0) = 1$.

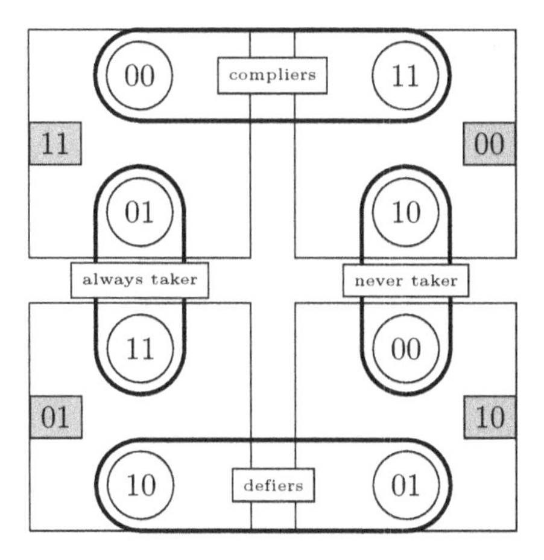

Figure 8.2: The four categories of individuals

There are four observable categories of individuals, represented by the four squares and named by the two digits (the observed values of w and d) indicated in a gray square. The counterfactuals are represented by two circles inside the square. For example, the square called 00 contains the individuals for which $w = 0$ and $d = 0$. For those individuals the unobserved counterfactual is $w = 1$ and d either equal to 0 or 1. If the counterfactual is $w = 1, d = 0$, the individual is a never-taker, $d = 0$ whatever the value of w. If the counterfactual is $w = 1, d = 1$, the individual is a complier: with $w = 0$, $d = 0$, but a change of w from 0 to 1 results in a change of d from 0 to 1. With the always takers and the never takers, the causal effect of w on y in Equation 8.5 is zero because the causal effect of w on d is zero. The monotonicity assumption rules out the existence of deniers. Therefore, the causal effect of w on d reduces to the treatment effect for the compliers.

Consider now the instrumental variable estimator. It estimates, in the general case, the following population estimand: $\frac{\mbox{cov}(y,w)}{\mbox{cov}(d,w)}$, which, for the case where both d and w are binary, reduces to:

$$\frac{\mbox{E}\left(y(1,d(1)) - y(0,d(0))\right)}{\mbox{E}\left(d(1) - d(0)\right)} \tag{8.6}$$

The numerator is the average causal effect of w on y for the compliers and the denominator is the share of compliers in the population. Angrist, Imbens, and Rubin (1996) call Equation 8.6 the **local average treatment effect** (**LATE**) to stress the fact that what is estimated using the instrumental variable estimator is an average treatment effect for only a subset of the population called the compliers, ie those for which a change in the value of w results in a change in the value of d. The numerator is also called the **reduced form** equation, as it measures the effect of the instrument on the outcome. The denominator is called the **intention to treat** equation, it measures the effect of the instrument on the treatment dummy.

As an example, we use the data set of Angrist et al. (2002) called `paces`. The authors investigate the effect of a large school voucher program in Columbia called PACES. This voucher covers more than half of the cost of private secondary school and may induce parents to enroll their children in private schools, which are known to provide much better service than public schools. In this case, w is a dummy for children who receive the voucher and d is a dummy for enrollment in private school.

```
paces %>% print(n = 2)
```

```
# A tibble: 1,577 x 18
     id privsch educyrs voucher pilot housvisit city   phone   age
  <dbl>   <dbl>   <dbl>   <dbl> <dbl>     <dbl> <fct>  <dbl> <dbl>
1     3       1       8       1     0         0 bogota     1    14
2     4       1       8       0     0         0 bogota     1    14
# i 1,575 more rows
# i 9 more variables: sex <fct>, strata <fct>, smpl <fct>,
#   month <fct>, married <dbl>, finish8 <dbl>, repetitions <dbl>,
#   in_school <dbl>, year <fct>
```

The treatment variable is `privsch` and the instrument is `voucher`. Several outcomes are considered by the authors; we'll consider only the number of years of finished education `educyrs`. The OLS estimate of the treatment is just the difference between the mean of the outcome for the two subsamples defined by the treatment variable:

```
lm(educyrs ~ privsch, paces) %>% gaze
##         Estimate Std. Error t value Pr(>|t|)
## privsch   0.2916     0.0557    5.23  1.9e-07
```

Therefore, the number of years of education is higher by about 0.29 of a year for pupils enrolled in private schools, and the effect is highly significant. We then compute the effect of the instrument on the treatment (intention to treat) and on the outcome (reduced form):

```
lm(privsch ~ voucher, paces) %>% gaze(coef = 2)
##           Estimate Std. Error t value Pr(>|t|)
## voucher     0.6421     0.0188    34.1   <2e-16
lm(educyrs ~ voucher, paces) %>% gaze(coef = 2)
##           Estimate Std. Error t value Pr(>|t|)
## voucher     0.1079     0.0553    1.95    0.051
```

The voucher has a large effect on enrolling in private school, the intention to treat effect being: 0.642. The effect of the instrument on the outcome is 0.108. The IV estimator is the ratio of these two effects, which is: 0.168. Therefore, in this example, we get an IV estimator of the treatment effect much smaller than the OLS estimator, which may be the symptom that unobserved determinants of the outcome are positively correlated with the enrollment in private school. Covariates can easily be added to the analysis. In their study, Angrist et al. (2002) use two covariates that indicate the type of survey (`pilot` is one if the individual was surveyed during the "pilot" survey, and `housvisit` is one if the survey was conducted in person and not by phone), `smpl` is a factor indicating the three subsamples (Bogota in 1995, Bogota in 1997 and Djamunid in 1993), `phone` is a dummy for owning a phone, `age` and `sex` are the age and the sex of the pupil and `strata` is a strata of residence. We then compute the OLS and IV estimators:

```
ols <- lm(educyrs ~ privsch + pilot + housvisit + smpl +
                phone + age + sex + strata + month, data = paces)
iv <- ivreg::ivreg(educyrs ~ privsch + pilot + housvisit + smpl +
                       phone + age + sex + strata + month | . - privsch +
                       voucher, data = paces)
ols %>% gaze(coef = 2)
##           Estimate Std. Error t value Pr(>|t|)
## privsch     0.1408     0.0424    3.32  0.00092
iv %>% gaze(coef = 2)
##           Estimate Std. Error t value Pr(>|t|)
## privsch     0.1342     0.0651    2.06    0.039
```

The two estimators are now almost identical: introducing the covariates reduces by almost one half the value of the OLS estimator and has a small effect on the IV estimator without covariates.

8.3 Regression discontinuity

Eligibility to a program is sometimes based on the value of an observable variable (called the **forcing variable**), and more precisely, on the fact that the value of this variable, for an individual is below or over a given threshold. Individuals just below and just over the threshold therefore constitute two groups of individuals who are very similar, except that the first group receives the treatment and the second group doesn't. This is called a **regression discontinuity** (**RD**) design.

8.3.1 Sharp and Fuzzy

Two variants of regression discontinuity designs can be considered:

- **sharp discontinuity**, there is a one-to-one correspondence between eligibility and treatment,
- **fuzzy discontinuity**, the probability of being treated is very different just below and just over the threshold.

We'll consider, in this chapter, two data sets. The first, called `probation`, is from Lindo, Sanders, and Oreopoulos (2010) who studied the effects of academic probation on academic performance in Canadian universities. If a student's grade point average (GPA) is below a certain threshold, he is placed on academic probation. GPA is between 0 and 4.5, and there are three campuses in the sample, denoted by `"1"`, `"2"` and `"3"`; for the first two, the threshold is 1.5 and for the third one it is 1.6. We first compute the distance to the threshold using the `gpa` and the `campus` variables:

```
probation <- probation %>%
    mutate(distcut = gpa - ifelse(campus == "3", 1.6, 1.5))
```

We then compute the joint frequency table for the two variables of eligibility and probation.

```
probation %>% mutate(eligible = as.numeric(distcut < 0)) %>%
  count(eligible, probation) %>%
  group_by(eligible) %>%
  mutate(n = n / sum(n)) %>%
  pivot_wider(names_from = eligible, values_from = n)
```

```
# A tibble: 2 x 3
  probation         `0`     `1`
  <fct>           <dbl>   <dbl>
1 no          1.00      0.00671
2 yes         0.0000806 0.993
```

Although the discontinuity is sharp, the probability of being on probation is not exactly equal to 1 for eligible students and equal to 0 for ineligible students because of administrative errors in data reporting.[5] We remove these misleading observations:

```
probation <- probation %>%
    filter( (probation == "yes" & distcut < 0) |
            (probation == "no" & distcut >= 0))
```

Buser (2015) analyzed the effect of income on religiousness. He considered the BDH (Bono Desarollo Humano) program in Ecuador which consists of a cash transfer for the poorest families. The eligibility is based on a wealth index called Selben, households in the four first deciles being eligible. The data were collected in 2011 and are available as `bdh`. The

[5] see Lindo, Sanders, and Oreopoulos (2010), footnote 20, page 104.

key variables are the **selben** index (the threshold has been removed so that eligible households are those for which **selben** is negative), **treated**, which is a dummy for households which receive the grant and **attendmonth**, which is the monthly frequency of visits to the church. First, we investigate whether the discontinuity is sharp or fuzzy, by computing the percentage of recipients for households under and over the threshold:

```
bdh <- bdh %>%
    mutate(eligible = ifelse(selben < 0, "yes", "no"))
bdh %>% count(eligible, treated) %>%
  group_by(eligible) %>%
  mutate(n = n / sum(n)) %>%
  pivot_wider(names_from = eligible, values_from = n)
```

```
# A tibble: 2 x 3
  treated     no   yes
    <dbl>  <dbl> <dbl>
1       0 0.971  0.144
2       1 0.0292 0.856
```

The discontinuity is therefore fuzzy, as about 3% of the eligible households don't receive the grant and, on the contrary, about 14% of the recipients are not eligible.

8.3.2 Plotting the discontinuity

The plot is obtained by defining bins for the forcing variable, computing the mean of the outcome for each bin, plotting the points and adding a smoothing line. The **micsr** package provides the convenient **geom_binmeans** function to compute and plot the mean of the response for bins of x (see Figure 8.3).

```
probation %>% filter(abs(distcut) < 1.5) %>%
    ggplot(aes(distcut, gpa2)) +
    geom_binmeans(aes(size = after_stat(n)), shape = 21,
                  center = 0, width = 0.2) +
    geom_smooth(aes(linetype = probation),
                method = "lm", se = FALSE, color = "black")
```

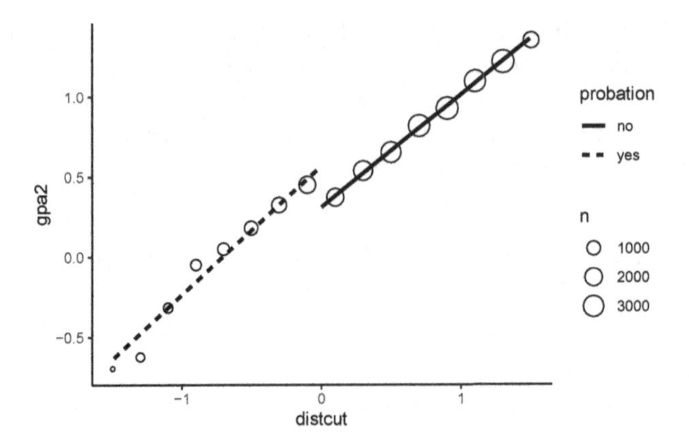

Figure 8.3: Discontinuity for probation

Two arguments of `geom_binmeans` are used: `center` to indicate the cutoff and `width` the width of the bins. The internal variable `n` is used to visualize the number of observations in every bin. Note the use of `after_stat`: `geom_binmeans` uses as stat `bins`, ie, it creates bins and counts the number of observations in each bin. This count (called `n`) is used in `geom_smooth` after the stat has been computed. For `geom_smooth` we use `probation` for the `linetype` aesthetic, so that two different smoothing lines are plotted on both sides of the cutoff. The effect of the treatment is then the difference of intercepts for the two regression lines. It is about 0.25 points of gpa in our example.

The `rdrobust::rdplot` function performs the same task (see Figure 8.4). There are numerous options available to customize the graphic, described in details in Calonico, Cattaneo, and Titiunik (2015).

```
library(rdrobust)
with(probation, rdplot(y = gpa2, x = distcut, nbins = 10))
```

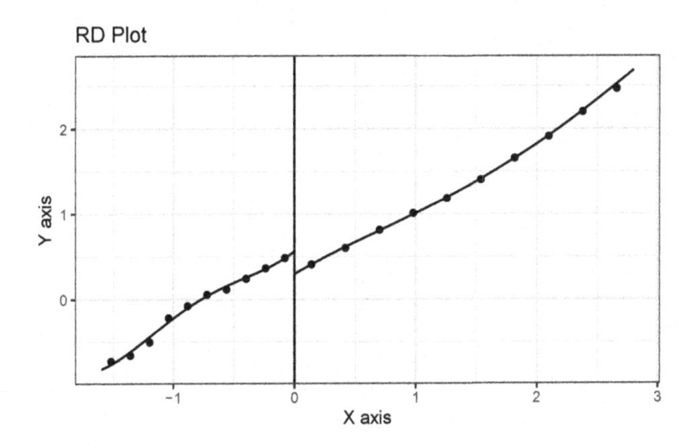

Figure 8.4: Discontinuity for probation using `rdplot`

Placebo tests consist of applying the same visual techniques to variables that shouldn't be impacted by the discontinuity. In Figure 8.5, we plot the high school grade percentile and we can see that there is no clear discontinuity for this variable around the threshold.

```
with(probation, rdplot(y = hsgrade, x = distcut, nbins = 10))
```

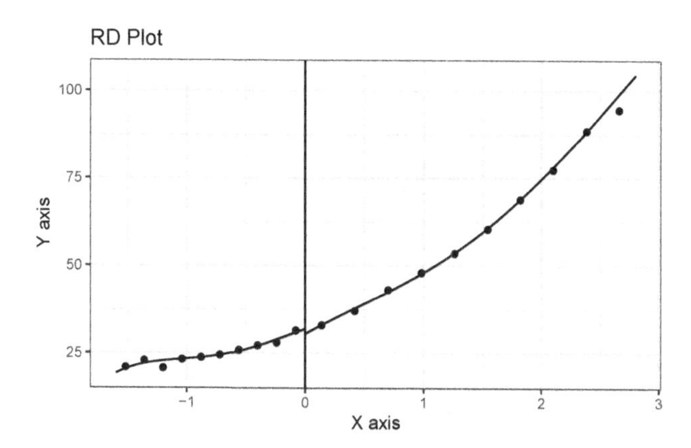

Figure 8.5: Placebo test for probation

rdrobust's functions don't have a data argument. Therefore, series should be provided, and we could have written the previous expression as:

```
rdplot(y = probation$hsgrade, x = probation$distcut, nbins = 10)
```

Using `with(data, expression)`, series of `data` can be called directly in `expression` without prefixing them by `probation$`.

8.3.3 Computing the effect of the treatment

As seen previously, the effect of the eligibility (and the effect of the treatment if the discontinuity is sharp) is the difference between the intercepts of two smoothing lines, which can be obtained using OLS or some more complicated fitting tools like local polynomials. In the simplest case, it is given by the estimation of β_t in the following regression, t_n being the treatment variable and x_n the forcing variable:

$$y_n = \beta_0 + \beta_t t_n + \beta_x x_n + \beta_{xt} x_n t_n + \epsilon_n$$

Note that the presence of β_{xt} allows to have different slopes on both sides of the cutoff. Two crucial choices have to be made while estimating the effect of the treatment: the **bandwidth**, i.e., the range of values of the forcing variable used in the estimation and the degree of smoothness of the fitting line. A larger bandwidth leads to a more efficient estimator (as the number of observations is higher), but the estimation may be biased as the sample contains observations not close enough to the cutoff. We first regress the outcome on the forcing variable interacted with the treatment variable (note the use of the `*` operator in the formula), with a bandwidth of 1.5:

```
lm(gpa2 ~ distcut * probation, probation,
   subset = abs(distcut) < 1.5) %>% gaze
```

```
                       Estimate Std. Error t value Pr(>|t|)
distcut                  0.7036     0.0119   59.05   <2e-16
probationyes             0.2554     0.0199   12.86   <2e-16
distcut:probationyes     0.0997     0.0306    3.25   0.0011
```

The coefficient of `probationyes` is highly significant; students just under the cutoff and who therefore benefit from probation have a subsequent GPA higher by one-quarter of a point. The interaction term is also positive, indicating that the slope of the regression line is higher on the right of the cutoff than on the left. To check the robustness of this result, we change the bandwidth and add a quadratic term in the forcing variable.

```
lm(gpa2 ~ distcut * probation, probation,
   subset = abs(distcut) < 0.5) %>% gaze(coef = "probationyes")
##              Estimate Std. Error t value Pr(>|t|)
## probationyes   0.2266     0.0342    6.62  3.7e-11
lm(gpa2 ~ (distcut + I(distcut ^ 2)) * probation, probation,
   subset = abs(distcut) < 1.5) %>% gaze(coef = "probationyes")
##              Estimate Std. Error t value Pr(>|t|)
## probationyes   0.1941     0.0289    6.71    2e-11
lm(gpa2 ~ (distcut + I(distcut ^ 2)) * probation, probation,
   subset = abs(distcut) < 0.5) %>% gaze(coef = "probationyes")
##              Estimate Std. Error t value Pr(>|t|)
## probationyes   0.2197     0.0512    4.29  1.8e-05
```

Depending on the specification, the coefficient varies from 0.19 to 0.26. The **rdrobust** package provides the `rdrobust` function which computes the treatment effect using local polynomials. The degree of the polynomial is controlled with the `p` argument (the default is 1) and the bandwidth with the `h` argument. If the bandwidth is not indicated, it is automatically computed internally using the `rdbwselect` function.

```
with(probation, rdrobust(gpa2, distcut)) %>% gaze
```

```
Bandwidth (N used)        : 0.457 (8592)
coefficient (se)          : -0.222 (0.039)
Conv. stat (p-value)      : -5.707 (0.000)
Robust. stat (p-value) : -4.595 (0.000)
```

The bandwidth is 0.457 (as the initial range of the forcing variable is $[-1.6, 2.8]$) and 8592 observations are used out of 44311 in the `probation` data set. Besides the standard t statistic, `rdrobust` computes a robust statistic, which is slightly lower than the conventional statistic. When the discontinuity is fuzzy, the effect of the treatment is computed using two stage least squares, the eligibility status being an instrument for treatment.

```
bdh <- bdh %>% mutate(eligible = as.numeric(selben < 0))
ivreg::ivreg(attendmonth ~ selben * treated | selben * eligible,
             data = bdh) %>% gaze(coef = "treated")
##         Estimate Std. Error t value Pr(>|t|)
## treated    1.722      0.598    2.88    0.004
```

The effect of receiving the grant increases monthly church attendance by 1.72, and the effect is significant. The `rdrobust` function can be used in a fuzzy discontinuity context using the `fuzzy` argument to indicate the treatment variable. A matrix of supplementary covariates can be added using the `covs` argument.

```
with(bdh, rdrobust(y = attendmonth, x = selben, fuzzy = treated,
                   covs = cbind(age, hsize, schooling))) %>% gaze
```

```
Bandwidth (N used)        : 1.851 (1114)
coefficient (se)          : 2.072 (0.918)
Conv. stat (p-value)      : 2.257 (0.024)
Robust. stat (p-value) : 2.124 (0.034)
```

As the `h` argument is not specified, the bandwidth is automatically computed and is 1.851, and the number of observation used is 1114 out of 2645 in the original data set. The coefficient of `treatment` is 2.07 (slightly higher than the one obtained previously), and the conventional and the robust tests indicate that the coefficient is significant at the 5% level.

8.3.4 Manipulation test

A critical hypothesis of regression discontinuity designs is that individuals are set randomly on both sides of the cutoff. On the contrary, individuals being aware of the existence of the treatment and of the value of the cutoff may manipulate their value of the forcing variable in order to be on one specific side of the cutoff. In this case, the distribution of the forcing variable should also exhibit a discontinuity at the cutoff. The first manipulation test was proposed by McCrary (2008). The `rddensity:rddensity` function performs an extension of this test.

```
library(rddensity)
dens_test <- probation %>% pull(distcut) %>% rddensity
dens_test %>% gaze
```

```
Bandwith (left-right)       : 0.609-0.827
Observations (left-right) : 4448-12049
Statistic (p-value)         : 0.764 (0.445)
```

The p-value being equal to 0.45, the absence of manipulation is not rejected. The `rddensity:rdplotdensity` function plots the density on the left and on the right of the cutoff, with a confidence interval; if the two confidence intervals overlaps, the hypothesis of no manipulation is not rejected. This is the case with the `probation` data set, as shown in Figure 8.6.

```
ra <- rdplotdensity(dens_test, probation$distcut)
```

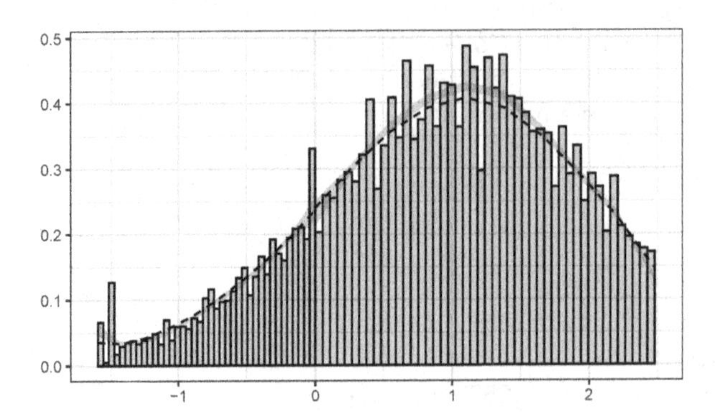

Figure 8.6: Density estimation on both sides of the cutoff for the `probation` data set

8.4 Difference-in-differences

Sometimes, the outcome is observed for two periods. In the first period, the treatment hasn't been implemented. For the second period, some individuals have been treated (the treatment group), as some individuals haven't (the control group). The effect of the treatment can then be estimated by:

$$\frac{\sum_{n\in T}(y_{n2}-y_{n1})}{N_T}-\frac{\sum_{n\in C}(y_{n2}-y_{n1})}{N_C} \tag{8.7}$$

Each term is the mean difference of the outcome between the two periods for the two groups, and the estimator is the difference of these differences. Consider as an example that the treatment is a job-training program implemented in 2021 and that the outcome is wage observed in 2022. If the first term of Equation 8.7 is \$1000, this means that the average annual wage in the treatment group increased by \$1000 in 2022 compared to 2021. This would be a relevant estimator of the effect of the program if nothing had changed on the labor market in 2022 compared to 2021. But if the economic situation improved, then the average wages will increase even for those who haven't been treated. This is measured by the second term in Equation 8.7 (say \$600). Therefore, the effect of the treatment is the difference between the two terms, which is \$400.

As an example, Di Tella and Schargrodsky (2004) sought to estimate the causal effect of police on crime. This task is difficult using non-experimental data, as there may be a reverse causality relationship between police and crime: more police reduces crime (negative causal relationship), but an increase in crime may lead authorities to increase police (positive reverse causal relationship). In July 18, 1994, a terrorist attack destroyed the main Jewish-owned center in Buenos Aires, Argentina, killing 85 people, and the federal government decided to provide 24 hour police protection to Jewish-owned institutions. Di Tella and Schargrodsky (2004) collected data on three "barrios" in Buenos Aires at the block level on a monthly basis. The outcome of interest is the number of car thefts. Denoting x as a dummy equal to 1 if the block contains or is close to a Jewish institution and p a dummy equal to 1 after the attack, the treatment effect is, in a regression context, the estimate of θ on the following equation:

$$y_{nt} = \beta_0 + \beta_x x_n + \beta_p p_{nt} + \theta x_n p_{nt} + \epsilon_{nt}$$

with $t = 1, 2$, (the periods before and after the attack). The `car_thefts` data set contains repeated observations of car thefts (`thefts`) for 876 blocks. Each block is observed 10 times on a monthly basis, and the month of the attack (July) is split into two half-month observations.

```
car_thefts %>% print(n = 3)
```

```
# A tibble: 8,760 x 8
  block date       barrio calle   distance thefts period  days
  <dbl> <date>     <chr>  <chr>   <fct>     <dbl> <fct>  <dbl>
1   870 1994-04-01 Once   Cordoba one           0 before    30
2   851 1994-04-01 Once   Tucuman two           0 before    30
3   843 1994-04-01 Once   Lavalle same          0 before    30
# i 8,757 more rows
```

We first compute the total number of thefts (`thefts`) before and after the attack for all the blocks. As the two periods are of unequal length (3.5 and 4.5 months), we divide the number of thefts for the two periods by the corresponding number of days and we multiply by 30.5 to get a monthly value.

```
two_obs <- car_thefts %>%
    group_by(block, period) %>%
    summarise(thefts = sum(thefts) / sum(days) * 30.5,
              .groups = "drop")
mean(two_obs$thefts)
## [1] 0.09328
```

The number of monthly theft per block is about 0.09. `distance` is a factor indicating the distance from the block to the nearest Jewish-owned institution: its levels are `"same"` (same block), `"one"`, `"two"` and `">2"` (one block, two blocks or more than two blocks). We add this variable to `two_obs` by selecting `distance` and `block` in the original data frame, selecting only the distinct rows (one per block) and joining it to `two_obs`:

```
two_obs <- two_obs %>%
  left_join(distinct(car_thefts, block, distance))
```

We then compute the regression by coercing the `distance` to a dummy for the same block:

```
two_obs %>% mutate(distance = ifelse(distance == "same", 1, 0)) %>%
    lm(formula = thefts ~ period * distance) %>% gaze
```

```
                        Estimate Std. Error t value Pr(>|t|)
periodafter              0.01055    0.00645    1.64    0.102
distance                 0.00898    0.02218    0.40    0.686
periodafter:distance    -0.07232    0.03137   -2.31    0.021
```

The effect of police on thefts is -0.07, which is very high as the average number of monthly theft per block is 0.09 and it is significant. The same difference-in-differences estimator can be obtained by computing a t-test of equality of the two means. We first reshape the data set in order to have one line per block, and we compute the after-before difference:

```
diffs <- two_obs %>%
    mutate(distance = ifelse(distance == "same", "yes", "no")) %>%
    pivot_wider(names_from = period, values_from = thefts) %>%
    mutate(dt = after - before)
diffs %>% print(n = 3)
```

```
# A tibble: 876 x 5
  block distance before  after      dt
  <dbl> <chr>     <dbl>  <dbl>   <dbl>
1     1 no        0      0       0
2     2 no        0      0.224   0.224
3     3 no        0.290 0.0449 -0.246
# i 873 more rows
```

and we then use the `t.test` function:

```
t.test(dt ~ distance, diffs, var.equal = TRUE) %>% gaze
## diff = 0.072 (0.026), t = 2.739, df: 874, pval = 0.006
```

The difference-in-differences can be extended to the case where the observation units before and after the implementation of the treatment are not the same. This is the case in Hong (2013)', which studied the impact of the introduction of Napster on music expenditure. The study is based on the Consumer Expenditure Survey, which is performed on a quarterly basis, and the data set is called `napster`. Napster was introduced in June 1999 and became the dominant file-sharing service. Households with (without) internet access constitute the treatment (control) group. From the `date` series, we construct a `period` variable using June 1999 as the cutoff:

```
napster <- napster %>%
    select(date, expmusic, internet, weight) %>%
    mutate(period = ifelse(date < ymd("1999-06-01"), "before", "after"),
           period = factor(period, levels = c("before", "after")))
napster %>% print(n = 3)
```

```
# A tibble: 107,650 x 5
  date       expmusic internet weight period
  <date>        <dbl> <fct>     <dbl> <fct>
1 1997-05-01        0 no       27471. before
2 1997-06-01        0 yes      23567. before
3 1997-06-01        0 no       26211. before
# i 107,647 more rows
```

we then proceed to the estimation, with `expmusic`, the expenditures on recorded music as a response:

```
lm(expmusic ~ period * internet, napster, weight = weight) %>% gaze
```

```
                        Estimate Std. Error t value Pr(>|t|)
periodafter               -1.749      0.253   -6.91    5e-12
internetyes               14.781      0.432   34.23   <2e-16
periodafter:internetyes   -4.589      0.552   -8.32   <2e-16
```

`internetyes` has a strong positive effect on expenditure. The interaction term between `period` and `internet` indicates that, the deployment of Napster led to a significant reduction of the expense for the "treated" individuals (those who have internet access) of \$4.6.

8.5 Matching

The fundamental problem of estimating the treatment effect on observational data is that the treatment and the control sample are not drawn from the same population, so that average (observable or non-observable) characteristics are not the same. The idea of matching is to select, for each treated observation, an observation in the control group as similar as possible, based on observed characteristics. If it is possible to do it for all the observations in the treatment group, the resulting sample would be similar to experimental data, i.e., a sample with treated and control observations drawn from the same population. When there are few covariates which take a small number of different values, it is possible to find in the control group an observation which has exactly the same observed characteristics than a treated observation (for example, a man aged 25 with 12 years of education). On the contrary, with numerous and/or continuous covariates, it is impossible to match a treated observation with an exactly similar observation in the control group.

In this case, the dimension of the problem is reduced to one using a **propensity score** estimator. The propensity score is the conditional probability (given x) of being treated. It can be fitted using for example a logit or a probit, the response being a dummy for treated individuals. Then, each treated individual is matched with the observation in the control group which has the closest propensity score. Finally, the estimation of the treatment effect is performed on the subset of the sample composed by all the observations of the treatment group and those of the control group that match these observations.

The matching estimator was first proposed by Rosenbaum and Rubin (1983) and the interest in matching methods in econometrics really started with the articles of Dehejia and Wahba (1999, 2002). These two articles revisited the results of LaLonde (1986) who showed that the estimation of treatment effect with observational data can be seriously biased. Using Lalonde's data, they showed that matching methods can be used to estimate consistently treatment effects on observational data. They also proposed an algorithm, implemented in **Stata** by Ichino (2002):

- first compute the propensity scores using a probit or a logit model with a rich set of covariates, eventually with squares and interactions between covariates,
- start with a small number of strata, for example 5 (0-0.2, 0.2-0.4, ..., 0.8-1), and test for each strata the hypothesis that the means of the scores are equal in the two groups,

- if the test fails for one strata, split it in two (for example, the 0.2-0.4 strata is decomposed in two stratas 0.2-0.3 and 0.3-0.4) until the equal mean hypothesis is not rejected for every strata,
- then compute the same test for each covariate,
- if the test fails for some covariates, estimate a more flexible propensity score model, adding squares and interactions between covariates.

Once the stratas are computed, a natural estimator of the treatment effect is: $\sum_k(\bar{y}_k^t - \bar{y}_k^c)f_k$, where $\bar{y}_k^t$ and $\bar{y}_k^c$ are the mean values of the outcome in the treatment and in the control group and $f_k = t_k/N_T$ is the frequency of strata k for the treatment group (with t_k the number of treated individuals in strata k and N_T the total number of treated individuals). The variance of the estimated treatment effect can be computed with or without the hypothesis of equal variance between stratas, between groups or both.

As an example, we replicate the results of Ichino, Mealli, and Nannicini (2008) who studied the effect of temporary work agency (TWA) jobs on the probability of finding a stable job. The data set, called `twa`, contains 2030 observations (511 treated and 1519 untreated) for two regions, Tuscany and Sicily. We restrict the sample to Tuscany:

```
tuscany <- twa %>% filter(region == "Tuscany")
```

There are 281 observations in the treatment group and 628 in the control group. The group variable `group` is a factor with levels `"control"` and `"treatment"`, and the outcome is also a factor indicating the employment status one year after the program. Its levels are `none` (no job), `other`, `fterm` for fixed-term contract and `perm` for a permanent contract. Following the authors, we define the outcome of interest as a dummy for a permanent contract:

```
tuscany <- tuscany %>% mutate(perm = ifelse(outcome == "perm", 1, 0))
```

To get a first idea of the treatment effect, we compute the mean of `perm` for the two groups:

```
tuscany %>% group_by(group) %>%
    summarise(n = n(), outcome = mean(perm))
```

```
# A tibble: 2 x 3
  group       n outcome
  <fct>   <int>   <dbl>
1 control   628   0.166
2 treated   281   0.313
```

The proportion of individuals who have a permanent job is 31.3% for the treated group and 16.6% for the control group and the apparent treatment effect is therefore 14.7%. The `micsr::pscore` function implements the algorithm previously described. The first two arguments are `formula` and `data`. The formula should have two variables on the left-hand side, the first indicating the outcome and the second the group. The group variable can be either a dummy or a factor with two levels, the second indicating the treated individuals. To estimate the propensity score, we use the same formula as Ichino, Mealli, and Nannicini (2008), and in particular, we use a square term for the distance to the next agency and an interaction between self-employed status and the city of Livorno:

```
tuscany <- tuscany %>%
    mutate(dist2 = dist ^ 2,
           livselfemp = I((city == "livorno") * (occup == "selfemp")),
           perm = ifelse(outcome == "perm", 1, 0))
ftusc <- perm + group ~ city + sex + marital + age +
    loc + children + educ + pvoto + training +
    empstat + occup + sector + wage + hour + feduc + femp + fbluecol +
    dist + dist2 + livselfemp
ps <- pscore(ftusc, tuscany)
```

Three supplementary arguments of `pscore` can be used: the maximum number of iterations (default 4), the tolerance level for the t-tests (default 0.005) and the link for the binomial model (default to logit). For the tolerance level, Ichino (2002) advised to use a low level with the following argument: with for example 20 covariates, using a 5% level, if the tests are mutually independent, the probability that one of the tests rejects the balancing property, although it holds, is 37%. `pscore` returns a `pscore` object which contains three tibbles.

- `strata` contains information about the stratas (the frequencies, the average propensity scores and the probability value of the hypothesis of no difference of the propensity scores in the two groups),
- `cov_balance` has a line for every covariate and contains the strata for which the probability value is the smallest,
- `model` contains the original data sets with some supplementary columns:
  - `pscore` contains the propensity score for every observation,
  - `.gp` is a factor with levels `"control"` and `"treated"`,
  - `.cs` is a boolean indicating whether the propensity score for an observation lies in the interval of scores for the treated,
  - `.resp` contains the response,
  - `.cls` indicates the strata for the observation.

A `summary` method is provided and the `print` method for `summary.pscore` objects has a `step` argument that allows to print the result of each step of the estimator. To get the information about the stratas, we use:

```
ps %>% summary %>% print(step = "strata")
```

strata	n_treated	n_control	ps_treated	ps_control	p.value
[0,0.1)	11	217	0.065	0.053	0.122
[0.1,0.2)	24	138	0.153	0.151	0.843
[0.2,0.4)	60	118	0.291	0.294	0.783
[0.4,0.6)	60	81	0.497	0.501	0.632
[0.6,0.7)	35	21	0.668	0.646	0.006
[0.7,0.8)	56	14	0.754	0.745	0.292
[0.8,1)	35	3	0.837	0.850	0.375

Two of the initial stratas were cut in halves (0-0.2 and 0.6-0.8). The control subsample is restricted to the range of the values of the propensity score for the treated; therefore, only 592 observations of the control group out of 628 are used. With `step = "covariates"`, we get a (long) table indicating, for each covariate, the results of the balance test between the treatment and the control group:

```
ps %>% summary %>% print(step = "covariates")
```

Finally, `step = "atet"` gives the values of the estimated ATET (average treatment effect of the treated) and different estimates of its standard deviation:

```
ps %>% summary %>% print(step = "atet")
```

```
ATET                          : 0.1769
sd: equal variance
  - within groups             : 0.04603
  - within strata             : 0.04879
  - within groups and strata: 0.04713
  - no                        : 0.03543
```

The estimated treatment effect is 0.177, which is slightly higher than the treatment effect computed with the whole sample which was 14.7%. It is highly significant with the different flavors of the standard deviations ranging from 0.035 to 0.049.

An alternative to using strata and computing the ATET as a weighted average of the difference of the mean of the outcome between treated and control observations in each stratum is to match each treated observation to one or several observations in the control group. The simplest algorithm consists of selecting, for every treated observation, the control observation which has the closest value of propensity score. Using the `model` tibble returned by `pscore`, we begin with constructing two tibbles, one for the treatment group and one for the control group, containing only the index of the observation and the value of the propensity score:

```
tusc_tr <- ps$model %>% filter(group == "treated") %>%
    select(id_tr = id, ps_tr = pscore)
tusc_ctl <- ps$model %>% filter(group == "control") %>%
    select(id_ctl = id, ps_ctl = pscore)
tusc_tr %>% print(n = 2)
## # A tibble: 281 x 2
##    id_tr ps_tr
##    <dbl> <dbl>
## 1    214 0.148
## 2    310 0.134
## # i 279 more rows
```

We then need to join the two tables, and **dplyr** provides functions that perform mutating joins. Among them, the `dplyr::left_join` function is appropriate, as it returns all the rows of the "left" table (the treatment group `tusc_tr`) and only those of the "right" table that match. Standard use of mutating joins are based on equality of one or several joining variables, but **dplyr** (since its version 1.1.0) performs also **inequality** and **rolling** joins.

With inequality joins, one can for example match a treated observation with all the observations in the control group with higher propensity scores. We do it below only for the first two treated observations:

```
tusc_tr %>% slice(1:2) %>%
    left_join(tusc_ctl, join_by(ps_tr <= ps_ctl)) %>%
    count(id_tr)
```

```
# A tibble: 2 x 2
  id_tr     n
  <dbl> <int>
1   214   309
2   310   331
```

and they are matched to respectively 309 and 331 observations in the control group. With **rolling** joins, one can select only one observation, the closest one, using the `closest` function:

```
match_sup <- tusc_tr %>%
    left_join(tusc_ctl, join_by(closest(ps_tr <= ps_ctl))) %>%
    rename(ps_sup = ps_ctl, id_sup = id_ctl)
match_sup %>% print(n = 3)
```

```
# A tibble: 281 x 4
  id_tr  ps_tr id_sup ps_sup
  <dbl>  <dbl>  <dbl>  <dbl>
1   214 0.148    4098 0.149
2   310 0.134    4594 0.135
3   332 0.0789   4765 0.0789
# i 278 more rows
```

this time one row is returned for every treated observation, and the same control observation can be matched with several treated observations. Next, we match with the closest lower propensity score control:

```
match_inf <- tusc_tr %>%
    left_join(tusc_ctl, join_by(closest(ps_tr >= ps_ctl))) %>%
    rename(ps_inf = ps_ctl, id_inf = id_ctl)
```

We then join the two tables (by the index for treated observations `id_tr`) and select the control observation which is the closest:

```
match_nearest <- match_sup %>% select(- ps_tr) %>%
    left_join(match_inf, by = "id_tr") %>%
    mutate(ps_sup = ifelse(is.na(ps_sup), 2, ps_sup),
           id_ctl = ifelse( (ps_tr - ps_inf) < (ps_sup - ps_tr),
                           id_inf, id_sup),
```

```
            ps_ctl = ifelse(id_ctl == id_inf, ps_inf, ps_sup)) %>%
    select(id_tr, id_ctl, ps_tr, ps_ctl)
```

For a couple of treated observations, the propensity score is greater than the highest propensity score in the control group. Therefore, **ps_sup** is **NA**, and we set it to an arbitrary high value so that the **id_inf** observation is selected. We can then compute the number of observations in the control group that are used:

```
match_nearest %>% pull(id_ctl) %>% unique %>% length
## [1] 146
```

and select control observations which are the most often used to match treated observations:

```
match_nearest %>% count(id_ctl) %>% top_n(2, n)
```

```
# A tibble: 2 x 2
  id_ctl     n
   <dbl> <int>
1   4211    13
2   4935    16
```

For example, observation 4935 in the control group matches 16 observations in the treatment group. For some observations, the algorithm may result in a poor match for some treated observations if the difference between the probability scores of this observation and the matched control observation is high.

```
match_nearest %>% mutate(diff = ps_tr - ps_ctl) %>% top_n(3, diff)
```

```
# A tibble: 3 x 5
  id_tr id_ctl ps_tr ps_ctl   diff
  <dbl>  <dbl> <dbl>  <dbl>  <dbl>
1   375   4935 0.809  0.786 0.0225
2    89   4038 0.907  0.881 0.0258
3   412   4935 0.809  0.786 0.0225
```

In our sample, the highest difference is about 2.6%. The sample can be reduced to observations for which the difference is lower than a given value. This is called **caliper matching**. For example, to restrict the sample to treated observations for which the propensity score difference with its matched control observation is lower than 1%, we would use:

```
mathc_nearest <- match_nearest %>% filter(abs(ps_tr - ps_ctl) < 0.01)
```

We then lose 25 treated observations. To compute the treatment effect, we pivot the tibble in "long" format (one line for each observation) and we join it to **tuscany** to get the response (**perm**) for each observation:

```
match_smpl <- match_nearest %>%
  select(- ps_tr, - ps_ctl) %>%
  pivot_longer(1:2) %>%
  separate(name, into = c(".id", "group")) %>%
  select(id = value, gp = group) %>%
  left_join(select(tuscany, id, perm), by = "id")
match_smpl %>% print(n = 2)
```

```
# A tibble: 562 x 3
     id gp     perm
  <dbl> <chr> <dbl>
1   214 tr        0
2  4098 ctl       0
# i 560 more rows
```

The treatment effect is then:

```
means <- match_smpl %>% group_by(gp) %>% summarise(perm = mean(perm))
atet <- means$perm[2] - means$perm[1]
c(treatment = means$perm[2], control = means$perm[1], atet = atet)
## treatment   control      atet
##    0.3132    0.1246    0.1886
```

The ATET (0.1886) is close to the one obtained previously using the algorithm based on stratas (0.1769).

Several packages have implemented the matching techniques previously described and some more advanced methods. We'll describe here the **MatchIt** package. The `MatchIt::matchit` function performs the matching, with a formula-data interface. It has numerous extra arguments, we set here the `replace` argument to `TRUE`, so that one observation in the control group can match several observations in the treatment group.

```
library(MatchIt)
ftusc2 <- update(ftusc, group ~ .)
mtch <- matchit(ftusc2, tuscany, replace = TRUE)
```

The `"nearest"` method is used (the default) and the number of matched observations is 427 (as previously), the 281 treated observations and 146 observations of the control group. A `match.data` function is provided in order to extract the data frame restricted to the treated observations and the subset of observations of the control group that match.

```
mtch %>% match.data %>% print(n = 2)
```

```
# A tibble: 427 x 31
     id   age sex    marital children feduc fbluecol  femp  educ pvoto
  <dbl> <dbl> <fct>  <fct>      <dbl> <dbl>    <dbl> <dbl> <dbl> <dbl>
1  4862    32 female married        0     5        0     1    18    98
2  3756    33 female single         0     5        0     0    13    80
```

```
# i 425 more rows
# i 21 more variables: training <dbl>, dist <dbl>, nyu <dbl>,
#   hour <dbl>, wage <dbl>, hwage <dbl>, contact <dbl>, region <fct>,
#   city <fct>, group <fct>, sector <fct>, occup <fct>,
#   empstat <fct>, contract <fct>, loc <fct>, outcome <fct>,
#   perm <dbl>, dist2 <dbl>, livselfemp <I<int>>, distance <dbl>,
#   weights <dbl>
```

A `weight` column is added to the data frame that contains weights to be used in subsequent treatment. There are 427 observations, the 281 treated observations and 146 observations of the control group. The weight for the treated observations is 1. For an observation of the control group that matches only one treated observation, the weight is $146/281 = 0.52$ and, for example, for an observation of the control group that matches four treated observations, the weight is :$146/281 \times 4 = 2.08$. Caliper matching is performed using the `caliper` argument. The value indicated is by default a share of the standard deviation of the propensity score, but a value can be indicated in the scale of the propensity score if `std.capiler` is set to `FALSE`. Using the same value of 1% as previously, we would use:

```
mtch_cap <- matchit(ftusc2, tuscany, replace = TRUE,
                    caliper = 0.01, std.caliper = FALSE)
```

MatchIt proposes more enhanced matching method; we won't describe them here. Once the matching has been performed, the quality of the balancing process can be assessed. The `summary` method prints, for each covariates, the mean in both groups and several statistics, especially the standardized mean differences, which should be close to 0. The `plot` method draws a Love plot (see Figure 8.7); for every covariate, two standardized mean differences are plotted, one for the raw data set and one for the balanced one.

```
mtch %>% summary %>% plot
```

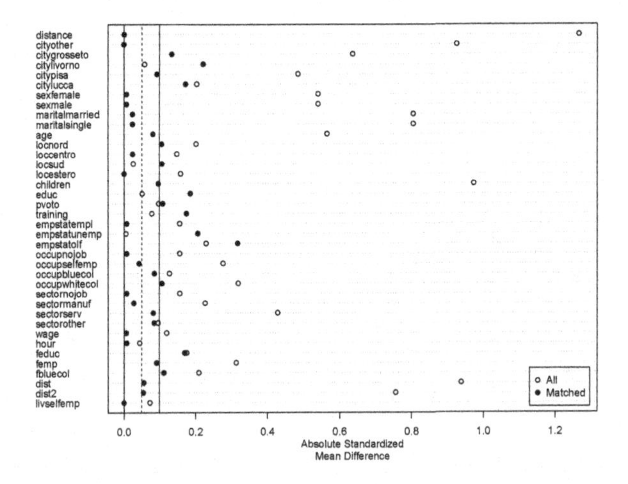

Figure 8.7: Love plot for the `twa` data set

We can see the effectiveness of matching, as the mean difference after matching (the black dots) are considerably lower than before matching (the white dots).

8.6 Synthetic control

Consider the case where units are aggregate units, as regions or countries and the treatment is an historical event or a policy intervention that affects one of them. Comparative case studies can then be used to estimate the causal effect of the treatment, comparing the situation of the treated unit after the treatment to the situation of one or several comparable units which haven't been treated. Synthetic control methods, introduced by Abadie and Gardeazabal (2003), are refinement of comparative case studies, where the comparison untreated unit is a weighted average of some units taken from a set of units called the **donor pool**. The example used by Abadie and Gardeazabal (2003) is the effect of terrorism in the Basque Country: the donor pool is the 17 Spanish regions, and the **synthetic control** of the Basque Country is a weighted average of two regions, Catalonia (85%) and Madrid (15%). The construction of the synthetic control is data-driven; it doesn't rely on any subjective thought of the researcher of which Spanish regions look much like the Basque Country before the beginning of terrorism. Examples of the application of synthetic control methods are the analysis of the effect of:

- state fragmentation (Reynaerts and Vanschoonbeek 2022),
- the Brexit (Douch and Edwards 2022),
- a tobacco control program in California, (Abadie, Diamond, and Hainmueller 2010),
- the reunification of Germany (Abadie, Diamond, and Hainmueller 2015),
- mafia in southern Italy (Pinotti 2015; Becker and Klosner 2017) ,
- the Legal Arizona Workers Act (Bohn, Lofstrom, and Raphael 2014).

Consider a set of $N+1$ units, the first being the treated unit. Variables for these units are observed for T periods and the treatment that affects unit 1 happened at the end of period T_0. Therefore, $t = 1, \ldots, T_0$ are pre-treatment periods, and $t = T_0 + 1, \ldots T$ are post-treatment periods. y is the outcome and x is a set of covariates (that are supposed to explain the variations of y). y^T_{nt} is the value of the outcome for unit n at period t if the unit were treated at this period and y^C_{nt} is the value without treatment. Of course, as usual in treatment effect analysis, for any n and t, y^T_{nt} and y^C_{nt} are never both observed. More precisely, y^T_{nt} is only observed for $n = 1$ and $t > T_0$. The effect of the treatment that we seek to estimate is:

$$y^T_{1t} - y^C_{1t} \; \forall \, t > T_0$$

The unknown y^C_{1t} is replaced by:

$$\hat{y}^C_{1t} = \sum_{n=2}^{N+1} w_n y_{nt}$$

where w_n are unknown weights such that $w_n \geq 0$ and $\sum_{n=2}^{N+1} w_n = 1$. These weights define the **synthetic control** for unit 1. The weights have to be chosen so that synthetic control of unit 1 is as similar as possible to unit 1 in the pre-treatment period. This similarity is evaluated using a set of covariates that are assumed to have a causal effect on the outcome. The time dimension is not taken into account, so that the value of the covariate for a unit is typically the mean or the median of this covariate for the whole pre-treatment period or its value for a given period. Note that the outcome in the pre-treatment period can be included in the set of covariates. For each covariate and for a given set of weights, the difference between the actual value of x_{1k} and the one of its synthetic control is:

$$x_{1k} - \sum_{n=2}^{N+1} w_n x_{nk}$$

and the synthetic control is defined by the set of weights w_n that minimize:

$$\sum_{k=1}^{K} v_k \left(x_{1k} - \sum_{n=2}^{N+1} w_n x_{nk} \right)^2 \tag{8.8}$$

where v is a second set of weights for the covariates. The simplest choice for v_k is $1/\hat{\sigma}_k^2$, with $\hat{\sigma}_k^2$ the empirical variance of covariate k in the sample, so that Equation 8.8 becomes:

$$\sum_{k=1}^{K} \left(x_{1k}/\hat{\sigma}_k - \sum_{n=2}^{N+1} w_n x_{nk}/\hat{\sigma}_k \right)^2$$

which is the sum of squares of the differences of the standardized covariates values for the treated observation and its synthetic control. More generally, for a given value of v, we can define $w(v)$ as the solution of the minimization of the function given by Equation 8.8:

$$\sum_{t=1}^{T_0} \left(y_{1t} - \sum_{n=2}^{N+1} w_n(v) y_{nt} \right)^2 \tag{8.9}$$

Abadie (2021) discusses in length different choices for covariates' weights v. The synthetic control method is implemented in **R** in two packages: **Synth** (Abadie, Diamond, and Hainmueller 2011) and **tidysynth** (Dunford 2021). We'll describe the use of the latter as it is written in the tidyverse style.

```
library(tidysynth)
```

The data set used is called `basque_country`. It is just the `basque` data set shipped with the **Synth** package, with a few modifications.

```
basque_country %>% print(n = 2)
```

```
# A tibble: 731 x 16
     id region    year gdpcap agriculture energy industry construction
  <dbl> <chr>    <dbl>  <dbl>       <dbl>  <dbl>    <dbl>        <dbl>
```

```
1      2 Andaluc~  1955   1.69           NA      NA        NA            NA
2      2 Andaluc~  1956   1.76           NA      NA        NA            NA
# i 729 more rows
# i 8 more variables: services <dbl>, administration <dbl>,
#   illit_educ <dbl>, prim_educ <dbl>, medium_educ <dbl>,
#   high_educ <dbl>, popdens <dbl>, invest <dbl>
```

It is a panel for 17 Spanish regions for 43 years (from 1955 to 1993). The first function to use is `synthetic_control` which defines the outcome, the unit and the time variables, the treated unit and the treatment period. In this study, the authors investigate the effect of terrorism in the Basque Country, which started in 1969, on GDP per capita:

```
bc <- basque_country %>%
    synthetic_control(outcome = gdpcap,
                      unit = region,
                      time = year,
                      i_unit = "Basque",
                      i_time = 1969,
                      generate_placebo = TRUE)
bc
```

```
# A tibble: 34 x 6
  .id       .placebo .type    .outcome .original_data      .meta
  <chr>        <dbl> <chr>    <list>   <list>              <list>
1 Basque           0 treated  <tibble> <tibble [43 x 16]>  <tibble>
2 Basque           0 controls <tibble> <tibble [688 x 16]> <tibble>
3 Andalucia        1 treated  <tibble> <tibble [43 x 16]>  <tibble>
4 Andalucia        1 controls <tibble> <tibble [688 x 16]> <tibble>
5 Aragon           1 treated  <tibble> <tibble [43 x 16]>  <tibble>
# i 29 more rows
```

We explicitly set the value `generate_placebo` to `TRUE`, even if it is the default value. If `generate_placebo` is set to `FALSE`, the result is a tibble which contains two lines for the treated unit (Basque Country): the first contains the information for the treated unit, the second contains the information for the donor set, i.e., all the Spanish regions except the Basque Country. The `.outcome` column contains the pre-period (15 years) values of the outcome for the treated (first line) and the donor set (second line). With `generate_placebo = TRUE`, there are $2 \times 16 = 32$ more lines, i.e., a couple of lines for every untreated unit. All the analysis is in this case performed not only for the treated unit, but also for all the units of the donor pool. As we'll see later, this placebo analysis enables to perform some inference about the efficiency of the treatment for the treated unit. Next, we define the covariates, using the `generate_predictor` function. Its syntax is similar to `mutate` or `summarise`, but the second argument of the function is `time_window` which indicates on which subperiod the computation has to be made. We compute the mean of the shares of education and investment for the 1964-1969 period, the population density in 1969, the GDP per capita for the 1960-1969 period and the sector shares for the 1961-1969 period.

```
bc <- bc %>%
    generate_predictor(time_window = 1964:1969,
```

```
                      illit_educ = mean(illit_educ, na.rm = TRUE),
                      prim_educ = mean(prim_educ, na.rm = TRUE),
                      medium_educ = mean(medium_educ, na.rm = TRUE),
                      high_educ = mean(high_educ, na.rm = TRUE),
                      invest = mean(invest, na.rm = TRUE)) %>%
    generate_predictor(time_window = 1969,
                      popdens = popdens) %>%
    generate_predictor(time_window = 1960:1969,
                      gdpcap = mean(gdpcap, na.rm = TRUE)) %>%
    generate_predictor(time_window = seq(1961, 1969, 2),
                      agriculture = mean(agriculture, na.rm = TRUE),
                      energy = mean(energy, na.rm = TRUE),
                      industry = mean(industry, na.rm = TRUE),
                      construction = mean(construction, na.rm = TRUE),
                      services = mean(services, na.rm = TRUE),
                      administration =
                        mean(administration, na.rm = TRUE))
```

A `.predictors` column is then added to the tibble. For the first line, the value is a tibble with 13 lines (the number of covariates) and two columns (the name of the variable and the value for the Basque Country). For the second line, there are 17 columns (the name of the variable and the values for the 16 regions of the donor set). Then the weights are computed using the `generate_weights` function. The second argument of this function is `optimization_window` which indicates the subset of the pre-treatment period which is used to compute the weights:

```
bc <- bc %>% generate_weights(optimization_window = 1960:1969)
```

`.unit_weights`, `.predictor_weights` and `.loss` columns are added. The first two contain respectively the values of w and v; the last one contains two mean square prediction errors, one for the variable and one for the unit. Finally, the synthetic control is computed using the `generate_control` function:

```
bc <- bc %>% generate_control
```

All the results can conveniently be extracted and plotted using a set of functions called `grab_###` and `plot_###`. `grab_balance_table` computes a balance table for the treated unit, its synthetic control and the whole sample for the covariates.

```
grab_balance_table(bc) %>% print(n = Inf)
```

```
# A tibble: 13 x 4
   variable      Basque synthetic_Basque donor_sample
   <chr>          <dbl>            <dbl>        <dbl>
 1 high_educ       3.26             3.10         2.68
 2 illit_educ      3.32             7.65        11.0
 3 invest         24.6             21.6         21.4
 4 medium_educ     7.46             6.92         5.41
 5 prim_educ      86.0             82.3         80.9
```

```
 6 popdens         247.            196.          99.4
 7 gdpcap            5.29            5.27         3.58
 8 administration    4.07            5.37         7.11
 9 agriculture       6.84            6.18        21.4
10 construction      6.15            6.95         7.28
11 energy            4.11            2.76         5.31
12 industry         45.1            37.6         22.4
13 services         33.8            41.1         36.5
```

The value of the covariate for the Basque Country may be very different from the mean of the values for the donor pool, but it should be close to the one of its synthetic control. It is particularly the case here for `agriculture` and `popdens`. To get the unit weights and only select those which are not almost zero, we can use:

```
grab_unit_weights(bc) %>% filter(weight > 1E-05)
```

```
# A tibble: 2 x 2
  unit     weight
  <chr>     <dbl>
1 Cataluna  0.851
2 Madrid    0.149
```

As stated previously, the synthetic control of the Basque Country consists of 85% of Cataluna and 15% of Madrid. It is a common feature of this method that a small number of units in the donor pool are really used. To get the covariates weights:

```
grab_predictor_weights(bc) %>% arrange(desc(weight))
```

```
# A tibble: 13 x 2
  variable       weight
  <chr>           <dbl>
1 popdens        0.339
2 gdpcap         0.199
3 industry       0.133
4 administration 0.107
5 agriculture    0.0937
# i 8 more rows
```

Population density and pre-treatment average of the GDP par capita are the two covariates with the largest weights. Figure 8.8's plots represent these weights using the `plot_weights` function.

```
plot_weights(bc)
```

The real series of GDP per capita for the Basque Country and its synthetic control are obtained using the `grab_synthetic_control` function. Figure 8.9 presents the two series and is obtained using the `plot_trends` function.[6]

[6]Note the use of `labs`; **tidysynth** uses **ggplot** so that the plots can be customized with any function of the **ggplot2** package.

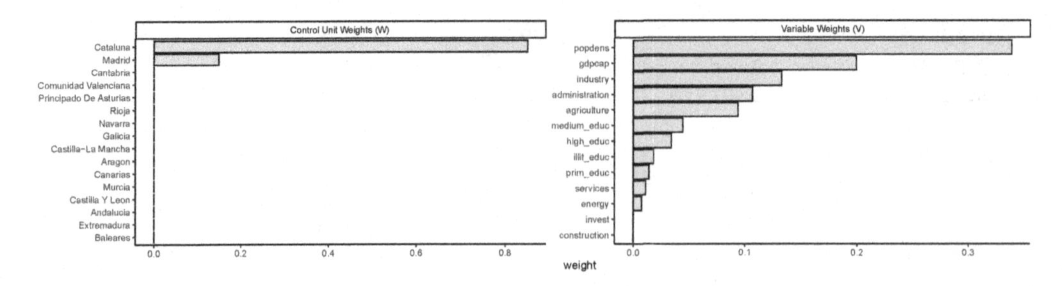

Figure 8.8: Weights for the units and the predictors

```
plot_trends(bc) + labs(title = NULL, y = "GDP per capita")
```

```
# A tibble: 43 x 3
  time_unit real_y synth_y
      <dbl>  <dbl>   <dbl>
1      1955   3.85    3.70
2      1956   3.95    3.85
3      1957   4.03    4.00
4      1958   4.02    4.03
5      1959   4.01    4.06
# i 38 more rows
```

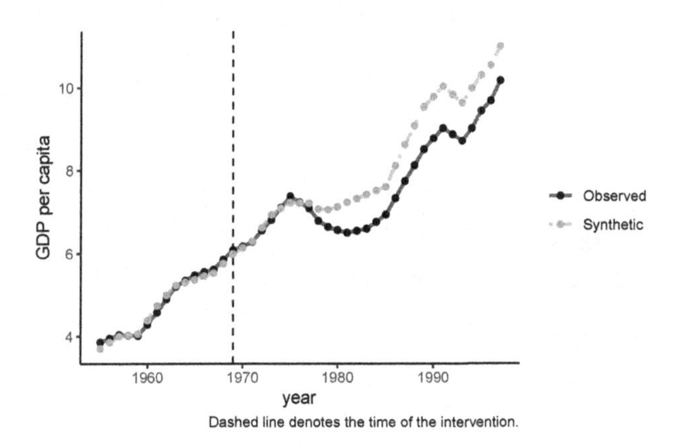

Figure 8.9: GDP per capita for the Basque Country and its synthetic control

Instead of plotting the GDP per capita for the real and the synthetic Basque Country, one can also plot the difference between both series (see Figure 8.10). The difference should be small and erratic before the treatment, and large and negative after the treatment.

```
plot_differences(bc) + theme_get()
```

The quality of the fit can be evaluated using **grab_loss**:

```
bc %>% grab_loss
```

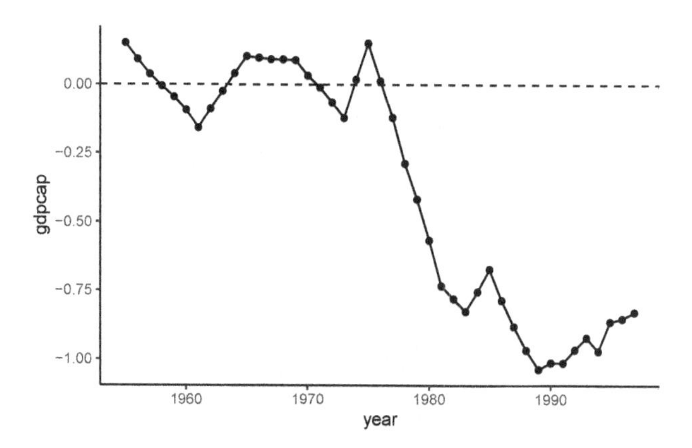

Figure 8.10: Difference of GDP per capita for the Basque Country and its synthetic control

```
# A tibble: 17 x 4
  .id       .placebo variable_mspe control_unit_mspe
  <chr>        <dbl>         <dbl>             <dbl>
1 Basque           0       0.00886            0.249
2 Andalucia        1       0.00850            0.0254
3 Aragon           1       0.00372            0.0313
4 Baleares         1       1.09               0.300
5 Canarias         1       0.0594             0.121
# i 12 more rows
```

The column `variable_mspe` indicates the value of Equation 8.8 and `control_unit_mspe` the value of Equation 8.9. The efficiency of the treatment and the ability of the method to estimate it can be assessed by computing the mean square prediction error for the outcome before and after the treatment:

$$m_n^{\text{pre}} = \frac{\sum_{t=1}^{T_0} (y_{nt} - \hat{y}_{nt})^2}{T_0} \text{ and } m_n^{\text{post}} = \frac{\sum_{t=T_0+1}^{T} (y_{nt} - \hat{y}_{nt})^2}{T - T_0}$$

The synthetic control should be very close to the treated unit before the treatment, so that m_n^{pre} should be low. On the contrary, if the treatment is efficient, the treated unit and its synthetic control should be very different after the treatment, so that m_n^{post} should be high. Therefore, the ratio $r_n = m_n^{\text{post}} / m_n^{\text{pre}}$ should be high if the treatment is efficient and if the synthetic control method is relevant. Moreover, it should be much higher than the ones obtained for untreated units. The significance of the treatment can be established using the distribution of r for all the units (the treated one and the untreated ones). Denoting $\bar{r}$ and $\hat{\sigma}_r$, as the mean and the standard deviation of r, we can define $z_n = (r_n - \bar{r})/\hat{\sigma}_r$ which is asymptotically distributed as a standard normal deviate. These indicators can be retrieved using the `grab_significance` function:

```
bc %>% grab_significance
```

```
# A tibble: 17 x 8
  unit_name                type     pre_mspe post_mspe mspe_ratio  rank
```

```
  <chr>                    <chr>      <dbl>      <dbl>       <dbl> <int>
1 Basque                   Treated  0.00821      0.493        60.1     1
2 Principado De Asturias   Donor    0.0157       0.650        41.3     2
3 Andalucia                Donor    0.00741      0.205        27.7     3
4 Castilla-La Mancha       Donor    0.0162       0.206        12.7     4
5 Navarra                  Donor    0.0224       0.224        10.0     5
# i 12 more rows
# i 2 more variables: fishers_exact_pvalue <dbl>, z_score <dbl>
```

The highest ratio is actually the one for the treated unit, i.e., the Basque Country. The mean and the standard deviations of this ratio are respectively 11.95 and 16.23. The z value for the Basque Country is 2.97, much higher than the critical value for a normal at the 1% level. The second highest value is for Asturias, with a z value of 1.81 lower than the critical value at the 5% level. The MSPE ratios are represented in Figure 8.11, using the `plot_mspe_ratio` function:

```
plot_mspe_ratio(bc) + labs(title = NULL)
```

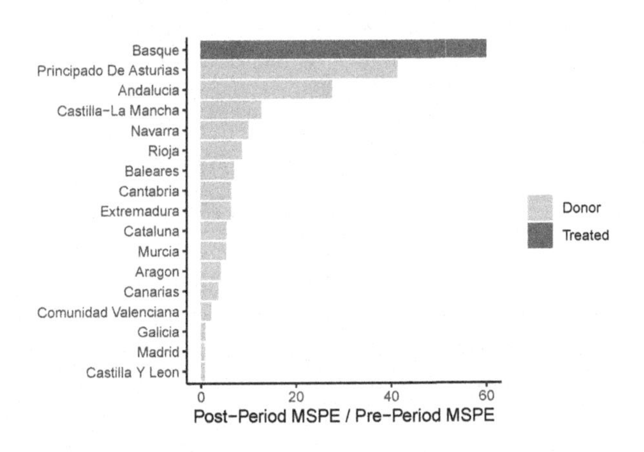

Figure 8.11: Plot of the MSPE ratio

Finally, a popular graphic in synthetic control analysis consists of plotting the difference between the real series of any unit and its synthetic control. The difference should be sharp for the post-treatment period for the treated unit and small for the other units which are not treated. This so-called placebos plot is obtained using the `plot_placebos` function and is presented in Figure 8.12

```
plot_placebos(bc)
```

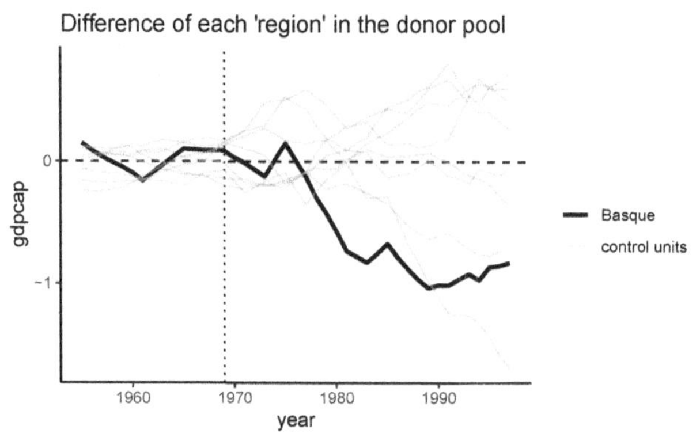

Figure 8.12: Placebos plot

9

Spatial econometrics

Spatial econometrics is a very dynamic field in modern econometrics. Geolocated data are now frequently available, even when the data doesn't concern geographic entities like countries, regions or towns, but households or firms. From their geographic coordinates, one is able to define for every observation a set of neighbors, and the interactions between neighbors can then be taken into account. There are two very different types of geographical data: vectors and rasters. A vector is a point or a set of points that define a line. A raster is a grid that contains the value of one variable (for example, a numeric indicating the elevation or a factor indicating land use). Relatively recently, two **R** packages have emerged that provide plenty of function that enables to deal easily with vectors and rasters. These are respectively **sf** (**simple feature**) and **terra**.[1] In this chapter, we'll consider only vectors.

The first two sections are devoted to simple features; Section 9.1 presents the structure of simple features objects and Section 9.1.2, using the example of a spatial RD design, illustrates how to deal with simple features in a statistical analysis. Section 9.3 deals with the detection and the measurement of spatial correlation. Finally, Section 9.4 presents some popular spatial models, namely the spatial error and the spatial autoregressive models.

9.1 Simple features

In a spatial statistical analysis, the first task is to get geographical information about the observations. This is usually done by importing external files in **R**, for example shapefiles. This task is performed by the **sf** library and the result is an `sf` object.

9.1.1 Structure of a simple feature

To understand what a simple feature is, it is best to construct a small one "by hand". The geographical representation of an observation is an `sfg` for simple feature geometry. It can be a point, a set of points that form a line or a polygon if the line ends where it started. We first load the **sf** package:

```
library(sf)
```

```
Linking to GEOS 3.10.2, GDAL 3.4.1, PROJ 8.2.1; sf_use_s2() is TRUE
```

[1]**sf** (Pebesma and Bivand 2023) and **terra** (Hijmans 2023) respectively supersede the **sp** and the **raster** packages.

The message indicates that **sf** uses three important external libraries:

- **GEOS** to compute topological operations,
- **GDAL** to read external files with a large variety of formats,
- **PROJ** to transform the coordinates in a specific **CRS** (coordinate reference system).

We consider in this example the three main cities in France. For Paris, the latitude[2] is 48.87 and the longitude[3] is 2.33. Most of **sf**'s functions start with `st_`. The `st_point` function can be used to construct a `sfg` (simple feature geometry) for Paris, the argument being a numeric vector containing the longitude and the latitude.

```
paris <- st_point(c(2.33, 48.87))
```

We then perform the same operation for Lyon and Marseilles:

```
lyon <- st_point(c(4.85, 45.76))
marseilles <- st_point(c(5.37, 43.30))
```

We then construct a `sfc` (simple feature column) which is a set of `sfg`s, using the `st_sfc` function. The different elements are entered as unnamed arguments of the function and a `crs` argument can be used to specify the **CRS**. Three formats exist to describe the **CRS**: **proj4string** (a character string), **EPSG** (an integer), and **WKT**, for well known text (a list). We can use here either `"+proj=longlat +datum=WGS84 +no_defs"` or `4326`. The datum is the representation of the earth from which the latitudes and the longitudes are computed.

```
cities_coords <- st_sfc(paris, lyon, marseilles, crs = 4326)
cities_coords
```

```
Geometry set for 3 features
Geometry type: POINT
Dimension:     XY
Bounding box:  xmin: 2.33 ymin: 43.3 xmax: 5.37 ymax: 48.87
Geodetic CRS:  WGS 84
```

Printing the `sfc` gives interesting information, as the bounding box (the coordinates of the rectangle that contains the data) and the **CRS**. Finally, we can construct a `sf` object using `st_sf` by binding a data frame containing information about the cities and the `sfc`.

```
cities_data <- tibble(pop = c(2.161, 0.513, 0.861),
                      area = c(105.4, 47.87, 240.6))
cities <- st_sf(cities_data, cities_coords)
cities
```

[2]The latitude is a coordinate that specifies the north/south position of a point on earth. It is an angular value equal to 0 at the equator and to -/+ 90 at the pole. The value is positive in the northern hemisphere and negative in the southern hemisphere.

[3]The longitude is a coordinate that specifies the east/west position of a point on earth. Meridians are lines that connect the two poles, and the longitude is the angular value of the position of a point to the reference (Greenwich) meridian.

```
Simple feature collection with 3 features and 2 fields
Geometry type: POINT
Dimension:     XY
Bounding box:  xmin: 2.33 ymin: 43.3 xmax: 5.37 ymax: 48.87
Geodetic CRS:  WGS 84
    pop   area      cities_coords
1 2.161 105.40 POINT (2.33 48.87)
2 0.513  47.87 POINT (4.85 45.76)
3 0.861 240.60  POINT (5.37 43.3)
```

A `sf` is just a data frame with a specific column that contains the coordinates of the observations. This column (called the geometry column) can be extracted using the `st_geometry` function:

```
cities %>% st_geometry
```

and is "sticky", which means that if some columns are selected, the geometry column is always returned along with the selected columns, i.e.:

```
cities %>% select(pop)
```

returns `pop` **and** `cities_coords`, even if the latter colon was not selected. We can then plot our three cities, along with a map of France, which is obtained using the `geodata::gadm` function.[4]

```
france <- geodata::gadm("FR", 1, ".")
```

`france` is not an object of class `sf`,[5] so we first coerce it to a `sf` using the `st_as_sf` function, and then we extract the geometry and the series called `NAME_1` which contains the name of the regions:

```
france <- france %>% st_as_sf %>%
    select(region = NAME_1)
```

Imported vector data are often large:

```
france %>% object.size %>% format("MB")
## [1] "3.6 Mb"
```

and they can be simplified using the `rmapshaper::ms_simplify` function, with a `keep` argument which indicates the degree of simplification (0.01 means that we seek to obtain a `sf` 100 times lighter than the initial one).

[4]This function is an interface to the http://www.gadm.org site which provides the boundaries of all countries in the world with different administrative level. The second argument set to 1 means that we want the boundaries of French regions, and the third argument set to "." means that the file is stored in the current directory.

[5]It is actually an object of class `SpatVector`, used by the **terra** package.

```
france <- france %>%
    rmapshaper::ms_simplify(keep = 0.01)
france %>% object.size %>% format("MB", digits = 2)
## [1] "0.07 Mb"
```

sf provides a `plot` method to get quickly a map of the data. Note that a thematic map is plotted for all the series of the `sf`, and it is therefore recommended to select first a unique series. If one is only interested in the vectors, they can be extracted before plotting using `st_geometry`:

```
france %>% st_geometry %>% plot
```

For more advanced maps, several specialized packages are available, in particular **tmap** (Tennekes 2018) and **mapsf** (Giraud 2023). In this chapter, we'll only use **ggplot**, which provides a `geom_sf` function. This geom is very special compared to other geoms, as the kind of geom that is plotted depends on the geometry column of the `sf`: if `cities` is provided as the data argument, points will be plotted; but if `france` is provided, polygons will be drawn. In the following code, we start with `france` as the data argument of `ggplot` and then the call to `geom_sf` results in the drawing of the administrative borders of French regions. Then we use a second time `geom_sf` with this time `cities` as data argument, so that points are drawn for the three cities. We use here `aes(size = pop)` so that the size of the points is related to the population of the cities. The result is presented in Figure 9.1.

```
france %>% ggplot() + geom_sf() +
    geom_sf(data = cities, aes(size = pop))
```

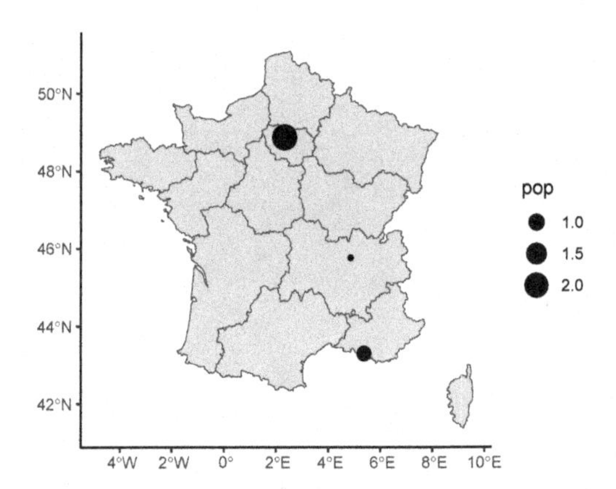

Figure 9.1: Map of France with its three main cities

sf provides several functions to compute values of interest. For example, to get the distance from Paris to Lyon and Marseilles, we would use:

```
st_distance(cities[1, ], cities[2:3, ])
## Units: [m]
```

```
##          [,1]    [,2]
## [1,] 394508 662105
```

`st_distance` always returns a matrix, the number of elements of the first (second) argument being the number of rows (columns) of the matrix. If only one argument is supplied, the result is a square matrix with 0 on the diagonal. Note that the numbers have a unit of measurement, which is here meters. To define a unit and to convert from one unit to another, the **units** package provides the `set_units` function. Consider the following example: we first provide a numeric (42.195), we define its unit as kilometers, and then we convert it to miles:

```
library(units)
d <- 42.195
dkm <- d %>% set_units(km)
dkm
## 42.2 [km]
dkm %>% set_units(miles)
## 26.22 [miles]
```

Even when the conversion is simple, it is advisable to use `set_units` instead of applying the conversion "by hand":

```
dm <- dkm %>% set_units(m)
dm
## 42195 [m]
dkm * 1000
## 42195 [km]
```

The numerical values are the same, but in the second case the unit hasn't changed and is still kilometers. `st_area` computes the area of a polygon:

```
st_area(france) %>% units::set_units(km ^ 2)
## Units: [km^2]
##  [1] 71023 47924 27507 39465  8705 57530 31917 12030 30044 84593 73135
## [12] 32267 31837
```

9.1.2 Computation on sf objects

The regression discontinuity framework can be adapted to consider geographic discontinuities. Some entities are considered on both sides of a border and the forcing variable is then the distance to the border.[6] The `us_counties` data set contains the borders of US counties:

```
us_counties %>% print(n = 3)
```

```
Simple feature collection with 3141 features and 4 fields
```

[6] With the convention that the sign of the distance is different on both sides of the border.

```
Geometry type: MULTIPOLYGON
Dimension:     XY
Bounding box:  xmin: -19940000 ymin: 2147000 xmax: 20010000 ymax: 11520000
Projected CRS: WGS 84 / Pseudo-Mercator
# A tibble: 3,141 x 5
  fips  gid        state   county                                  geometry
* <chr> <chr>      <chr>   <chr>                         <MULTIPOLYGON [m]>
1 01001 USA.1.1_1 Alabama Autauga (((-9675520 3850831, -9652842 38505~
2 01003 USA.1.2_1 Alabama Baldwin (((-9798375 3599675, -9797718 36046~
3 01005 USA.1.3_1 Alabama Barbour (((-9508473 3713483, -9545463 37133~
# i 3,138 more rows
```

Kumar (2018) investigates the effect of a legal restriction on home equity extraction which is specific to Texas on mortgage defaults. Kumar measures mortgage default rates in Texas and in the bordering states and compares mortgage default rates for counties which are closed to the Texas border (on both sides). Let's first plot the map of the counties, using **ggplot** and **geom_sf** (see Figure 9.2):

```
us_counties %>%
    filter(! state %in% c("Alaska", "Hawaii")) %>%
    ggplot() + geom_sf()
```

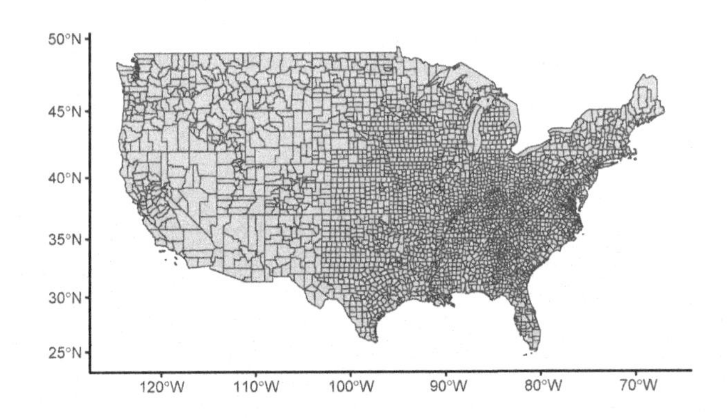

Figure 9.2: Map of American counties

The polygons can be merged using the **group_by** / **summarise** functions of **dplyr**. If the **sf** is grouped by states, the statistics computed by the **summarise** function is performed at the state level and the **geometry** now contains the coordinates of the states. With a void call to **summarise**, we get only this new **geometry**, and we can use it to plot a map of states (see Figure 9.3):

```
states <- us_counties %>%
    filter(! state %in% c("Alaska", "Hawaii")) %>%
    group_by(state) %>%
    summarise()
states %>% ggplot() + geom_sf()
```

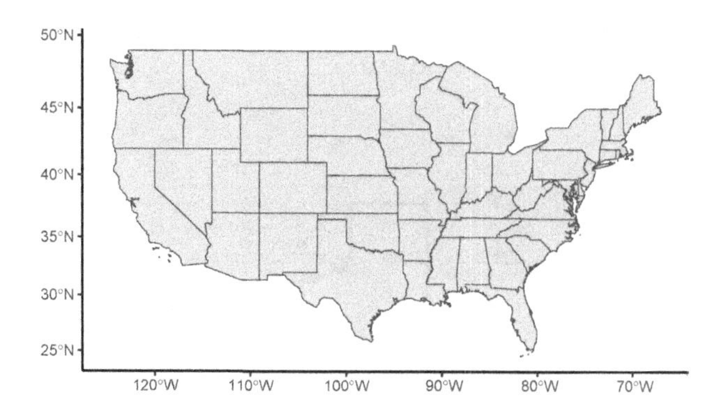

Figure 9.3: Map of American states

To get the border states of Texas, we use spatial indexation, i.e., we use the `[` operator. The first argument is used to select rows; it is normally a vector of integers (the positions of the lines to extract), a vector of characters (the names of the lines to extract) or a logical vector. Here we can index an `sf` by another `sf`, and the result is (by default) a new `sf` containing the elements of the first one which has common points with the second one:

```
border_states <- states[filter(states, state == "Texas"), ]
border_states
```

```
Simple feature collection with 5 features and 1 field
Geometry type: GEOMETRY
Dimension:     XY
Bounding box:  xmin: -12140000 ymin: 2979000 xmax: -9918000 ymax: 4439000
Projected CRS: WGS 84 / Pseudo-Mercator
# A tibble: 5 x 2
  state                                                           geometry
  <chr>                                                     <GEOMETRY [m]>
1 Arkansas   POLYGON ((-10407328 3897869, -10410758 3897882, -1044339~
2 Louisiana  MULTIPOLYGON (((-10253638 3470493, -10257225 3475979, -1~
3 New Mexico POLYGON ((-11868806 3763464, -11868752 3752373, -1187066~
4 Oklahoma   POLYGON ((-10942378 4046699, -10950200 4049735, -1095627~
5 Texas      MULTIPOLYGON (((-11360369 3475920, -11367496 3476824, -1~
```

We then compute the Texas border. It is defined by the common points of the borders of Texas and its border states. This is obtained using the `st_intersection` function which returns four lines (one for each border state), and the border is then obtained by merging these four lines using the `st_union` function. The result is presented in Figure 9.4.

```
texas <- filter(border_states, state == "Texas") %>% st_geometry
border <- filter(border_states, state != "Texas") %>% st_geometry
border <- st_intersection(texas, border) %>%  st_union
border_states %>% ggplot() +
    geom_sf() +
```

```
    geom_sf(data = border, linewidth = 1) +
    geom_sf_label(aes(label = state))
```

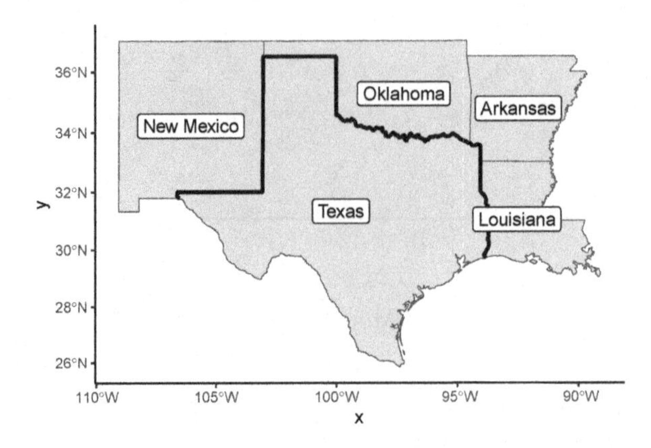

Figure 9.4: Border states and Texas borders

Note the use of **geom_sf_label** function which, is a specialized version of **geom_label** that is suitable for writing labels on a map. We then select the counties that belong to any of these states and we compute the distance to the Texas border. It is defined as the distance between the centroid of the county and the closest point of the border. The centroids are obtained using the **st_centroid** function. We've already used **st_distance** to comp ute the distance between two points. It can also be used to compute the (shortest) distance between a point and a line:

```
border_counties <- us_counties %>%
    filter(state %in% pull(border_states, state))
centroids <- border_counties %>% st_geometry %>% st_centroid
dists <- st_distance(centroids, border)[, 1] %>% set_units(miles)
border_counties <- border_counties %>% add_column(dists)
head(dists)
## Units: [miles]
## [1] 194.7 157.4 259.2 212.0 241.6 130.3
```

st_distance returns a matrix of distance, each column corresponding to a line of the second argument. As here there is only one line, we convert this matrix to a vector by taking its first column. The unit of the returned values is meters; we convert it to miles, like in the original article and we add this column to the sf. As in the original article, we fill with different colors in Figure 9.5 counties that are less than 25, 50, 75 and 100 miles from the border:

```
border_counties %>%
    mutate(dist_class = cut(dists, c(0, 25, 50, 75, 100))) %>%
  ggplot() +
  geom_sf(aes(fill = dist_class)) +
```

```
    scale_fill_grey(na.translate = FALSE) +
    geom_sf(data = border)
```

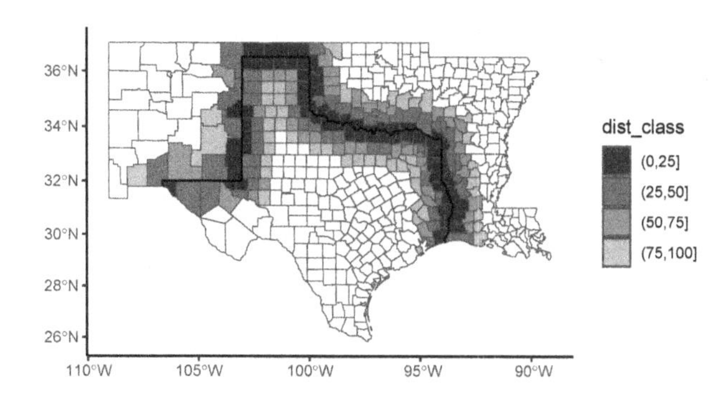

Figure 9.5: Counties close to the Texan border

Finally, we select the relevant series from the data set of the paper, called `mortgage_default`, and we merge it with `border_counties`:

```
mortdef_sf <- border_counties %>%
  right_join(mortgage_defaults, by = "fips") %>%
  mutate(dists = ifelse(state == "Texas", dists, - dists))
```

Note the use of the quite unusual `dplyr::right_join` function. By using `right_join` with `border_counties` as the first argument and `mortgage_defaults` as the second argument, we get an object of the class of the first argument (therefore a `sf`) and we get all the lines of the `mortgage_defaults` tibble and only those of `us_counties` that match. We use the convention that the distance is positive for Texan counties and negative for neighboring states. Finally, we plot the discontinuity in Figure 9.6.

```
mortdef_sf %>%
    filter(abs(dists) < 50) %>%
    mutate(state = ifelse(state == "Texas", "Texas", "other")) %>%
    ggplot(aes(dists, default)) +
    micsr::geom_binmeans(aes(size = after_stat(n)), shape = 21) +
    geom_smooth(aes(linetype = state, weight = loans),
                method = "lm", se = FALSE, color = "black")
```

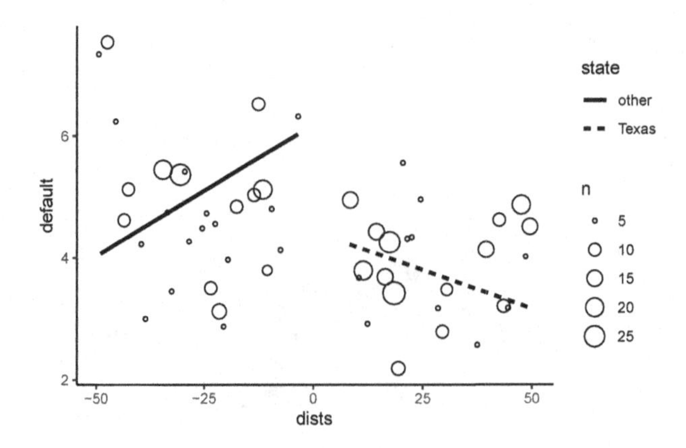

Figure 9.6: Discontinuity of the mortgage defaults data set

The intercept is a bit more than 7% outside the border and about 4.5% inside the border, so that the effect of the Texas specific regulation seems significant (a reduction of about 2.5% of the mortgage default rate).

9.2 Two examples

To illustrate the techniques presented in the subsequent sections, we'll use two data sets. The first one, from Wheeler (2003), is called `agglo_growth` and deals with the topic of economies and diseconomies of agglomeration. The second one, called `sp_growth` is from Ertur and Koch (2007) and is used to fit an extension of Solow's growth model.

9.2.1 Agglomeration economies and diseconomies

The `agglo_growth` contains data about US counties in 1990, identified by their fips code.

```
agglo_growth
```

```
# A tibble: 3,106 x 14
  fips  emp_gr pop_gr    emp    pop college manuf  unemp income
  <chr>  <dbl>  <dbl>  <dbl>  <dbl>   <dbl> <dbl>  <dbl>  <dbl>
1 01001  0.203 0.0591 12591. 32259.  0.121  0.248 0.0900  5774.
2 01003  0.367 0.224  29807. 78556.  0.121  0.206 0.0686  5960.
3 01005  0.177 0.0264  8642. 24756.  0.0923 0.344 0.0874  4544.
4 01007  0.224 0.0528  5377. 15723.  0.0490 0.424 0.158   4859.
5 01009  0.248 0.0737 13713. 36459.  0.0529 0.287 0.0892  5213.
# i 3,101 more rows
# i 5 more variables: educ_sh <dbl>, hw_sh <dbl>, pol_sh <dbl>,
#   notwhite <dbl>, type <fct>
```

emp_gr and pop_gr are the employment and the population growth in each county between 1980 and 1990 and emp and pop the level of employment and of the population in 1980. Wheeler (2003) investigates the existence of:

- agglomeration economies, which implies that the growth of a given territory will be positively correlated with its size,
- agglomeration diseconomies for large cities that experience congestion, crime, pollution, etc.

The hypothesis is therefore that the relation between size and growth should be inverted U shaped, which means that, for small territories, the agglomeration economies effect is dominant; as for large territories, the agglomeration diseconomies becomes dominant. We'll reproduce some of the results using the counties of Louisiana. We first join the tibble to the sf called us_counties which contains the geometries of the counties that we've already used in the previous section:

```
louisiana <- us_counties %>%
    right_join(agglo_growth, by = "fips") %>%
    filter(state == "Louisiana")
```

We first plot a thematic map (Figure 9.7), with colors for counties related to the growth of the population (pop_gr). It is recommended to create first a categorical variable using base::cut, because it is easier to visualize a discrete palette of colors than a continuous one. We use scale_fill_brewer to use one of the palettes provided by the **RColorBrewer** package, which provides sequential, diverging and qualitative palettes. The first two are suitable to represent the values of numerical variables. Sequential palettes use a color, light for low values of the variable and dark for high values. Divergent palettes use two colors, with light colors for middle range values and dark colors for low and high values. All the available palettes can be displayed using RColorBrewer::display.brewer.all(). We use here the Oranges sequential palette.

```
louisiana %>%
    mutate(pop_gr = cut(pop_gr, (-3:3) / 10)) %>%
    ggplot() + geom_sf(aes(fill = pop_gr)) +
    scale_fill_brewer(palette = "Oranges")
```

The inverted U shaped hypothesis between the log of the initial population and the growth rate can be tested by regressing the growth rate on the log population and its square. The later coefficient should then be negative.

```
mod1 <- lm(pop_gr ~ poly(log(pop), 2, raw = TRUE), louisiana)
mod1 %>% gaze
```

```
                                Estimate Std. Error t value Pr(>|t|)
poly(log(pop), 2, raw = TRUE)1  0.68298    0.19333    3.53  0.00079
poly(log(pop), 2, raw = TRUE)2 -0.02974    0.00883   -3.37  0.00132
```

The coefficient of the square term is negative and significant and we get a maximum growth for the fitted model for a value of log(pop) equal to $-0.683/(-0.0297 \times 2) = 11.4819$, i.e.,

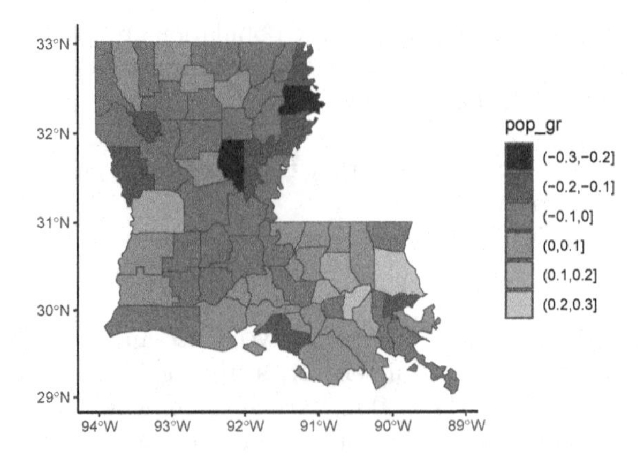

Figure 9.7: Population growth and initial population in Louisiana

a population of about 97 thousand inhabitants. This result is illustrated in Figure 9.8, a scatterplot representing the relationship between the logarithm of the initial population and its growth rate and a fitting line with a second degree polynomial:

```
louisiana %>% ggplot(aes(pop, pop_gr)) +
    geom_point() +
    geom_smooth(method = "lm", formula = y ~ poly(x, 2, raw = TRUE),
                se = FALSE, color = "black") +
  scale_x_continuous(trans = "log10")
```

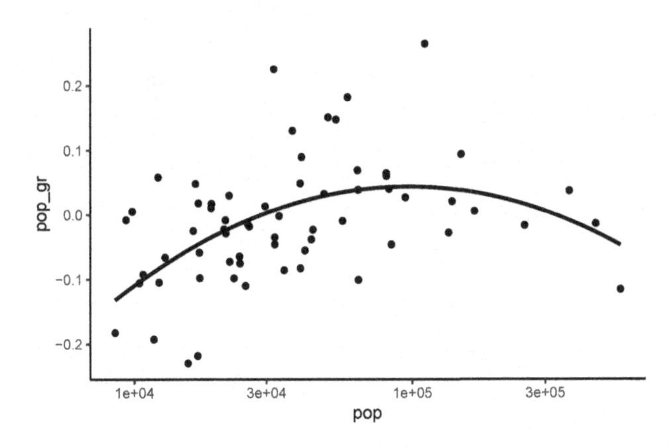

Figure 9.8: Relation between initial population and population growth in Louisiana

9.2.2 Solow model

The second example is from Ertur and Koch (2007) who estimated a growth model for the countries of the world taking spatial correlation into account. The `sp_solow` data set is provided by the **necountries** package:

```
data("sp_solow", package = "necountries")
sp_solow
```

```
# A tibble: 91 x 6
  name      code   gdp60  gdp95 saving labgwth
  <chr>     <chr>  <dbl>  <dbl>  <dbl>   <dbl>
1 Angola    AGO    5136.  2629. 0.0736 0.0233
2 Argentina ARG   18733. 24738. 0.178  0.0165
3 Australia AUS   26480. 45331. 0.247  0.0210
4 Austria   AUT   15283. 45023. 0.261  0.00291
5 Burundi   BDI     889.  1339. 0.0521 0.0168
# i 86 more rows
```

There are 91 countries in the `sp_solow` data set, the series being the gdp in 1960 and 1995, the saving rate (`saving`) and the growth of the labor force (`labgwth`). As in Section 3.1.2, we denote `i` as the saving rate and `v` as the sum of the growth rate of the labor force and 0.05.[7] We also compute the annual growth rate (`growth`) as the difference of the logs of the GDP for 1995 and 1960 divided by 35.

```
sp_solow <- sp_solow %>%
  rename(i = saving) %>%
  mutate(v = labgwth + 0.05,
         growth = (log(gdp95) - log(gdp60)) / 35)
```

We then need to join this data frame to a sf containing the administrative boundaries of the countries and the coordinate of their capital. Several packages enable to get an sf of the world. For example, the **spData** package has a `world` sf which is obtained from Natural Earth,[8] **geodata** has a `world` function that enables to download from the gadm website[9] an object of class `SpatVector` (than can be easily converted to a `sf`, as shown in Section 9.1.1). We use here the convenient **necountries** package that also uses Natural Earth and provides a `countries` function with different arguments to select a subset of countries of the world. By default, `countries` returns 199 lines, the 193 United Nations' recognized countries, the two observer countries (Palestine and Vatican) and four not or not fully recognized countries (Kosovo, Somaliland, Northern Cyprus and Taiwan). Each line includes the geometry of the "main" part of the countries. Some countries have "parts" or "dependencies" that can be included using `part = TRUE` and `dependency = TRUE`. A "part" is an area which has the same political status as the rest of the country, but is far from the main part of the country. Examples of parts include Alaska and Hawaii for the United States, Martinique and Guadeloupe for France and Canaries for Spain. A "dependency" is an area with a specific political status. Examples of dependencies are Greenland for Denmark, New Caledonia for France and Gibraltar for the United Kingdom.

`sp_solow` has columns that contain the names and the iso-codes of the countries (respectively `name` and `code`). Any of them can be used to join `sp_solow` with the `countries`' object, but it is much safer to use `code`, as it avoids the problem of small differences in

[7]0.05 being the sum of the rate of depreciation and the rate of technical progress, assumed to be the same for all countries.

[8]<https://www.naturalearthdata.com/>

[9]<https://gadm.org/>

countries' names. A lot of countries of the world are not present in `sp_solow` (especially most of the communist countries). We check whether all the countries of `sp_solow` are present in the `countries` object, with the `check_join` function; the `by` argument indicates the series in `sp_solow` that identifies the country:

```
library(necountries)
countries() %>% check_join(sp_solow, by = "code")
##
## Countries in the external tibble not in the countries' sf:
##  HKG, ZAR
```

The two problems are that the D.R. of Congo (`iso3` code `COD`) used to be called Zaire (`iso3` code `ZAR`) and that Hong Kong, which is a part of China for the **necountries** package, was considered as a sovereign country in Ertur and Koch (2007) study. We then correct the code for the D.R. of Congo, we add Hong Kong using the `include` argument and we remove `Antarctica` with the exclude argument:

```
sp_solow <- sp_solow %>%
  mutate(code = ifelse(code == "ZAR", "COD", code))
sps <- countries(include = "Hong Kong", exclude = "Antarctica") %>%
    select(iso3, country, point) %>%
    left_join(sp_solow, by = "code") %>% select(- name)
sps
```

```
Simple feature collection with 200 features and 8 fields
Active geometry column: polygon
Geometry type: GEOMETRY
Dimension:     XY
Bounding box:  xmin: -180 ymin: -55.67 xmax: 180 ymax: 83.12
Geodetic CRS:  WGS 84
# A tibble: 200 x 10
  iso3  country              point                                   polygon
  <chr> <chr>          <POINT [°]>                        <MULTIPOLYGON [°]>
1 AFG   Afghanist~  (69.18 34.52) (((74.54 37.02, 74.39 36.99, 74.09 ~
2 ALB   Albania     (19.82 41.33) (((20.57 41.87, 20.5 41.73, 20.53 4~
3 DZA   Algeria     (3.049 36.77) (((-4.822 25, -5.516 25.42, -6.14 2~
4 AND   Andorra     (1.527 42.51) (((1.707 42.5, 1.64 42.47, 1.448 42~
5 AGO   Angola     (13.23 -8.836) (((13.07 -4.635, 13.03 -4.677, 12.8~
# i 195 more rows
# i 6 more variables: gdp60 <dbl>, gdp95 <dbl>, i <dbl>,
#   labgwth <dbl>, v <dbl>, growth <dbl>
```

Note that the resulting `sf` has two geometry columns, polygons for the borders of the countries (`polygon`) and points for the capitals (`point`). Note also that the active geometry column is `polygon`. We then draw in Figure 9.9 a world map with the color of the countries related to the annual growth during the 1960-95 period and a point with a size related to the initial (1960) GDP. **necountries** provides a `plot` method which draws a basic map using **ggplot2**. A basic thematic map can be drawn using the `fill` argument to fill the areas of the countries and the `centroid` or the `capital` arguments to draw a point at the position of the centroid or the capital of each country. Any **ggplot2** functions can be used to customize the graphic:

```
sps %>% plot(fill = "growth", centroid = "gdp60", palette = "Blues") +
  scale_size_continuous(range = c(0.5, 3)) +
  theme(legend.position = "bottom") +
  guides(size = guide_legend(nrow = 3, reverse = TRUE),
         fill = guide_legend(nrow = 3, byrow = TRUE)) +
  labs(fill = "Growth rate (1960-95)", size = "GDP in 1960")
```

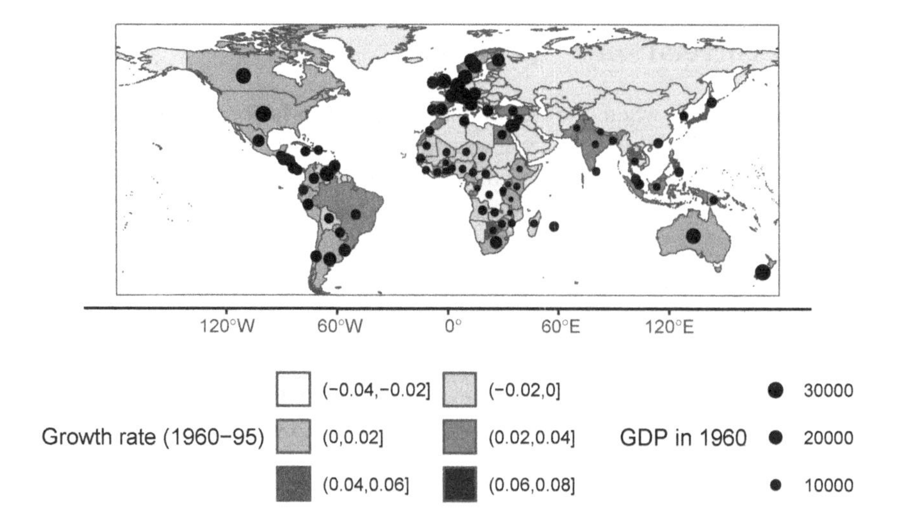

Figure 9.9: Growth and initial GDP, 1960-1995

The basic Solow model, which was given in Equation 3.3, is then estimated:

```
lm(log(gdp95) ~ log(i) + log(v), sp_solow) %>% gaze
```

```
       Estimate Std. Error t value Pr(>|t|)
log(i)    1.276      0.125   10.19   <2e-16
log(v)   -2.709      0.642   -4.22    6e-05
```

Remember that the structural model implies that $\beta_s = -\beta_v = \kappa/(1-\kappa)$. This hypothesis can be tested using a reparametrization of the model:

```
lm(log(gdp95) ~ log(i / v) + log(v), sp_solow) %>% gaze
##          Estimate Std. Error t value Pr(>|t|)
## log(i/v)    1.276      0.125    10.2   <2e-16
## log(v)     -1.433      0.681    -2.1    0.038
```

for which the hypothesis is that the coefficient of `log(v)` equals 0. This hypothesis is rejected at the 5% level (but not at the 1% level). Imposing this hypothesis, we get:

```
lm(log(gdp95) ~ log(i / v), sp_solow) %>% gaze
##             Estimate Std. Error t value Pr(>|t|)
## log(i/v)       1.379      0.117    11.8   <2e-16
```

which implies a value of κ (the share of the capital in the GDP) equal to $\hat{\beta}_i/(1+\hat{\beta}_i) = 0.58$ which is implausibly high.

9.3 Spatial correlation

Spatial correlation occurs when one can define a distance relationship between observations. The notion of distance is broad, but we'll consider in this section only geographical distance. For each observation, one can in this case define a set of neighbors. If the geographical representation of an observation is a polygon (which is the relevant choice for countries or regions), two observations for example can be said to be neighbors if they have a common border. If the geographical representation of an observation is a point (for example for a city), two observations are neighbors if the distance between them is less than, say, 100 kilometers. Once the set of neighbors have been defined for every observations, weights can be computed. The weights can be equal or may depend on the distance between the observation and its neighbors. These two operations of defining the set of neighbors and their weights are performed by the **spdep** package (Pebesma and Bivand 2023).

9.3.1 Contiguity and weights

To get the matrix of contiguity for the counties of Louisiana, we use the `spdep::poly2nb` function:

```
library(spdep)
nb_louis <- poly2nb(louisiana)
nb_louis
```

```
Neighbour list object:
Number of regions: 64
Number of nonzero links: 322
Percentage nonzero weights: 7.861
Average number of links: 5.031
```

The print method indicates the number of contiguity links (322). Instead of storing these links in a full matrix (with 1 for contiguous counties and 0 otherwise) which would have $64^2 = 4096$ cells with only 322 of them with a value of 1 (about 7.9%), an `nb` object is returned. It is a list of 64 vectors which contains, for each observation, the positions of its neighbors. For example:

```
nb_louis %>% head(3)
```

```
[[1]]
[1] 16 23 40 48 56

[[2]]
[1]  6 16 33 41 48

[[3]]
[1]  4 20 26 55 62 64
```

The first two counties have five neighbors and the third one has six. A `summary` method provides more details. `poly2nb` has a `queen` argument which is `TRUE` by default. Queen contiguity means that two polygons which have only one common point are neighbors. Setting `queen` to `FALSE` implies that rook continuity is used. In this case, only counties that have a common border are neighbors.

```
nb_louis_rook <- poly2nb(louisiana, queen = FALSE)
```

Printing `nb_nc_rook`, one can check that the number of contiguity links (318) is slightly lower than previously (322). `nb` objects can't be plotted as is with **ggplot2**. We provide a convenient `st_nb` function which performs this task, as it returns a `sf` object (see Figure 9.10):

```
louisiana %>%
    ggplot() +
    geom_sf(fill = NA) +
    geom_sf(data = st_nb(louisiana), linetype = "dotted") +
    geom_sf(data = st_nb(louisiana, queen = FALSE))
```

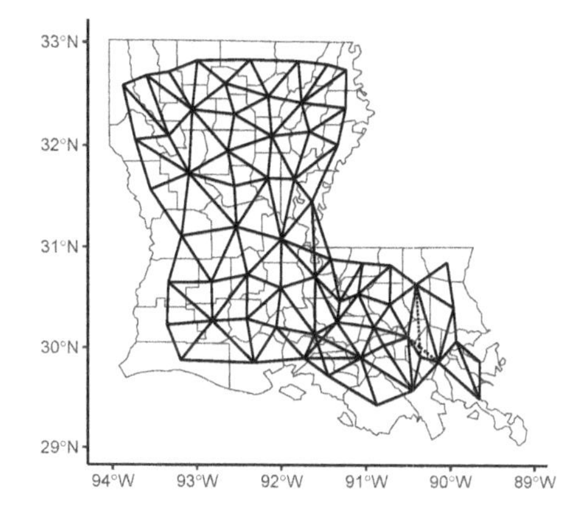

Figure 9.10: Queen and rook links for the counties of Louisiana

We first plot queen contiguity links with dotted lines and then rook contiguity with plain lines, so that the specific queen contiguity links appear as dotted segments. A matrix of weights is obtained by using the `nb2listw` function:

```
W_louis <- nb_louis %>% nb2listw
```

A `listw` object is returned, which contains two lists: the first (`neighbours`) is the same as the one of the `nb` object. The second contains weights:

```
W_louis$weights %>% head(3)
```

```
[[1]]
[1] 0.2 0.2 0.2 0.2 0.2

[[2]]
[1] 0.2 0.2 0.2 0.2 0.2

[[3]]
[1] 0.1667 0.1667 0.1667 0.1667 0.1667 0.1667
```

We can see that weights of neighbors for a given observation are all equal (1/5 for the first two observations which have five neighbors and 1/6 for the third one which has six neighbors) and sum to one. Therefore, premultiplying a vector $(y - \bar{y})$ by W results in a vector with typical value $\sum_m w_{nm} y_m - \bar{y} = \tilde{y}_n - \bar{y}$. `nb` and a `listw` objects can be coerced to matrices using the `nb2mat` and `listw2mat` functions.

```
W_louis %>% listw2mat
```

Instead of defining the neighborhood as common borders, one can consider distance between points, for example the capitals or the centroids of the countries using the `sp_solow` data set. Remember that `sps` contains two `sf`, `polygon` and `point` and that the active geometry is `polygon`. We first specify `point` as being the active geometry, using the `st_set_geometry` function, and we remove all countries for which the data are not available:

```
sp_solow2 <- sps %>% na.omit %>% st_set_geometry("point")
```

Then, we use the `dnearneigh` function to compute the set of neighbors for each country: the first argument is the `sf` and the next two arguments, called `d1` and `d2` are mandatory and should be set to the minimum and the maximum distances that should be used to define the neighbors. Note that this distance should be indicated in kilometers if the CRS is geographical, which is the case here.

```
d <- dnearneigh(sp_solow2, 0, 1000)
d
```

```
Neighbour list object:
Number of regions: 91
Number of nonzero links: 238
Percentage nonzero weights: 2.874
Average number of links: 2.615
18 regions with no links:
3 8 10 17 27 41 44 49 50 55 62 63 64 65 72 74 79 91
29 disjoint connected subgraphs
```

With a distance of 1000 km, there are 238 links and an average of 2.62 neighbors per country. Note that 18 countries have no neighbors. We then compute the weights, using `nb2listwdist`. The first two argument are a `nb` and a `sf` objects. The `type` argument indicates how the weights should be computed. Denoting d_{nm} the distance between the two unit n and m, with `type = "idw"`, we get $w_{nm} = d_{nm}^{-\alpha}$. With `type = "exp"`, the weights are $w_{nm} = \exp(-\alpha d_{nm})$. The `alpha` argument controls the value of α which is 1 by default. For example, if `type = "idw"` and `alpha = 2`, the weights are the inverse of the square of the distance. Once the weights have been computed, they can be normalized in different ways, using the `style` argument. If `type = "raw"` (the default), no normalization is performed. A usual choice is `"W"` where the weights of a unit are normalized so that they sum to 1:

```
w <- nb2listwdist(d, sp_solow2, type = "idw", alpha = 1,
                  style = "W", zero.policy = TRUE)
```

Note the use of `zero.policy`: if `TRUE`, the matrix of weights is computed even if, as it is the case here, some countries have no neighbors.

9.3.2 Tests for spatial correlation

The spatial correlation hypothesis can be tested either for a specific variable or using the residuals of a linear regression. Two main tests have been proposed. The first is Moran's I test: denoting W as the weights matrix where the weights sum to one for a given line, the statistic is defined as:

$$I = \frac{(y-\bar{y})^\top W (y-\bar{y})}{(y-\bar{y})^\top (y-\bar{y})} = \frac{\sum_n (y_n-\bar{y})(\tilde{y}_n-\bar{y})}{\sum_n (y_n-\bar{y})^2}$$

which is simply the ratio of the covariance between y and $\tilde{y}$ and the variance of y. It therefore can be obtained as the coefficient of y in a linear regression of $\tilde{y}$ on y. If the variance of $\tilde{y}$ and y were equal, it would also be the coefficient of correlation between the value of y for a unit and its neighbors. The Moran test is computed using the `spdep::moran.test` function which takes as argument a series and a `listw` object:

```
moran.test(louisiana$pop_gr, W_louis) %>% gaze
## Moran I = 0.306 (-0.016, 0.078), z = 4.147, pval = 0.000
```

The value of the statistic is 0.306. Under the hypothesis of no correlation, the expected value of this statistic is $-1/(N-1) = -0.016$ and the standard deviation is 0.078.[10] The standardized statistic (4.147) is asymptotically normal and the hypothesis of no spatial correlation is rejected. An alternative to Moran's I test is Geary's C test, which is defined as:

$$C = \frac{N-1}{2N}\frac{\sum_n \sum_m w_{nm}(y_n-y_m)^2}{\sum_n (y_n-\bar{y})^2}$$

Introducing deviations from the mean in the previous expression and developing, we get:

[10]The expression of the variance of the Moran statistic can be found in R. S. Bivand and Wong (2018).

$$C = \frac{1}{2}\frac{N-1}{N}\left(1 - 2I + \frac{\sum_n (y_m - \bar{y})^2 \sum_m w_{mn}}{\sum_n (y_n - \bar{y})^2}\right)$$

If the weight matrix were symmetric, as $\sum_n w_{nm} = 1$, we also would have $\sum_m w_{mn} = 1$, so that the last term is one and $C = \frac{N-1}{N}(1 - I)$. Therefore, we can expect C to be close to $(1 - I)$. Geary's test is implemented in the `spdep::geary.test` function

```
geary.test(louisiana$pop_gr, W_louis) %>% gaze
## Geary C = 0.681 (1.000, 0.083), z = 3.824, pval = 0.000
```

These tests are unconditional tests of spatial correlation. The same tests can be performed on the residuals of linear models.

```
model_louisiana <- lm(pop_gr~poly(log(pop),2,raw =TRUE), louisiana)
lm.morantest(model_louisiana, W_louis) %>% gaze
## Moran I = 0.136 (-0.020, 0.077), z = 2.026, pval = 0.021
```

The hypothesis of no spatial correlation is still rejected at the 5% level, but the p-value is much higher than the one associated with the unconditional test.

Ertur and Koch (2007) computed the Moran test for the residuals of the OLS estimation of the standard Solow's growth model. The set of neighbors is obtained with setting an infinite maximum distance, so that all countries are neighbors to each other. Two matrices of weights are computed: one with $w_{nm} = d_{nm}^{-2}$ (`W1`) and the other with $w_{nm} = \exp(-2d_{nm})$ (`W2`):

```
lm_solow <- lm(log(gdp95) ~ log(i) + log(v), sp_solow2)
d <- dnearneigh(sp_solow2, 0, Inf)
W1 <- nb2listwdist(d, sp_solow2, type = "idw", alpha = 2,
                   style = "W", zero.policy = TRUE)
W2 <- nb2listwdist(d, sp_solow2, type = "exp", alpha = 2,
                   style = "W", zero.policy = TRUE)
lm.morantest(lm_solow, W1) %>% gaze
## Moran I = 0.431 (-0.020, 0.065), z = 6.927, pval = 0.000
lm.morantest(lm_solow, W2) %>% gaze
## Moran I = 0.560 (-0.022, 0.125), z = 4.675, pval = 0.000
```

Whatever the weighting matrix, the hypothesis of no spatial correlation is rejected.

9.3.3 Local spatial correlation

A first glance of local spatial correlation can be obtained using Moran's plot (see Figure 9.11), implemented in the `spdep::moran.plot` function, where $y_n - \bar{y}$ is on the x axis and $\tilde{y}_n - \bar{y}$ on the y axis. Therefore, both variables have zero mean and the intercept of the fitting line is therefore 0, the slope being Moran's I statistic:

```
moran.plot(louisiana$pop_gr, W_louis)
```

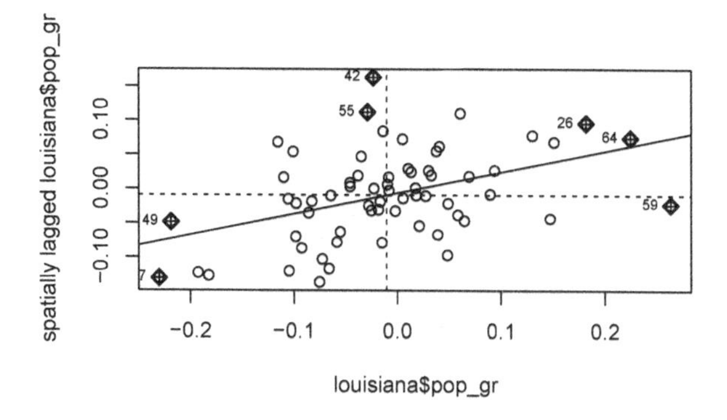

Figure 9.11: Moran plot

Each observation is situated in one of the four quarters of the plane. Observations in the upper-right quarter are *"high-high"* observations, which means that the value of y for observation n and its neighbors are higher than the sample mean. Similarly, the lower-left quarter contains *"low-low"* observations, i.e., observations for which own values of y and values of its neighbors are lower than the sample mean. The upper-left and the lower-right quarters contain respectively the *"low-high"* and the *"high-low"* observations. In case of no spatial correlation, the points should be randomly disposed around the origin and the slope of the regression line should be 0. On the contrary, in case of positive spatial correlation, a majority of points should be of the *"low-low"* or *"high-high"* category and the regression line should have a positive slope.

Anselin (1995) proposed local versions of Moran's I and Geary's C statistics. We then have for each observation I_n and G_n so that $\sum_n I_n$ and $\sum_n G_n$ are respectively proportional to the global Moran and Geary statistics. Local Moran's statistics are defined as:

$$I_n = (y_n - \bar{y}) \sum_m w_{nm}(y_m - \bar{y}) = (y_n - \bar{y})(\tilde{y}_n - \bar{y})$$

and their sum is $(y_n - \bar{y})(\tilde{y}_n - \bar{y})$, which is the numerator of Moran's I statistic. Therefore, we have:

$$I = \frac{\sum_n I_n}{\sum_n (y_n - \bar{y})^2}$$

Local Moran's statistics are obtained using the `spdep::localmoran` function:

```
locmor <- localmoran(louisiana$pop_gr, W_louis)
```

which returns a matrix, with a line for every observation and column containing the local Moran values, their expectation, variance, the standardized statistic and the p-value. It's easier to coerce this matrix to a tibble in order to extract extreme values of the statistic:[11]

[11]Note that we also rename for convenience the fifth column which has a non-standard name (`Pr(z != E(Ii))`) to `pval`.

```
locmor %>% as_tibble %>% rename(pval = 5) %>% filter(pval < 0.01)
```

```
# A tibble: 11 x 5
  Ii          E.Ii        Var.Ii      Z.Ii        pval
  <localmrn>  <localmrn>  <localmrn>  <localmrn>  <localmrn>
1  1.17852    -1.603e-02 0.154640     3.038      0.002384
2  0.93029    -7.606e-03 0.090380     3.120      0.001810
3 -0.03674    -2.119e-05 0.000175    -2.776      0.005510
4  2.31977    -6.666e-02 0.744962     2.765      0.005694
5  3.00564    -8.712e-02 1.210852     2.811      0.004945
# i 6 more rows
```

There are therefore 11 out of 64 (17%) observations for which the p-value is lower than 1%, which confirms the presence of spatial correlation. The local Moran statistic can also be represented in a map (Figure 9.12), in order to identify the "hot spots":[12]

```
z_locmor <- locmor %>% as_tibble %>% pull(Ii) %>% as.numeric
z_locmor <- z_locmor %>%
  cut(c(-1, - 0.5, 0, 0.5, 1, 1.5, 2, 2.5, Inf))
louisiana %>% add_column(z = z_locmor) %>%
  mutate(z = fct_rev(z)) %>%
  ggplot + geom_sf(aes(fill = z)) +
  scale_fill_grey()
```

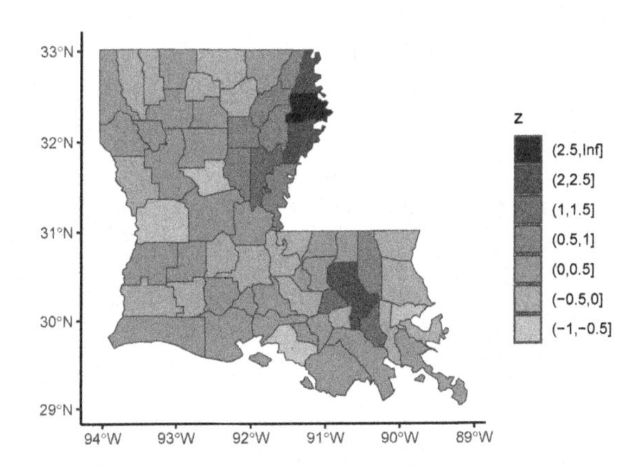

Figure 9.12: Map of local Moran statistic

Two hot spots are then identified, in the north-west and the south-west of the state.

[12]Note that the orders of the levels are reversed using `forcats::fct_rev`, so that the high values are represented in dark gray.

9.4 Spatial econometrics

9.4.1 Spatial models and tests

We consider here linear gaussian models that are extended in order to integrate the spatial features of the sample, using the weighting matrix described in the previous section. Two main models can be considered. The first one is the **spatial error model** (**SEM**) which can be written in matrix form as:

$$y = \alpha\iota + X\beta + \epsilon \text{ with } \epsilon = \rho_\epsilon W\epsilon + \nu \tag{9.1}$$

Therefore, the error for observation n is linearly related to the errors of its neighbors $\tilde{\epsilon}_n$. OLS estimation gives unbiased and consistent estimators but, as always with non-spherical errors, it is inefficient and the estimated covariance matrix of the coefficients based on the simple formula $(\sigma^2(\tilde{X}^\top\tilde{X})^{-1})$ is biased.

The second model is called the **spatial autoregressive model** (**SAR**).[13]. It extends the basic gaussian linear model by adding as a regressor the mean value of the response for the neighbor units. For one observation, the model writes $y_n = \alpha + \beta^\top x_n + \rho_y\tilde{y}_n + \epsilon_n$ or, in matrix form:

$$y = \alpha j + X\beta + \rho_y W y + \epsilon = Z\gamma + \rho_y W y + \epsilon$$

The reduced form of the model is:

$$y = (I - \rho_y W)^{-1} Z\gamma + (I - \rho_y W)^{-1}\epsilon \tag{9.2}$$

Therefore, the values of y depend on all the values of ϵ, and $\tilde{y}_n$ is therefore correlated with ϵ_n. The OLS estimator is therefore biased and inconsistent. Moreover, in Equation 9.2, the spatial dependence in the parameter ρ_y feeds back (R. Bivand, Millo, and Piras 2021, 6), which is not the case for Equation 9.1. This means that for the SEM model, the marginal effect of a covariate is the corresponding coefficient, but this is not the case for the SAR model. Moreover, the matrix $(I - \rho_y W)^{-1}$ can be written as an infinite series:

$$(I - \rho_y W)^{-1} = I + \rho_y W + \rho_y^2 W^2 + \rho_y^3 W^3 + ...$$

Consider the case of three observations, France (1), Italy (2) and Spain (3). The matrix of contiguity is:

$$W = \begin{pmatrix} 0 & 0.5 & 0.5 \\ 1 & 0 & 0 \\ 1 & 0 & 0 \end{pmatrix}$$

France has two neighbors (Italy and Spain) and Italy and Spain only one (France). The weights are such that they sum to 1 for each line. Consider a variation of the unique covariate in Spain (Δx_3) and denote β the corresponding coefficient. The direct effect on the response for the three countries is obviously $\Delta y_0^\top = (0, 0, \beta\Delta x_3)$. This increase of y in Spain implies an

[13] This model is also known as the spatial lag model.

increase of y in the neighboring country, France. Therefore, $\Delta y_1^\top = (0.5\rho_y\beta\Delta x_3, 0, 0)$. This increase of y in France implies an increase of y in the neighboring countries, Italy and Spain: $\Delta y_2^\top = (0, 0.5\rho_y^2\beta\Delta x_3, 0.5\rho_y^2\beta\Delta x_3)$. This increase of y in Italy and Spain implies an increase of y in the neighboring country of Italy and Spain, which is France: $\Delta y_3^\top = (0.5\rho_y^3\beta\Delta x_3, 0, 0)$, etc. To take a numerical example, consider $\beta\Delta x_3 = 1$ and $\rho_y = 0.3$. Then, $\Delta y_0^\top = (0, 0, 1)$, $\Delta y_1^\top = (0.15, 0, 0)$, $\Delta y_2^\top = (0, 0.045, 0.045)$ and $\Delta y_3^\top = (0.0135, 0, 0)$. In total, we have $\sum_{i=0}^{3} \Delta y_i^\top = (0.177, 0.045, 1.045)$. These four effects and their sum is computed below:

```
W <- matrix(c(0, 1, 1, 0.5, 0, 0, 0.5, 0, 0), nrow = 3)
bdx <- c(0, 0, 1)
rho <- 0.3
dy0 <- diag(3) %*% bdx
dy1 <- rho * W %*% bdx
dy2 <- rho ^ 2 * W %*% W %*% bdx
dy3 <- rho ^ 3 * W %*% W %*% W %*% bdx
dy_03 <- cbind(dy0, dy1, dy2, dy3)
dy_03
```

```
     [,1] [,2]  [,3]   [,4]
[1,]    0 0.15 0.000 0.0135
[2,]    0 0.00 0.045 0.0000
[3,]    1 0.00 0.045 0.0000
```

```
apply(dy_03, 1, sum)
```

```
[1] 0.1635 0.0450 1.0450
```

The total effect of Δx_3 is obtained using the formula: $\Delta y = (I - \rho_y W)^{-1}\Delta x$:

```
solve(diag(3) - rho * W) %*% bdx
```

```
        [,1]
[1,] 0.16484
[2,] 0.04945
[3,] 1.04945
```

which is very close to the sum of the direct effect and the first three indirect effects computed previously.

Spatial models are usually estimated by maximum likelihood, by assuming a multivariate normal distribution for the iid errors. For the SEM model, the idiosyncratic errors can be written as a function of the response:

$$\nu = (I - \rho_\epsilon W)y - (I - \rho_\epsilon W)Z\gamma \tag{9.3}$$

so that the Jacobian of the transformation is

$$\left|\frac{\partial \nu}{\partial y}\right| = |I - \rho_\epsilon W|$$

.

the likelihood is similar to the one of the linear gaussian model except that an extra term, which is the log of the Jacobian, should be added:

$$-N/2(\ln 2\pi + \ln \sigma^2) + \ln |I - \rho_\epsilon W| - \frac{1}{\sigma^2}\nu^\top \nu$$

where ν is given by Equation 9.3. For the **SAR** model, Equation 9.2 indicates that the Jacobian is the same, so that the log-likelihood is:

$$-N/2(\ln 2\pi + \ln \sigma^2) + \ln |I - \rho_y W| - \frac{1}{\sigma^2}\epsilon^\top \epsilon^2$$

with $\epsilon = y - Z\gamma - \rho_y Wy$.

These two models can be augmented by spatial lags of the covariates, they are in this case called Durbin's models. There is an inserting relationship between Durbin's SAR model and the SEM model. The former can be written:

$$y = \rho_y Wy + Z\gamma + WZ\theta + \epsilon$$

with ϵ iid errors. The **common factor** hypothesis is that $\theta = -\rho_y \gamma$. In this case, we have:

$$(I - \rho_y W)y = (I - \rho_y W)Z\gamma + \epsilon$$

Or finally:

$$y = Z\gamma + (I - \rho_y W)^{-1}\epsilon$$

which is the SEM model, as denoting $\eta = (I - \rho_y W)^{-1}\epsilon$ the errors of this model, we have $y = Z\gamma + \eta$, with $\eta = \rho_y W\eta + \epsilon$.

Once it has been established that there is some spatial dependence, one has to choose the "right" specification. Several tests have been proposed, based on the three test principles. Lagrange multiplier tests proposed by Anselin et al. (1996) are popular, as they only require the OLS estimation. Different flavors of tests have been proposed, testing $\rho_y = 0$, $\rho_\epsilon = 0$ or $\rho_y = \rho_\epsilon = 0$. The basic version of, for example the first test ($\mathrm{H}_0 : \rho_y = 0$) has, as a maintained hypothesis that $\rho_\epsilon = 0$. A robust version of the test has been proposed which doesn't impose this maintained hypothesis. The common factor hypothesis can easily be tested using a likelihood ratio test. The Wald statistic is less easy to compute, as a set of non-linear hypotheses $\theta = -\rho_y \beta$ should be tested.

9.4.2 Application to the growth model

To overcome the irrelevant empirical results of the standard Solow model, Ertur and Koch (2007) consider an endogenous growth model with spillovers. The production function is as usual a Cobb-Douglas:

$$Y_n(t) = A_n(t)K_n^{\kappa}(t)L_n^{1-\kappa}(t)$$

κ being the elasticity of the production with the capital and also the share of the profits in the national product. The level of technology is given by:

$$A_n(t) = \Omega(0)e^{\mu t}k_n^{\phi}(t)\prod_{m\neq n}^{N} A_n^{\gamma w_{nm}}(t)$$

where $k_n(t) = K_n(t)/L_n(t)$ is the physical capital per worker, $\Omega(0)$ is the initial level of the technology and μ is a constant rate of technological growth, as in the Solow model. The next term takes into account the spillover effect of domestic investment, following Romer (1986), and the strength of this spillover effect is measured by ϕ. The last term is specific to the model of Ertur and Koch (2007), for which the spillovers are not restricted to domestic investment, but concerns also the technology of neighboring countries. The effect of the technology of country m on the technology of country n is the product of a constant parameter γ and a specific weight, w_{nm}, which is a decreasing function of the distance between both countries. Ertur and Koch (2007) showed (equation 10, page 1038) that, at the steady state, the output per capita is:

$$\begin{array}{rcl}\ln y_n^* & = & \frac{1}{1-\kappa-\phi}\ln\Omega + \frac{\kappa+\phi}{1-\kappa-\phi}\ln i_n - \frac{\kappa+\phi}{1-\kappa-\phi}\ln v_n \\ & - & \frac{\gamma\kappa}{1-\kappa-\phi}\sum_{m\neq n}^{N} w_{nm}\ln i_m + \frac{\gamma\kappa}{1-\kappa-\phi}\sum_{m\neq n}^{N} w_{nm}\ln v_n \\ & + & \frac{\gamma(1-\kappa)}{1-\kappa-\phi}\sum_{m\neq n}^{N} w_{nm}\ln y_m^*\end{array}$$

In matrix form, denoting $y = \ln y^*$, $X = (\ln i \ln v)$, $\beta = \begin{pmatrix}(\kappa+\phi)/(1-\kappa-\phi) \\ -(\kappa+\phi)/(1-\kappa-\phi)\end{pmatrix}$, $\theta = \begin{pmatrix}\gamma\kappa/(1-\kappa-\phi) \\ -\gamma\kappa/(1-\kappa-\phi)\end{pmatrix}$ and $\rho_y = \gamma(1-\kappa)/(1-\kappa-\phi)$, the model can be written as:

$$y = \alpha i + X\beta + WX\theta + \rho_y Wy + \epsilon$$

if $\gamma = 0$, the model reduces to the model of Romer (1986) and the last two terms disappear. Note that in this case κ and ϕ are not identified, but only their sum. Moreover, if ϕ is also 0, we get the Solow model. Whether ϕ equals zero or not, the resulting model is a standard linear model already estimated in Section 9.2.2:

```
ols_solow <- lm(log(gdp95) ~ log(i) + log(v), sp_solow2)
```

Moreover, imposing the theoretical constraint of the growth model, we must have the sum of the two coefficients of `log(i)` and `log(v)` equal to 0. Therefore, the constrained model can be estimated using the difference of the logarithms of the two covariates as the unique regressor. For convenience, we call the ratio of the two covariates `i` and we store it in a new data frame called `sp_solow3`. The constrained OLS model can then be estimated:

```
sp_solow3 <- sp_solow2 %>% mutate(i = i / v)
ols_solowc <- lm(log(gdp95) ~ log(i), sp_solow3)
```

The hypothesis of $\gamma = 0$ can be tested using Lagrange multipliers tests, testing either the alternative that there is a spatial lag or correlated errors. `spdep::lm.LMtests` provides different flavors of these tests. Its two main arguments are a `lm` model and a matrix of spatial weights. The `test` argument can be either `"LMerr"` and "`LMlag`" for the standard tests of uncorrelated errors and absence of spatial lag, `"RLMerr"` and `"RLMlag"` for the robust versions of these two tests and `"SARMA"`, suitable when the alternative hypothesis is the presence of both correlation of the errors and a spatial lag. `test = "all"` enables to perform the five tests:

```
lm.LMtests(ols_solow, W1, test = "all") %>% gaze
## RSerr    : 39.874, df = 1, p-value =  0.000
## RSlag    : 51.467, df = 1, p-value =  0.000
## adjRSerr:  3.004, df = 1, p-value =  0.083
## adjRSlag: 14.596, df = 1, p-value =  0.000
## SARMA    : 54.471, df = 2, p-value =  0.000
```

All the tests reject the hypothesis of uncorrelated errors or/and spatial lag, except the robust test of uncorrelated errors. Therefore, these tests seem to indicate that the correct model is a spatial lag model. We then estimate the three spatial models using the **spatialreg** package (Pebesma and Bivand 2023) which provides three functions: `lagsarlm` for SAR models, `sacsarlm` for SEM model and `errorsarlm` for SAR-SEM models. The three first arguments are `formula`, `data` and `listw`, which should be a `listw` object containing the weights. The `Durbin` argument enables to estimate Durbin's versions of the model; this can be done either by setting it to `TRUE` (a spatial lag is then added for all the covariates) or a formula indicating the subset of covariates for which spatial lags have to be added. We estimate also, as for the OLS estimator, the variant of the model that imposes the theoretical restrictions on the coefficients. The results are presented in Table 9.1.

```
library(spatialreg)
m1 <- lagsarlm(log(gdp95) ~ log(i) + log(v),
               sp_solow2, W1, Durbin = TRUE)
m2 <- errorsarlm(log(gdp95) ~ log(i) + log(v),
                 sp_solow2, W1, Durbin = FALSE)
m3 <- sacsarlm(log(gdp95) ~ log(i) + log(v),
               sp_solow2, W1, Durbin = TRUE)
m1c <- update(m1, . ~ . - log(v), data = sp_solow3)
m2c <- update(m2, . ~ . - log(v), data = sp_solow3)
m3c <- update(m3, . ~ . - log(v), data = sp_solow3)
```

We can first test the theoretical restrictions using either a likelihood ratio or a Wald test:

```
library(lmtest)
library(car)
linearHypothesis(ols_solow, "log(i) + log(v) = 0",
                 test = "Chisq") %>% gaze
## Chisq = 4.427, df: 1, pval = 0.035
lrtest(ols_solow, ols_solowc) %>% gaze
## Chisq = 4.467, df: 1, pval = 0.035
linearHypothesis(m1, c("log(i) + log(v) = 0",
                       "lag.log(i) + lag.log(v) = 0")) %>% gaze
## Chisq = 2.382, df: 2, pval = 0.304
```

Table 9.1: Results of the growth models

	OLS	SAR	SEM	SARMA	OLSc	SARc	SEMc
$\ln i$	1.276	0.837	0.843	0.876	1.379	0.863	0.863
	(0.125)	(0.099)	(0.099)	(0.102)	(0.117)	(0.099)	(0.099)
$\ln v$	−2.709	−1.636	−1.848	−1.737			
	(0.642)	(0.555)	(0.550)	(0.562)			
$W \ln i$		−0.338		−0.325		−0.278	
		(0.182)		(0.392)		(0.179)	
$W \ln v$		0.566		0.076			
		(0.851)		(1.398)			
ρ_y		0.746		0.633		0.740	
		(0.070)		(0.338)		(0.070)	
ρ_ϵ			0.823	0.317			0.835
			(0.053)	(0.509)			(0.050)
Num.Obs.	91	91	91	91	91	91	91
Log.Lik.	−102.177	−73.451	−76.569	−72.912	−104.410	−74.640	−78.156
F	74.025				138.297		

```
lrtest(m1, m1c) %>% gaze
## Chisq = 2.377, df: 2, pval = 0.305
```

As seen previously, the theoretical restrictions are rejected at the 5% level using a Wald test for the Solow model. This is not the case for the spatial lag model. Note also that the Wald and the likelihood ratio tests give very similar results. An interesting feature of the model of Ertur and Koch (2007) is that their specification enables the identification of all the structural parameters of the model, κ, ϕ and γ. If $\phi = 0$, $\theta = \gamma\beta$ and $\rho_y = \gamma$, so that: $y = \alpha j + (I - \gamma W)X\beta + \gamma y + \epsilon$, the reduced form of the model is, in matrix form:

$$y = \alpha j + X\beta + (I - \gamma W)^{-1}\epsilon$$

which is a SEM model without Durbin terms. Therefore, the hypothesis that $\phi = 0$ is in the context of this model the hypothesis of common factors and can easily be tested using a likelihood ratio test:

```
lrtest(m1c, m2c) %>% gaze
## Chisq = 7.033, df: 1, pval = 0.008
```

The hypothesis of common factor is rejected at the 1% level. Our preferred specification is therefore a spatial lag model imposing the theoretical restrictions. The structural parameters of the model (s) can be computed using the fitted reduced form parameters (r). The relation between r and s is:

$$r^\top = (\beta, \theta, \rho) = \left(\frac{\kappa + \phi}{1 - \kappa - \phi}, -\gamma\frac{\kappa}{1 - \kappa - \phi}, \gamma\frac{1 - \kappa}{1 - \kappa - \phi}\right)$$

Inversing this relation, we get the structural parameters s as a function of the reduced form parameters:

$$s^\top = (\kappa, \phi, \gamma) = \left(\frac{\theta}{\theta-\rho}, \frac{\beta}{1+\beta} - \frac{\theta}{\theta-\rho}, \frac{\rho-\theta}{1+\beta}\right)$$

For our model, we get:

```
beta <- unname(coef(m1c)["log(i)"])
theta <- unname(coef(m1c)["lag.log(i)"])
rho <- unname(coef(m1c)["rho"])
kappa <- theta / (theta - rho)
phi <- beta / (1 + beta) - theta / (theta - rho)
gamma <- (rho - theta) / (1 + beta)
s <- c(kappa = kappa, phi = phi, gamma = gamma)
s
## kappa    phi  gamma
## 0.2729 0.1902 0.5463
```

To apply the delta method, we compute the matrix of derivatives of s with respect to r:

$$\Gamma(r) = \frac{\partial s}{\partial r^\top}(r) = \begin{pmatrix} 0 & -\frac{\rho}{(\theta-\rho)^2} & \frac{\theta}{(\theta-\rho)^2} \\ \frac{1}{(1+\beta)^2} & \frac{\rho}{(\theta-\rho)^2} & -\frac{\theta}{(\theta-\rho)^2} \\ -\frac{\theta-\rho}{(1+\beta)^2} & -\frac{1}{1+\beta} & \frac{1}{1+\beta} \end{pmatrix}$$

and the asymptotic variance of s is then:

$$\hat{V}(\hat{r}) = \Gamma(\hat{r})\hat{V}(\hat{s})\Gamma(\hat{r})^\top$$

```
Vr <- vcov(m1c)[c("log(i)", "lag.log(i)", "rho"),
                c("log(i)", "lag.log(i)", "rho")]
G <- c(0,
       - rho / (theta - rho) ^ 2, theta / (theta - rho) ^ 2,
       1 / (1 + beta) ^ 2, rho / (theta - rho) ^ 2,
       - theta / (theta - rho) ^ 2,
       (theta - rho) / (1 + beta) ^ 2,
       - 1 / (1 + beta), 1 / (1 + beta))
G <- matrix(G, nrow = 3, byrow = TRUE)
V <- G %*% Vr %*% t(G)
```

We finally compute the table of the structural parameters:

```
stdev <- V %>% diag %>% sqrt
z <- s / stdev
p <- pnorm(abs(z), lower.tail = FALSE) * 2
struct_par <- cbind(Estimate = s, "Std. Error" = stdev,
                    "z value" = z, "Pr(>|z|" = p)
rownames(struct_par) <- names(s)
struct_par
```

```
      Estimate Std. Error z value   Pr(>|z|
kappa   0.2729     0.1182   2.309 2.095e-02
phi     0.1902     0.1071   1.775 7.587e-02
gamma   0.5463     0.1159   4.712 2.453e-06
```

The value of κ is compatible with the observable share of the profits in the national product. ϕ is significant only at the 10% level. Finally, γ is highly significant, which is a confirmation of the relevance of the model of Ertur and Koch (2007).

Part III

Special responses

The models analyzed in the first two parts of the book were (with very few exceptions) characterized by a response that was a real number defined on the whole real line. When the response was a strictly positive real number, it was possible to coerce it to a real number by taking its logarithm. In this part, we'll consider models with "special responses", i.e., responses that are not defined on the whole real line. We'll then encounter responses that are either real but defined only on a part of the real line, integers or categorical. Most of the time, the models described in this part will be estimated by maximum likelihood, which means that they are highly parametrical as the distribution law of the response should be completely specified.

A fundamental difference with the linear model is that the conditional expectation of the response is no longer a linear function of the covariates. More specifically, in the first two parts of the book, we used a linear index function: $\mu_n = \alpha + \beta^\top x_n = \gamma^\top z_n$ and we had $\mathrm{E}(y \mid x_n) = \mu_n$. Therefore, $\beta_k = \frac{\partial \mathrm{E}(y|x_n)}{\partial x_{nk}}$ so that the marginal effect of the kth covariate was the corresponding coefficient. In the chapters of this part, we'll denote $\eta_n = \alpha + \beta^\top x_n = \gamma^\top z_n$ the linear index function and μ_n a function of interest which will be most of the time the conditional expectation of the response. The two variables are related in the following way: $\eta_n = g(\mu_n)$, where the g function is called the **link**. Therefore, we now have: $\frac{\partial \mu_n}{\partial x_{nk}} = \beta_k \frac{\partial g^{-1}}{\partial \eta_n}(\eta_n)$ so that the marginal effect of the kth covariate is now proportional (but not equal) to the corresponding coefficient.

Chapter 10 is devoted to binomial responses, i.e., responses that take only two values. Chapter 11 presents truncated responses, for which the real value of the variable is only observed in a certain range. Count responses, i.e., responses that take non-negative integer values are presented in Chapter 12. Models that deal with duration responses are presented in Chapter 13. Finally, Chapter 14 is devoted to random utility models for multinomial responses.

10

Binomial models

Binomial responses can take only two mutually exclusive values that can be, without loss of generality coded as 1 and 0. Common examples are transport mode choice (car vs. public transit), working force participation for women, union membership, etc. For this kind of responses, the statistical distribution is obviously a binomial distribution with one trial. This is a major difference with the estimators that will be reviewed in the other chapters of this part, for which assuming a distribution function for the response is a crucial choice. Denoting 1 as "a success" and 0 as "a fail", this distribution is fully characterized by a unique parameter, μ, which is the probability of success and is also the expected value of the variable, as: $\mbox{E}(y) = (1-\mu)\times 0 + \mu \times 1 = \mu$. The variance of the distribution is: $\mbox{V}(y) = (1-\mu)(0-\mu)^2 + \mu(1-\mu)^2 = \mu(1-\mu)$. It is therefore inversely U-shaped, has a maximum for $\mu = 0.5$ (with a value of 0.25) and is symmetric around this value. As μ tends to 0 or to 1, the variance of y obviously tends to 0 as almost all the values in a given sample will be either equal to 0 or 1.

To get a regression model for a binomial response, we first define an **index function**, also called the **linear predictor**: $\eta_n = \gamma^\top z_n = \alpha + \beta^\top x_n$. Then, a function F is chosen that relates this index function to the unique parameter of the binomial distribution: $\mu_n = F(\eta_n)$. Different choices of F leads to different binomial models.

Section 10.1 will present the three most common choices for F which result in the linear probability, the logit and the probit models. Section 10.2 will present two distinct structural models that can justify the use of these models. Section 10.3 presents the generalized linear models from which the binomial model is a special case. Section 10.4 is devoted to the estimation, the evaluation and the testing of binomial models. Section 10.5 presents relevant estimators when some covariates are endogenous. Finally, Section 10.6 presents the ordered model.

10.1 Functional form and the linear-probability, probit and logit model

The most obvious choice for F is the identity function, so that $\mu_n = \eta_n = \alpha + \beta^\top x_n$. Therefore, the parameter of the binomial distribution is assumed to be a linear function of the covariates. On the one hand, this choice has several interesting features. It is very simple to estimate, as it is a linear model and, moreover, it can be simply extended to IV estimation. As a linear model, $\frac{\partial \mu_n}{\partial x_{kn}} = \beta_k$, so that the estimated parameters can be interpreted as the (constant) marginal effects of the corresponding covariate on the probability of success. On the other hand, it has two serious drawbacks. Firstly, the residuals are $y_n - \hat{\mu}_n$ but, as y_n is either 0 or 1, the residuals are respectively $-(\hat{\alpha} + \hat{\beta}^\top x_n)$ or $1 - (\hat{\alpha} + \hat{\beta}^\top x_n)$ and therefore depends on the values of x_n. The linear-probability model, estimated by least-squares is therefore inefficient, as the residuals are heteroskedastic and the standard deviations reported by a least squares program are biased. As usual, the solution would be either to estimate the linear-probability model by GLS or to use heteroskedasticity-robust estimator for the covariance matrix of the estimators. Secondly, as the fitted probabilities of success are linear functions of the covariates, they are not bounded by 0 and 1 and, therefore, it is possible that the model will predict, for some observations, probabilities that would be either negative or greater than 1.

Therefore, it is customary to use a functional form F which has the following properties:

- $F(z)$ is increasing in z,
- $\lim_{z\rightarrow -\infty} F(z) = 0$,
- $\lim_{z\rightarrow +\infty} F(z) = 1$.

which are the features of any cumulative density function for continuous variables defined on the whole real line support. Two common choices are the normal (Φ) and the logistic (Λ) distributions:

$$\left\{\begin{array}{rcl}\Phi(z) & = & \int_{-\infty}^{z}\phi(t)dt = \int_{-\infty}^{z}\frac{1}{\sqrt{2\pi}}e^{-\frac{1}{2}t^2}dt \\ \Lambda(z) & = & \frac{e^z}{1+e^z}\end{array}\right.$$

which lead respectively to the **probit** and **logit** models. The density function for the logistic distribution (obtained by taking the derivative of Λ) is $\lambda(z) = \frac{e^z}{(1+e^z)^2}$. Both density functions are symmetric around 0 and are "bell-shaped", but they have two important differences, as illustrated in Figure 10.1:

- the variance of the standard normal distribution is 1 and is equal to $\pi^2/3$ for the logistic distribution,
- the logistic distribution has much heavier tails than the normal density.

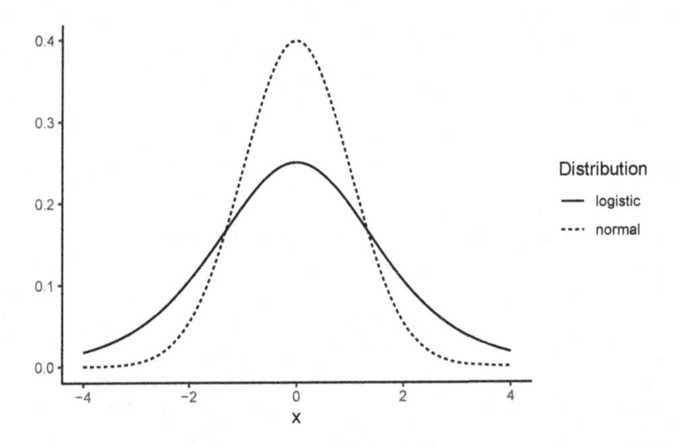

Figure 10.1: Logistic and normal densities

As $\mu_n = F(\eta_n)$ (with $\eta_n = \alpha + \beta^\top x_n$), the marginal effect of the kth covariate on the probability is:

$$\frac{\partial \mu_n}{\partial x_{nk}} = \beta_k f(\eta_n) \tag{10.1}$$

where f is the first derivative of F, which is respectively, for the probit and logit models, ϕ and λ, the normal and logistic densities. Therefore, the marginal effect is obtained by multiplying the coefficient by $f(\eta_n)$ which depends on the value of the covariates for a given observation. Therefore, the marginal effect is observation-dependent, but the ratio of two marginal effects for two covariates is not, as it is obviously, from Equation 10.1, equal to the ratio of the two corresponding coefficients. As the coefficient of proportionality is the normal/logistic density, the maximum marginal effect is for $\eta_n = 0$, which results in a probability of success of 0.5. The corresponding values of the densities are 0.4 and 0.25 for the normal and logistic densities. Therefore, a rule of thumb to interpret coefficients is to multiply them respectively by 0.4 and 0.25 for the probit and logit model to get an estimation of the maximum marginal effect.

The coefficients of the logit and probit can therefore not be compared. This is due to the fact that they are scaled differently, as the standard deviation of the logistic distribution is $\pi/\sqrt{3} \approx 1.81$, compared to 1 for the normal distribution. Therefore, it would be tempting to multiply the probit coefficients by 1.81 to compare them to the logit coefficients, but Amemiya (1981) showed that, empirically, the value of 1.6 performs better.

As an example, we consider the data set used by Horowitz (1993) which concerns the transport mode chosen for work trips by a sample of 842 individuals in Washington DC in the late sixties. The response `mode` is 1 for car and 0 for transit. The covariates are the in- and out-vehicle times (`ivtime` and `ovtime`) and the cost differences between car and transit. Therefore, a positive value indicates that car trip is longer/more expensive than the corresponding trip using public transit. We multiply the cost by 8.42 to obtain dollars in 2022 (the CPI for 2022 is 842 with a 100 base in 1967). The generalized cost of a trip is the sum of the monetary cost and the value of the time spent in the transport. We use two-thirds of the minimum hourly wage (about \$1.4 in the US in the late sixties, which is about \$8 in dollars of 2022) to valuate an hour of transport:

```
mode_choice <- mode_choice %>%
    mutate(cost = cost * 8.42,
           gcost = (ivtime + ovtime) * 8 + cost)
```

To fit the three models, we use the `micsr::binomreg` function, which has a `link` argument which enables to estimate the three models.

```
lp_m <- binomreg(mode ~ gcost, mode_choice, link = "identity")
pt_m <- update(lp_m, link = "probit")
lt_m <- update(lp_m, link = "logit")
gaze(lp_m)
##        Estimate Std. Error z-value Pr(>|z|)
## gcost   0.02255    0.00236    9.58   <2e-16
gaze(pt_m)
##        Estimate Std. Error z-value Pr(>|z|)
## gcost    0.1129     0.0128    8.79   <2e-16
gaze(lt_m)
```

```
##        Estimate Std. Error z-value Pr(>|z|)
## gcost    0.2112     0.0248    8.52   <2e-16
```

The coefficient of `gcost` for the linear-probability model is 0.0226, which means that a one dollar increase of the generalized cost differential will increase the probability of using the car by 2.26 percentage points. If we use the previously described rule of thumb to multiply the probit/logit coefficients by 0.4/0.25 in order to have an upper limit for the marginal effect, we get 5.28 and 4.51 percentage points, which are much higher values than for the linear probability model. This is because the coefficient of the linear model estimates the marginal effect at the sample mean. In our sample, the mean value of the covariate is 2.9. To get comparable marginal effects for the probit/logit models, we should first compute $\hat{\alpha} + \hat{\beta}\bar{x}$ (1.15 and 2 respectively for the probit and logit models) and use these values with the relevant densities ($\phi(1.15) = 0.206$ and $\lambda(2) = 0.105$). At the sample mean, the marginal effects are then 0.023 and 0.022 and are therefore very close to the linear probability model coefficient. The scatterplot and the fitted probability curves are presented o in Figure 10.2.

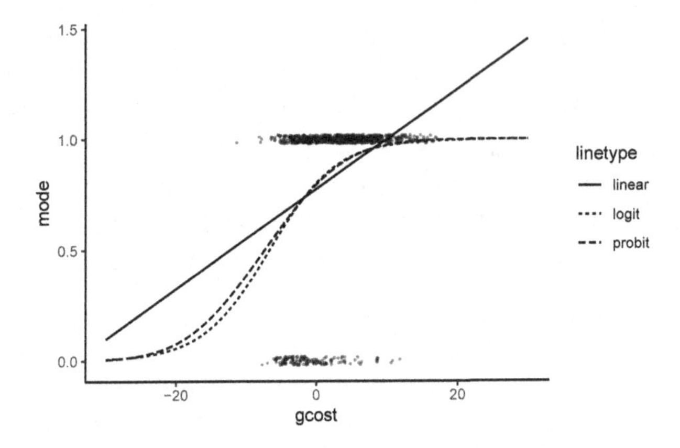

Figure 10.2: Fitted probabilities

The fitted probabilities are given by a straight line for the linear probability model and by an S curve for the probit and logit models. These last two curves are very similar except for low values of the covariate. Note also that at the sample mean ($x = 2.9$), the slopes of the three curves are very similar. This illustrates the fact that the three models result in similar marginal effects around the mean value of the covariate. The linear probability model is $\hat{\mu} = 0.774 + 0.023 \times x$. Therefore, $\hat{\mu} < 0$ for $x < -0.774/0.023 = -34.33$ and $\hat{\mu} > 1$ for $x > (1 - 0.774)/0.023 = 10.007$. In this sample, there are no observations for which $\hat{\mu} < 0$ but, for 83 out of 842 observations, $\hat{\mu} > 1$. Finally, the ratio of the logit and probit coefficients is $0.211/0.211 = 1.871$, which is a bit larger than the value of 1.6 suggested by Amemiya (1981).

We now consider a second data set called `airbnb`, used by Edelman, Luca, and Svirsky (2017). The aim of their study is to analyze the presence of racial discrimination on the Airbnb platform. The authors create guest accounts that differ by the first name chosen. More specifically, the race of the applicant is suggested by the choice of the first name, either a "white" (Emily, Sarah, Greg) or an "African American" (Lakisha or Kareem) first name. The response is acceptance and is 1 if the host gave a positive response and 0 otherwise. In our simplified example, we use only three covariates, guest's race suggested by the first name `guest_race`, the price `price` (in logs) and `city`, the cities where the experience took place

(Baltimore, Dallas, Los Angeles, St. Louis and Washington, DC). Note that the mean of the response is 0.45 which is a distinctive feature of this data set compared to the previous one. As the mean value of the probability of success is close to 50% we can expect that the rule of the thumb which consists of multiplying the logit/probit coefficients by 0.25/0.4 would give an estimated value for the marginal effect close to the one directly obtained in the linear probability model.

```
airbnb <- airbnb %>%
    mutate(acceptance = ifelse(acceptance == "no", 0, 1)) %>%
    filter(! is.na(price), ! is.na(city))
lp_a <- binomreg(acceptance ~ guest_race + log(price) + city,
                 airbnb, link = "identity")
pt_a <- update(lp_a, link = "probit")
lt_a <- update(lp_a, link = "logit")
```

To summarize the results, we print in Table 10.1 the coefficients of the linear probability model, those of the logit multiplied by 0.25, those of the probit multiplied by 0.4 and the ratio of the logit and the probit coefficients.

Table 10.1: Comparison of the coefficients for the Airbnb data set

	linear	probit	logit	logit / probit
(Intercept)	0.708	0.214	0.215	1.604
guest_raceblack	-0.084	-0.085	-0.085	1.602
log(price)	-0.045	-0.047	-0.047	1.603
cityDallas	0.023	0.023	0.023	1.593
cityLos-Angeles	0.015	0.016	0.016	1.590
citySt-Louis	0.010	0.010	0.010	1.573
cityWashington	-0.037	-0.037	-0.037	1.611

With this rescaling, the three models give similar results. For a 100% increase of the price, the probability of acceptance reduces by 4.16 percentage points. The estimated marginal effect for black guests is about -8.5 percentage points. However, computing a derivative is not relevant in this case as the covariate is a dummy. We should therefore better compute the difference between the probabilities of acceptance, everything other being equal, which means here for a given price of the property and for the reference city, which is Baltimore. The average price being equal to \$182 in our sample, we have, for the probit model: $\Phi(0.497-0.213-0.107\times\ln 182)-\Phi(0.497-0.213-0.107\times\ln 182)=-0.084$ and for the logit model: $\Lambda(0.796-0.341-0.171\times\ln 182)-\Lambda(0.796-0.171\times\ln 182)=-0.084$, which means that, at least in this example, the previous computation of the derivative gives an extremely accurate approximation of the effect of this dummy covariate. Finally, note that the ratio of the logit and probit coefficients is very close to the value of 1.6 advocated by Amemiya (1981).

10.2 Structural models for binomial responses

Two structural models have been proposed to give a theoretical foundation to the probit/logit models. Without loss of generality, we'll present these two models for the case where there is a unique covariate.

10.2.1 Latent variable and index function

We observe that y is equal to 0 or 1, but we now assume that this values is related to a latent continuous variable (called y^*) which is unobserved. $y = 1$ will result for "high" values of y^* and $y = 0$ for low values of y^*. More specifically we'll assume that the observation rule is:

$$\left\{ \begin{array}{lll} y=0 & \text{if} & y^* \leq \psi \\ y=1 & \text{if} & y^* > \psi \end{array} \right.$$

where ψ is an unknown threshold. Now assume that the value of y^* is partly explained by an observable covariate x, the unexplained part being modelized by a random error ϵ. We then have: $y^* = \alpha + \beta x + \epsilon$, so that the observation rule becomes:

$$\left\{ \begin{array}{lll} y=0 & \text{if} & \epsilon \leq \psi - \alpha - \beta x \\ y=1 & \text{if} & \epsilon > \psi - \alpha - \beta x \end{array} \right.$$

This observation rule depends on $\psi - \alpha$ and not on the separate values of ψ and α. Therefore, ψ can be set to any arbitrary value, for example 0. Then, the probability of success is: $1 - F(-\alpha - \beta x)$ where F is the cumulative density function of ϵ. For example, if $\epsilon \sim \mathcal{N}(0, \sigma)$, $\mbox{P}(y = 1 \mid x) = 1 - \Phi(-(\alpha - \beta x)/\sigma)$. We can see from this expression that only α/σ and β/σ can be identified, so that, σ can be set to any arbitrary value, for example 1. Moreover, by the symmetry of the normal distribution, we have $1 - \Phi(-z) = \Phi(z)$, so that the probability of success becomes $F(y = 1 \mid x) = \Phi(\alpha + \beta^\top x)$, which defines the probit model. Assuming that the distribution of ϵ is logistic, we have a probability of success equal to $1 - \Lambda(-\alpha - \beta x)$ which reduces, as the logistic distribution is also symmetric, to: $F(y = 1 \mid x) = \Lambda(\alpha + \beta x) = e^{\alpha + \beta x}/(1 + e^{\alpha + \beta x})$.

10.2.2 Random utility model

Consider now that we can define a utility function for the two alternatives that correspond to the two values of the binomial response. As an example, y equals 1 or 0 if car or public transit is chosen, and the only covariate x is the generalized cost. The utility of choosing a transport mode doesn't depend only on the generalized cost, but also on some other unobserved variables. The effect of these variables are modelized as the realization of a random variable ϵ. We can therefore define the following random utility functions:

$$\left\{ \begin{array}{lll} U_0 & = & \alpha_0 + \beta x_0 + \epsilon_0 \\ U_1 & = & \alpha_1 + \beta x_1 + \epsilon_1 \end{array} \right.$$

where β is the marginal utility of \$1. The choice of the individual is deterministic. They will choose the car if the utility of this mode is greater than the utility of public transit. Therefore, we have the following observation rule:

$$\begin{cases} y = 0 & \text{if} \quad \epsilon_1 - \epsilon_0 \leq -(\alpha_1 - \alpha_0) - \beta(x_1 - x_0) \\ y = 1 & \text{if} \quad \epsilon_1 - \epsilon_0 > -(\alpha_1 - \alpha_0) - \beta(x_1 - x_0) \end{cases}$$

Denoting $\epsilon = \epsilon_1 - \epsilon_0$ the error difference, $\alpha = \alpha_1 - \alpha_0$ and $x = x_1 - x_0$ the difference of generalized cost for the two modes, we have:

$$\begin{cases} y = 0 & \text{if} \quad \epsilon \leq -(\alpha + \beta x) \\ y = 1 & \text{if} \quad \epsilon > -(\alpha + \beta x) \end{cases}$$

The probability of "success" (here choosing the car) is therefore $\mathrm{P}(y = 1 \mid x) = 1 - F(-\alpha - \beta x)$, with F as the cumulative density of ϵ. If the distribution is symmetric, this probability reduces once again to $\mathrm{P}(y = 1 \mid x) = F(\alpha + \beta x)$, and the probit or logit models are obtained by choosing either the normal or the logistic distribution.

10.3 Binomial model as a generalized linear model

The estimation of binomial models with **R** is performed using the `stats::glm` function, which stands for a **generalized linear model**. It is therefore important to have at least some basic knowledge about generalized linear models to understand the output of the fitted models.

10.3.1 Generalized linear models

The generalized linear models (**GLM**) are a wide family of models that are intended to extend the linear model. These models have the following components:

- a *random component* which specifies the distribution of the response, as a member of the exponential family, and in particular the expected value $\mathrm{E}(y) = \mu$,
- a *systematic component*: some covariates $x_1, x_2, \dots x_m$ produce a linear predictor $\eta_n = \alpha + \beta^\top x_n$,
- the *link* function g which specifies the relation between the random and the systematic components: $\eta = g(\mu)$.

The exponential family is defined by the following density function:

$$f(y; \theta, \phi) = e^{(y\theta - b(\theta)) / \phi + c(y, \phi)} \tag{10.2}$$

θ and ϕ being respectively a position and a scale parameter. Linear models are actually a specific case of generalized linear models with a normal distribution and an identity link. We have in this case the following density function:

$$\phi(y; \mu, \sigma) = \frac{1}{\sqrt{2\pi}\sigma} e^{-\frac{1}{2}\frac{(y-\mu)^2}{\sigma^2}} = e^{\frac{y\mu - 0.5\mu^2}{\sigma^2} - 0.5y^2/\sigma^2 - 0.5\ln(2\pi\sigma^2)}$$

which is a member of the exponential family with $\theta = \mu$, $\phi = \sigma^2$, $b(\theta) = 0.5\theta^2$ and $c(y, \phi) = -0.5(y^2/\phi + \ln(2\pi\phi))$. The first two derivatives of Equation 10.2 with respect to θ are:

$$\left\{\begin{array}{lcl} \frac{\partial l}{\partial \theta} & = & \frac{1}{\phi}(y - b'(\theta)) \\ \frac{\partial^2 l}{\partial \theta^2} & = & -\frac{1}{\phi}b''(\theta) \end{array}\right.$$

As $\mathrm{E}\left(\frac{\partial l}{\partial \theta}\right) = 0$, we have $\mathrm{E}(y) = b'(\theta)$. Moreover, by the information matrix equality: $\mathrm{E}\left(\frac{\partial^2 l}{\partial \theta^2}\right) + \mathrm{E}\left({\frac{\partial l}{\partial \theta}}^2\right) = 0$, so that $\mathrm{V}(y) = \phi b''(\theta)$.

Going back to the normal (or gaussian) model with an identity link, we have, for a given set of estimates (which leads to the **proposed model**): $\hat{\mu}_n = \hat{\eta}_n = \hat{\alpha} + \hat{\beta}^\top x_n$ and the log-likelihood function is:

$$\ln L(y, \hat{\mu}) = -\frac{N}{2}\ln(2\pi + \sigma^2) - \frac{1}{2\sigma^2}\sum_{n=1}^{N}(y_n - \hat{\mu}_n)^2$$

For a hypothetical "perfect" or **saturated model** with a perfect fit, we would have $\hat{\mu}_n = y_n$, so that the log-likelihood would be $-\frac{N}{2}\ln(2\pi + \sigma^2)$. Minus two times the difference of these two values of the log likelihood function is called the **scaled deviance** of the proposed model:

$$D^*(y; \hat{\mu}) = \sum_{n=1}^{N}\frac{(y_n - \hat{\mu}_n)^2}{\sigma^2}$$

and the deviance is obtained by multiplying the scaled deviance by σ^2 (or more generally by the scale parameter ϕ):

$$D(y; \hat{\mu}) = \sum_{n=1}^{N}(y_n - \hat{\mu}_n)^2$$

which is simply, for the linear model, the sum of square residuals. For the binomial model, the probability mass function is given by the probability of success μ if $y = 1$ and by the probability of failure $1-\mu$ if $y = 0$. This probability can be compactly written as $\mu^y(1-\mu)^{1-y}$ or as:

$$f(y; \mu) = e^{y\ln\mu + (1-y)\ln(1-\mu)} = e^{y\ln\frac{\mu}{1-\mu} + \ln(1-\mu)} = e^{y\theta - \ln(1+e^\theta)}$$

which is a member of the exponential family with: $\theta = \ln\frac{\mu}{1-\mu}$, $b(\theta) = \ln(1 + e^\theta)$, $c(\theta, y) = 0$ and $\phi = 1$. The model is fully characterized once the link is specified. For the logit model, we have $\mu = \frac{e^\eta}{1+e^\eta}$, so that $\eta = \ln\frac{\mu}{1-\mu} = g(\mu)$. We then have $\theta = \eta$, so that the logit link is called the **canonical** link for binomial models[1]. As the density for the binomial model returns a probability, the log-likelihood for the saturated model is zero. Therefore, the deviance is:

$$D(y;\hat{\mu}) = 2\sum_{n=1}^{N}(y_n \ln \hat{\mu}_n + (1-y_n)\ln(1-\hat{\mu}_n)) \tag{10.3}$$

The **null model** is a model with only an intercept. In this case, $\hat{\mu}_n = \hat{\mu}_0$ and the maximum likelihood estimator of μ_0 is $\sum_{n=1}^{N} y_n/N$, i.e., the share of success in the sample. The deviance of this model is called the **null deviance**. An alternative to the deviance as a measure of the fit of the model is the **generalized Pearson statistic**, defined as:

$$X^2 = \sum_{n=1}^{N}\frac{(y_n-\hat{\mu}_n)^2}{\mathrm{V}(\hat{\mu}_n)} = \sum_{n=1}^{N}\frac{(y_n-\hat{\mu}_n)^2}{\hat{\mu}_n(1-\hat{\mu}_n)} \tag{10.4}$$

In the linear model, residuals have several interesting properties:

- they are homoskedastic (or at least they may be homoskedastic if the variance of the conditional distribution of the response is constant),
- they have an intuitive meaning, as they are the difference between the actual and the fitted values of the response, the latter being an estimate of the conditional mean of the response,
- they are related to the value of the objective function, which is the sum of square residuals.

The most obvious definition of the residuals for binomial models is the **response residuals**, which are simply the difference between the response and the prediction of the model (the fitted probability of success $\hat{\mu}$). However, these residuals $(y_n - \hat{\mu}_n)$ are necessarily heteroskedastic, as the variance of y_n is $\mu_n(1-\mu_n)$. Scaling the response residuals by their standard deviation leads to **Pearson's residuals**: $(y_n - \hat{\mu}_n)/\sqrt{\hat{\mu}_n(1-\hat{\mu}_n)}$. The sum of squares of Pearson's residuals is the generalized Pearson statistic given by Equation 10.4. The **deviance residuals** are such that the sum of their squares equals the deviance statistic D. They are therefore defined by:

$$(2y_n - 1)\sqrt{2}\sqrt{y_n \ln \hat{\mu}_n + (1-y_n)\ln(1-\hat{\mu}_n)}$$

the term $2y_n - 1$ gives a positive sign for the residuals of observations for which $y_n = 1$ and a negative sign for $y_n = 0$, as for the two other types of residuals.

10.3.2 Estimation with `stats::glm`

The estimation of probit/logit models is performed using `glm`. The interface of `glm` is very similar to `lm`, but it has a supplementary argument called `family` which indicates the distribution of the response.[2] The family argument can be either a character string or a function. In the latter case, an argument called `link` can be specified, which indicates how the linear predictor $\eta_n = \alpha + \beta^\top x_n$ is related to the parameter of the distribution μ_n . If

[1]For every member of the exponential family, there is one canonical link, see McCullagh and Nelder (1989), page 30.

[2]Note that `family` is the second argument of `glm`, and `data` is the third.

we use `family = binomial(link = "probit")`, then $\mu_n = \Phi(\eta_n)$. The default choice is `"logit"` (the canonical link), so that the logit model can be obtained using either:

```
lgt <- glm(mode ~ gcost, data = mode_choice,
           family = binomial(link = 'logit'))
glm(mode ~ gcost, data = mode_choice, family = binomial)
glm(mode ~ gcost, data = mode_choice, family = binomial())
glm(mode ~ gcost, data = mode_choice, family = "binomial")
```

Remember that, while estimating a generalized linear model, three models are considered:

- the saturated model, for which there is one parameter for every observation and a perfect fit; therefore the log-likelihood, the deviance and the number of degrees of freedom are 0,
- the null model, with only one estimated coefficient and $N-1$ degrees of freedom,
- the proposed model, with $K+1$ estimated parameters, and therefore $N-K-1$ degrees of freedom.

A call to `glm` results in an object of class `glm` which inherits from class `lm`. As for `lm`, the `summary` method computes detailed results for the model and, if not saved in an object, these results are printed:

```
summary(lgt)
```

```

Call:
glm(formula = mode ~ gcost, family = binomial(link = "logit"),
    data = mode_choice)

Coefficients:
            Estimate Std. Error z value Pr(>|z|)
(Intercept)   1.3905     0.0998   13.94   <2e-16 ***
gcost         0.2112     0.0248    8.52   <2e-16 ***
---
Signif. codes:  0 '***' 0.001 '**' 0.01 '*' 0.05 '.' 0.1 ' ' 1

(Dispersion parameter for binomial family taken to be 1)

    Null deviance: 741.33  on 841  degrees of freedom
Residual deviance: 647.39  on 840  degrees of freedom
AIC: 651.4

Number of Fisher Scoring iterations: 5
```

The output indicates the deviance of the null and the proposed model, along with their respective degrees of freedom ($N-1=841$ and $N-K-1=840$). The latter is called the **residual deviance**. This information is elements of the object returned by `stats::glm` and can be extracted directly:

```
lgt$deviance
lgt$null.deviance
lgt$df.residual
lgt$df.null
```

or, for the fitted model, using the corresponding functions:

```
deviance(lgt)
df.residual(lgt)
```

We can check that the null deviance can be obtained by fitting a model with only an intercept:

```
update(lgt, . ~ 1)$deviance
## [1] 741.3
```

The residuals can be extracted from the fitted model using `resid`. The `resid` method for `glm` objects has a `type` argument which can be equal to `"response"`, `"pearson"` and `"deviance"`.

```
resid(lgt, "response") %>% head
##        1        2        3        4        5        6
##  0.03651 -0.55944  0.23718 -0.66782  0.04603 -0.80629
resid(lgt, "pearson") %>% head
##       1       2       3       4       5       6
##  0.1947 -1.1269  0.5576 -1.4179  0.2197 -2.0402
resid(lgt, "deviance") %>% head
##       1       2       3       4       5       6
##  0.2728 -1.2804  0.7358 -1.4846  0.3070 -1.8118
```

The fitted values of the model can be expressed on the scale of the linear predictor or the response. They are available in the returned object as `linear.predictors` and `fitted.values`:

```
lgt$linear.predictors %>% head
##      1      2      3      4      5      6
## 3.2729 0.2389 1.1682 0.6983 3.0313 1.4261
lgt$fitted.values %>% head
##      1      2      3      4      5      6
## 0.9635 0.5594 0.7628 0.6678 0.9540 0.8063
```

The latter can also be obtained using the `fitted` function:

```
fitted(lgt)
```

The `predict` method returns by default the fitted values but can also compute the predicted values for a new data frame. For example, if the difference of generalized cost is increased by 10%:

```
mode_choice2 <- mode_choice %>% mutate(gcost2 = gcost * 1.1)
```

The predictions can be computed in the scale of the linear predictors or of the response by setting the `type` argument to `"link"` (the default) or `"response"`:

```
predict(lgt, newdata = mode_choice2, type = "link") %>% head
##      1      2      3      4      5      6
## 3.2729 0.2389 1.1682 0.6983 3.0313 1.4261
predict(lgt, newdata = mode_choice2, type = "response") %>% head
##      1      2      3      4      5      6
## 0.9635 0.5594 0.7628 0.6678 0.9540 0.8063
```

10.4 Model estimation, evaluation and testing

10.4.1 Estimation

The `stats::glm` function uses an iterative weighted least squares method to fit all the flavors of GLM's models. However, probit and logit models are usually estimated by maximum likelihood. With $\eta_n = \alpha + \beta^\top x_n = \gamma^\top z_n$ the linear predictor, the individual contribution to the likelihood is $F(\eta_n)$ if $y_n = 1$ and $1 - F(\eta_n)$ if $y_n = 0$. The log-likelihood is then:

$$\ln L = \sum_{n=1}^{N} y_n \ln F(\eta_n) + (1 - y_n) \ln(1 - F(\eta_n))$$

The first-order condition for a maximum is that the vector of the first derivatives:

$$\frac{\partial \ln L}{\partial \gamma} = \sum_{n=1}^{N} \frac{y_n}{F(\eta_n)} f(\eta_n) z_n - \frac{1 - y_n}{1 - F(\eta_n)} f(\eta_n) z_n = \sum_{n=1}^{N} \frac{y_n - F_n}{F_n(1 - F_n)} f_n z_n = 0 \tag{10.5}$$

is zero, where we defined for convenience $F_n = F(\eta_n)$ and $f_n = f(\eta_n)$.

$\psi_n = \frac{y_n - F_n}{F_n(1-F_n)} f_n$ is called the **generalized residual**[3]. Generalized residuals have the same property as standard residuals in the linear regression model, they are orthogonal to all the covariates, and they sum to 0 if the regression contains an intercept. The hessian matrix of the second derivatives is:

$$\frac{\partial^2 \ln L}{\partial \gamma \partial \gamma^\top} = -\sum_{n=1}^{N} \left(\frac{y_n(1 - F_n)^2 + (1 - y_n) F_n^2}{F_n^2 (1 - F_n)^2} f_n^2 - \frac{y_n - F_n}{F_n(1 - F_n)} f'_n \right) z_n z_n^\top \tag{10.6}$$

with f'_n the derivative of f_n. Taking the expectation, we obtain a much simpler expression: as $E(y_n) = F_n$, the second term in brackets disappears and the first one simplifies to:

[3] Gourieroux et al. (1987).

$$\mathrm{E}\left(\frac{\partial^2 \ln L}{\partial\gamma\partial\gamma^\top}\right) = -\sum_{n=1}^{N} \frac{f_n^2}{F_n(1-F_n)} z_n z_n^\top \tag{10.7}$$

For the logit model, the density is: $\lambda_n = e^{\eta_n}/(1+e^{\eta_n})^2 = \Lambda_n(1-\Lambda_n)$ and Equation 10.5 reduces to:

$$\frac{\partial \ln L}{\partial \gamma} = \sum_{n=1}^{N} (y_n - \Lambda_n)\, z_n = 0$$

This expression is particularly appealing as the generalized residual $y_n - \Lambda_n$ is the response residual. Moreover, the expression of the matrix of second derivatives is particularly simple:

$$\frac{\partial^2 \ln L}{\partial\gamma\partial\gamma^\top} = -\sum_{n=1}^{N} \lambda_n z_n z_n^\top = -\sum_{n=1}^{N} \Lambda_n(1-\Lambda_n) z_n z_n^\top$$

and is equal to its expectation, as it doesn't depend on y. For the probit model, the vector of response residuals $(y_n - \Phi_n)$ is not orthogonal to the covariates. Moreover, the formula of the hessian is rather complicated and depends on y. However, its expectation (Equation 10.7) can be expressed compactly in terms of the inverse mills ratio, defined by: $r(z) = \phi(z)/\Phi(z)$. Noting that $\phi(z)/(1-\Phi(z)) = \phi(-z)/\Phi(-z) = r(-z)$ by symmetry of the normal distribution, Equation 10.7 simplifies to:

$$\mathrm{E}\left(\frac{\partial^2 \ln L}{\partial\beta\partial\beta^\top}\right) = -\sum_{n=1}^{N} r(\eta_n) r(-\eta_n) z_n z_n^\top \tag{10.8}$$

The generalized residuals are:

$$\psi_n = \frac{\phi_n(y_n - \Phi_n)}{\Phi_n(1-\Phi_n)} \tag{10.9}$$

and they are related to the latent variable y^* used in Section 10.2.1. Remember that $y^*_n = \mu_n + \epsilon_n$, with $\mu_n = \alpha + \beta^\top x_n$ and $\epsilon_n \sim \mathcal{N}(0, \sigma)$. Then, considering the latent variable, the residual can be defined as $\hat{\epsilon}_n = y^*_n - \hat{\mu}_n$. This residual can't be computed because y^*_n is unobserved, we only observe $y_n = 1$ if $y^*_n > 0$ and $y_n = 0$ if $y^*_n \leq 0$. However, its expectation can be computed. For $y_n = 1$:

$$\mathrm{E}(\hat{\epsilon} \mid x, y^* > 0) = \frac{\int_0^{+\infty} y^* \phi(y^* - \hat{\mu}) dy^*}{\int_0^{+\infty} \phi(y^* - \hat{\mu}) dy^*} - \hat{\mu} = \frac{\int_{-\hat{\mu}}^{+\infty} (\hat{\mu} + v)\phi(v) dv}{\int_{-\hat{\mu}}^{+\infty} \phi(v) dv} - \hat{\mu} = \frac{\phi(\hat{\mu})}{\Phi(\hat{\mu})}$$

Similarly, $\mathrm{E}(\hat{\epsilon} \mid x, y^* \leq 0) = -\frac{\phi(\hat{\mu})}{1-\Phi(\hat{\mu})}$, so that:

$$\mathrm{E}(\hat{\epsilon}_n \mid x_n) = y_n \mathrm{E}(\hat{\epsilon}_n \mid x_n, y^*_n > 0) + (1 - y_n)\mathrm{E}(\hat{\epsilon}_n \mid x_n, y^*_n \leq 0) = \frac{\phi(\hat{\mu}_n)(y_n - \Phi(\hat{\mu}_n))}{\Phi(\hat{\mu}_n)(1 - \Phi(\hat{\mu}_n))}$$

which is the generalized residual defined in Equation 10.9.

Table 10.2: Logit and probit models for the mode choice data set

	logit		probit	
	unconstrained	constrained	unconstrained	constrained
(Intercept)	1.062	1.390	0.664	0.821
	(0.195)	(0.100)	(0.110)	(0.056)
cost	0.156		0.086	
	(0.036)		(0.020)	
ivtime	0.545		0.308	
	(0.455)		(0.248)	
ovtime	4.760		2.380	
	(0.958)		(0.494)	
gcost		0.211		0.113
		(0.025)		(0.013)
Num.Obs.	842	842	842	842
AIC	642.6	651.4	645.5	652.5
BIC	661.6	660.9	664.5	662.0
Log.Lik.	−317.322	−323.694	−318.769	−324.270
Deviance	634.64	647.39	637.54	648.54
Deviance Null	741.33	741.33	741.33	741.33

The three estimators of the covariance matrix of the estimators can be used. The outer product of the gradient estimator is based on Equation 10.5:

$$\hat{V}_G(\hat{\gamma}) = \sum_{n=1}^{N} \left(\frac{y_n - F_n}{F_n(1 - F_n)} f_n\right)^2 z_n z_n^\top$$

The hessian based estimator is obtained by taking the inverse of the opposite of the hessian given by Equation 10.6. Finally the information-based estimator is obtained by taking the inverse of the opposite of the matrix given by Equation 10.7:

$$\hat{V}_I(\hat{\gamma}) = \left(\sum_{n=1}^{N} \frac{f_n^2}{F_n(1 - F_n)} z_n z_n^\top\right)^{-1}$$

As the hessian for the logit model doesn't depend on y, the last two estimators are the same for this model. We consider as an example two variants of the mode choice model: the first uses as distinct covariate the monetary cost, in- and out-vehicle time; the second uses as unique covariate the generalized cost.

```
lgt_unconst <- binomreg(mode ~ cost + ivtime + ovtime,
                        data = mode_choice, link = "logit")
lgt_const <- binomreg(mode ~ gcost, data = mode_choice, link = "logit")
pbt_unconst <- update(lgt_unconst, link = "probit")
pbt_const <- update(lgt_const, link = "probit")
```

We present in Table 10.2 the results of the logit and probit models. By default, the standard deviations are computed using the information-based estimation of the covariance matrix of the estimates. The hessian and the outer-product of the gradient estimators are obtained by setting the `vcov` argument of `vcov` or of `summary` to respectively `"hessian"` or `"opg"`. For example, to get the gradient-based estimator of the covariance matrix:

```
vcov(pbt_unconst, vcov = "opg")
```

The `micsr::stder` function enables to compute the different flavors of the estimated standard errors, which are obtained by taking the square roots of the diagonal elements of the covariance matrix:

```
vcov(pbt_unconst, vcov = "opg") %>% diag %>% sqrt
## (Intercept)        cost      ivtime      outime
##     0.11392     0.02111     0.32552     0.47637
stder(pbt_unconst, vcov = "opg")
## (Intercept)        cost      ivtime      outime
##     0.11392     0.02111     0.32552     0.47637
```

The sandwich estimator is obtained using the **micsr**'s method for `sandwich::vcovHC`.

```
sandwich::vcovHC(pbt_unconst) %>% diag %>% sqrt
## (Intercept)        cost      ivtime      outime
##     0.10864     0.01863     0.18973     0.54202
stder(pbt_unconst, vcov = vcovHC)
## (Intercept)        cost      ivtime      outime
##     0.10864     0.01863     0.18973     0.54202
```

The four estimators of the standard errors are presented in Table 10.3 for the unconstrained probit model. The first three give very similar estimates. The sandwich estimator gives slightly different results, especially a larger value for out-vehicle time and a smaller value for in-vehicle time.

Table 10.3: Estimation of the standard deviations of the estimates

	information	hessian	gradient	sandwich
(Intercept)	0.1092	0.1100	0.1139	0.1086
cost	0.0195	0.0198	0.0211	0.0186
ivtime	0.2382	0.2483	0.3255	0.1897
ovtime	0.4952	0.4938	0.4764	0.5420

10.4.2 Evaluation

Once several models are estimated, the evaluation and the selection process of one of them is based on several indicators. The first indicator is the value of the objective function, which is the log-likelihood. Closely related to the log-likelihood is the deviance, which is the opposite of twice the log-likelihood. Both measures are reported in Table 10.2. These measures favor lightly the logit models compared to the probit models and indicate an important difference between the constrained and unconstrained model. However, the comparison between the constrained and unconstrained models is spurious, because adding further covariates, even if they are irrelevant, necessarily increases the fit of the model. Therefore, we need indicators that penalize highly parametrized models. The two most popular indicators are the Akaike and the Bayes information criteria (**AIC** and **BIC**) which are respectively defined by AIC $= -2\ln L + 2K$ and BIC $= -2\ln L + K\ln N$. They are therefore obtained by augmenting the deviance by a term which is a multiple of the number of fitted parameters: 2 times for the AIC and $\ln N$ times for the BIC. The rule being to select the model for which the statistic is lower, we can see from Table 10.2 that the AIC leads to the choice of the unconstrained model, and the BIC leads to the choice of the constrained model. This is because the penalization in the BIC is higher, as $\ln 842 = 6.7$. These statistics can be extracted from the fitted model using the `logLik`, `deviance`, `BIC` and `AIC` methods for `micsr` objects, for example:

```
AIC(pbt_unconst)
## [1] 645.5
```

The `logLik` method for `micsr` objects has a `type` argument which enables to extract the value for the proposed model (`type = "mode"`, the default), the null model or the saturated model:

```
logLik(pbt_unconst)
## 'log Lik.' -318.8 (df=4)
logLik(pbt_unconst, type = "model")
## 'log Lik.' -318.8 (df=4)
logLik(pbt_unconst, type = "null")
## 'log Lik.' -370.7 (df=1)
logLik(pbt_unconst, type = "saturated")
## 'log Lik.' 0 (df=842)
```

In linear model, a popular indicator of the quality of a model is the coefficient of determination, called R^2. For linear models: $\sum(y_n - \bar{y})^2 = \sum(\hat{y}_n - \bar{y})^2 + \sum\hat{\epsilon}_n^2$ because the vectors of fitted values and residuals are orthogonal. The R^2 can therefore be defined using three equivalent formulas:[4]

$$R^2 = \frac{\sum(\hat{y}_n - \bar{y})^2}{\sum(y_n - \bar{y})^2} = 1 - \frac{\sum\hat{\epsilon}_n^2}{\sum(y_n - \bar{y})^2} = \hat{\rho}^2_{y,\hat{y}} \qquad (10.10)$$

The first formula is particularly appealing, as it indicates the share of the variance of the response that is explained by the model: it is therefore bounded by 0 and 1. It is 0 if the model has no explanatory power, which means that the fit is equivalent to the null model, i.e., the model with no covariates. It is one for a "perfect" model, i.e., a model for which the vector of residuals is 0.

The three formulas are not equivalent for binomial models and, therefore, there is no unambiguous formula for the R^2 for these models. A lot of different formulas have been proposed in the literature.[5] The **micsr** package provides an `rsq` function which has a type argument. By setting `type` to `"ess"`, `"rss"` and `"cor"`, we get the three versions of the R^2 described in Equation 10.10. The `"rss"` version is often called the Efron's R^2 (Efron 1978) and was previously proposed by Lave (1970).

[4]See Section 1.4.3.
[5]Useful surveys are Magee (1990), Windmeijer (1995) and Veall and Zimmermann (1996).

```
rsq(pbt_unconst, type = "ess")
## [1] 0.1164
rsq(pbt_unconst, type = "rss")
## [1] 0.1336
rsq(pbt_unconst, type = "cor")
## [1] 0.1342
```

As seen in Section 5.3.1.2, in the linear model, the R^2 is related to statistics that test the hypothesis that all the coefficients of the model except the intercept are 0. This leads to pseudo-R^2 that are obtained using any of the three classical tests statistic:

```
rsq(pbt_unconst, type = "wald")
## [1] 0.08751
rsq(pbt_unconst, type = "lr")
## [1] 0.116
rsq(pbt_unconst, type = "score")
## [1] 0.1111
```

Tjur (2009) proposed an R^2 that he called the coefficient of discrimination. This coefficient is the difference between the probability of success for the subsample for which $y = 1$ and the subsample for which $y = 0$. Tjur's measure is interestingly related to the ESS, the RSS and the correlation measure of the R^2. More precisely:

$$R^2 = \frac{1}{2}\left(R^2_{\text{ess}} + R^2_{\text{rss}}\right) = \sqrt{R^2_{\text{ess}} R^2_{\text{cor}}}$$

```
rsq(pbt_unconst, type = "tjur")
## [1] 0.125
```

It summarizes the difference in the distribution of the fitted values for the two subsamples defined by $y = 1/0$. The `plot` method for `binomreg` objects draws these two distributions as an histogram and indicates the average fit for the two groups by a dot on the horizontal axis. Tjur's R^2 is then simply the distance between these two points. For the probit unconstrained model of mode choice, the result is represented in Figure 10.3 and the same plot is presented for the `airbnb` data set in Figure 10.4.

Figure 10.4 reveals a very poor fit, as the two points are very close. This can be checked by computing the R^2:

```
rsq(pt_a, type = "tjur")
## [1] 0.0127
```

Estrella (1998) proposed a R^2 based on the likelihood ratio statistic comparing the proposed model and the null model, for which only one parameter is estimated. The average likelihood ratio statistic is:

$$A_{LR} = \frac{2}{N}\left(\ln L - \ln L_0\right)$$

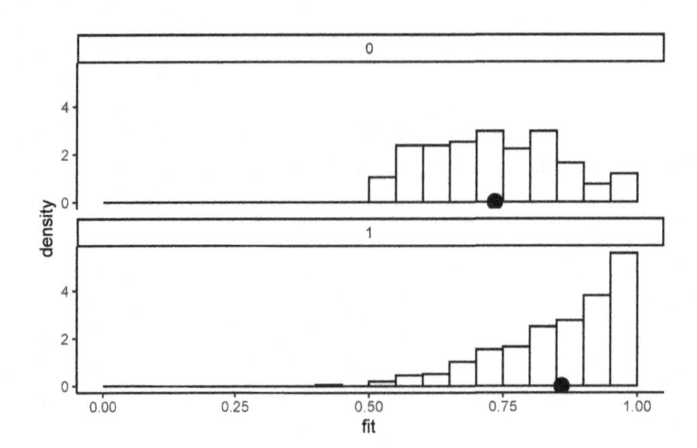

Figure 10.3: Histogram of the distribution of the fitted values for the probit mode choice model

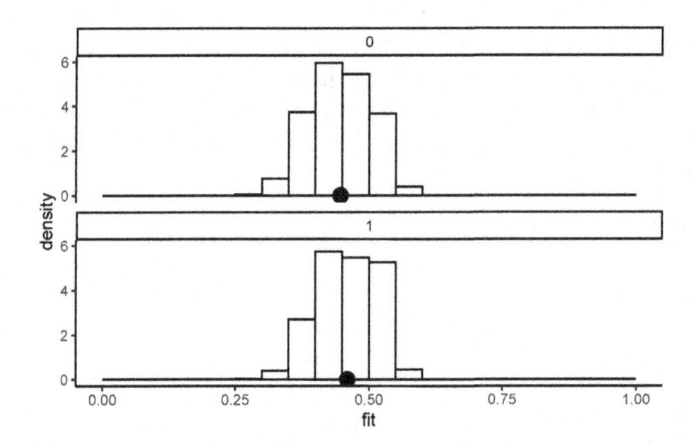

Figure 10.4: Histogram of the distribution of the fitted values for the Airbnb probit model

If the model has no explanatory power, $\ln L = \ln L_0$ so that the minimum value of A_{LR} is 0. For the saturated model, $L = 1$, so that the maximum value of A_{LR} is $B = -\frac{2}{N}\ln L_0$. The proposed R^2 follows the following differential equation:

$$\frac{dR^2}{1 - dR^2} = \frac{dA}{1 - A/B}$$

which means that the relative change of the R^2 should be equal to the relative change of the average likelihood ratio. The solution to this differential equation is: $1 - (1 - A/B)^B$, so that the R^2 is:

$$R^2 = 1 - \left(\frac{\ln L}{\ln L_0}\right)^{-\frac{2}{N}\ln L_0}$$

```
rsq(pbt_unconst, type = "estrella")
## [1] 0.1244
```

McFadden (1973) proposed the very popular pseudo-R^2:

$$R^2 = 1 - \frac{\ln L_0}{\ln L}$$

```
rsq(pbt_unconst, type = "mcfadden")
## [1] 0.14
```

McKelvey and Zavoina (1975) proposed a R^2 based on the latent variable. Denoting $\hat{y}^*_n = \hat{\alpha} + \hat{\beta}^\top x_n$ as the fitted values and $\bar{y}^*$ as the sample mean, the explained sum of squares is: $\sum(\hat{y}^*_n - \bar{y}^*)^2$ and the residuals sum of squares is not estimated, but its expected value is N times the variance of the errors, which is 1 for a probit and $\pi^2/3$ for a logit. The R^2 is then obtained by dividing the explained sum of squares by the sum of the explained sum of squares and either N or $N\pi^2/3$ respectively for the probit and logit models.

```
rsq(pbt_unconst, type = "mckel_zavo")
## [1] 0.2726
```

10.4.3 Testing

10.4.3.1 Tests of nested models

To test nested models, the three tests described in Section 3.6.3 and Section 5.3.1 are available. In Section 10.1, we estimated a model with the generalized cost as a unique covariate, which was computed as: $g_n = c_n + 8(i_n + o_n)$, where c, i, and o are the differences in monetary cost, in-vehicle time and out-vehicle time, based on the hypothesis that time value was \$8 per hour. The unconstrained model is:

$$P(y_n = 1) = \Phi(\alpha + \beta_c c_n + \beta_i i_n + \beta_o o_n)$$

The constrained model implies the two following hypotheses: $H_O : \beta_o = \beta_i = 8\beta_c$. It is more convenient to rewrite the model so that, under H_0, a subset of the parameters are 0:

$$\begin{aligned} P(y_n = 1) &= \Phi\left(\alpha + \beta_c\left(c_n + 8(i_n + o_n)\right) + (\beta_i - 8\beta_c)i_n + (\beta_o - 8\beta_c)o_n\right) \\ &= \Phi(\alpha + \beta_c g_n + \beta_i' i_n + \beta_o' o_n) \end{aligned}$$

where $\beta_i' = (\beta_i - 8\beta_c)$ and $\beta_o = (\beta_o - 8\beta_c)$ are the reduced form parameters of the binomial regression with the generalized cost, the in-vehicle and out-vehicle time as covariates. With this parametrization, the set of hypotheses is simply $\beta_i' = \beta_o' = 0$.

```
pbt_unconst2 <- binomreg(mode ~ gcost + ivtime + ovtime,
                         data = mode_choice, link = "probit")
```

Tests can be computed using several functions in the **lmtest** package (Zeileis and Hothorn 2002). The likelihood ratio test can easily be computed "by hand", as it is twice the difference of the log-likelihood functions of the unconstrained and constrained models:

```
as.numeric(2 * (logLik(pbt_unconst) - logLik(pbt_const)))
## [1] 11
```

but `lmtest::lrtest` is a convenient function which computes the statistic and the probability value:

```
lmtest::lrtest(pbt_unconst, pbt_const) %>% gaze
## Chisq = 11.002, df: 2, pval = 0.004
```

The Wald test can be computed using either `lmtest::waldtest` or `car::linearHypothesis`. `lmtest::waldtest` provides two possible syntaxes: two fitted models, as `lmtest::lrtest` or the fitted unconstrained model and a formula describing the constrained model. `car::linearHypothesis`, already described in Section 3.6.4.1, uses a character vector to indicate the hypothesis:

```
lmtest::waldtest(pbt_unconst2, pbt_const) %>% gaze
## Chisq = 10.987, df: 2, pval = 0.004
lmtest::waldtest(pbt_unconst2, . ~ . - ivtime - ovtime) %>% gaze
## Chisq = 10.987, df: 2, pval = 0.004
car::linearHypothesis(pbt_unconst,
                      c("ivtime = 8 * cost", "ovtime = 8 * cost")) %>%
  gaze
## Chisq = 10.987, df: 2, pval = 0.004
```

Finally, score tests are provided by the `micsr::scoretest` function. Its first argument is the constrained fitted model and the second one a formula that describes the unconstrained model:

```
scoretest(pbt_const , . ~ . + ivtime + ovtime) %>% gaze
## chisq = 10.231, df: 2, pval = 0.006
```

The three statistics are very close and the joint hypothesis is rejected at the 1% level.

10.4.3.2 Conditional moment test

The conditional moment tests have been presented in Section 5.3.2 and the relevant moments are given by Equation 5.44 (for the heteroskedasticity test), Equation 5.43 (for the normal test) and Equation 5.45 (for the omitted variable test). The empirical moments use the powers (up to 4) of the residuals. As the residuals are unobserved, ϵ_n^k is replaced by their expectations, i.e., by the uncentered moments of the residuals:

$$m_k = \mathrm{E}(\epsilon_n^k \mid x_n) = (1 - y_n)\mathrm{E}(\epsilon_n^k \mid x_n, y_n^* \leq 0) + y_n \mathrm{E}(\epsilon_n^k \mid x_n, y_n^* > 0) \tag{10.11}$$

For the probit model, the first four moments of the truncated normal distribution should be computed. A recursive formula for the moments of a normal variable $x \sim \mathcal{N}(\eta, \sigma)$ with $l \leq x \leq u$ is:[6]

$$m_k = (k-1)\sigma^2 m_{k-2} + \eta m_{k-1} - \sigma \frac{u^{k-1}\phi\left(\frac{u-\eta}{\sigma}\right) - l^{k-1}\phi\left(\frac{l-\eta}{\sigma}\right)}{\Phi\left(\frac{u-\eta}{\sigma}\right) - \Phi\left(\frac{l-\eta}{\sigma}\right)} \tag{10.12}$$

with $m_{-1} = 0$ and $m_0 = 1$. For a residual of the probit model, we have $\eta = 0$ and $\sigma = 1$ and the truncature is $-\mu_n$. We then obtain, for $y_n^* \leq 0$ ($l = -\infty$ and $u = -\mu_n$) and $y_n^* > 0$ ($l = -\mu_n$ and $u = +\infty$):

$$\left\{ \begin{array}{lcl} \mathrm{E}(\epsilon_n^k \mid x_n, y_n^* \leq 0) & = & (k-1)m_{k-2} - (-\mu_n)^{k-1}\frac{\phi(\mu_n/\sigma)}{1-\Phi(\mu_n/\sigma)} \\ \mathrm{E}(\epsilon_n^k \mid x_n, y_n^* > 0) & = & (k-1)m_{k-2} + (-\mu_n)^{k-1}\frac{\phi(\mu_n/\sigma)}{\Phi(\mu_n/\sigma)} \end{array} \right.$$

Using Equation 10.11, the recursive formula for the kth moment of ϵ is simply:

$$m_k = (k-1)m_{k-2} + (-\mu_n)^{k-1}\frac{(y_n - \Phi(\mu_n/\sigma))\phi(\mu_n/\sigma)}{\Phi(\mu_n/\sigma)(1-\Phi(\mu_n/\sigma))} = (k-1)m_{k-2} + (-\mu)^{k-1}\psi_n$$

where ψ_n is the generalized residual defined by Equation 10.9.

The omitted variable test uses the first moment, the homoskedasticity test the second and the normality test the third and fourth. The moments of ϵ_n, the theoretical moments for the three hypotheses and their empirical counterparts are presented in the following table:[7]

k	hypothesis	$\mathrm{E}(\epsilon_n^k \mid x_n)$	theor. moment	emp. moment
1	omit. var.	ψ_n	$\mathrm{E}(\epsilon_n w_n \mid x_n) = 0$	$\frac{1}{N}\sum_n \psi_n w_n$
2	homosc.	$1 - \mu_n \psi_n$	$\mathrm{E}((\epsilon_n^2 - 1)w_n \mid x_n) = 0$	$\frac{1}{N}\sum_n \mu_n \psi_n w_n$
3	asymetry	$(2 + \mu_n^2)\psi_n$	$\mathrm{E}(\epsilon_n^3 \mid x_n) = 0$	$\frac{1}{N}\sum_n \mu_n^2 \psi_n$
4	kurtosis	$3 - (3\mu_n + \mu_n^3)\psi_n$	$\mathrm{E}(\epsilon_n^4 - 3 \mid x_n) = 0$	$\frac{1}{N}\sum_n \mu_n^3 \psi_n$

(10.13)

[6] Unpublished note by Eric Orjebin, 2014, founded on the Wikipedia page entitled "Truncated normal distribution".

[7] Note that the empirical moments simplify because $\sum_n \psi_n = 0$ and $\sum_n \psi_n \mu_n = 0$, see Equation 10.5.

The `miscr::cmtest` function computes the conditional moment test; the first argument is a fitted model the second one is `test` which can be equal to `"normality"`, `"reset"` (for tests for omitted variables) or `"heterosc"`. The two joint hypothesis corresponding to the normality hypothesis can be tested one by one by setting `test` either to `"skewness"` or `"kurtosis"`. For the homoskedasticity tests, the set of variables can be selected using the `heter_cov` argument. By default, all the covariates used in the model are selected. By default, tests are performed using the hessian, but the outer product of the gradient form of the test can be computed by setting `opg` to `TRUE`.

```
cmtest(pbt_unconst, test = "normality") %>% gaze
## chisq = 4.700, df: 2, pval = 0.095
cmtest(pbt_unconst, test = "heterosc") %>% gaze
## chisq = 4.129, df: 3, pval = 0.248
cmtest(pbt_unconst, test = "reset") %>% gaze
## chisq = 3.218, df: 2, pval = 0.200
```

Our probit model seems to be correctly specified, as the three hypotheses are not rejected.

10.5 Endogeneity

We now consider the case where some of the covariates are endogenous. In a linear model, the solution is to use an instrumental variable estimator, which can be estimated using the 2SLS approach and therefore by using only the `lm` function. We treat in this section the case where the response is binomial, and we consider that the realization of y is related to the value of a latent variable y^*_n, with the usual observation rule: ($y = 0$ if $y^* \leq 0$ and $y = 1$ if $y^* > 0$). y^* is a linear function of a set of K_1 exogenous (x_1) and G endogenous (e) covariates:

$$y^*_n = \alpha + \beta^\top x_{1n} + \delta^\top e_n + \epsilon_n = \gamma^\top z_n + \epsilon_n$$

with $\gamma^\top = (\alpha, \beta^\top, \delta^\top)$ and $z_n^\top = (1, x_{1n}^\top, e_n^\top)$.

The reduced form equation for each endogenous variable is:

$$e_{gn} = \pi_g^\top w_n + \nu_{gn}$$

where $w_n^\top = (1, x_{1n}^\top, x_{2n}^\top)$, x_{2n} being a vector of K_2 external instruments. It is assumed that $K_2 \geq G$. The joint distribution of y^*_n and w_n is normal:

$$\left(\begin{array}{c} y^* \\ e \end{array}\right) \sim N\left(\left(\begin{array}{c} \gamma^\top z_n \\ \Pi w_n \end{array}\right); \left(\begin{array}{cc} \sigma_\epsilon^2 & \sigma_{\epsilon\nu}^\top \\ \sigma_{\epsilon\nu} & \Sigma_\nu \end{array}\right)\right)$$

where Π is an $(K_1 + K_2 + 1) \times G$ matrix with the gth line equal to $\pi_g^\top$, Σ_ν is the $G \times G$ matrix of covariance of ν and $\sigma_{\epsilon\nu}$ is a vector of length G containing the covariances between ϵ and ν. Conditional on w_n, the distribution of y^*_n is also normal:

$$y_n^* \mid e_n \sim N\left(\gamma^\top z_n + \sigma_{\epsilon\nu}^\top \Sigma_\nu^{-1}(e_n - \Pi w_n), \sigma_\epsilon^2 - \sigma_{\epsilon\nu}^\top \Sigma_\nu^{-1}\sigma_{\epsilon\nu}\right) \tag{10.14}$$

Let $\rho = \Sigma_\nu^{-1}\sigma_{\epsilon\nu}$ and $\sigma^2 = \sigma_\epsilon^2 - \sigma_{\epsilon\nu}^\top \Sigma_\nu^{-1}\sigma_{\epsilon\nu}$. The conditional mean of y_n^* is then $\theta^\top u_n$, with $\theta^\top = (\gamma^\top, \rho^\top)$ and $u_n^\top = (z_n^\top, \nu_n^\top)$ and its conditional variance is σ^2.

10.5.1 Maximum likelihood estimation

The joint density of y_n^* and e_n can be written as the product of the conditional density of y_n^* and the marginal density of e_n, which is multivariate normal:

$$\ln g(e_n) = -\frac{1}{2}\left(G \ln 2\pi + \ln \mid \Sigma_\nu \mid + \nu_n^\top \Sigma_\nu^{-1}\nu_n\right) \tag{10.15}$$

We consider here the case where y_n^* is not observed, but only its sign. Therefore, we observe y_n equal to 0 or 1, or $q_n = 2y_n - 1$ equal to -1 or $+1$. The log-likelihood is then $\ln L = \sum_{n=1}^N \ln g(e_n) + \ln f(y_n \mid e_n)$, where $g(e)$ is given by Equation 10.15 and:

$$f(y_n \mid d_n) = \Phi\left(q_n \frac{\theta^\top u_n}{\sigma}\right) = \Phi\left(q_n \frac{\gamma^\top z_n + \sigma_{\epsilon\nu}^\top \Sigma_\nu^{-1}(e_n - \Pi w_n)}{\sqrt{\sigma_\epsilon^2 - \sigma_{\epsilon\nu}^\top \Sigma_\nu^{-1}\sigma_{\epsilon\nu}}}\right) \tag{10.16}$$

The computation of the estimator is simplified by the use of the Cholesky decomposition of Σ_ν^{-1}, i.e., by considering the upper triangular matrix C such that $C^\top C = \Sigma_\nu^{-1}$. Then, the determinant of Σ_ν^{-1} is simply the product of the squares of the diagonal elements of C. Therefore, $\ln \mid \Sigma_\nu^{-1} \mid = 2\sum_g \ln C_{gg}$ and $\ln \mid \Sigma_\nu \mid = -2\sum_g \ln C_{gg}$. Denoting $\Pi^* = C\Pi$, $w_n^* = Cw_n$ and $\rho^* = C\sigma_{\epsilon\nu}$ we have : $\theta^\top u_n = \gamma^\top z_n + \rho^{*\top}(e_n^* - \Pi^* w_n)$ and $\sigma^2 = \sigma_\epsilon^2 - \rho^{*\top}\rho^*$. The marginal density of e_n and the conditional density of y_n are then:

$$\left\{\begin{array}{rcl} \ln g(e_n) & = & -\frac{1}{2}G \ln 2\pi + \sum_{g=1}^G \ln C_{gg} - \frac{1}{2}(e_n^* - \Pi^* w_n)^\top (e_n^* - \Pi^* w_n) \\ \ln f(y_n \mid e_n) & = & \Phi\left((2y_n - 1)\frac{\gamma^\top z_n + \rho^{*\top}(e_n^* - \Pi^* w_n)}{\sqrt{\sigma_\epsilon^2 - \rho^{*\top}\rho^*}}\right) \end{array}\right.$$

The maximum likelihood estimator is then obtained by maximizing the log-likelihood function $\ln L = \sum_{n=1}^N (\ln g(e_n) + \ln f(y_n \mid e_n))$ with respect to γ, ρ^*, Π^* and C. σ_ϵ is not identified and can be set to 1.

10.5.2 Two-step estimator

From Equation 10.14, we have $y_n^* \sim N(\gamma^\top z_n + \rho^\top \nu_n, \sigma^2)$. If y_n^* and ν_n were observed, the model could be consistently estimated by regressing y_n^* on z_n and ν_n. ν_n is actually unknown, but it can be consistently estimated using the estimation of $\hat{\Pi}$ obtained by maximizing $\sum_{n=1}^N \ln g(e_n)$. This is a seemingly unrelated regression problem, and it is well known that, for the special case where the set of covariates is the same for all the equations, the estimator can be obtained using OLS independently on each equation. From this first step, we obtain $\hat{\nu}_n = e_n - \hat{\Pi} w_n$ and, in the second step, $\hat{\gamma}$ and $\hat{\rho}$ are obtained by regressing y_n^* on z_n and $\hat{\nu}_n$.

Regressing y_n^* on a vector of 1, x_{1n}, e_n and $\hat{\nu}_n$ is one way to obtain the instrumental variable estimator. This approach (called **control function**) is identical to the 2SLS estimator but provides supplementary estimates ($\hat{\rho}$ associated with $\hat{\nu}_n$) that can be used to test the hypothesis of exogeneity. If $G = 1$, the test can be performed using the Student statistic. If $G > 1$, the joint hypothesis that $\rho = 0$ can be tested using a Wald test, the statistic being a χ^2 with G degrees of freedom under the null hypothesis of exogeneity.

This two-step instrumental variable estimator and test has been extended for the case where y_n^* is only partially observed by Smith and Blundell (1986) and Rivers and Vuong (1988) (respectively for the tobit and the probit models). It can be computed as follows:

- compute the OLS estimator of π_g for the G endogenous variables and retrieve the residuals $\hat{\nu}_{gn}$,
- estimate $\theta^\top = (\gamma^\top, \rho^\top)$ using a probit model with z_n, and $\hat{\nu}_n$ as covariates,
- test the hypothesis that $\rho = 0$.

As it is customary for two-step estimators, the covariance matrix returned by the probit model is inconsistent because it doesn't take into account the fact that ν_n is unknown and is replaced by a consistent estimator. Denoting $\pi = \mathrm{vec}\,\Pi$, the first-order approximation of the vector of scores is:

$$\frac{\partial \ln L}{\partial \theta}(\hat{\theta}, \hat{\pi}) \approx \frac{\partial \ln L}{\partial \theta} + \frac{\partial \ln^2 L}{\partial \theta \partial \theta^\top} \times (\hat{\theta} - \theta) + \frac{\partial \ln^2 L}{\partial \theta \partial \pi^\top} \times (\hat{\pi} - \pi)$$

Taking expectation and solving for $\hat{\theta} - \theta$, we get:

$$\hat{\theta} - \theta = \mathrm{E}\left(-\frac{\partial^2 \ln L}{\partial \theta \partial \theta^\top}\right)^{-1} \left(\frac{\partial \ln L}{\partial \theta} + \mathrm{E}\left(\frac{\partial^2 \ln L}{\partial \theta \partial \pi^\top}\right)(\hat{\pi} - \pi)\right) = A^{-1}\left(\frac{\partial \ln L}{\partial \theta} + B(\hat{\pi} - \pi)\right)$$

As the two terms in the brackets are uncorrelated and using the information matrix equality, we get:

$$\hat{V}(\hat{\theta}) = A^{-1} + A^{-1} B \hat{V}(\hat{\pi}) B^\top A^{-1}$$

A and B contain the second derivatives of the individual contribution to the log-likelihood function for the probit model. These are, defining: $\eta_n = \gamma^\top z_n + \rho \hat{\nu}_n$ and $r_n = \phi(\eta_n)/(1 - \Phi(\eta_n))$:

$$\frac{\partial^2 \ln l_n}{\partial \eta_n \partial \eta_n^\top} = -r_n (r_n + \eta_n) = -\psi_n$$

More precisely, $A = \sum_{n=1}^N \psi_n \hat{u}_n \hat{u}_n^\top$ and $B = \sum_{n=1}^N \rho^\top \otimes \psi_n \hat{u}_n w_n^\top$ or, defining Ψ a diagonal matrix of dimension N containing ψ_n: $A = \hat{U}^\top \Psi \hat{U}$ and $B = \hat{U}^\top \Psi W$, with $\hat{U}$ the $N \times (K_1 + 2G + 1)$ matrix with rows $(1, x_{1n}^\top, e_n^\top, \hat{\nu}_n^\top)$ and W the $N \times (K_1 + K_2 + 1)$ matrix with rows $w_n^\top = (1, x_{1n}^\top, x_{2n}^\top)$. As $\hat{V}(\hat{\pi})$ is the variance of the SUR estimator with identical covariates, $\hat{V}(\hat{\pi}) = \hat{\Sigma}_\nu \otimes (W^\top W)^{-1}$ and the expression further simplifies to:

$$\hat{V}(\hat{\theta}) = (\hat{U}^\top \Psi \hat{U})^{-1} + (\hat{\rho}^\top \hat{\Sigma} \hat{\rho}) \times (\hat{U}^\top \Psi \hat{U})^{-1} (\hat{U}^\top \Psi W)(W^\top W)^{-1}(W^\top \Psi \hat{U})(\hat{U}^\top \Psi \hat{U})^{-1}$$

Smith and Blundell (1986) proposed an endogeneity test based on the two-step estimator. As for the linear instrumental variable estimator, it is a Wald test that $\rho = 0$, but it uses the simple probit estimation of the covariance matrix.

10.5.3 Minimum χ^2 estimator

Newey (1987) argued that the minimum chi-square estimator (Amemiya 1978's) can be used in this context and is more efficient than the two-step estimator. This estimator is computed in five steps:

1. compute the OLS estimator of π_g for the G endogenous variables and compute the fitted values $\hat{d}_{gn}$ and the residuals $\hat{\nu}_{gn}$ of these regressions,
2. using a probit regress y on the whole set of exogenous variables $w_n^\top = (1, x_{n1}^\top, x_{n2}^\top)$ and on the previously computed residuals $\hat{\nu}_{gn}$. Save the coefficients of v_n $(\hat{\alpha})$, of $\hat{\nu}_n$ $(\hat{\lambda})$ and the part of the covariance matrix that corresponds to α $\hat{\Sigma}_1$,
3. using a probit, regress y on $\hat{u}_n^\top = (1, x_1^\top, \hat{\nu}_n^\top)$ and on the fitted values $\hat{d}_{gn}$ computed on the first step; save the coefficients of the fitted values $\hat{\delta}$ and compute $\hat{\rho} = \hat{\lambda} - \hat{\delta}$,
4. regress $\hat{\rho}^\top e_n$ on the whole set of exogenous variables w_n, save the covariance matrix $\hat{\Sigma}_2$ and compute $\hat{\Omega} = \hat{\Sigma}_1 + \hat{\Sigma}_2$,
5. compute the minimum χ^2 estimator $\hat{\gamma}$ and its variance $\hat{V}(\hat{\gamma})$:

$$\hat{V}(\hat{\gamma}) = (Z^\top W)(W^\top W)^{-1} \hat{\Omega}^{-1} (W^\top W)^{-1} (W^\top Z)$$

$$\hat{\gamma} = \hat{V}(\hat{\gamma})(Z^\top W)(W^\top W)^{-1} \hat{\Omega}^{-1}$$

10.5.4 Application

Adkins, Carter, and Simpson (2007) and Adkins (2012) analyzed the effect of managerial incentives on the use of foreign-exchange derivatives for hedging by U.S. bank holding companies, for the 1996-2000 period. The dependent variable `federiv` is 1 if the bank uses foreign-exchange derivatives.

The first set of covariates concerns ownership. When managers have a higher ownership position in the bank, their behavior is more in line with the preferences of shareholders, and they therefore have an incentive to take risk: the logarithm of the percentage of total shares outstanding that are owned by officers and directors (`linsown`) should therefore have a negative effect on the probability of using foreign-exchange derivatives. However, incentives provided by regulation may dominate the expected incentive relation and lead to a negative effect on the probability. On the contrary, institutional blockholders have imperfect information and, therefore, the logarithm of the percentage of total shares outstanding that are owned by all institutional investors (`linstown`) should have a negative effect on the probability of using foreign-exchange derivatives.

The second set of covariates concerns CEO compensation. Value of option awards (`optval`) should induce managers to take more risk and therefore should have a negative effect on the probability. On the contrary, cash bonus (`bonus`) may increase the probability of hedging in order to decrease variability in the firm's cash flows.

The other covariates are the leverage (`eqrat`), the logarithm of total assets (`ltass`), the return on equity (`roe`), the market to book ratio (`mktbt`), the foreign to total interest income ratio (`perfor`), a derivative dealer activity dummy (`dealdum`), dividends paid (`div`) and the year from 1996 to 2000 (`year`).

Three covariates are suspected to be endogenous: the leverage (`eqrat`), the option awards `optval` and the bonus, (`bonus`). The external instruments are the number of employees (`no_emp`), of subsidiaries (`no_subs`) and of officies (`no_off`), the CEO age (`ceo_age`), the 12 month maturity mismatch (`gap`) and the ratio of cash flow to total assets (`cfa`).

The instrumental variable probit estimator is obtained using `binomreg` with `link = "probit"` and a two-part formula, containing the covariates and the instruments. The method argument (the default `"ml"`, `"twosteps"` and `"min"`) indicates the estimation method. The results for the three estimators are presented in Table 10.4. To save place, we only present the coefficients for the covariates of main interest.

```
form <- federiv ~ eqrat + optval + bonus + ltass +
                    linsown + linstown + roe + mktbk +
                    perfor + dealdum + div + year |
                      . - eqrat - bonus - optval + no_emp +
                    no_subs + no_off + ceo_age + gap + cfa
bank_msq <- binomreg(form, data = federiv, link = "probit",
                   method = "minchisq")
bank_ml <- update(bank_msq, method = "ml")
bank_2st <- update(bank_msq, method = "twosteps")
```

The coefficients of `linstown`, `bonus` and `optval` have the expected sign. `linsown` has a positive sign, which must be driven by the strength of the regulatory constraints. Note that the standard deviations are much smaller for the highly parametrized ML estimator, compared to the two-step and the minimum χ^2 estimator.

To perform the endogeneity test, we first estimate the two-step estimator with the probit estimator of the covariance matrix of the coefficients. This is performed by setting the supplementary argument `robust` to `FALSE`:

```
bank_2s_nr <- binomreg(form, data = federiv, link = "probit",
                       method = "twosteps", robust = FALSE)
```

Then, a Wald test that $\rho = 0$ can be performed, using for example `car::linearHypothesis`:

```
bank_2s_nr %>% car::linearHypothesis(c("rho_eqrat = 0",
                                       "rho_optval = 0",
                                       "rho_bonus = 0")) %>% gaze
## Chisq = 7.547, df: 3, pval = 0.056
```

or more simply using the `miscsr::endogtest` function:

Table 10.4: IV probit models for the bank data set

	minchisq	two-step	ML
linsown	0.259	0.257	0.145
	(0.116)	(0.116)	(0.062)
linstown	0.370	0.372	0.202
	(0.135)	(0.135)	(0.106)
eqrat	21.775	21.825	12.491
	(13.386)	(13.424)	(5.354)
optval	−0.088	−0.087	−0.051
	(0.053)	(0.053)	(0.017)
bonus	1.757	1.735	1.018
	(0.888)	(0.888)	(0.187)
rho_eqrat		−25.506	−0.224
		(13.835)	(0.087)
rho_optval		0.096	0.347
		(0.049)	(0.075)
rho_bonus		−1.672	−0.730
		(0.875)	(0.140)
Num.Obs.	794	794	794

```
endogtest(form, federiv) %>% gaze
## chisq = 7.547, df: 3, pval = 0.056
```

The hypothesis of no endogeneity is rejected at the 10%, but not at the 5% level.

10.6 Ordered models

An ordered model is a model for which the response can take J distinct values (with $J > 2$). The construction of the model is very similar to the one of the binomial model. We consider, as in Section 10.2.1, a latent variable equal to the sum of a linear combination of the covariates and an error:

$$y^* = \alpha + \beta^\top x + \epsilon$$

Denoting $\mu = (\mu_0, \mu_1, \mu_1, \dots, \mu_J)$ a vector of parameters, with $\mu_0 = -\infty$ and $\mu_J = +\infty$, the rule of observation for the different values of y is then:

$$\left\{\begin{array}{lllllllll} y & = & 1 & \Leftrightarrow & \mu_0 & \leq & \alpha+\beta^\top x+\epsilon & \leq & \mu_1 \\ y & = & 2 & \Leftrightarrow & \mu_1 & \leq & \alpha+\beta^\top x+\epsilon & \leq & \mu_2 \\ \vdots & & & \vdots & & \vdots & & \vdots & \\ y & = & J-1 & \Leftrightarrow & \mu_{J-2} & \leq & \alpha+\beta^\top x+\epsilon & \leq & \mu_{J-1} \\ y & = & J & \Leftrightarrow & \mu_{J-1} & \leq & \alpha+\beta^\top x+\epsilon & \leq & \mu_J \end{array}\right.$$

For $y = 1$, subtracting α from the three terms of the inequality, we get: $\mu_0 - \alpha \leq \beta^\top x + \epsilon \leq \mu_1 - \alpha$. Therefore, the observation rule doesn't depend on μ_0, μ_1 and α, but on $\mu_0 - \alpha$ and $\mu_1 - \alpha$. Therefore, μ_0, μ_1 and α are not identified and, for example, α can be set to 0. The same reasoning applies to the other values of y. Denoting F the cumulative density of ϵ, the probability for a given value j of y is:

$$\mathrm{P}(y_n = j) = F(\mu_j - \beta^\top x_n) - F(\mu_{j-1} - \beta^\top x_n)$$

The probabilities are represented in Figure 10.5 for $J = 3$ by the areas under the density curve of ϵ between two consecutive values of $\mu_j - \beta^\top x$ (with $\mu_0 = -\infty$ and $\mu_3 = +\infty$).

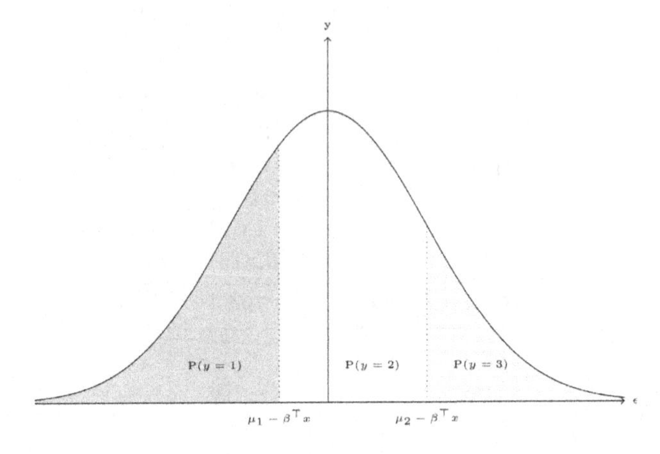

Figure 10.5: Probabilities for an ordered model

Consider now an increase of a covariate x_k and suppose that the associated coefficient β_k is positive. Then, all the values of $\mu_j - \beta^\top x$ decrease by the same amount $-\beta_k \Delta x_k$. The new probabilities are represented in Figure 10.6.

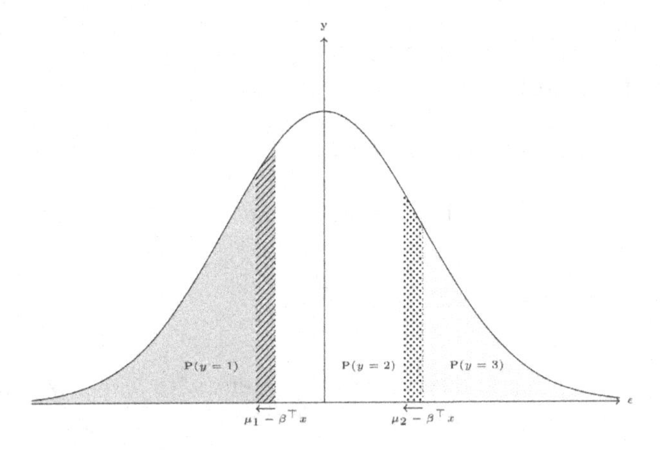

Figure 10.6: Marginal effects for an ordered model

The dashed area represents an increase of the probability that $y = 2$ and a reduction of the probability that $y = 1$, and the dotted area a reduction of the probability that $y = 2$ and an increase of the probability that $y = 3$. Therefore, the marginal effect of x_k is positive for $\mathrm{P}(y = 3)$, negative for $\mathrm{P}(y = 1)$ and ambiguous for $\mathrm{P}(y = 2)$. For small changes of x_k the dashed and dotted areas are proportional to the densities for the two limits of the range of ϵ for which $y = 2$. Therefore $\mathrm{P}(y = 2)$ increases if the density for the lower limit $(\mu_1 - \beta^\top x)$ is greater than the density for the upper limit $(\mu_2 - \beta^\top x)$, which is the case here, and decrease otherwise. More generally, when y takes J values, the effect on the probabilities is unambiguous only for $\mathrm{P}(y = 1)$ and $\mathrm{P}(y = J)$. The probability of the outcome can be written compactly using the $\mathbf{1}(x)$ function which equals 1 if x is true and 0 otherwise:

$$\mathrm{P}(y_n) = \sum_{j=1}^{J} \mathbf{1}(y_n = j) \left[F(\mu_j - \beta^\top x_n) - F(\mu_{j-1} - \beta^\top x_n)\right] \tag{10.17}$$

For a sample of size N, the log-likelihood function is obtained by summing the logarithms of Equation 10.17 for all the observations:

$$\ln L = \sum_{n=1}^{N} \sum_{j=1}^{J} \mathbf{1}(y_n = j) \ln \left[F(\mu_j - \beta^\top x_n) - F(\mu_{j-1} - \beta^\top x_n)\right]$$

As for the binomial model, the most common choices for the distribution of ϵ are the normal and the logistic distributions, which lead respectively to the ordered probit and logit models.

As an example, we consider the article of Abiad and Mody (2005) who study the determinants of financial reform. The data set, called `fin_reform`, is a panel of 35 countries for 24 years (from 1973 to 1996). The variable of interest `fli` is an index of financial liberalization, which takes integer values from 0 to 18. Denote I_{nt} the value of this variable for country n on year t divided by 18, so that I_{nt} equal to 0 or to 1 indicates respectively no and perfect financial liberalization. It is assumed that the yearly variation of the index is given by the following equation, denoting I^*_{nt} the desired value of I_{nt}:

$$\Delta I_{nt} = \alpha(I^*_{nt} - I_{n(t-1)}) + \epsilon_{nt}$$

I^* is unobserved and is supposed to be equal to 1, so that the target is perfect financial liberalization. α is an adjustment factor and is supposed to be equal to $\alpha = \theta I_{n(t-1)}$, so that the resistance to reform is a function of the state of liberalization. We then have:

$$\Delta I_{nt} = \theta I_{n(t-1)}(1 - I_{n(t-1)}) + \epsilon_{nt}$$

The response is therefore the change in the index and the main covariate is `indxl` $\times$ `(1 - indxl)` where `indxl` is the lagged value of the index on the 0-1 scale. The computation of these variables are presented below; note the use of the `lag` function on the data frame grouped by `country`. This ensures that the lag value for the first year (1973) for the second country (Australia) is `NA` and not the value of the first country (Argentina) for the last year (1996).

```
fin_reform <- fin_reform %>%
    group_by(country) %>%
    mutate(dindx = fli - lag(fli),
           indx = fli / 18,
           indxl = lag(indx),
           rhs1 = indxl * (1 - indxl)) %>%
    ungroup
```

The authors test the possibility of regional diffusion: countries within a region would then be induced to catch up with the highest level of liberalization observed within the region. Therefore, they introduce a covariate which is equal to the difference between the previous value of the index and the maximum value in the group of country:

```
g <- fin_reform %>% group_by(region, year) %>%
  summarise(max_indxl = max(indxl))
fin_reform <- fin_reform %>% left_join(g) %>%
  mutate(catchup = max_indxl - indxl)
```

Other covariates are the political orientation of the government (`pol`, a factor with levels `center`, `left` and `right`) and dummies for first year of office (`dum_1yofc`), for balance of payments and bank crises in the first two previous years (`dum_bop` and `dum_bank`) and for recession `recession` (growth rate `gdpg` negative) and high inflation `hinfl` (inflation rate `infl` greater than 50%). The relevant dummies are computed below:

```
fin_reform <- fin_reform %>%
    mutate(dum_bop = ifelse(bop | lag(bop) | (! is.na(lag(bop, 2)) &
                                               lag(bop, 2)), 1, 0),
           dum_bank = ifelse(bank | lag(bank) | (! is.na(lag(bank, 2)) &
                                                  lag(bank, 2)), 1, 0),
           dum_1yofc = ifelse(!is.na(yofc) & yofc == 1, 1, 0),
           recession = ifelse(gdpg <= 0, 1, 0),
           hinfl = ifelse(infl > 50, 1, 0))
```

Ordered models can be fitted using `MASS::polr`, which has a `method` argument similar to the `link` argument of the `binomial` function used as a `family` argument in `glm`. It can be set to `logistic` (called `logit` for binomial) , `probit`, `loglog`, `cloglog` and `cauchy`. As for binomial models, `probit` and `logit` are by far the most used links. Another implementation of the ordered model is `micsr::ordreg`. Table 10.5 presents the three specifications presented by Abiad and Mody (2005), table 7, page 78.

```
mass1 <- MASS::polr(factor(dindx) ~ rhs1 + catchup, fin_reform,
                    method = "logistic")
mod1 <- ordreg(factor(dindx) ~ rhs1 + catchup, fin_reform,
               link = "logit")
mod2 <- update(mod1, . ~ . + dum_bop + dum_bank + recession + hinfl)
mod3 <- update(mod2, . ~ . + dum_1yofc + imf + usint + pol + open)
```

The first column presents the result of the model with the resistance to reform and the diffusion covariates which have both of the expected signs and are significant. In the second column, the shocks covariates are added, and only the balance of payments and the bank crisis dummies are significant. Finally, in the third column, the political and economic structure covariates are added, but none of them are significant.

Table 10.5: Ordered logit models for the financial reform data set

	(1)	(2)	(3)
rhs1	4.001***	4.652***	4.188***
	(1.008)	(1.031)	(1.048)
catchup	0.842**	0.897**	0.993**
	(0.310)	(0.321)	(0.353)
dum_bop		0.526**	0.439*
		(0.183)	(0.187)
dum_bank		−1.025***	−0.993***
		(0.279)	(0.280)
recession		−0.071	−0.056
		(0.260)	(0.262)
hinfl		−0.161	−0.264
		(0.298)	(0.307)
dum_1yofc			0.194
			(0.208)
imf			0.326+
			(0.195)
usint			−0.066+
			(0.034)
polleft			0.242
			(0.239)
polright			0.169
			(0.208)
open			−0.001
			(0.002)
Num.Obs.	805	805	805
AIC	1555.3	1542.1	1544.4
BIC	1625.7	1631.2	1661.7
Log.Lik.	−762.662	−752.052	−747.202

+ $p < 0.1$, * $p < 0.05$, ** $p < 0.01$, *** $p < 0.001$

11

Censored and truncated models

We'll discuss in this chapter models for which the value of the response is continuous and observed only in a certain range. These variables are truncated for a certain value, which can be on the left side of the distribution (l), on the right side (u) or on both sides. Therefore, the distribution of such a variable is a mix of a discrete and a continuous distribution:

- the value of y is continuous on the $]l, u[$ interval, and its distribution can be described by a density function $f(y)$,
- there is a mass of probability on l or/and on u, which is described by a probability $P(y = l)$ or/and $P(y = u)$.

Section 11.1 describes the situations where such responses occur. The Tobit model is presented in details in Section 11.2, the relevant estimation methods in Section 11.3 and their implementation in **R** in Section 11.4. Section 11.5 presents different tools useful to evaluate fitted models. Finally, Section 11.6 presents the two-equations tobit model.

11.1 Truncated response, truncated and censored samples

Truncated responses are observed in different contexts in economics. The first is a corner solution, the second is a problem of missing data, also called censoring, and the last is a problem of selection, also called incidental truncation.

11.1.1 Corner solution

Consider a consumer who can buy two goods, food (z) and vacations (y). Denoting q_z and q_y as the quantities of the two goods, assume that the preferences of the consumer can be represented by the following utility function:

$$U(q_y, q_z) = (q_y + \mu)^\beta q_z^{1-\beta}$$

where $0 < \beta < 1$ and $\mu > 0$. The consumer seeks to maximize their utility subject to their budget constraint, which writes $x = p_y q_y + p_z q_z$, where x is the income and p_y and p_z are the unit prices of the two goods. For an interior solution, the consumer should equate the marginal rate of substitution to the price ratio:

$$\frac{\beta}{1-\beta}\frac{q_z}{q_y + \mu} = \frac{p_y}{p_z}$$

We therefore have $p_z q_z = \frac{1-\beta}{\beta} p_y (q_y + \mu)$. Replacing in the budget constraint and solving for q_z and then for q_y, we finally get the demand functions.

$$\left\{ \begin{array}{rcl} q_z & = & (1-\beta)\frac{x}{p_z} + (1-\beta)\frac{p_y}{p_z}\mu \\ q_y & = & \beta\frac{x}{p_y} - (1-\beta)\mu \end{array} \right.$$

Note that the demand function for y can return negative values, which of course is impossible. Therefore, the pseudo-demand function previously written are only suitable for an interior solution, i.e., when both goods are consumed. This is only the case for a sufficient level of income, namely $\bar{x} = \frac{1-\beta}{\beta} p_y \mu$. For a lower level of income, we have $q_y = 0$ and therefore $q_z = x/p_z$.

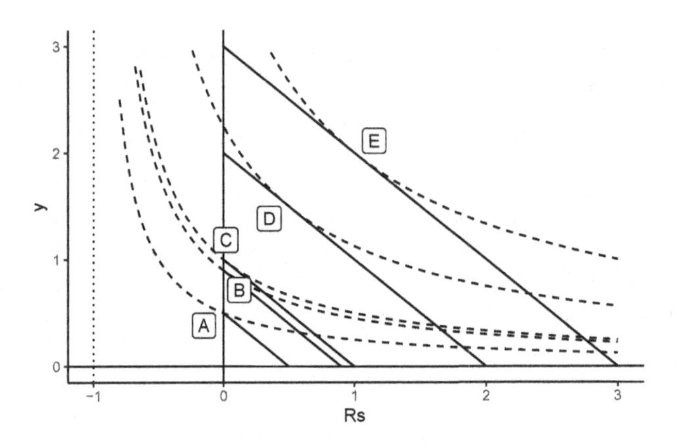

Figure 11.1: Internal and corner solution

This situation is depicted on Figure 11.1. Points D and E correspond to interior solutions for large values of the income. On the contrary, A and B are corner solutions for low income households. We then have $q_y = 0$ and the value of the marginal rate of substitution (the slope of the indifference curve) is lower than the price ratio. Point C corresponds to the level of income that leads to a corner solution but for which the marginal rate of substitution equals the price ratio. The consumption of y starts when the income is greater than this level. The expression simplifies by taking as the response the expense for the good, $y = p_y q_y$ and not the quantity. We then have: $y = \beta x - (1-\beta) p_y \mu$ or, replacing $p_y \mu$ by its expression in terms of $\bar{x}$:

$$y = -\beta\bar{x} + \beta x = \alpha + \beta x$$

In a linear regression context, the slope is therefore the marginal propensity to consume the specific good (for 1 more dollar of income, the expense increases by β dollars) and the intercept is the opposite of β times the minimum income $\bar{x}$ for which the consumption is positive.

The expense on good y is an example of a truncated variable, more precisely a left-zero truncated variable. Note that, as stressed by Wooldridge (2010), pp.517-520, 0 is a relevant value for the variable (this is not the case for the situation described in the next section), but all the values of 0 don't have the same meaning. For example, at point B, y equal 0, but a small increase of the income would lead the household to start consuming the good.

On the contrary, at point A, y also equals 0, but even with a large increase of income, the household would still consume only good z and not good y.

11.1.2 Data censoring and truncation

Data censoring occurs when the value of the variable is reported only in a certain range $]l, u[$ and is set to l or u otherwise. For example, De Crombrugghe, Palm, and Urbain (1997) estimate the demand for food using household survey data in the Netherlands. For the upper 5 percentiles (13030 Dfl), the expenditure is not reported, but it is replaced by the average value (17670 Dfl). Therefore, the response is right-truncated with $u = 13030$. In this case, the data censoring process takes the **top-coding** form and 13030 is not a relevant value of the covariate. Sometimes, the censoring process leads to a truncation sample, which means that only observations for which the response is in the continuous observed range are selected. A classic example is Hausman and Wise (1976, 1977) who used data from the New Jersey negative income tax experiment, for which families with income above 1.5 times the poverty level were excluded.

11.1.3 Sample selection

The last data-generating process is the one of sample selection. It means that the observation of the response in a particular sample is not random because of a self-selection process. This kind of process was first analyzed by Gronau (1973) in the context of women participation in the labor force: the response is the wage offered to women, but it is only observed for women who participate in the labor market. Although it is clearly a different form of truncation compared to the case of a corner solution or data censoring, the models that deal with these kinds of responses are much alike; it therefore makes sense to consider these two cases in the same chapter.

11.1.4 Truncated and censored samples

Responses considered in this chapter are **truncated variables**, but the sample used can be either truncated or censored. For the demand for vacations:

- a **censored sample** consists of households for which the expenditure on vacations is positive and on household for which this expenditure is 0,
- a **truncated sample** consists only of households for which the expenditure is strictly positive.

Samples used in consumer expenditure surveys are censored. A representative sample of households is surveyed, and this includes households that don't have any expenditure on vacations during the survey. On the contrary, samples that consist of individual surveyed in a travel agency or in an airport are truncated samples, i.e., sample for which the variable of interest (vacation expenditure) is strictly positive. Estimation of models using censored samples are called **censored regression model** or **tobit** models. The tobit name comes from James Tobin, who is the first economist who proposed this model in econometrics (Tobin 1958), and it was proposed by Goldberger (1964) because of its similarity to the probit model. Estimation on truncated samples leads to the **truncated regression model** (Cragg 1971; and Hausman and Wise 1976, 1977).

In a classic paper, Amemiya (1984) surveyed different flavors of the tobit model and proposed a typology of five categories, tobit1, tobit2, ..., tobit5. We'll concentrate in this chapter on the first two categories:

- the **tobit-1** model is a model with one equation which explains jointly the probability that the value of the response is in the observable range and the value of the response if it is observed,
- the **tobit-2** model, which is a bivariate model, with the first equation indicating whether the response is in the observable range or not, and the second one indicating the value of the response when it is observed.

11.2 Tobit-1 model

We'll denote by tobit-1 (or tobit, for short) a linear model of the usual form: $y_n = \alpha + \beta^\top x_n + \epsilon_n = \gamma^\top z_n + \epsilon_n$, where y is only observed in a certain range, say $y \in]l, u[$. In general, the tobit name is restricted to models estimated in a censored sample. We'll treat in the same section the case where the estimation is performed in a truncated sample. In a semi-parametric setting, no hypotheses are made on the distribution of ϵ_n. On the contrary, a fully parametric model will specify the distribution of ϵ; for example, it will suppose that $\epsilon_n \sim \mathcal{N}(0, \sigma_\epsilon^2)$, i.e., that the errors of the model are normal and homoskedastic. In the context of the linear regression model, violation of these assumptions is not too severe, as the estimator is still consistent. This is not the case for the model studied in this chapter, as wrong assumptions of homoskedasticity and normality will lead to biased and inconsistent estimators.

11.2.1 Truncated normal distribution, truncated and censored sample

Early models assume that the (conditional) distribution of the response is normal. But the fact that the response is truncated implies that the distribution of y is truncated normal. This distribution is represented in Figure 11.2. Starting from a normal distribution $\frac{1}{\sigma}e^{-\frac{1}{2}\left(\frac{y-\mu}{\sigma}\right)^2}$, we first compute the probability that $l < y < u$ and we divide the normal density by this probability, which is: $\Phi\left(\frac{u-\mu}{\sigma}\right) - \Phi\left(\frac{l-\mu}{\sigma}\right)$. The density of y is therefore:

$$f(y) = \frac{1}{\sigma}\frac{\phi\left(\frac{y-\mu}{\sigma}\right)}{\Phi\left(\frac{u-\mu}{\sigma}\right) - \Phi\left(\frac{l-\mu}{\sigma}\right)}$$

so that $\int_l^u f(y)dy = 1$. As y is truncated, its expected value and its variance are not μ and σ^2. More precisely, left(-right) truncation will lead to an expected value greater(-lower) than μ. In Figure 11.2, the expected value is greater than μ because the truncation is more severe on the left. Obviously, reducing the range of the values of y implies a reduction of the variance, so that $\mathrm{V}(y) < \sigma^2$. To compute the first two moments of this distribution, it's easier to consider first a truncated standard normal deviate z. The three following results will be used: $\phi(z)' = -z\phi(z)$, $[\Phi(z) - z\phi(z)]' = z^2\phi(z)$ and $\lim\limits_{z \to \pm\infty} z\phi(z) = 0$.

For the left-truncated case, the expectation and the variance are:

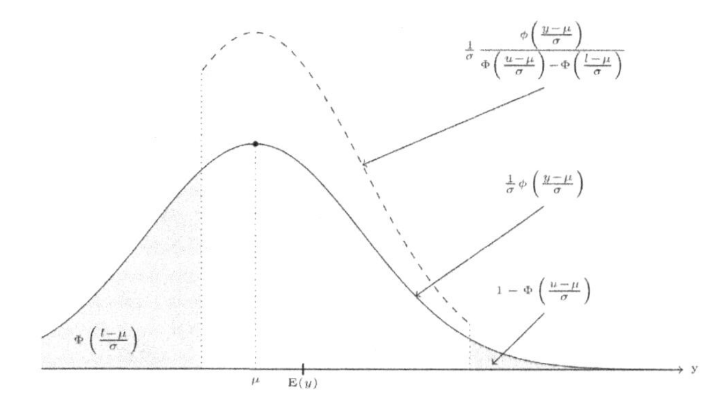

Figure 11.2: Truncated normal distribution

$$\left\{\begin{array}{lcl} E(z|z>l) & = & \dfrac{\displaystyle\int_l^{+\infty} z\phi(z)dz}{1-\Phi(l)} = \dfrac{\left[-\phi(z)\right]_l^{+\infty}}{1-\Phi(l)} = \dfrac{\phi(l)}{1-\Phi(l)} = \lambda_l \\ V(z|z>l) & = & \dfrac{\displaystyle\int_l^{+\infty} z^2\phi(z)dz}{1-\Phi(l)} - \lambda_l^2 = \dfrac{\left[\Phi(v) - v\phi(v)\right]_l^{+\infty}}{1-\Phi(l)} - \lambda_l^2 = 1 - \lambda_l\left[\lambda_l - l\right] \end{array}\right.$$

where λ_l is called the **inverse mills ratio**. For a general normal variable $y \sim \mathcal{N}(\mu, \sigma)$, denoting $\tilde{l} = (l-\mu)/\sigma$, the expectation is:

$$E(y|y>l) = \frac{\displaystyle\int_l^{+\infty} y\phi\left(\frac{y-\mu}{\sigma}\right)/\sigma dy}{1-\Phi(\tilde{l})} = \frac{\displaystyle\int_{\tilde{l}}^{+\infty} (\mu+\sigma z)\phi(z)dz}{1-\Phi(\tilde{l})} = \mu + \sigma\lambda_{\tilde{l}} \tag{11.1}$$

and the variance is:

$$\begin{array}{lcl} V(z|z>l) & = & \dfrac{\displaystyle\int_l^{+\infty} \left(y-\mu-\sigma\lambda_{\tilde{l}}\right)^2 \phi\left(\frac{y-\mu}{\sigma}\right)/\sigma dy}{1-\Phi(\tilde{l})} \\ & = & \dfrac{\displaystyle\int_{\tilde{l}}^{+\infty} \sigma^2(z-\lambda_{\tilde{l}})^2\phi(z)dz}{1-\Phi(\tilde{l})} = \sigma^2\left[1-\lambda_{\tilde{l}}(\lambda_{\tilde{l}}-\tilde{l})\right] \end{array} \tag{11.2}$$

Similarly, for the right-truncated case, denoting $\tilde{u} = (u-\mu)/\sigma$ and $\lambda_{\tilde{u}} = -\phi(\tilde{u})/\Phi(\tilde{u})$, $\mathrm{E}(y \mid y<u) = \mu + \sigma\lambda_{\tilde{u}}$ and $\mathrm{V}(y \mid y<u) = \sigma^2\left[1-\lambda_{\tilde{u}}(\lambda_{\tilde{u}}-\tilde{u})\right]$.

Consider now the special (and very common) case where the distribution of y is normal left-truncated at $l=0$, with untruncated mean and variance equal to $\mu_n = \alpha + \beta^\top x_n = \gamma^\top z_n$ and σ_ϵ^2. Then, $\tilde{l} = -\mu_n/\sigma$ and the inverse mills ratio is:[1]

[1]By symmetry of the normal distribution, $\phi(-z) = \phi(z)$ and $1-\Phi(z) = \Phi(-z)$.

$$\lambda_{\tilde{0}} = \frac{\phi(-\mu_n/\sigma)}{1 - \Phi(-\mu_n/\sigma)} = \frac{\phi(\mu_n/\sigma)}{\Phi(\mu_n/\sigma)}$$

Let $r(x) = \frac{\phi(x)}{\Phi(x)}$; then $\lambda_{\tilde{0}} = r(\mu_n/\sigma)$. The derivative of r is: $r'(x) = -r(x)\left[r(x) + x\right]$. $r(x)$ is represented in Figure 11.3. It is a decreasing function, with $\lim_{x \to -\infty} r(x) + x = 0$ and $\lim_{x \to +\infty} r(x) = 0$.

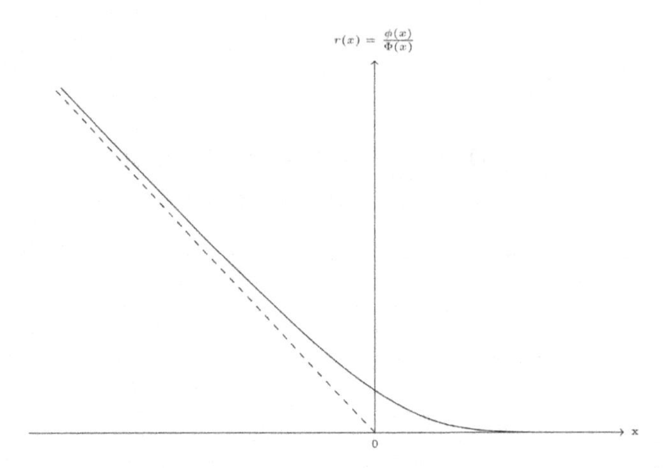

Figure 11.3: Inverse Mills ratio

The expectation and the variance of y left-truncated at 0 can then be written, using Equation 11.1 and Equation 11.2:

$$\left\{ \begin{array}{lcl} \mathrm{E}(y \mid x_n, y > 0) & = & \mu_n + \sigma r(\mu_n/\sigma) \\ \mathrm{V}(y \mid x_n, y > 0) & = & \sigma^2 \left[1 + r'(\mu_n/\sigma)\right] \end{array} \right.$$

Therefore, truncation has two consequences for the linear regression model:

- the conditional variance depends on x so that the errors of the model are heteroskedastic,
- the conditional expectation of y is no longer equal to $\mu_n = \alpha + \beta^\top x_n$, but to $\mu_n + \sigma r(\mu_n/\sigma)$ or, stated differently, the errors of the model are correlated with the covariate as $\mathrm{E}(\epsilon \mid x) = \sigma r(\mu_n/\sigma)$.

The first point implies that the OLS estimator is inefficient, the second one that it is biased and inconsistent. For the case where there is only one covariate and $\beta > 0$, this correlation is illustrated in Figure 11.4 which presents the distribution of y for different values of x. The mode of the distribution is $\alpha + \beta x$ (and it would also be $\mathrm{E}(y \mid x)$ if the response weren't truncated). $\mathrm{E}(y \mid x)$ is obtained by adding $\mathrm{E}(\epsilon \mid x)$ to $\alpha + \beta x$.

As x increases, $\alpha + \beta x$ increases, which reduces $\mathrm{P}(y < 0)$ and makes the truncated normal density closer to the untruncated one. As we can see in Figure 11.4, the distance between the mode of the distribution $\alpha + \beta x$ and $\mathrm{E}(y|x)$, which is $\mathrm{E}(\epsilon \mid x)$ decreases with higher values of x. This situation is illustrated, using simulated data on Figure 11.5. The plain line is defined by $\alpha + \beta x$. The dotted line is the regression line for this sample. Its slope is slightly lower than β, which illustrates the fact that the OLS estimator of the slope is downward biased (if $\beta > 0$, which is the case here). The dashed line depicts $\mathrm{E}(y \mid x, y > 0)$.

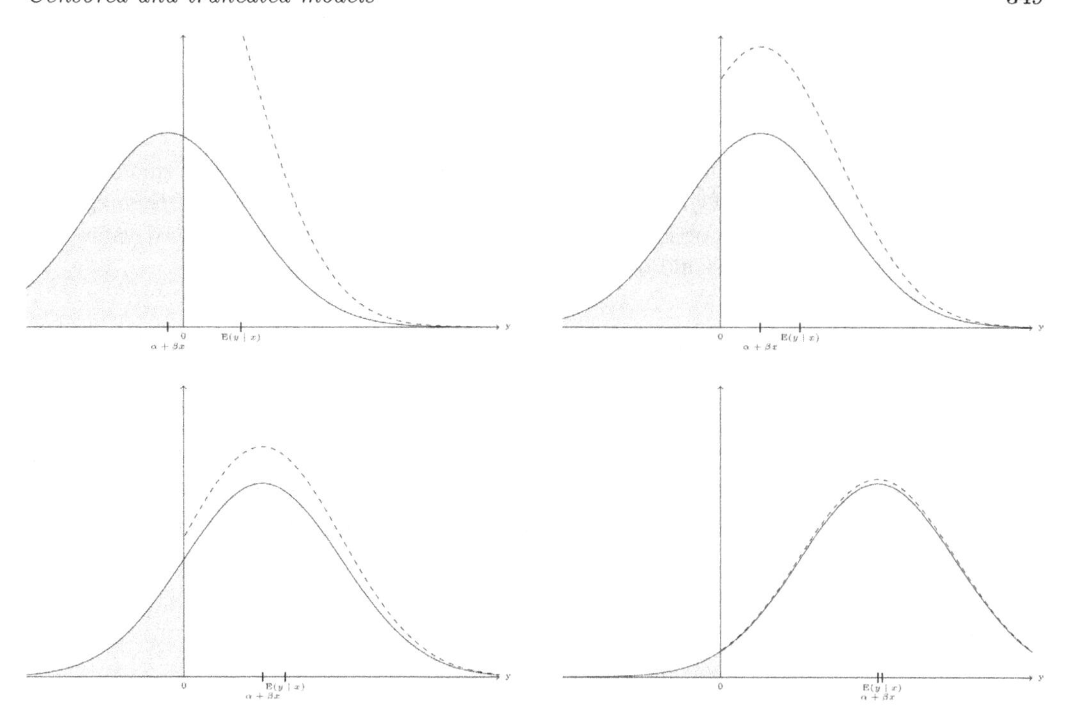

Figure 11.4: Truncated normal distribution for y

For large values of x, it is almost the same as the black line, which indicates that in this range of values of x, $\mathrm{E}(y \mid x, y > 0)$ is almost equal to $\alpha + \beta x$, which means than the correlation between x and ϵ almost vanishes. Conversely, for low values of x, the gap between $\mathrm{E}(y \mid x, y > 0)$ and $\alpha + \beta x$ increases. This gap is $\mathrm{E}(\epsilon \mid x, y > 0)$, it is positive and is particularly high for very low values of x. As $x \to -\infty$, $\mathrm{E}(y \mid x, y > 0) \to 0$ and therefore $\mathrm{E}(\epsilon \mid x, y > 0) \to -(\alpha + \beta x)$.

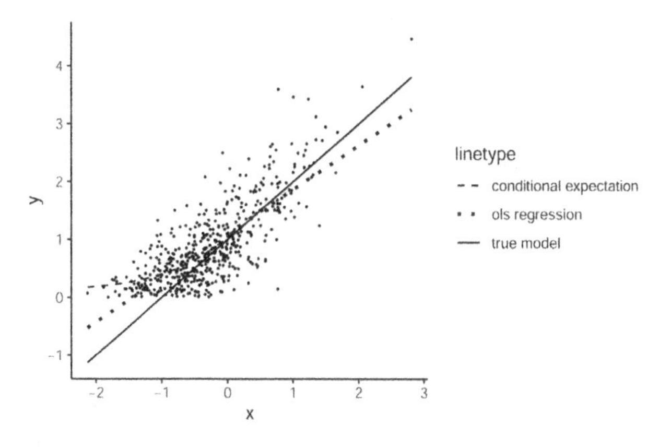

Figure 11.5: OLS bias in a truncated sample

By now, we have considered a truncated sample, which is a sample containing only observed values of y. Consider now that the underlying variable is: $y^* \mid x \sim \mathcal{N}(\mu, \sigma)$ with the following rule of observation:

$$\left\{\begin{array}{lcll} y & = & 0 & \text{if} \quad y^* < 0 \\ y & = & y^* & \text{if} \quad y^* \geq 0 \end{array}\right. \tag{11.3}$$

The observed response y is therefore either y^* if positive, or 0 if $y^* \leq 0$. In this case, the conditional expected value of y can be computed as the weighted average of the expected value given that y is greater or lower than 0, the first one being the expected value of y left-truncated at 0 and the second one being 0. With $\mu_n = \alpha + \beta x_n$:

$$\begin{array}{rcl} \text{E}(y \mid x_n) & = & \left[1 - \Phi\left(\frac{\mu_n}{\sigma}\right)\right] \times 0 + \Phi\left(\frac{\mu_n}{\sigma}\right) \times \text{E}(y \mid x, y > 0) \\ & = & \mu_n \Phi\left(\frac{\mu_n}{\sigma}\right) + \sigma\phi\left(\frac{\mu_n}{\sigma}\right) \end{array}$$

As for the previous case, the conditional expected value of y is not μ_n, which implies that the OLS estimator is biased and inconsistent. Least squares estimation is illustrated, using simulated data, on Figure 11.6. The downward bias of the slope seems more severe than for the truncated sample because there are much more observations for very low values of x, i.e., in the range of the values of x where the correlation between x and ϵ is severe.

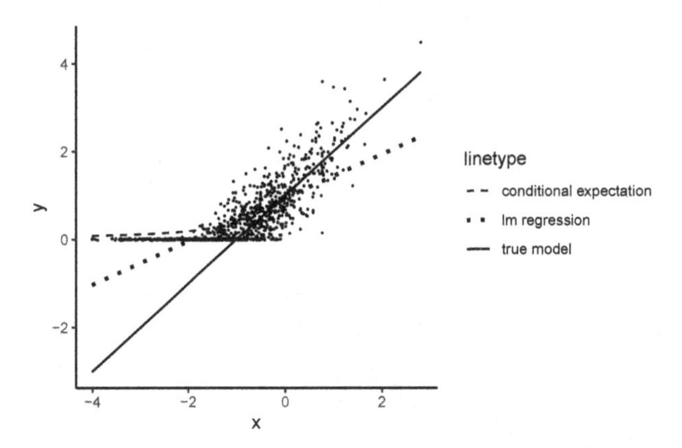

Figure 11.6: OLS bias in a censored sample

The asymptotic bias of the OLS estimator has been computed by Goldberger (1981) for a truncated sample and by Greene (1981) for a censored sample, with the hypothesis that y and x follow a jointly normal distribution. In both cases, we have, denoting $\hat{\beta}$ the OLS estimator: plim $\hat{\beta} = \theta\beta$, with $0 < \theta < 1$. Therefore, the bias of the OLS estimator is an attenuation bias, which means that the OLS estimator converges in absolute value to a value lower than the true parameter. Moreover, for the censored sample case:

$$\text{plim } \hat{\beta} = \Phi(\mu_y/\sigma)\beta \tag{11.4}$$

Therefore, in this case, θ is the probability of observing a positive value of y, which can be consistently estimated by the share of positive observations in the sample.

11.2.2 Interpretation of the coefficients

This section concerns only the case of corner solution and not the case of data censoring (like top-coding). In both cases, the regression function: $\mu_n = \alpha + \beta^\top x_n$ returns the mean of the distribution of the untruncated distribution of y. In the data censoring case, which is just a problem of missing values of the response, this is the relevant distribution to consider and therefore β_k is the marginal effect of covariate x_k that we have to consider. On the contrary, for corner solution models, the relevant distributions that we have to consider is on the one hand the probability of $y > 0$ and on the other hand the zero left-truncated distribution of y. Therefore, μ_n is the mean of an untruncated latent variable, β_k is the marginal effect of x_k on this latent variable and none of these values are particularly meaningful. For a corner solution model, the effect of a change in x_k is actually twofold:

- firstly, it changes the probability that the value of y is positive: $\mathrm{P}(y > 0 \mid x)$,
- secondly, it changes the expected value of y if it is positive: $\mathrm{E}(y \mid x, y > 0)$.

The probability that y is positive and the conditional expectation for positive values of y are, denoting as usual $\mu_n = \alpha + \beta^\top x_n$:

$$\left\{\begin{array}{rcl} \mathrm{P}(y_n > 0 \mid x_n) & = & \Phi\left(\frac{\mu_n}{\sigma}\right) \\ \mathrm{E}(y_n \mid x_n, y_n > 0) & = & \mu_n + \sigma r\left(\frac{\mu_n}{\sigma}\right) \end{array}\right.$$

and the unconditional expectation of y is just the product of these two expressions:

$$\mathrm{E}(y_n \mid x_n) = \mathrm{P}(y_n > 0 \mid x_n) \times \mathrm{E}(y_n \mid x_n, y_n > 0)$$

Its derivative with respect to x_k gives:

$$\begin{array}{rcccl} \frac{\partial \mathrm{E}(y_n \mid x_n)}{\partial x_{nk}} & = & \frac{\partial \mathrm{P}(y_n > 0 \mid x_n)}{\partial x_{nk}} & \times & \mathrm{E}(y_n \mid x_n, y_n > 0) \\ & + & \mathrm{P}(y_n > 0 \mid x_n) & \times & \frac{\partial \mathrm{E}(y_n \mid x_n, y_n > 0)}{\partial x_{nk}} \end{array}$$

with:

$$\left\{\begin{array}{rcl} \frac{\partial \mathrm{P}(y_n > 0 \mid x_n)}{\partial x_{nk}} & = & \frac{\beta_k}{\sigma}\phi\left(\frac{\mu_n}{\sigma}\right) \\ \frac{\partial \mathrm{E}(y_n \mid x_n)}{\partial x_{nk}} & = & \beta_k\left[1 + r'\left(\frac{\mu_n}{\sigma}\right)\right] \end{array}\right.$$

The effect of a change of a covariate is represented in Figure 11.7 for the simple covariate case, with $\beta > 0$. When the value of x increases from x_1 to x_2, the untruncated normal density curve moves to the right, the mode increasing from $\mu_1 = \alpha + \beta x_1$ to $\mu_2 = \alpha + \beta x_2$. The increase of the probability that $y > 0$ is represented by the gray area, as it is the area between the two density curves from 0 to $+\infty$, which reduces to the area between the two curves between $\mu_1 + \mu_2$ and $+\infty$.[2] This is the first source of change of $\mathrm{E}(y \mid x)$ which, for a small variation of x is equal to $\Delta\Phi\left(\frac{\alpha+\beta x}{\sigma}\right) \times \mathrm{E}(y \mid x, y > 0)$. The second source of change is the increase of the conditional expectation of x, which is multiplied by the probability that y is observed: $\Delta\mathrm{E}(y \mid x, y > 0) \times \Phi\left(\frac{\alpha+\beta x}{\sigma}\right)$. The first one can be considered as an increase of y on the **extensive margin**, i.e., due to the fact that for more people, we observe $y > 0$. The second one is an increase of y on the **intensive margin**, which means that people for

[2]The area between 0 and the intersection point $\frac{\mu_1+\mu_2}{2}$ and the one between $\frac{\mu_1+\mu_2}{2}$ and $\mu_1 + \mu_2$ are the same with the opposite sign.

which y was already positive, the value of y increases.[3] The sum of these two components gives the marginal effect of a variation of x on the unconditional expected value of y, which is simply:

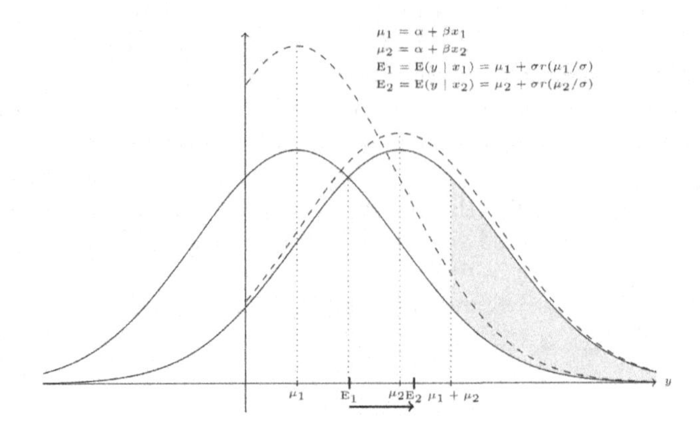

Figure 11.7: Effect of a change of x

$$\frac{\partial \mathrm{E}(y_n \mid x_n)}{\partial x_{nk}} = \beta_k \Phi\left(\frac{\mu_n}{\sigma}\right)$$

Note (Equation 11.4) that it is exactly the probability limit of the OLS estimator.

11.3 Methods of estimation

Several consistent estimators are available for the truncated and the censored model. We'll start by inefficient estimators (non-linear least squares, probit and two-step estimators). We'll then present the maximum likelihood estimator which is asymptotically efficient if the conditional distribution of y is normal and homoskedastic. We'll finally develop the symmetrically trimmed least squares estimator, which is consistent even if the distribution of y is not normal and heteroskedastic.

11.3.1 Non-linear least squares

The conditional expected value of y: $\mathrm{E}(y \mid x) = \gamma^\top z + \sigma r\left(\frac{\gamma^\top z}{\sigma}\right)$ is non-linear in x. Therefore, the parameters can be consistently estimated using non-linear least squares, by minimizing:

$$\sum_{n=1}^{N} \left[y_n - \gamma^\top z_n - \sigma r\left(\frac{\gamma^\top z_n}{\sigma}\right)\right]^2$$

[3]This decomposition of marginal effects for tobit models was first proposed by McDonald and Moffitt (1980)

11.3.2 Probit and two-step estimators

The probability that y is positive is $\Phi\left(\frac{\gamma^\top z_n}{\sigma}\right)$, therefore, a probit model can be used to estimate the vector of coefficients $\frac{\gamma}{\sigma}$. σ is not identified, and each element of γ is only estimated up to a $1/\sigma$ factor. To estimate the probit model, we first have to compute a binary response from the observed response which is equal to 1 if $y > 0$ and 0 if $y = 0$. We then obtain a vector of estimated coefficients $\hat{\delta}$ which are related to the structural coefficients of the model by the relation $\delta = \frac{\gamma}{\sigma}$. Obviously, the probit estimation can only be performed for a censored sample, and not a truncated sample for which all the values of y are positive. Remember that the expected value of y is: $\mathrm{E}(y_n \mid x_n) = \gamma^\top z_n + \sigma r\left(\gamma^\top z_n/\sigma\right)$. If γ/σ were known and denoting $r_n = r(\gamma^\top z_n/\sigma)$, estimating the equation: $y_n = \gamma^\top z_n + \sigma r_n + \nu_n$ by least squares would lead to consistent estimates of γ and σ as $\mathrm{E}(y_n \mid x_n, r_n) = \gamma^\top x_n + \sigma r_n$ or $\mathrm{E}(\nu_n \mid x_n, r_n) = 0$. r_n is obviously unknown, as it depends on the parameters we seek to estimate, but it can be consistently estimated, using the probit estimator, by $\hat{r}_n = r\left(\hat{\delta}^\top z_n\right)$. This idea leads to the **two-step estimator** first proposed by Heckman (1976):

- first estimate the coefficient of the probit model $\hat{\delta}$ and estimate r_n by $\hat{r}_n = r(\hat{\delta}^\top z_n)$,
- then regress y on x and $\hat{r}$ and estimate $\hat{\gamma}$ and $\hat{\sigma}$.

Denote $W = (Z, \hat{r}) = (1, X, \hat{r})$ the matrix of covariates for the second step and $\lambda^\top = (\gamma^\top, \sigma)$ the associated vector of parameters. The covariance matrix of the parameters reported by the OLS estimation is $\hat{\sigma}_\epsilon^2(W^\top W)^{-1}$. It is inconsistent for two reasons:

- the errors of the model are heteroskedastic, their variance being $\mathrm{V}(\epsilon_n) = \sigma^2(1 + r'(\gamma^\top z_n/\sigma))$,
- the supplementary covariate $\hat{r}(\delta^\top z_n)$ differs from the true value of $r(\delta^\top z_n)$, which inflates the variance of the estimators. A consistent estimate of the covariance matrix of the two-step estimator is[4]:

$$\hat{\sigma}^2(W^\top W)^{-1} W\left[\Sigma + (I - \Sigma) Z \hat{\mathrm{V}}_{\text{probit}} (I - \Sigma) Z^\top\right] W^\top (W^\top W)^{-1}$$

Σ is a matrix that takes into account the heteroskedasticity, it is a diagonal matrix that contains either $\hat{\sigma}^2(1 + r'(\hat{\delta}^\top z_n))$ or, following the argument of White (1980), the square of the residual $y_n - \hat{\gamma}^\top z_n - \hat{\sigma}\hat{r}_n$. The second matrix, which uses the covariance matrix of the first stage probit regression, takes into account the fact that $\hat{r}_n$ is introduced in place of r_n.

11.3.3 Maximum Likelihood estimation

Without loss of generality, we'll assume that $y = 0$ for observations from 1 to N_o and $y > 0$ for those from $N_o + 1$ to N are not. Estimating the model on the truncated sample, we obtain the likelihood by multiplying the truncated density of y for all the individuals from $N_o + 1$ to N:

$$L^T(\gamma, \sigma \mid y, x) = \prod_{n=N_o+1}^{N} \frac{1}{\sigma \Phi\left(\frac{\gamma^\top z_n}{\sigma}\right)} \phi\left(\frac{y_n - \gamma^\top z_n}{\sigma}\right)$$

[4]See for example Amemiya (1984), equation 28, page 13.

or, taking the logarithm:

$$\ln L^T(\gamma, \sigma \mid y, x) = -\frac{N - N_o}{2}(\ln \sigma^2 + \ln 2\pi) - \frac{1}{2\sigma^2} \sum_{n=N_o+1}^{N} (y_n - \gamma^\top z_n)^2 - \sum_{n=N_o+1}^{N} \ln \Phi\left(\frac{\gamma^\top z_n}{\sigma}\right)$$

Note that, except for the last term, this is the log-likelihood of the normal gaussian model. For the censored sample, the individual contribution to the likelihood sample will depend on whether $y = 0$ or not:

- if $y = 0$, the contribution is the probability that $y = 0$, which is $1 - \Phi\left(\frac{\gamma^\top z}{\sigma}\right)$,
- if $y > 0$, the contribution is the product of the probability that $y > 0$ and the density of the truncated distribution of y, which is: $\Phi\left(\frac{\gamma^\top z}{\sigma}\right)\frac{1}{\sigma\Phi\left(\frac{\gamma^\top z}{\sigma}\right)}\phi\left(\frac{y-\gamma^\top z}{\sigma}\right)$.

The likelihood function is therefore:

$$\begin{array}{rcl} L^C(\gamma, \sigma | y, x) & = & \prod_{n=1}^{N_o}\left[1 - \Phi\left(\frac{\gamma^\top z_n}{\sigma}\right)\right]\prod_{n=N_o+1}^{N}\Phi\left(\frac{\gamma^\top z_n}{\sigma}\right) \\ & \times & \prod_{n=N_o+1}^{N}\frac{1}{\sigma\Phi\left(\frac{\gamma^\top z_n}{\sigma}\right)}\phi\left(\frac{y_n - \gamma^\top x_n}{\sigma}\right) \end{array} \quad (11.5)$$

which is simply the product of:

- the likelihood of a probit model which explains that $y = 0$ or $y > 0$ (the first line of Equation 11.5),
- the likelihood of y for the truncated sample (the second line of Equation 11.5).

Denoting L^P the likelihood of the probit model, we then have:

$$L^C(\gamma, \sigma \mid y, x) = L^P(\gamma, \sigma \mid y, x) \times L^T(\gamma, \sigma \mid y, x)$$

Taking logs and rearranging terms, we finally get:

$$\begin{array}{rcl} \ln L^C(\gamma, \sigma \mid y, x) & = & \sum_{n=1}^{N_o}\ln\left[1 - \Phi\left(\frac{\gamma^\top z_n}{\sigma}\right)\right] - \frac{N-N_o}{2}\left(\ln \sigma^2 + \ln 2\pi\right) \\ & - & \frac{1}{2\sigma^2}\sum_{n=N_o+1}^{N}\left(y_n - \gamma^\top z_n\right)^2 \end{array}$$

Denoting $d_n = \mathbf{1}(y_n > 0)$, the first derivatives with γ are, denoting: $\mu_n = \gamma^\top z_n$ and $r_n = \frac{\phi(\mu_n/\sigma)}{1-\Phi(\mu_n/\sigma)}$:

$$\frac{\partial L^C}{\partial \gamma} = \frac{1}{\sigma^2}\sum_{n=1}^{N}\left(-(1 - d_n)\sigma r_n + d_n(y_n - \mu_n)\right) z_n = \frac{1}{\sigma^2}\sum_{n=1}^{N}\psi_n z_n$$

For the maximum likelihood estimator, the vector of generalized residuals is then:[5]

$$\psi_n = -(1 - d_n)\sigma r_n + d_n(y_n - \mu_n) \quad (11.6)$$

[5]See Section 10.4.1.

and it is orthogonal to all the regressors. Note that, for positive values of y, the generalized residual is just the standard residual. For null values of y, which means negative values of y^*, the generalized residual is: $\mathrm{E}(y_n^* - \mu_n \mid x, y_n^* \leq 0) = -\sigma r_n$. As y_n^* and therefore the residuals for null observations are unobserved; they are simply replaced by their expectations. The hessian is rather tricky, but its expression can be greatly simplified using a reparametrization, due to Olsen (1978): $\delta = \gamma/\sigma$ and $\theta = 1/\sigma$. Olsen (1978) showed that the log-likelihood function of the censored model expressed in terms of δ and θ is globally concave and therefore admits a unique optimum which is a maximum.

11.3.4 Semi-parametric estimators

In a semi-parametric approach, only the regression function, i.e., $\mathrm{E}(y \mid x) = \mu_n = \alpha + \beta^\top x$, is parametrically specified, while the rest of the model (especially the conditional distribution of y) is not. This approach is therefore much more generally applicable. Compared to the estimators presented in the previous two sections, which are only consistent if the conditional distribution of y is normal and homoskedastic, the semi-parametric estimator presented in this section (Powell 1986) is consistent in a much broader context, as it requires only the symmetry of the conditional distribution of the response.

For the zero left-truncated response case, the OLS estimator is biased because the conditional distribution of y is asymmetric, as the observations on the lower tail of the distribution ($y < 0$) are either missing (the case of a truncated sample) or set to 0 (the case of a censored sample). For the case of a truncated sample, trimming the observations for which $y_n > 2\mu_n$, i.e., observations that lie in the upper tail, would restore the symmetry, and OLS estimation on this trimmed sample would be consistent. This situation is depicted in Figure 11.8.

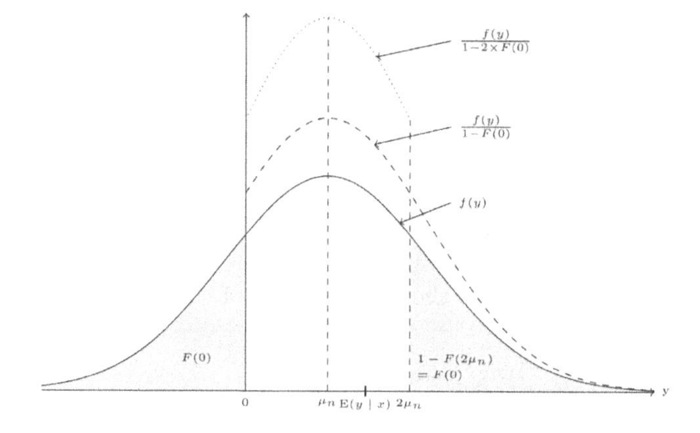

Figure 11.8: Symmetrically trimmed truncated distribution

The plain line represents the distribution of the untruncated response, the dashed line the corresponding distribution of the left zero-truncated response. This distribution is asymmetric, and the expected value of y for $x = x_n$ is above μ_n because of the left-truncation. The dotted line represents the two-sided truncated distribution of y, truncated at 0 on the left side and at $2\mu_n$ on the right side. As the untruncated distribution of y is symmetric, so is the two-sided truncated distribution, for which the conditional expected value of y is now equal to μ_n. Therefore, if we were able to remove from the sample all the observations for which $y_n > 2\mu_n$, the OLS estimator on this trimmed sample would be consistent. Of course, the problem is that the right-truncation point is unknown for every observation, as it depends on γ which is the parameter vector that we seek to estimate. The subset of observations that should be kept are such that $0 < y_n < 2\gamma^\top z_n$, or $-\gamma^\top z_n < \epsilon_n < \gamma^\top z_n$, the first inequality being always verified for a truncated sample. Note that this condition will remove all the observations for which $\gamma^\top z_n < 0$.

The first-order conditions to minimize the sum of squares of the residuals for the trimmed sample is then similar to the normal equations ($\sum_{n=1}^N (y_n - \gamma^\top z_n) z_n = 0$) on the relevant subset of the sample:

$$\sum_{n=1}^N \mathbf{1}(y_n < 2\gamma^\top z_n)(y_n - \gamma^\top z_n) z_n = 0 \tag{11.7}$$

Note that the left-hand side of this equation is a discontinuous function of γ, as even a small change of γ may, for some observations, turn the condition $y_n < 2\gamma^\top z_n$ from true to false (or the opposite). Moreover, note that $\gamma = 0$ is a trivial solution as, in this case, $\mathbf{1}(y_n < 2\gamma^\top x_n) = 0 \; \forall n$. Therefore, it is safer to consider the estimator as the result of a minimization problem instead of solving the set of non-linear equations. By direct integration, we can check that minimizing the following function:

$$R_T = \sum_{n=1}^N \left[y_n - \max\left(\frac{y_n}{2}, \gamma^\top z_n\right)\right]^2$$

leads to Equation 11.7. This **symmetrically truncated least squares** estimator easily extends to the case of censored samples. In this case, negative values of y are unobserved, and the value of the response is set to 0. A symmetrically censored sample is obtained by setting the response, for observations whose the observed value of y is greater than $2\mu_n$ to $2\mu_n$. This means that as the response is zero left-censored, we also right-censor it, with a truncation value that is specific to the observation, and that depends on the set of unknown parameters γ we seek to estimate. The resulting first-order condition to minimize the sum of squares of the "symmetrically censored sample" is:

$$\sum_{n=1}^N \mathbf{1}(y_n < 2\gamma^\top z_n) \left[\min\left(y_n, 2\gamma^\top z_n\right) - \gamma^\top z_n\right] z_n = 0$$

which is the first-order conditions of the minimization of the following function:

$$\begin{array}{rcl} R_C & = & \sum_{n=1}^N \left[y_n - \max\left(\frac{y_n}{2}, \gamma^\top z_n\right)\right]^2 \\ & + & \sum_{n=1}^N \mathbf{1}(y_n < 2\gamma^\top z_n) \left[\left(\frac{y_n}{2}\right)^2 - \max(0, \gamma^\top z_n)^2\right] \end{array}$$

11.4 Estimation of the tobit-1 model with R

The estimation of the tobit-1 model is available in functions of different packages: `AER::tobit`, `censReg::censReg` and `micsr::tobit1`. All these functions use the usual formula-data interface to describe the model to be estimated and also a `left` and a `right` argument to indicate the truncation points. The default values of these last two arguments are 0 and $+\infty$, which correspond to the most usual zero left-truncated case. The `micsr::tobit1` function allows to use either a censored or a truncated sample by setting the `sample` argument either to `"censored"` or `"truncated"`. The other two functions only allow the estimation of the censored regression model. The truncated regression model can also be estimated using the `truncreg::truncreg` function. `micsr::tobit1` also has the advantage of providing several different estimators, selected using the `method` argument: `"ml"` for maximum likelihood, `"lm"` for linear model, `"twostep"` for the two-step estimator, `"trimmed"` for the trimmed estimator and `"nls"` for the non-linear least squares estimator, the two other functions only providing the maximum likelihood estimator. We'll present three examples of application, with respectively a left-truncated, a right-truncated and a two-sided truncated response.

11.4.1 Left-truncated response

The `charitable` data set is used by Wilhelm (2008) and concerns charitable giving.

```
charitable %>% print(n = 3)
```

```
# A tibble: 2,384 x 7
  donation donparents education        religion   income married south
     <dbl>      <dbl> <fct>            <fct>       <dbl>   <dbl> <dbl>
1      335       5210 less_high_school other      21955.       0     0
2       75      13225 high_school      protestant 22104.       0     0
3     6150.      3375 some_college     catholic   50299.       0     0
# i 2,381 more rows
```

The response is called `donation`; it measures annual charitable giving in US dollars. This variable is left-censored for the value of 25, as this value corresponds to the item "less than \$25 donation". Therefore, for this value, we have households who didn't make any charitable giving and some who made a small giving (from \$1 to \$25). The covariates used are the donations made by the parents (`donparents`), two factors indicating the educational level and religious beliefs (respectively `education` and `religion`), annual income (`income`) and two dummies for living in the south (`south`) and for married couples (`married`). Wilhelm (2008) considers the value of the donation in logs and subtracts from it $\ln 25$, so that the response is 0 for households who gave no donation or a small donation.

```
charitable <- charitable %>% mutate(logdon = log(donation) - log(25))
```

The model can either be estimated using `logdon` as the response and the default values of `left` or `right` or by using `log(donation)` as the response and setting `left` to `log(25)`.

```
char_form <- logdon ~ log(donparents) + log(income) +
    education + religion + married + south
ml_aer <- AER::tobit(char_form, data = charitable)
ml_creg <- censReg::censReg(char_form, data = charitable)
ch_ml <- tobit1(char_form, data = charitable)
```

Table 11.1: Estimation of charitable giving models

	OLS	2-steps	ML	SCLS
(Intercept)	−10.071***	−9.485***	−17.618***	−15.388***
	(0.556)	(1.081)	(0.898)	(1.472)
log(donparents)	0.135***	0.109***	0.200***	0.167***
	(0.017)	(0.014)	(0.025)	(0.035)
log(income)	0.941***	0.963***	1.453***	1.320***
	(0.056)	(0.072)	(0.087)	(0.120)
married	0.562***	0.614***	0.767***	0.702***
	(0.079)	(0.062)	(0.117)	(0.169)
south	0.111	0.128**	0.113	0.064
	(0.071)	(0.043)	(0.105)	(0.130)
σ		0.656*	2.114***	
		(0.276)	(0.041)	
Num.Obs.	2384	2384	2384	2384
AIC			8038.5	
BIC			8119.4	
Log.Lik.			−4005.274	

The three functions return identical results, except that they are parametrized differently: `micsr::tobit1` estimates σ as the two other functions estimate $\ln \sigma$. Using `micsr::tobit1`, we also estimate the two-step, the **SCLS** (symmetrically censored least squares) and the OLS estimators.

```
ch_twostep <- update(ch_ml, method = "twostep")
ch_scls <- update(ch_ml, method = "trimmed")
ch_ols <- update(ch_ml, method = "lm")
```

The results of the three models are presented in Table 11.1.[6] The last two columns of Table 11.1 match the first two columns of table 3 of Wilhelm (2008), page 577. Note that the OLS estimators are in general lower in absolute values than those of the three other estimators, which illustrates the fact that OLS estimators are biased toward zero when the response is censored. More precisely Equation 11.4 indicates that the OLS estimator converges to $\Phi(\mu_y/\sigma)\beta$. $\Phi(\mu_y/\sigma)$ is the probability of $y > 0$ (y being in our example the log of the donation minus $\ln 25$) and can be consistently estimated by the share of uncensored observations in our sample, which is about two-thirds. Therefore, the ratio of the OLS estimator and one of a consistent estimator, for example the ML estimator, should be approximately equal to two-thirds. This is actually the case for the first two covariates: $0.135/0.2 = 0.672$ and $0.941/1.453 = 0.648$.

11.4.2 Right-truncated response

The `food` data set is used by De Crombrugghe, Palm, and Urbain (1997) to estimate the demand for food in the Netherlands.

[6] To save space, the coefficients of the levels of the `education` and `religion` covariates are omitted.

```
food %>% print(n = 3)
```

```
# A tibble: 4,611 x 7
   year weights hsize  ageh income  food midage
  <dbl>   <dbl> <dbl> <dbl>  <dbl> <dbl>  <dbl>
1  1980   0.615     2     1  55900  5280      0
2  1980   1.52      1     1  10600  1700      0
3  1980   1.16      2     5  34800  6730      0
# i 4,608 more rows
```

Two surveys are available, for 1980 and 1988, we'll use only the first one. Food expenses are top-coded for the top 5 percentiles, which corresponds to an expense of 13030 Dfl. The value reported for these observations is 17670 Dfl, which is the mean value of the expense for the top 5 percentile.

```
food %>% filter(year == 1980) %>%
    summarise(n = sum(food == 17670),
              f = mean(food == 17670))
```

```
# A tibble: 1 x 2
      n      f
  <int>  <dbl>
1   127 0.0453
```

The percentage of censored observations is not exactly 5%, because some observations have been excluded due to missing values for some covariates. The response being expressed in logarithms, the right threshold is set to `log(13030)` and the left one to `- Inf` as the response is not left-truncated.

```
food_tobit <- tobit1(log(food) ~ log(income) + log(hsize) + midage,
                     data = food, subset = year == 1980,
                     left = -Inf, right = log(13030))
food_tobit %>% gaze
```

```
            Estimate Std. Error z-value Pr(>|z|)
log(income)  0.34005    0.01856    18.3  < 2e-16
log(hsize)   0.47304    0.01588    29.8  < 2e-16
midage       0.09501    0.01462     6.5  8.1e-11
sigma        0.36716    0.00507    72.4  < 2e-16
```

The main coefficient of interest is the one associated with the `log(income)` covariate. Remember that in this data censoring case, the coefficient is the marginal effect, in the present context the income elasticity of food which is equal to 0.34.

11.4.3 Two-sided tobit models

Hochguertel (2003) estimates the share of riskless assets, which can be either an internal solution, or a corner solution with the share equal to 0 or 1 ($l = 0$ and $u = 1$). Hochguertel therefore estimates a two-sided tobit model. The paper seeks to explain the low share of risky assets in portfolios of Dutch households. The data set is called `portfolio`. This data set is a panel of annual observations from 1993 to 1998. In his paper, Hochguertel (2003) used panel and cross-section estimators (the latter being obtained on the whole data set by pooling the six time-series). The response is `share`, and the two covariates of main interest are `uncert` and `expinc`.

- `uncert` indicates the degree of uncertainty felt by the household; it is a factor with levels `low`, `moderate` and `high`,
- `expinc` indicates the prediction of the household concerning the evolution of their income in the next 5 years; it is a factor with levels `increase`, `constant` and `decrease`.

We'll use a simpler specification than the one used by the author, which contains a lot of covariates; we only add to the two covariates previously described the net worth, the age of household's head and its square and a dummy for households whose head is a woman. We use the `micsr::tobit1` function and we set the `left` and the `right` arguments respectively to 0 and 1.

```
portfolio <- portfolio %>% mutate(agesq = age ^ 2 / 100)
prec_ml <- tobit1(share ~ uncert + expinc + networth +
                        age + agesq + female,
                      left = 0, right = 1, data = portfolio)
prec_ml %>% gaze
```

```
             Estimate Std. Error z-value Pr(>|z|)
uncertmod     0.04410    0.01489    2.96  0.00307
uncerthigh    0.08155    0.01756    4.64  3.4e-06
expinccst     0.04059    0.01184    3.43  0.00061
expincdecr    0.04631    0.01582    2.93  0.00343
networth     -0.02940    0.00122  -24.19  < 2e-16
age          -0.02848    0.00269  -10.58  < 2e-16
agesq         0.03055    0.00257   11.88  < 2e-16
female        0.14153    0.01447    9.78  < 2e-16
sigma         0.45578    0.00423  107.64  < 2e-16
```

As expected, high uncertainty and pessimistic expectations about future income increase the share of riskless assets. Net worth has a negative effect on the share of riskless assets, and households headed by a woman have a higher share of riskless assets. Finally the effect of age is U-shaped. Actually, Hochguertel (2003) doesn't estimate this model, as he suspected the presence of heteroskedascticity. Therefore, he estimates a model for which σ is replaced by $\sigma_n = e^{\delta^\top w_n}$, where w_n are a set of covariates and δ a set of further parameters to be estimated. The `crch::crch` function (Messner et al. 2014) enables the estimation of such models. The model is described using a two-part formula, the second one containing the covariates that are used to estimate σ_n. By default, a logistic link is used ($\ln \sigma_n = \gamma^\top z_n$), but other links can be selected using the `link.scale` argument. Moreover, departure from normality can be taken into account using the `dist` argument by, for example, switching from a gaussian distribution (the default) to a Student or a logistic. For the skedasticity function, we use all the covariates used in the main equation except `uncert` and `expinc` which appear to be insignificant.

```
prec_ht <- crch::crch(share ~ uncert + expinc + networth +
          age + agesq + female | networth +
          age + agesq + female, left = 0, right = 1,
          data = portfolio)
summary(prec_ht)
```

```
Call:
crch::crch(formula = share ~ uncert + expinc + networth + age +
    agesq + female | networth + age + agesq + female, data = portfolio,
    left = 0, right = 1)

Standardized residuals:
   Min     1Q Median     3Q    Max
-2.112 -0.888 -0.258  0.407  2.776

Coefficients (location model):
             Estimate Std. Error z value Pr(>|z|)
(Intercept)  2.66330    0.08142   32.71  < 2e-16 ***
uncertmod    0.03474    0.01385    2.51   0.0122 *
uncerthigh   0.05365    0.01639    3.27   0.0011 **
expinccst    0.02645    0.01069    2.47   0.0133 *
expincdecr   0.04223    0.01461    2.89   0.0038 **
networth    -0.14347    0.00435  -32.96  < 2e-16 ***
age         -0.02202    0.00308   -7.15  8.6e-13 ***
agesq        0.02669    0.00303    8.80  < 2e-16 ***
female       0.09196    0.01542    5.96  2.5e-09 ***

Coefficients (scale model with log link):
             Estimate Std. Error z value Pr(>|z|)
(Intercept)  0.41494    0.12607    3.29   0.0010 ***
networth    -0.09291    0.00255  -36.37   <2e-16 ***
age         -0.01575    0.00511   -3.08   0.0021 **
agesq        0.02267    0.00497    4.56    5e-06 ***
female       0.07027    0.02796    2.51   0.0120 *
---
Signif. codes:  0 '***' 0.001 '**' 0.01 '*' 0.05 '.' 0.1 ' ' 1

Distribution: gaussian
Log-likelihood: -5.96e+03 on 14 Df
Number of iterations in BFGS optimization: 29
```

Without conducting a formal test, it is clear that the heteroskedastic specification is supported by the data, the log-likelihood value of the second specification being much larger than the one of standard tobit model. The value of some coefficients are strikingly different. For example, the coefficient of `networth` is -0.029 for the tobit model, but -0.1435 for the heteroskedastic model, as this covariate also has a huge effect on the conditional variance of the response.

11.5 Evaluation and tests

11.5.1 Conditional moment tests

The most popular method of estimation for the tobit-1 model is the fully parametric maximum likelihood method. Contrary to the OLS model, the estimator is only consistent if the GDP process is perfectly described by the likelihood function, i.e., if $\epsilon_n \sim \mathcal{N}(0, \sigma)$. In particular, the consistency of the estimator rests on the hypothesis of normality and homoskedasticity. The conditional moment tests have been presented in Section 5.3.2, they use different powers of the residuals. We have seen in Section 10.4.3.2 for the probit model how to compute these tests when the residuals are not observable: ϵ_n^k is then replaced by $\mathrm{E}(\epsilon_n^k \mid x_n)$. The computation of the conditional moment test for the tobit model is quite similar, except that the residuals are partially observed (when $d_n = \mathbf{1}(y_n > 0) = 1$). Then, we'll use ϵ_n^k if $d_n = 1$ and $\mathrm{E}(\epsilon_n^k \mid x_n)$ if $d_n = 0$. From Equation 10.12, the moments of $\epsilon \sim \mathcal{N}(0, \sigma_\epsilon)$ right-truncated at $-\mu_n$ are, denoting $r_n = \frac{\phi(\mu_n/\sigma_\epsilon)}{1-\Phi(\mu_n/\sigma_\epsilon)}$:

$$\mathrm{E}(\epsilon_n^k \mid x_n, y_n^* \leq 0) = (k-1)\sigma_\epsilon^2 m_{k-2} - \sigma_\epsilon(-\mu_n)^{k-1} r_n$$

For example, $E(\epsilon_n | x_n, y_n^* \leq 0) = -\sigma_\epsilon r_n$, which is the generalized residual for $d_n = 0$, see Equation 11.6. Then a similar table as the one given in Equation 10.13 can be written for the tobit model:

k	hypothesis	$\mathrm{E}(\epsilon_n^k \mid x_n \mid \epsilon_n \leq -\mu_n)$	emp. moment
1	omit. variable	$-\sigma r_n$	$\frac{1}{N}\sum_n \left[-(1-d_n)r_n + d_n\epsilon_n)\right] w_n$
2	homosc.	$\sigma^2(1 + z_n r_n)$	$\frac{1}{N}\sum_n \left[(1-d_n)\sigma^2 z_n r_n + d_n(\epsilon_n^2 - \sigma^2)\right] w_n$
3	asymetry	$-\sigma^3 r_n(2 + z_n^2) r_n$	$\frac{1}{N}\sum_n \left[-(1-d_n)\sigma^3 r_n(2+z^2) + d_n\epsilon_n^3\right] w_n$
4	kurtosis	$\mathrm{E}(\epsilon_n^4 - 3 \mid x_n) = 0$	$\frac{1}{N}\sum_n \left[(1-d_n)\sigma^4 z_n r_n(3+z_n^3) + d_n(\epsilon_n^4 - 3\sigma^2)\right] w_n$

(11.8)

Conditional moment tests can be computed using the `micsr::cmtest` function, which can take as input a model fitted by either `AER::tobit`, `censReg::censReg` or `micsr::tobit1`. To test respectively the hypothesis of normality and of homoskedasticity, we use:

```
cmtest(ch_ml, test = "normality") %>% gaze
## chisq = 116.351, df: 2, pval = 0.000
cmtest(ch_ml, test = "heterosc") %>% gaze
## chisq = 103.592, df: 12, pval = 0.000
```

Normality and heteroskedasticity are strongly rejected. The values are different from Wilhelm (2008), as he used the "outer product of the gradient" form of the test. These versions of the test can be obtained by setting the `opg` argument to `TRUE`.

```
cmtest(ch_ml, test = "normality", opg = TRUE) %>% gaze
## chisq = 200.117, df: 2, pval = 0.000
cmtest(ch_ml, test = "heterosc", opg = TRUE) %>% gaze
## chisq = 127.308, df: 12, pval = 0.000
```

Non-normality can be further investigated by testing separately the fact that the skewness and kurtosis indicators are respectively different from 0 and 3.

```
cmtest(ch_ml, test = "skewness") %>% gaze
## z = 10.393, pval = 0.000
cmtest(ch_ml, test = "kurtosis") %>% gaze
## z = 2.329, pval = 0.020
```

The hypothesis that the conditional distribution of the response is mesokurtic is not rejected at the 1% level and the main problem seems to be the asymmetry of the distribution, even after taking the logarithm of the response. This can be illustrated (see Figure 11.9) by plotting the (unconditional) distribution of the response (for positive values) and adding to the histogram the normal density curve.

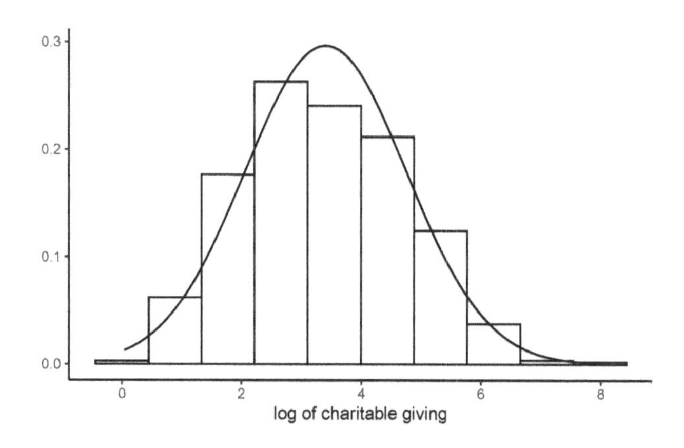

Figure 11.9: Empirical distribution of the response and normal approximation

11.5.2 Endogeneity

We have seen in Section 10.5 how endogeneity can be taken into account in a probit model. The analysis is almost the same for the probit model. The joint density of the distribution of the latent variable y^* and the endogenous covariates w is assumed to be normal and can be written as the product of the marginal density of w (Equation 10.15) and of the conditional density of y^* (Equation 10.16). For the tobit model, the former is exactly the same, but the latter is different because, for the probit model, only the sign of y^* is observed. This is not the case for the tobit model, the value of the response being observed if $y^* > 0$. Therefore, Equation 10.16 becomes:

$$\begin{array}{rcl} f(y_n \mid w_n) & = & (1-d_n)\left[1-\Phi\left(\frac{\gamma^\top z_n+\sigma_{\epsilon\nu}^\top\Sigma_\nu^{-1}(e_n-\Pi v_n)}{\sqrt{\sigma_\epsilon^2-\sigma_{\epsilon\nu}^\top\Sigma_\nu^{-1}\sigma_{\epsilon\nu}}}\right)\right] \\ & - & \frac{1}{2}d_n\left[\ln 2\pi+\ln(\sigma_\epsilon^2-\sigma_{\epsilon\nu}^\top\Sigma_\nu^{-1}\sigma_{\epsilon\nu})+\frac{y_n-\gamma^\top z_n-\sigma_{\epsilon\nu}^\top\Sigma_\nu^{-1}(e_n-\Pi w_n)}{(\sigma_\epsilon^2-\sigma_{\epsilon\nu}^\top\Sigma_\nu^{-1}\sigma_{\epsilon\nu}))}\right] \end{array} \quad (11.9)$$

where, as usual, $d_n = \mathbf{1}(y_n^* > 0)$. Note that, as the value of the response is partly observed, σ_ϵ is identified, contrary to the probit case where it has been arbitrarily set to 1. As for the probit model, several estimators are available: the maximum likelihood, the two-step and the minimum χ^2 estimators. Moreover, the hypothesis of exogeneity can be performed by computing a Wald test on the two-step estimator.

In the protection for sale model of Grossman and Helpman (1994), empirically tested by Goldberg and Maggi (1999), trade protection is determined by capital owners' lobbying. Matschke and Sherlund (2006) extend this model by introducing the potential effect of trade unions' lobbying on trade protection. The response is non-tariff barriers coverage ratio τ, transformed as $\frac{\tau}{1+\tau}$. The two covariates of Grossman and Helpman (1994) are the inverse of the import penetration ratio[7] divided by the importation demand elasticity x_1 and this variable in interaction with a dummy for capital owners' lobbying x_2. The supplementary covariate in the model of Matschke and Sherlund (2006) x_3 takes into account trade union's lobbying. The data set is called `trade_protection`. We first compute the response and the covariates using raw series:[8]

```
trade_protection <- trade_protection %>%
  mutate(y = ntb / (1 + ntb),
         x1 = vshipped / imports / elast,
         x2 = cap * x1) %>%
  rename(x3 = labvar)
```

The theoretical model of Matschke and Sherlund (2006) results in the following structural equation:

$$\frac{\tau}{1+\tau} = -\frac{\Theta}{\Theta + a}x_1 + \frac{1}{\Theta + a}x_2 + \frac{1}{\Theta + a}x_3$$

Therefore, denoting $(\alpha, \beta_1, \beta_2, \beta_3)$ the reduced form parameters, the model of Matschke and Sherlund (2006) implies that $\alpha = 0$ and that $\beta_2 = \beta_3$ as, in the model of Grossman and Helpman (1994) $\alpha = \beta_3 = 0$. Matschke and Sherlund (2006) suspect that the three covariates may be endogenous and therefore use the model of Smith and Blundell (1986). Three models are estimated:

- Grossman and Helpman's model (GH), i.e., x_3 is omitted,
- Matschke and Sherlund's "full" model, that doesn't impose that $\beta_2 = \beta_3$,
- Matschke and Sherlund's "short" model, that imposes that $\beta_2 = \beta_3$,

The hypothesis that $\alpha = 0$ is not imposed in the three models. Numerous instruments are used, as described in Matschke and Sherlund (2006), page 415 and are defined below as a one-sided formula:

```
inst <- ~ sic3 + k_serv + inv + engsci + whitecol + skill + semskill +
  cropland + pasture + forest + coal + petro + minerals + scrconc +
  bcrconc + scrcomp + bcrcomp + meps + kstock + puni + geog2 + tenure +
  klratio + bunion
```

[7]The value of gross imports divided by the value of shipments.

[8]The third covariate is quite complicated to compute and is directly available in `trade_protection` as `labvar`.

Table 11.2: Estimation results of the protection for sale model

	GH	Full	Short
α	0.0250	0.0109	0.0045
	(0.0226)	(0.0232)	(0.0219)
β_1	−0.0028*	−0.0024*	−0.0020**
	(0.0011)	(0.0010)	(0.0007)
β_2	0.0026*	0.0032**	
	(0.0011)	(0.0011)	
β_3		0.0023	
		(0.0014)	
$\beta_2 = \beta_3$			0.0031**
			(0.0010)
γ_1	0.0026+	0.0023	0.0029***
	(0.0016)	(0.0015)	(0.0008)
γ_2	−0.0025	−0.0042**	
	(0.0016)	(0.0016)	
γ_3		−0.0048**	
		(0.0017)	
$\gamma_2 = \gamma_3$			−0.0048***
			(0.0014)
σ	0.2103***	0.2033***	0.2043***
	(0.0172)	(0.0166)	(0.0167)
Num.Obs.	194	194	194

The formulas for the different models are, in terms of response and covariates only `y ~ x_1 + x_2` for the GH model, `y ~ x_1 + x_2 + x_3` for the full model and `y ~ x_1 + I(x_2 + x_3)` for the short model. To construct the relevant two-part formulas for the IV estimation, we use the **`Formula::as.Formula`** function that enables to construct a two-part formula, using a standard formula and a one-sided formula. As an example:

```
Formula::as.Formula(y ~ x1 + x2, ~ z1 + z2 + z3)
## y ~ x1 + x2 | z1 + z2 + z3
```

We then proceed to the estimation of the three models, using **`micsr::tobit1`** with the **`twostep`** method, as in the original paper:

```
GH <- tobit1(Formula::as.Formula(y ~ x1 + x2, inst),
             trade_protection, method = "twostep")
Short <- tobit1(Formula::as.Formula(y ~ x1 + I(x2 + x3), inst),
                trade_protection, method = "twostep")
Full <- tobit1(Formula::as.Formula(y ~ x1 + x2 + x3, inst),
               trade_protection, method = "twostep")
```

The results are presented in Table 11.2.[9] As in the original article, we use the name of the coefficients and not the name of the terms, which are listed below:

[9]See Matschke and Sherlund (2006), table 3, page 417.

```
names(coef(Full))
## [1] "(Intercept)" "x1"          "x2"          "x3"
## [5] "rho_x1"      "rho_x2"      "rho_x3"      "sigma"
```

For example, **x2** is replaced by β_2 and **rho_x2** by γ_2. Tests of exogeneity for the three models give:

```
endogtest(Formula::as.Formula(y ~ x1 + x2, inst),
          trade_protection, model = "tobit") %>% gaze
## chisq = 3.938, df: 2, pval = 0.140
endogtest(Formula::as.Formula(y ~ x1 + I(x2 + x3), inst),
          trade_protection, model = "tobit") %>% gaze
## chisq = 13.957, df: 2, pval = 0.001
endogtest(Formula::as.Formula(y ~ x1 + x2 + x3, inst),
          trade_protection, model = "tobit") %>% gaze
## chisq = 12.773, df: 3, pval = 0.005
```

The exogeneity hypothesis is not rejected for the GH model, but it is for the two versions of Matschke and Sherlund's model. The GH model, which is a special case of the Full model is relevant if the hypotheses $\beta_3 = \gamma_3 = 0$ are not rejected. Performing a Wald test, we get:

```
lmtest::waldtest(Full, GH) %>% gaze
## Chisq = 8.441, df: 2, pval = 0.015
```

and the hypothesis is therefore rejected at the 5% (but not at the 1%) level. Then, the hypothesis that $\beta_2 = \beta_3$ and $\gamma_2 = \gamma_3$ implied by Matschke and Sherlund's model can be tested:

```
car::linearHypothesis(Full, c("x2 = x3", "rho_x2 = rho_x3")) %>% gaze
## Chisq = 1.540, df: 2, pval = 0.463
```

and is not rejected.

11.6 Tobit-2 model

Tobit-2 models are bivariate models, the first equation being the **selection equation** and the second the **outcome equation**. The model can be written as follow:

$$\left\{ \begin{array}{rcl} y_1^* &=& \alpha_1 + \beta_1^\top x_1 + \epsilon_1 = \mu_1 + \epsilon_1 \\ y_2^* &=& \alpha_1 + \beta_2^\top x_2 + \epsilon_2 = \mu_2 + \epsilon_2 \end{array} \right.$$

and the observation rule is:

$$\left\{ \begin{array}{rcl} y &=& y_2^* \text{ if } y_1^* > 0 \\ y &=& 0 \text{ if } y_1^* \leq 0 \end{array} \right.$$

Note that the tobit-2 model reduces to the tobit-1 model if $x_1 = x_2$ and $\beta_1 = \beta_2$. In the outcome equation, y_2^* can be replaced by $\ln y_2^*$, so that predicted values of y_2 are necessarily positive. The tobit-2 model is more general than the tobit-1 model, as it allows the economic mechanisms that explain the fact that the response is observed to be different from those that explain the value of the response. The seminal papers about the tobit-2 model are Gronau (1973) with his model of female labor supply, and Heckman (1976, 1979) who precisely described the statistical properties of this model and proposed a two-step estimator. Hurdle models, proposed by Cragg (1971), can also be considered as tobit-2 models. These models describe the level of consumption of a good as a two-step process: first, the good should be selected (selection equation) and then the level of the consumption is set (outcome equation). The tobit-2 model is also widely used to measure the treatment effect of public programs. If the selection of individuals in a particular program is not purely random, the selection equation describes as a binomial model the process of getting hired in the program and the outcome equation measures the effectiveness of the program, e.g., the wage one year after leaving the program.

11.6.1 Two-part models

With the assumption of uncorrelation between ϵ_1 and ϵ_2 (which means uncorrelation of the unobserved part of the two responses), the tobit-2 model is called the **two-part** model, and the two equations can be estimated independently, the first by any binomial model (very frequently but not necessarily a probit) with the response equal to 0/1 for positive/negative values of y_1^* and the second by least squares (with either y_2 or $\ln y_2$ as the response). The main advantage of this model is that it can be estimated without further hypothesis about the conditional distribution of y_2 and can therefore be viewed as a semi-parametric estimator, which would be consistent in a wide variety of contexts (e.g., heteroskedasticity and non-normality).

11.6.2 Hurdle models

Hurdle models, proposed by Cragg (1971), share with the two-part models the fact that the errors of the two equations are uncorrelated. Cragg proposed three flavors of hurdle models:

- simple hurdle models with a log-normal or a truncated normal distribution for the outcome response,
- double hurdle models with a normal distribution for the outcome response.

The **log-normal simple hurdle** model is a two-part model for which the outcome equation consists of a linear regression of the logarithm of the outcome equation on the set of covariates x_2, which is consistent even without the hypothesis of normality and homoskedasticity.

The **truncated normal simple hurdle** model shares with two-part models the fact that the two equations can be estimated independently, but the outcome equation is estimated by maximum likelihood and therefore the consistency of the estimator relies on the hypothesis of normality and homoskedasticity.

Finally, the **normal double hurdle model** is not per se a tobit-2 model because zero observations may appear not only because $y_1^* < 0$, but also because $y_2^* < 0$. If the response is the expenditure for a given good, it is positive only if the good is selected by the household ($y_1^* > 0$) and if the solution of the consumer problem is not a corner solution ($y_2^* > 0$).

For this latter model, we have $\mathrm{P}(y_1^* > 0) = 1 - \Phi(-\mu_1) = \Phi(\mu_1)$, $\mathrm{P}(y_2^* > 0) = 1 - \Phi(-\mu_2/\sigma) = \Phi(\mu_2/\sigma)$, so that, given the hypothesis of independence of the two errors:

$$\mathrm{P}(y > 0) = \Phi(\mu_1)\Phi(\mu_2/\sigma)$$

The density of y for positive values of y is:

$$f(y \mid x_2, y > 0) = \frac{1}{\sigma}\frac{\phi\left(\frac{y-\mu_2}{\sigma}\right)}{\Phi\left(\frac{y-\mu_2}{\sigma}\right)}$$

So that finally, the likelihood is:

$$\begin{array}{rcl} L^{2H}(\gamma_1, \gamma_2, \sigma | y, X) & = & \prod_{n=1}^{N_o}\left[1 - \Phi(\mu_{n1})\Phi(\mu_{n2}/\sigma)\right]\prod_{n=N_o+1}^{N}\Phi(\mu_{n2}/\sigma)\Phi(\mu_{n1}) \\ & \times & \prod_{n=N_o+1}^{N}\frac{1}{\sigma}\frac{\phi(\frac{y_n-\mu_{n2}}{\sigma})}{\Phi(\mu_{n2}/\sigma)} \end{array}$$

This expression is very similar to Equation 11.5 which indicates that the likelihood for the tobit-1 model is the product of:

- the likelihood of a probit model which explains that $y = 0$ or $y > 0$,
- the likelihood of y for the truncated sample.

The second term is exactly the same, but the first one is different, as the probability of a positive value of y is now $\Phi(\mu_2/\sigma)\Phi(\mu_1)$. The likelihood can be simplified as:

$$L^{2H}(\gamma_1, \gamma_2, \sigma | y, X) = \prod_{n=1}^{N_o}\left[1 - \Phi(\mu_{n1})\Phi(\mu_{n2}/\sigma)\right]\prod_{n=N_o+1}^{N}\frac{1}{\sigma}\Phi(\mu_{n1})\phi\left(\frac{y_n - \mu_{2n}}{\sigma}\right)$$

Hurdle models can be estimated using the **mhurdle** package (Carlevaro and Croissant 2023).

11.6.3 Correlated models

Finally, we consider the case where the errors of the two equations are correlated. In this case, the model should be fully parametrized and a natural way to do it is to suppose that y_1^* and y_2^* (or $\ln y_2^*$) follow a bivariate normal distribution. The variance of y_1^* is arbitrarily set to 1, as only the sign of y_1^* is observed. Therefore, we can write the system of two equations as:

$$\left\{\begin{array}{rcl} y_1^* & = & \gamma_1^\top z_1 + \epsilon_1 = \mu_1 + u_1 \\ y_2^* & = & \gamma_2^\top z_2 + \epsilon_2 = \mu_2 + \sigma u_2 \end{array}\right.$$

where u_1 and u_2 are a couple of standard normal deviates. Then the joint distribution of the two latent variables is:

$$\left(\begin{array}{c} y_1^* \\ y_2^* \end{array}\right) \sim \mathcal{N}\left(\left(\begin{array}{c} \mu_1 \\ \mu_2 \end{array}\right), \left(\begin{array}{cc} 1 & \rho\sigma \\ \rho\sigma & \sigma^2 \end{array}\right)\right)$$

Two properties of the bivariate normal distribution should be remembered. For a couple of standard normal deviates (u_1 and u_2), the joint normal density can be written as the product of the marginal density of u_1 and the conditional density of u_2:

$$\phi_b(u_1, u_2, \rho) = \frac{1}{2\pi\sqrt{1-\rho^2}} e^{-\frac{1}{2}\left(\frac{u_1^2+u_2^2-2\rho u_1 u_2}{1-\rho^2}\right)} = \phi(u_1)\frac{1}{\sqrt{1-\rho^2}}\phi\left(\frac{u_2-\rho u_1}{\sqrt{1-\rho^2}}\right) \quad (11.10)$$

Then, the expectation of u_2 for u_1 left-truncated at l is:

$$\begin{aligned}
\mathrm{E}(u_2 \mid u_1 > l) &= \frac{\int_l^{+\infty}\int_{-\infty}^{+\infty} u_2\phi_b(u_1, u_2, \rho) du_1 du_2}{\int_l^{+\infty}\int_{-\infty}^{+\infty} \phi_b(u_1, u_2, \rho) du_1 du_2} \\
&= \frac{\int_l^{+\infty}\left[\int_{-\infty}^{+\infty} u_2\phi\left(\frac{u_2-\rho u_1}{\sqrt{1-\rho^2}}\right) du_2\right]\phi(u_1)du_1}{\sqrt{1-\rho^2}(1-\Phi(l))} \\
&= \frac{\int_l^{+\infty}\left[\int_{-\infty}^{+\infty}(v\sqrt{1-\rho^2}+\rho u_1)\phi(v)dv\right]\phi(u_1)du_1}{1-\Phi(l)} \\
&= \frac{\rho\int_l^{+\infty}\left[\int_{-\infty}^{+\infty}\phi(v)dv\right]u_1\phi(u_1)du_1}{1-\Phi(l)} \\
&= \rho\frac{\phi(l)}{1-\Phi(l)}
\end{aligned} \quad (11.11)$$

y_2 is observed if $y_1^* > 0$ or, equivalently, if $u_1 > -\mu_1$. As u_1 follows a standard normal distribution, the probability that y_2 is observed is:

$$\mathrm{P}(y_1^* > 0) = \mathrm{P}(u_1^* > -\mu_1) = 1 - \Phi(-\mu_1) = \Phi(\mu_1) \quad (11.12)$$

Denote $f(y_2)$ the marginal distribution of y_2: it is obtained by integrating out the joint distribution of y_1^* and y_2^* for all positive values of y_1^*. Using Equation 11.10 and Equation 11.12:

$$\begin{aligned}
f(y_2) &= \frac{1}{\sigma}\frac{\int_0^{+\infty}\phi_b(y_1^*-\mu_1,(y_2-\mu_2)/\sigma))dy_1^*}{\int_0^{+\infty}\phi(y_1^*-\mu_1)dy_1^*} \\
&= \frac{1}{\sigma}\frac{\int_{-\mu_1}^{+\infty}\phi_b(u_1,(y_2-\mu_2)/\sigma))du_1}{1-\Phi(-\gamma_1^\top z_1)} \\
&= \frac{\int_{-\mu_1}^{+\infty}\phi\left(\frac{u_1-\rho(y_2-\mu_2)/\sigma}{\sqrt{1-\rho^2}}\right)\phi(\frac{y_2-\mu_2}{\sigma})}{\sigma\sqrt{1-\rho^2}\Phi(\mu_1)} \\
&= \frac{\Phi\left(\frac{\mu_1+\rho(y_2-\mu_2)/\sigma}{\sqrt{1-\rho^2}}\right)}{\sigma\sqrt{1-\rho^2}\Phi(\mu_1)}\phi\left(\frac{y_2-\mu_2}{\sigma}\right)
\end{aligned} \quad (11.13)$$

The contribution of an observation to the likelihood is either $P(y_1^* < 0)$ (one minus the probability given by Equation 11.12) if y_2 is not observed and $P(y_1^* > 0)$ (Equation 11.12) times the density of y_2 (Equation 11.13) if it is; which leads to the following likelihood function:

$$L(\theta|y, X) = \prod_{n=1}^{N_o}[1-\Phi(\mu_{n1})]\prod_{n=N_o+1}^{N}\frac{1}{\sigma\sqrt{1-\rho^2}}\Phi\left(\frac{\mu_{n1}+\rho\frac{y_n-\mu_{n2}}{\sigma}}{\sqrt{1-\rho^2}}\right)\phi\left(\frac{y_n-\mu_{n2}}{\sigma}\right)$$

Consider now the conditional expectation of y if it is observed, i.e., if $y_1^* > 0$:

$$\mathrm{E}(y \mid x_2, y > 0) = \mathrm{E}(\mu_2 + \sigma_\epsilon u_2 \mid x_2, u_1^* > -\mu_1) = \mu_2 + \sigma_\epsilon \mathrm{E}(u_2 \mid u_1 > -\mu_1)$$

From Equation 11.11, the last term is just $\rho\frac{\phi(-\mu_1)}{1-\Phi(-\mu_1)}$, so that:

$$\mathrm{E}(y \mid x_2, x_1, y > 0) = \mu_2 + \sigma\rho\frac{\phi(\mu_1)}{\Phi(\mu_1)} = \gamma_2^\top z_2 + \sigma\rho r(\mu_1) \tag{11.14}$$

As for the tobit-1 model, the conditional expectation is not equal to $\mu_2 = \gamma^\top z_2$ because of the supplementary term $\sigma\rho(\gamma_1^\top x_1)$. The linear estimator is therefore biased if the omitted variable if $r(\gamma_1^\top z_1)$ is correlated with x_2, which is obviously the case if there are common covariates in x_1 and x_2. Equation 11.14 also directly leads to an alternative estimator. This is a two-step estimator, sometimes called the **heckit** estimator. It consists of first regressing $\mathbf{1}(y_1^* > 0)$ on x_1 using a probit, and then estimating $r(\gamma^\top z_1)$ by $r(\hat{\gamma}^\top z_1)$, $\hat{\gamma}^\top z_1$ being the linear predictor of the probit model. In a second step, regressing y_2 on x_2 and the supplementary covariate $r(\hat{\gamma}_1^\top z_1)$ leads to a consistent estimate of γ_2.

As previously seen (Figure 11.3), $r(z)$ is almost linear in z for a wide range of values of z. Therefore, if $x_1 = x_2$, the coefficients of the second steps of the heckit estimator are only identified by the non-linearity of r. The high correlation between x_2 and $r(\hat{\gamma}_1^\top z_1)$ will lead in this case to very imprecise estimates. It is therefore recommended to impose exclusion conditions, i.e., to exclude at least one covariate of the x_1 set for the second step of the estimator. This (or these) excluded covariates must be relevant for the selection equation, but not for the outcome equation. In practice, it is often difficult to provide theoretical bases to such exclusion conditions.

The relative merits of the two-part and the heckit models have been discussed in numerous articles, especially in the field of health economics. The arguments are presented in Jones (2000) and Dow and Norton (2003).

11.6.4 Application

Makowsky and Stratmann (2009) analyze the behavior of policemen in terms of issuing speeding tickets, because of excessive speed. A policeman has to make two decisions:

- whether to give a ticket or not,
- if a ticket is given, its amount should be set, with the recommendation of applying the following formula: \$50 + (speed - (max speed allowed + 10))

The authors make the hypotheses that a policeman's behavior will depend on their institutional and political environment. First, there are two kinds of policemen, belonging to the municipality or to the state of Massachusetts. Municipality agents are headed by a chief who is nominated by the municipal authorities and can be fired at any time. One can suppose, that contrary to the state policemen, these local policemen will act in the interest of the local authorities, for which the fees policy has two aspects:

- fees are a source of income, which can be particularly important for municipalities with budget problems,
- fees given to drivers from the municipality can make them unhappy and unwilling to vote for the current authorities, which is not an issue for drivers from other municipalities.

The data set, called `traffic_citations` includes:

- two responses: `fine` which indicates whether a ticket has been given or not, and `amount` which is the amount of the fine when it has been issued,
- covariates that describe local fiscal conditions: `prval` is the average property value in the municipality, which is the base of the main local tax, and `overloss` which indicates that an override referendum (which indicates that the municipality anticipates insufficient revenues) fails to pass,
- characteristics of the offender: `ethn` indicates the ethnicity of the driver (`"other"`, `"hispanic"` or `"black"`), the sex is indicated by a dummy `female` and `res` indicates whether the driver lives in the municipality (`"loc"`), in another municipality of the state (`"oto"`) or in another State (`"ost"`), `courtdist` is the distance to the courts where the driver can appeal the citation, `mph` is the difference between the speed of the driver and the legal limit and `cdl` is a dummy for a commercial driver's license.
- characteristics of the policemen: `stpol` is a dummy for state officers and is therefore 0 for local policemen.

We start by defining the selection and the outcome equations. The authors suppose that the probability of receiving a fine depends on the difference between the actual and the limit speed, the ethnicity, the sex and the age of the offender, the fact that the driver's license is commercial and covariates describing local fiscal conditions and place of residence of the offender in interaction with the dummy for state officers. As the log of age is introduced in interaction with the `female` dummy, we divide the age by the sample mean so that the coefficient of `female` is the effect for the mean age. For the outcome equation, the set of covariates is the same except that the dummy for commercial driver's license is removed:

```
traffic_citations <- traffic_citations %>% mutate(age = age / 35)
sel <- fine ~ log(mph) + ethn + female * log(age) +
    stpol * (res + log(prval) + oloss) + cdl
out <- update(sel, log(amount) ~ . - cdl)
traffic_citations <- traffic_citations %>% mutate(locpol = 1 - stpol)
sel_c <- fine ~ log(mph) + ethn + female * log(age) +
    locpol : (res + log(prval) + oloss) + cdl + locpol
out_c <- update(sel_c, log(amount) ~ . - cdl)
```

Therefore, the identification is performed using the hypothesis that `cdl` enters the selection equation, but not the outcome equation. The authors justify this hypothesis by the fact that a fine received by drivers with a commercial license may have important consequences for them, because the accumulation of points may lead to the suspension of their license and can cause the loss of their employment and affect their future income. Once the fine has been issued, there is no clear incentive for the officer to impose a lower fine (Makowsky and Stratmann 2009, 515–16).

The hypothesis is that local policemen have an incentive to fine when their town experiences fiscal problems and when the offender is not a resident of the town. Therefore, the coefficients on `prval`, `overloss`, `locoto` and `locost` should be negative for the first one and positive for the other three. Interacting these covariates with the dummy for state officers, an opposite sign should be observed. We can then compute the heckit estimator. First, we estimate the selection equation using a probit and we estimate the inverse mills ratio:

```
probit <- glm(sel, data = traffic_citations,
              family = binomial(link='probit'))
lp <- probit$linear.predictor
mls <- dnorm(lp) / pnorm(lp)
```

We then compute the second stage of the estimator by using OLS to fit the outcome equation, using the estimation of the inverse mills ratio as a supplementary covariate:

```
lm <- lm(out, traffic_citations)
heck <- update(lm, . ~ . + mls)
```

The **sampleSelection** package (Toomet and Henningsen 2008) is devoted to the estimation of sample selection models. The `sampleSelection::selection` function performs the estimation of such models; two formulas should be provided for the selection and for the outcome equations. The method of estimation is indicated using the `method` argument which could be either `"2step"` and `"ml"` respectively for the two-step and the maximum likelihood estimator. The results are presented in Table 11.3.

```
library(sampleSelection)
tc_heck <- selection(sel, out, data = traffic_citations,
                     method = "2step")
tc_ml <- update(tc_heck, method = "ml")
```

Remember that for the probit, the marginal effect is the coefficient times the probability that a fine has been issued. As the mean value of `fine` is close to 0.5, we can multiply probit's coefficients by 0.4 as an estimate of the marginal effect at the sample mean. Being a female reduces the probability of having a fine by about 8%. The probability is higher for Hispanic drivers (about 12%), there is not such effect for black drivers. State policemen are less severe than local policemen in terms of fine issuing, but when the fine is issuing; the amount is the same as the one set by local policemen. The coefficients of `log(prval)` and `oloss` have the expected sign and are highly significant. Moreover, as expected, the coefficients of these covariates in interaction with state policemen have the opposite sign and are significant. Out-of-town and out-of-state offenders have a higher probability to receive a fine and, if it is the case, the average amount is higher than the one for local offenders. The interaction term with the state policemen dummy is not significant for out-of-town offenders and is positive for out-of-state offenders. Finally, as expected, the probability of receiving a fine is lower for drivers with a commercial license. In the two-step model, the inverse mills ratio is positive and highly significant. The implied coefficient of correlation is 0.506 and the estimated value using ML is 0.360. Therefore, we can conclude that the unobserved parts of the selection and the outcome equations are positively correlated.

Table 11.3: Traffic citations

	maximum likelihood			heckit		
	selection	outcome	auxiliary	selection	outcome	auxiliary
(Intercept)	−0.091	2.246***		−0.005	2.231***	
	(0.148)	(0.056)		(0.148)	(0.057)	
log(mph)	1.654***	0.959***		1.615***	1.005***	
	(0.019)	(0.008)		(0.019)	(0.015)	
ethnhispanic	0.296***	0.035***		0.291***	0.043***	
	(0.029)	(0.009)		(0.029)	(0.009)	
ethnblack	−0.006	−0.022**		−0.009	−0.022*	
	(0.026)	(0.009)		(0.026)	(0.009)	
female	−0.193***	−0.036***		−0.194***	−0.042***	
	(0.011)	(0.004)		(0.011)	(0.005)	
log(age)	−0.412***	−0.023***		−0.407***	−0.035***	
	(0.018)	(0.006)		(0.018)	(0.007)	
stpol	−1.736***	−0.119		−1.548***	−0.213*	
	(0.294)	(0.087)		(0.295)	(0.092)	
resoto	0.275***	0.037***		0.276***	0.047***	
	(0.014)	(0.006)		(0.014)	(0.007)	
resost	0.532***	0.121***		0.533***	0.139***	
	(0.022)	(0.009)		(0.022)	(0.010)	
log(prval)	−0.436***	−0.035***		−0.434***	−0.050***	
	(0.012)	(0.005)		(0.012)	(0.007)	
oloss	0.672***	0.077***		0.670***	0.098***	
	(0.040)	(0.014)		(0.040)	(0.015)	
cdl	−0.293***			−0.315***		
	(0.032)			(0.033)		
female × log(age)	0.194***	0.016		0.188***	0.020+	
	(0.030)	(0.011)		(0.030)	(0.011)	
stpol × resoto	0.021	0.034*		0.017	0.032*	
	(0.046)	(0.014)		(0.046)	(0.014)	
stpol × resost	0.164**	0.038*		0.166**	0.036*	
	(0.052)	(0.016)		(0.052)	(0.016)	
stpol × log(prval)	0.252***	0.019*		0.235***	0.030***	
	(0.026)	(0.008)		(0.027)	(0.009)	
stpol × oloss	−0.325*	−0.054+		−0.346**	−0.072*	
	(0.134)	(0.028)		(0.134)	(0.029)	
sigma			0.338***			0.349
			(0.002)			
rho			0.360***			0.506
			(0.019)			
invMillsRatio						0.177***
						(0.017)
Num.Obs.	68 357			68 357		
R2				0.445		
R2 Adj.				0.445		
AIC	90 598.4					
BIC	90 918.0					
RMSE	0.34			0.33		

12

Count data

It is often the case in economics that the response is a count, i.e., a non-negative integer. In this case, fitting a linear model has a number of drawbacks:

- the integer nature of the response is not taken into account,
- the fitted model can predict negative values for some values of the covariates,
- if the distribution of the response is asymmetric, the logarithm transformation can't be used in the common case where the response is 0 for a subset of observations.

There is therefore the need for specific models that will be presented in this chapter. In Section 12.1, we'll illustrate the features of count data, using a survey of empirical studies. Section 12.2 is devoted to the benchmark count model, the Poisson model. Section 12.3 discusses two common features of count data: overdispersion and excess of zero. Finally Section 12.4 deals with the endogeneity problem with a count response.

12.1 Features of count data

12.1.1 An empirical survey

We start with the presentation of some data sets for which the response is a count. All these data are available in the **micsr.count** package and are summarized in Table 12.1.

About half of these data sets concern demand for health care. `doctor_aus` (Cameron and Trivedi 1986), `doctor_cal` (Gurmu 1997), `doctor_ger` (Santos Silva and Windmeijer 2001) and `doctor_us` (Mullahy 1998) concern annual doctor visits, respectively in Australia, California, Germany and the United States. `health_ins` (Deb and Trivedi 2002) focuses on the link between health insurance and doctor visits using a randomized experiment in the United States, as `health_reform` (Winkelmann 2004) analyzes the effect of a health insurance reform in Germany. `elderly` (Deb and Trivedi 1997) focus on the demand for health care by elderly in the US and `hospitalization` (Geil et al. 1997) uses as the response the number of stays in the hospital.

The response of `amc12` (Ellison and Swanson 2016) is the number of students per school who got a very high score at the AMC12 test (a test in mathematics). `asymptotic` (Koenker and Zeileis 2009) is a meta-analysis of wage equations reported in 156 papers and the analysis stresses on the hypothesis used to get the consistency of an estimator, i.e., that the number of regressors is fixed as the number of observations increases. `bids` (Cameron and Johansson 1997) analyzes the number of bids received by 126 firms that were targets of tender offers. `cartels` (Miller 2009) is a time series of bi-annual observations, the response being the number of cartels discoveries and the main covariate the pre- and post-period when

Table 12.1: Empirical survey of count data sets

data	μ	σ^2	$\frac{\sigma^2}{\mu}$	D9	V5	$P(Y=0)$	$\hat{P}(Y=0)$	$\frac{P(Y=0)}{\hat{P}(Y=0)}$
amc12	1	3.5	4.4	2.0	4.36	0.66	0.6	1.1
asymptotic	15	519.3	34.5	30.0	34.46	0.00	0.0	0.0
bids	2	2.1	1.2	3.0	1.18	0.07	0.2	0.3
cartels	5	7.8	1.5	9.1	1.51	0.03	0.0	1.6
cigmales	8	181.5	22.0	30.0	22.00	0.63	0.0	17.7
doctor.aus	0	0.6	2.1	1.0	2.11	0.80	0.8	1.0
doctor.cal	2	11.1	5.0	6.0	4.98	0.33	0.2	2.0
doctor.ger	1	11.3	9.2	3.0	9.16	0.68	0.4	1.7
doctor.us	5	163.9	33.4	12.0	33.38	0.24	0.0	6.0
elderly	6	45.7	7.9	13.0	7.91	0.16	0.0	12.7
health.ins	3	20.3	7.1	7.0	7.09	0.31	0.1	2.9
health.reform	3	18.1	7.1	6.0	7.15	0.35	0.2	2.2
hospitalization	0	0.4	3.3	1.0	3.32	0.90	0.9	1.0
majordrg	0	0.2	1.3	1.0	1.34	0.89	0.9	1.0
publications	2	3.7	2.2	4.0	2.19	0.30	0.2	1.4
somerville	2	39.6	17.6	6.0	17.64	0.63	0.4	1.5
strikes	8	16.1	2.1	13.0	2.06	0.01	0.0	8.8
trips	5	24.4	5.4	11.0	5.35	0.19	0.1	2.1

a new leniency program was introduced. `cigmales` (Mullahy 1997) measures the number of cigarettes smoked daily by males in the United States. `majordrg` (Greene 2001) contain data about the number of major derogatory reports by cardholders. `publications` (Long 1997) analyzes the scientific production (measured by the number of publications) of individuals who got a PhD in biochemistry. `somerville` (Seller, Stoll, and Chavas 1985) reports the number of recreational visits to a lake in Texas. `strikes` (Kennan 1985) is a time series that reports the number of strikes (and also their length) in the United States. `trips` (Terza 1998) contains the number of trips taken by members of households the day prior the survey interview.

As we have seen in Section 5.1.1, count data can be considered as a random variable that follows a Poisson distribution, which has a unique parameter, this parameter being the mean and the variance of the series. The ML estimator of this parameter is the sample mean, which is indicated in the first column (μ) of Table 12.1. The empirical variance (ω) is reported in the second column. Count data often take small values, and this is the case for almost all our data sets. We report in Table 12.1 the 9th decile, which is less than 10 for a majority of the data sets. Noticeable exceptions are `asymptotic` (the number of covariates in an econometric equation) and `cigmales` (the number of cigarettes smoked daily).

While comparing real count data to Poisson distribution, one often faces two problems:

- **overdispersion**, which means that the variance is much greater than the mean,
- **excess of zero**, which means that the mean probability of a zero value computed using the Poisson distribution is often much less than the observed share of zero in the sample.

The third column of Table 12.1 contains the mean variance ratio and shows that overdispersion seems to be the rule. The ratio is in general much larger than 1, noticeable exceptions being `bids`, `cartels` and `majordrg`. This overdispersion is important because we consider that the Poisson parameter is the same for every observation. Adding covariates will give specific values of the Poisson parameter for every observations and will therefore reduce the (conditional) variance. Anyway, we'll see in subsequent sections that the overdispersion problem is often still present even after introducing covariates, because of unobserved heterogeneity.

The excess of zero problem is difficult to distinguish from overdispersion, as zero is an extreme value of the distribution and therefore, an excess of zero leads to overdispersion. The excess of zero can in part be explained by the fact that 0 is a special value that may not be correctly explained by the same process that explains strictly positive values. An example is `cigmales`, where 0 means no smoking, and smoking vs. non-smoking is quite different in nature than smoking a few or a lot of cigarettes. The eighth column of Table 12.1 returns the ratio between the empirical share of zero and the Poisson probability of 0 and, unsurprisingly, this ratio is huge for `cigmales`.

12.1.2 Analyzing the number of trips

Throughout this chapter, we'll use the `trips` data set. It was used by Terza (1998) and the response is the number of trips taken the day before the interview. We first consider the unconditional distribution of the number of `trips`, computing the first two moments and the share of 0:

```
trips %>% summarise(m = mean(trips), v = var(trips),
                    s0 = mean(trips == 0))
## # A tibble: 1 x 3
##       m     v    s0
##   <dbl> <dbl> <dbl>
## 1  4.55  24.4 0.185
```

This series exhibits important overdispersion, as the variance is about 5 times larger than the mean. The ML estimator is the sample mean, and therefore, the predicted probability of 0 for the Poisson model is:

```
dpois(0, 4.551)
## [1] 0.01056
```

which is much lower than the actual percentage of 0 which is 18.5%. There is therefore a large excess of 0, which also can be associated with overdispersion as 0 is a value far from the mean of 4.551. The **topmodels** package[1] provides an interesting plotting function called `rootogram` to compare the actual and the fitted distribution of a count. The first is represented by bars and the second by a curve. The most simple representation is obtained by setting the `style` and the `scale` parameters respectively to `"standing"` and `"raw"` (see Figure 12.1). With the first, the bottom of all the bars is on the horizontal axes; with the second one, the absolute frequencies are displayed on the vertical axes. For example, for

[1]The **topmodels** package is not on CRAN; it can be installed using `install.packages("topmodels", repos="http://R-Forge.R-project.org")`.

$Y = 0$, the absolute frequency is 107 and, as seen previously, the fitted probability is 0.01056 which implies, as the number of observations is 577, a fitted frequency of $577 \times 0.01056 = 6.1$. The frequencies are clearly over-estimated for values of the response close to the mean (3, 4, 5, 6) and under-estimated for extreme values. This is particularly the case for 0, which indicates that a specific problem of excess of zero may be present.[2]

```
uncond_trips <- glm(trips ~ 1, data = trips, family = "poisson")
topmodels::rootogram(uncond_trips, plot = FALSE, breaks = (0:16) - 0.5,
                     style = "standing", scale = "raw", confint = FALSE) %>%
  autoplot + coord_cartesian(xlim = c(0, 15))
```

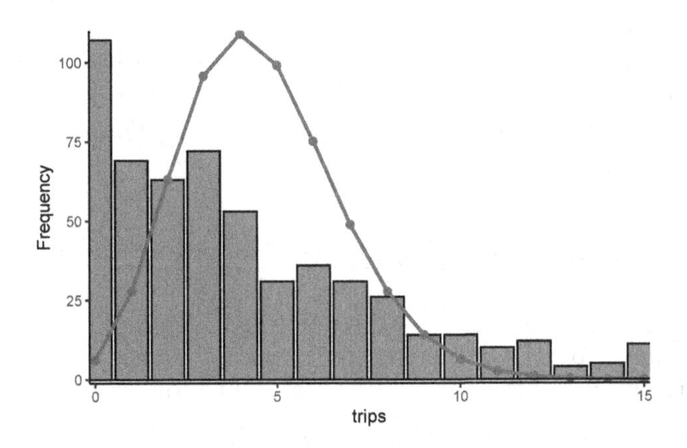

Figure 12.1: Unconditional distribution of trips

12.2 Poisson model

12.2.1 Derivation of the Poisson model

The basic count data model is the Poisson model. It relies on the hypothesis that the response is a Poisson variable and therefore that the probability of a given value y is:

$$P(Y = y, \mu) = \frac{e^{-\mu}\mu^y}{y!}$$

The Poisson distribution is defined by a unique parameter μ, which is the mean and the variance of the distribution. This is a striking feature of the Poisson distribution which differs for example from the normal distribution which has two separate position and dispersion parameters (respectively the mean and the standard deviation). The Poisson probability can also be written as:

$$P(Y = y, \mu) = e^{y \ln \mu - \mu - \ln y!}$$

[2]Note the use of `ggplot2::autoplot` that coerces a graphical object to a `ggplot` object.

and therefore, the Poisson model belongs to the generalized linear model family (see Section 10.3), with $\theta = \ln\mu$, $b(\theta) = \mu = e^{\theta}$, $\phi = 1$ and $c(y,\phi) = -\ln y!$. The mean is $b'(\theta) = e^{\theta} = \mu$ and the variance $\phi b''(\theta) = e^{\theta} = \mu$. The link defines the relation between the linear predictor η and the parameter of the distribution μ: $\eta = g(\mu)$. The canonical link is such that $\theta = \eta$ and it is therefore, for the Poisson model, the log link: $\eta = \ln\mu$. For a given set of covariates, the Poisson parameter for observation n is $\mu_n = g^{-1}(\eta_n)$, with $\eta_n = \alpha + \beta^{\top} x_n = \gamma^{\top} z_n$ the linear predictor. Therefore:

$$\mu_n = \mathrm{E}(y \mid x_n) = e^{\alpha+\beta^{\top}x_n} = e^{\gamma^{\top}z_n} \tag{12.1}$$

and the log-likelihood function is:

$$\ln L(\gamma, y, Z) = \sum_{n=1}^{N} \ln \frac{e^{-e^{\gamma^{\top}z_n}} e^{y_n\gamma^{\top}z_n}}{y_n!} = \sum_{n=1}^{N} -e^{\gamma^{\top}z_n} + y_n\gamma^{\top}z_n - \ln y_n!$$

The first-order conditions for a maximum are:

$$\frac{\partial \ln L}{\partial \gamma} = \sum_{n=1}^{N} \left(y_n - e^{\gamma^{\top}z_n}\right) z_n = 0 \tag{12.2}$$

which states, as in the OLS model, that the difference between the actual value of the response y and the prediction of the model (which can be called the residual) should be orthogonal to every covariate. It is a very important property, as it implies that the Poisson model is consistent if the conditional expectation function is correctly specified, whether or not the hypothesis of the Poisson distribution is correct. Moreover, as the first element of z_n is 1, the first element of Equation 12.2 implies that:

$$\bar{y} = \frac{1}{N}\sum_{n=1}^{N} e^{\hat{\gamma}^{\top}z_n} \tag{12.3}$$

which means that the mean of the predictions equals the mean value of y in the sample.

For the `trips` data set, the covariates are the share of trips for work or school (`workschl`), the number of individuals in the household (`size`), the distance to the central business district (`dist`), a dummy (`smsa`) for large urban area, the number of full-time workers in a household (`fulltime`), the distance from home to the nearest transit node (`distnode`), household income divided by the median income of the census tract (`realinc`), a dummy if the survey period is Saturday or Sunday (`weekend`) and a dummy for owning at least one car (`car`). The Poisson model is fitted using the `glm` function and setting the `family` argument to `poisson`. As for the binomial model, the `family` argument can be a character, a function without argument (in this case, the default link is chosen) or a function with a link argument.[3] For the sake of comparison, we also estimate a linear model using `lm`:

```
ftrips <- trips ~ workschl + size + dist + smsa + fulltime + distnod +
  realinc + weekend + car
```

[3]The other possible other links for the Poisson model are `"identity"` and `"sqrt"`, but the default `"log"` link is almost always used.

```
pois_trips <- glm(ftrips, family = poisson, trips)
lm_trips <- lm(ftrips, trips)
```

12.2.2 Interpretation of the coefficients

As for any non-linear models, the marginal effects are not equal to the coefficients. Taking the derivative of the conditional expectation function (Equation 12.1) with the kth covariate, we get the following marginal effect for observation n:

$$\frac{\partial \mathrm{E}(y \mid x_n)}{\partial x_{nk}} = \beta_k e^{\gamma^\top z_n}$$

which can be consistently estimated by $\hat{\beta}_k e^{\hat{\gamma}^\top z_n}$. Note that the ratio of the marginal effects of two covariates k and l is β_k/β_l and therefore doesn't depend on the values of the covariates for a specific observation. The average marginal effect (AME) for the N observations is:

$$\hat{\beta}_k \frac{\sum_{n=1}^N e^{\hat{\gamma}^\top z_n}}{N}$$

But, if the regression contains an intercept, the sample mean of the predictions is the sample mean of the response (see Equation 12.3). Therefore, the average marginal effect is $\hat{\beta}_k\bar{y}$. Therefore, we can for example compare the estimate slopes in a linear model (which are directly the marginal effects) to the Poisson estimators multiplied by the mean of the response.

We now print the coefficients obtained for the Poisson and for the linear model:

```
rbind(lm = coef(lm_trips), poisson = coef(pois_trips))
```

```
        (Intercept) workschl   size      dist     smsa fulltime
lm          -0.8799  -2.1504 0.8285 -0.012944 -0.04841   1.3331
poisson     -0.5703  -0.4556 0.1667 -0.002199 -0.03093   0.2482
         distnod realinc weekend   car
lm      0.028952 0.13339 -0.47650 2.219
poisson 0.004879 0.01882 -0.07382 1.413
```

An increase of 1 of `realinc` means that household's income increases by an amount equal to the local median income. The linear models indicates that the number of trips then increases by 0.133 trips (the sample mean for trips being 4.55). The Poisson model indicates that the relative increase of trips dy/y is equal to 0.019, i.e., 2%. At the sample mean, it corresponds to an increase of $0.019 \times 4.55 = 0.086$ trips.

If the interview concerns a weekend day, the OLS coefficient is -0.48, which indicates that the number of trips taken during the weekend are about one-half less, compared to a weekday. For such a dummy variable, denoting β_w and z_0 a given vector of covariates such that $x_w = 0$, $\gamma^\top z_0$ is the linear predictor for a weekday and $\gamma^\top z_0 + \beta_w$ the linear predictor for a week-end day. Therefore, the relative difference τ of the number of trips between a weekend day and a weekday is:

$$\tau = \frac{e^{\gamma^\top z_0 + \beta_w} - e^{\gamma^\top z_0}}{e^{\gamma^\top z_0}} = e^{\beta_w} - 1 = e^{-0.0738} - 1 = -0.071 = 7.1\%$$

The previous expression indicates also that $\beta_w = \ln(1+\tau) \approx \tau$. For small values of β_w, the coefficient can be interpreted as the relative difference of the response. In terms of absolute value, at the sample mean, the absolute difference is $-0.071 \times 4.55 = -0.323$, which is close to the OLS coefficient. Of course the approximation is only relevant for small values of the coefficient. If we consider the coefficient of `cars`, the value for the Poisson model is 1.413 which implies an increase of $e^{1.413} - 1 = 3.1 = 310\%$ much larger that the value of the coefficient.

12.2.3 Computation of the variance of the estimator

The matrix of the second derivatives is:

$$\frac{\partial \ln L}{\partial \gamma \partial \gamma^\top} = -\sum_{n=1}^{N} e^{\gamma^\top z_n} z_n z_n^\top$$

The information matrix is the opposite of the hessian (as it doesn't depend on y, it is equal to its expected value) and the variance of the gradient. The latter is:

$$\sum_{n=1}^{N} \mathrm{E}\left((y_n - e^{\gamma^\top z_n})^2 z_n z_n^\top\right) = \sum_{n=1}^{N} e^{\gamma^\top z_n} z_n z_n^\top$$

as $\mathrm{E}\left((y_n - e^{\gamma^\top z_n})^2\right)$ is the conditional variance which equals the conditional expectation for a Poisson distribution. The estimated variance of the estimator is therefore:

$$\hat{\mathrm{V}}(\hat{\beta}) = \left(\sum_{n=1}^{N} e^{\hat{\gamma}^\top z_n} z_n z_n^\top\right)^{-1} = \left(\sum_{n=1}^{N} \hat{\mu}_n z_n z_n^\top\right)^{-1}$$

This estimator is consistent only if the distribution of the response is Poisson, and therefore if the conditional variance equals the conditional mean. A more general estimator is based on the sandwich formula:

$$\left(\sum_{n=1}^{N} \hat{\mu}_n z_n z_n^\top\right)^{-1} \left(\sum_{n=1}^{N} (y_n - \hat{\mu}_n)^2 z_n z_n^\top\right) \left(\sum_{n=1}^{N} \hat{\mu}_n z_n z_n^\top\right)^{-1} \quad (12.4)$$

One alternative is to assume that the variance is proportional to the mean: $\mathrm{V}(y|x) = \phi \mathrm{E}(y|x)$. ϕ can then be consistently estimated by:

$$\hat{\phi} = \frac{1}{N} \sum_{n=1}^{N} \frac{(y_n - \hat{\mu}_n)^2}{\mu_n} \quad (12.5)$$

which leads to the third estimator of the covariance matrix:

$$\hat{\mathrm{V}}(\hat{\gamma}) = \hat{\phi}\left(\sum_{n=1}^{N} \hat{\mu}_n z_n z_n^\top\right)^{-1} \quad (12.6)$$

The "quasi-Poisson" model is obtained by using the log link and Equation 12.6 for the variance. It leads to the same estimator as the Poisson model and a variance that is greater (respectively lower) than the one of the Poisson model if there is overdispersion (respectively underdispersion). It can be fitted using `glm` by setting the `family` argument to `quasipoisson`:

```
qpois_trips <- update(pois_trips, family = quasipoisson)
```

Compared to the Poisson model, the standard deviations of the quasi-Poisson model are inflated by a factor which is the square root of the estimate of the variance-mean ratio. This factor can be estimated using Equation 12.5, which makes use of the residuals of the Poisson regression.

```
y <- pois_trips %>% model.frame %>% model.response
mu <- pois_trips %>% fitted
e_resp <- pois_trips %>% resid(type = "response")
e_pears <- pois_trips %>% resid(type = "pearson")
```

The response residual is $y_n - \hat{\mu}_n$, and we get the Pearson residual by dividing the response residual by the standard deviation, which is $\sqrt{\hat{\mu}_n}$. Therefore, the estimation of the variance-mean ratio is simply the sum of square of the Pearson residuals divided by the sample size (the number of degrees of freedom can be used instead).

```
phi <- sum(e_pears ^ 2) / df.residual(pois_trips)
c(phi, sqrt(phi))
## [1] 3.366 1.835
```

The estimated variance-mean ratio is more than 3, so that the standard deviations of the quasi-Poisson model are almost 2 times larger than those of the Poisson model (1.835). The sandwich estimator can be constructed by extracting from the fitted Poisson model the model matrix (Z), the fitted values ($\hat{\mu}_n$) and the response residuals ($y_n - \hat{\mu}_n$):

```
mu <- pois_trips %>% fitted
e <- pois_trips %>% resid(type = "response")
Z <- pois_trips %>% model.matrix
```

We can then compute the "bread" (`B`) and the "meat" (`M`):

```
B <- crossprod(mu * Z, Z)
M <- crossprod(e ^ 2 * Z, Z)
```

Applying Equation 12.4, and taking the square root of the diagonal elements, we get:

```
sand_vcov <- solve(B) %*% M %*% solve(B)
sand_vcov %>% stder %>% head(4)
## (Intercept)    workschl        size        dist
##    0.185364    0.117199    0.025192    0.001696
```

The **sandwich** package provides the `vcovHC` function which computes different flavors of the sandwich estimator of the variance. The one that corresponds to Equation 12.4 is obtained by setting `type` to `HC`.

```
sandwich::vcovHC(pois_trips, type = "HC") %>% stder %>% head(4)
## (Intercept)    workschl        size        dist
##    0.185363    0.117199    0.025191    0.001696
```

The sandwich standard errors can also be extracted from the Poisson model using `stder` with the `vcov` and `type` arguments (respectively equal to `sandwich::vcovHC` and `"HC"`):

```
stder(pois_trips, vcov = sandwich::vcovHC, type = "HC") %>% head(4)
## (Intercept)    workschl        size        dist
##    0.185363    0.117199    0.025191    0.001696
```

Taking the ratio of the simple standard deviations of the coefficients and the sandwich estimator, we get:

```
stder(sand_vcov) / stder(pois_trips)
```

```
(Intercept)    workschl        size        dist        smsa
      1.481       1.677       2.059       1.139       1.996
   fulltime    distnod    realinc     weekend         car
      1.695       1.931       2.888       2.023       1.415
```

As previously, the robust standard deviations are about twice the basic ones on average, but the ratio now depends on the coefficient.

12.2.4 Overdispersion

A more general expression for the variance is the following Negbin specification:

$$\sigma_n^2 = \mu_n + \alpha\mu_n^p$$

with two special cases:

- $p = 1$ (**NB1**) which implies that the variance is proportional to the expectation, $\sigma_n^2 = (1+\alpha)\mu_n$,
- $p = 2$ (**NB2**) which implies that the variance is quadratic in the expectation, $\sigma_n^2 = \mu_n + \alpha\mu_n^2$.

Tests for overdispersion can easily be implemented using one of the three test principles: Wald, likelihood ratio and score tests. The first two require the estimation of a more general model that doesn't impose the equality between the conditional mean and the conditional variance. The score test requires only the estimation of the constrained (Poisson) model. It can take different forms; we'll just present here the test advocated by Cameron and Trivedi (1986) who use an auxiliary regression.[4] The null hypothesis is:

$$\mathrm{E}\left((y_n - \mu_n)^2 - y_n\right) = \sigma_n^2 - \mu_n = 0$$

With the NB1 specification, the alternative is:

$$\mathrm{E}\left((y_n - \mu_n)^2 - y_n\right) = (1+\alpha)\mu_n - \mu_n = \alpha\mu_n$$

or:

$$\frac{\mathrm{E}\left((y_n - \mu_n)^2 - y_n\right)}{\mu_n} = \alpha \tag{12.7}$$

The hypothesis of equidispersion (i.e., $\alpha = 0$) can be tested by regressing $\frac{(y_n - \hat{\mu}_n)^2 - y_n}{\hat{\mu}_n}$ on a constant and comparing the Student statistic to the relevant critical value. With the NB2 specification, the alternative is:

$$\mathrm{E}\left((y_n - \mu_n)^2 - y_n\right) = \mu_n + \alpha\mu_n^2 - \mu_n = \alpha\mu_n^2$$

or:

$$\frac{\mathrm{E}\left((y_n - \mu_n)^2 - y_n\right)}{\mu_n} = \alpha\mu_n \tag{12.8}$$

The hypothesis of equidispersion (i.e., $\alpha = 0$) can then be tested by regressing $\frac{(y_n - \hat{\mu}_n)^2 - y_n}{\hat{\mu}_n}$ on $\hat{\mu}_n$ without an intercept and comparing the Student statistic to the relevant critical value. Note that the response in this auxiliary regression can be written as: $(y_n - \hat{\mu}_n)^2/\hat{\mu} - y_n/\hat{\mu}_n$, which is the difference between the square of the Pearson residual and the ratio of the actual and the fitted value of the response:

```
hmu <- fitted(pois_trips)
y <- model.response(model.frame(pois_trips))
e_pears <- resid(pois_trips, type = "pearson")
resp_areg <- e_pears ^ 2 - y / hmu
lm(resp_areg ~ 1) %>% gaze(coef = 1)
##               Estimate Std. Error t value Pr(>|t|)
## (Intercept)      2.314      0.406     5.7  1.9e-08
lm(resp_areg ~ hmu - 1) %>% gaze
##     Estimate Std. Error t value Pr(>|t|)
## hmu   0.4671     0.0765    6.11  1.9e-09
```

[4]Score tests that are not obtained from auxiliary regressions are presented by Dean (1992).

In both regressions, the t statistic is much higher than 1.96, the p-value is very close to zero so that the equidispersion hypothesis is strongly rejected.

To see to what extent the introduction of covariates has reduced the overdispersion problem, we can produce once again a rootogram, this time using the fitted Poisson model. We use the default style which is `"hanging"`, which means that the bars are vertically aligned on the points that display the fitted frequencies and the default scale which is `"sqrt"`; the vertical axis indicates the square root of the frequencies, to increase the readability of small frequencies. Rootograms for the Poisson model are presented on Figure 12.2 along with the previous rootogram for the unconditional distribution of `trips`. The two figures are first created without being plotted by setting the `plot` argument to `FALSE`, and are then assembled in one figure using `ggpubr::ggarrange`. The limits on the vertical axis are set to the same values for the two plots so that the frequencies for the two figures can be easily compared.

```
uncond_trips <- glm(trips ~ 1, data = trips, family = "poisson")
fig_raw <- topmodels::rootogram(uncond_trips,
                                plot = FALSE) %>%
    autoplot + coord_cartesian(xlim = c(0, 15), ylim = c(-3, 12))
fig_covar <- topmodels::rootogram(pois_trips, plot = FALSE) %>%
    autoplot + coord_cartesian(xlim = c(0, 15), ylim = c(-3, 12))
figure <- ggpubr::ggarrange(fig_raw, fig_covar,
                            labels = c("No covariates", "Covariates"),
                            ncol = 2, nrow = 1)
figure
```

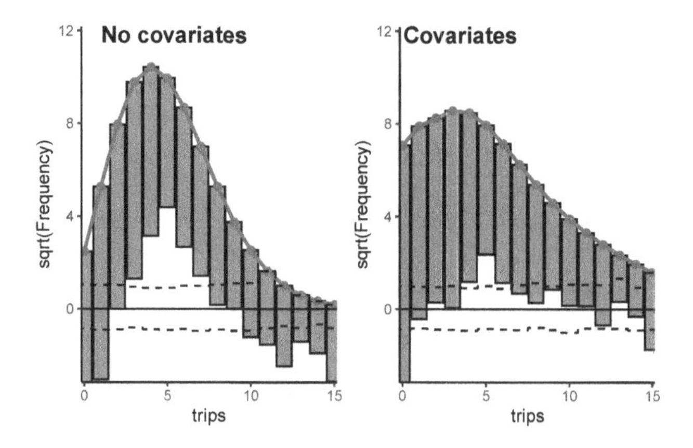

Figure 12.2: Rootograms for Poisson models with and without covariates

One can see that the bottom of the bars is closer to the horizontal axis for the right figure (Poisson model with covariates), which indicates that the inclusion of covariates reduces slightly the overdispersion problem, which anyway still remains. Moreover, the excess of zero largely remains, which suggests that a model that takes the specificity of zero values should be used. The overdispersion phenomenon can also be represented by plotting the square of the residuals against the fitted values, the straight line (NB1 hypothesis) and the parabola (NB2 hypothesis) that fit the data. The result is presented in Figure 12.3.

```
tb <- tibble(e2 = resid(pois_trips, type = "response") ^ 2,
             mu = fitted(pois_trips))
tb %>% ggplot(aes(mu, e2)) + geom_point() +
    coord_cartesian(xlim = c(0, 15), ylim = c(0, 150)) +
    geom_smooth(method = lm, se = FALSE, formula = y ~ x - 1,
                color = "red") +
    geom_smooth(method = lm, se = FALSE, formula = y ~ x + I(x ^ 2) - 1)
```

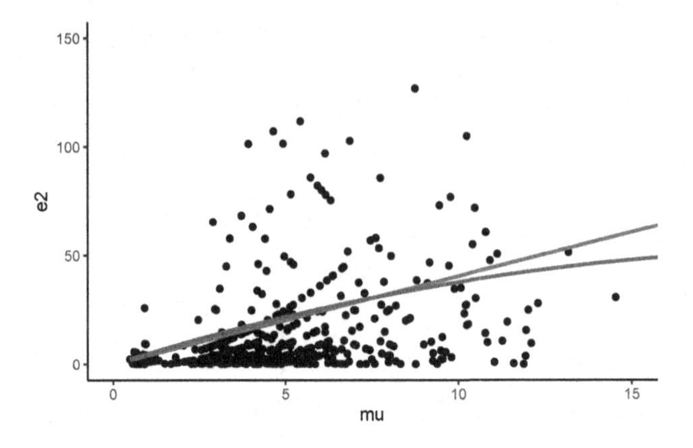

Figure 12.3: Squared residuals and fitted values

12.3 Overdispersion and excess of zero

We describe in this section models that have been proposed to overcome the limits of the Poisson model, namely the fact that it doesn't deal correctly with overdispersion and the excess of zero.

12.3.1 Mixing model

For the Poisson model with the log-link, the parameter of the Poisson distribution for individual n is $\mu_n = e^{\gamma^\top z_n}$. The problem of overdispersion occurs when the variance of y is greater than its expectation, although the hypothesis underlying the use of the Poisson model is that they are equal. One solution is to define the Poisson parameter for individual n as : $\lambda_n = \nu_n \mu_n = \nu_n e^{\gamma^\top z_n}$. ν_n is an unobserved positive term that inflates the variance of y. Actually, for the same set of covariates z, two individuals n and m will be characterized by two different values of their Poisson parameters if $\nu_n \neq \nu_m$. For a given value of ν, the probability of y is still Poisson:

$$P(y_n \mid x_n, \nu_n; \gamma) = \frac{e^{-\nu_n e^{\gamma^\top z_n}} \left(\nu_n e^{\gamma^\top z_n}\right)^{y_n}}{y_n!}$$

but, as ν_n is unobserved, this probability can't be used to estimate the parameters of the model. To get the distribution of y conditional on x only, we need to integrate out ν_n, assuming a specific distribution for ν, called the **mixing distribution**. Two popular choices are the gamma and the log-normal distributions. The gamma density is:

$$g(\nu;\delta,\gamma) = \nu^{\delta-1}e^{-\gamma\nu}\gamma^{\delta}/\Gamma(\delta)$$

where δ and γ are respectively called the shape and the intensity parameters and $\Gamma(z) = \int_0^{+\infty} t^{z-1}e^{-t}dt$ is the Gamma function. The expected value and the variance are respectively δ/γ and δ/γ^2. As γ contains an intercept α, α and the expected value of ν can't be identified and a simple choice of normalization is to set $\delta = \gamma$, so that the expected value of ν is set to 1 and the variance is then $1/\gamma$. The density of ν is then:

$$g(\nu;\delta) = \nu^{\delta-1}e^{-\delta\nu}\delta^{\delta}/\Gamma(\delta)$$

The distribution of y conditional on x_n only, called the **negative binomial** distribution, is then:

$$\begin{aligned} P(y_n \mid x_n;\gamma,\delta) &= \int_0^{+\infty} P(y_n \mid x_n, v;\gamma)g(\nu;\delta)d\nu \\ &= \int_0^{+\infty} \frac{e^{-\mu_n\nu}(\mu_n\nu)^{y_n}}{y_n!}\nu^{\delta-1}e^{-\delta\nu}\delta^{\delta}/\Gamma(\delta)d\nu \end{aligned}$$

Rearranging terms, we get:

$$P(y_n \mid x_n;\gamma,\delta) = \frac{\mu_n^{y_n}\delta^{\delta}}{y_n!\Gamma(\delta)}\int_0^{+\infty} e^{-(\mu_n+\delta)\nu}\nu^{y_n+\delta-1}d\nu$$

with the change of variable $t = (\mu_n + \delta)\nu$, we finally get:

$$P(y_n \mid x_n;\gamma,\delta) = \frac{\mu^{y_n}\delta^{\delta}}{y_n!\Gamma(\delta)}\frac{\Gamma(y_n+\delta)}{(\mu+\delta)^{y_n+\delta}}$$

and finally, as, for integer values of x: $\Gamma(x+1) = x!$:

$$P(y_n \mid x_n;\gamma,\delta) = \frac{\Gamma(y_n+\delta)}{\Gamma(y_n+1)\Gamma(\delta)}\left(\frac{\mu_n}{\mu_n+\delta}\right)^{y_n}\left(\frac{\delta}{\mu_n+\delta}\right)^{\delta}$$

Conditional on x_n and ν_n, $\mathrm{E}(y_n \mid x_n, \nu_n) = \mathrm{V}(y_n \mid x_n, \nu_n) = \nu_n\mu_n$. Conditional on x_n only, $\mathrm{E}(y_n|x_n) = \mathrm{E}_{\nu}(\nu\lambda_n) = \mu_n$ (as $\mathrm{E}(\nu) = 1$). To get the variance, we use the formula of variance decomposition:

$$\begin{aligned} \mathrm{V}(y_n|x_n) &= \mathrm{E}_{\nu}\left(\mathrm{V}(y_n|x_n,\nu_n)\right) + \mathrm{V}_{\nu}\left(\mathrm{E}(y_n|x_n,\nu_n)\right) = \mathrm{E}_{\nu}(\lambda_n) + \mathrm{V}_{\nu}(\lambda_n) \\ &= \mu_n + \mu_n^2\mathrm{V}(\nu_n) = \mu_n(1+\mu_n/\delta) \end{aligned}$$

Different parametrizations result in different flavors of the **Negbin model**. Replacing δ by $\sigma = 1/\delta$ we get the NB2 model, with a quadratic conditional variance function:

$$\mathrm{V}(y_n|x_n) = \mu_n + \sigma\mu_n^2 = (1+\sigma\mu_n)\mu_n \tag{12.9}$$

Replacing δ by μ_n/σ, we get the NB1 model, with a linear conditional variance function:

$$V(y_n \mid x_n) = (1+\sigma)\mu_n \tag{12.10}$$

Note that, as δ is positive, so is σ in the versions of the Negbin model so that the Negbin model exhibits necessarily overdispersion, with the special case of equidispersion if $\sigma = 0$. Actually, in this case, $\delta \rightarrow \infty$, so that the variance of ν tends to 0 and the Poisson model is therefore obtained as a limiting case.

Another choice for the mixing distribution is the log-normal distribution, with $\ln \nu \sim \mathcal{N}(0, \sigma)$. The first parameter of the distribution is set to 0. For a log-normal distribution, the exponential of this parameter is the median of the log-normal. Therefore, the choice of normalization for the log-normal distribution is to set the median (and not the mean as in the gamma distribution) to 0. The log-normal density is $\frac{1}{\sigma\nu}\phi(\ln \nu/\sigma)$ where $\phi()$ is the standard normal density. The probabilities are then given by:

$$P(y_n \mid x_n; \gamma, \delta) = \int_0^{+\infty} \frac{e^{-e^{\gamma^\top z_n}\nu_n}(e^{\gamma^\top z_n}\nu_n)^y}{y!} \frac{1}{\sqrt{2\pi}\sigma\nu} e^{-\frac{1}{2}\left(\frac{\ln \nu}{\sigma}\right)^2} d\nu$$

using the change of variable $t = \ln \nu / (\sqrt{2}\sigma)$, we get:

$$P(y_n \mid x_n; \gamma, \delta) = \frac{1}{\sqrt{\pi} y_n!} \int_{-\infty}^{+\infty} e^{-e^{\gamma^\top z_n + \sqrt{2}\sigma t}} e^{y(\gamma^\top z_n + \sqrt{2}\sigma t)} e^{-t^2} dt$$

There is no closed form for this integral, but it can be approximated using Gauss-Hermite quadrature:

$$\int_{-\infty}^{+\infty} f(t)e^{-t^2}dt \approx \sum_{r=1}^{R} \omega_r f(t_r)$$

where R is the number of points for which the function is evaluated, t_r are called the nodes and ω_r the weights. For a log-normal variable, $\mathrm{E}(x) = e^{\mu + 0.5\sigma^2}$ and $\mathrm{V}(x) = e^{2\mu+\sigma^2}(e^{\sigma^2} - 1)$. As, we set μ to 0, $\mathrm{E}(y_n \mid x_n) = \mathrm{E}_\nu\left(\mathrm{E}(y_n \mid, x_n, \nu)\right) = e^{\frac{1}{2}\sigma^2}\mu_n$. Moreover, $\mathrm{E}_\nu\left(\mathrm{V}(y_n \mid, x_n, \nu)\right) = e^{\frac{1}{2}\sigma^2}\mu_n$ and $\mathrm{V}_\nu(\mathrm{E}(y_n \mid, x_n, \nu)) = \mu_n^2 e^{\sigma^2}(e^{\sigma^2} - 1)$ so that:

$$\mathrm{V}(y_n \mid x_n) = e^{\frac{1}{2}\sigma^2}\mu_n + e^{\sigma^2}(e^{\sigma^2} - 1)\mu_n^2 = \left(1 + (e^{\sigma^2} - 1)\mathrm{E}(y \mid x_n)\right)\mathrm{E}(y_n \mid x_n) \tag{12.11}$$

We then estimate the Poisson model and the three mixed models we've just described. The `micsr::poisreg` function enables the estimation of this mixed models. It has a `mixing` argument. If `mixing = "none"` (the default), the basic Poisson model is estimated. Setting `mixing` to `"gamma"` and `"lognorm"` result respectively in the Negbin and the Log-normal-Poisson models. For the Negbin model, the two different flavors are obtained by setting `vlink` either to `"nb1"` (the default) or to `"nb2"` to get respectively the Negbin1 and the Negbin2 models.[5] The results are presented in Table 12.2.

[5]The NB2 model is implemented in `MASS::glm.nb`.

Table 12.2: Mixed models for the demand for trips

	poisson	Negbin1	Negbin2	Log-normal
(Intercept)	−0.570	−0.402	−0.628	−0.908
	(0.125)	(0.180)	(0.161)	(0.168)
workschl	−0.456	−0.205	−0.365	−0.296
	(0.070)	(0.107)	(0.136)	(0.132)
size	0.167	0.151	0.176	0.179
	(0.012)	(0.021)	(0.024)	(0.026)
dist	−0.002	−0.001	−0.002	−0.001
	(0.001)	(0.002)	(0.002)	(0.002)
smsa	−0.031	0.049	−0.030	0.017
	(0.044)	(0.075)	(0.081)	(0.085)
fulltime	0.248	0.251	0.318	0.327
	(0.026)	(0.044)	(0.056)	(0.055)
distnod	0.005	0.006	0.005	0.006
	(0.001)	(0.002)	(0.002)	(0.003)
realinc	0.019	0.015	0.020	0.015
	(0.006)	(0.011)	(0.013)	(0.017)
weekend	−0.074	−0.119	−0.018	−0.070
	(0.049)	(0.084)	(0.090)	(0.092)
car	1.413	1.195	1.304	1.314
	(0.123)	(0.174)	(0.155)	(0.159)
sigma		2.394	0.485	0.683
		(0.239)	(0.047)	(0.041)
Num.Obs.	577	577	577	577
AIC	3340.9	2780.9	2779.6	2780.1
BIC	3384.5	2828.9	2827.6	2828.1
Log.Lik.	−1660.440	−1379.459	−1378.811	−1379.061

```
pois <- poisreg(trips ~ workschl + size + dist + smsa + fulltime +
                  distnod + realinc + weekend + car, trips)
nb1 <- update(pois, mixing = "gamma", vlink = "nb1")
nb2 <- update(pois, mixing = "gamma", vlink = "nb2")
ln <- update(pois, mixing = "lognorm")
```

The most striking result is the improvement of the fit. The coefficient of σ in the three specifications is highly significant, and the AIC or the BIC conclude that the mixed models are much better than the Poisson model. Using these indicators, it is difficult to discriminate between these three models because the AIC and the BIC are almost the same. The standard errors are much higher for the mixed models compared to the Poisson model, but remember that, when overdispersion is present, the standard errors of the Poisson model are downward biased. Using respectively Equation 12.10, Equation 12.9 and Equation 12.11, we compute the ratio of the variance and the expectation at the sample mean for the three models, i.e., using $\bar{y}$ instead of $\mathrm{E}(y \mid x_n)$ in the three formulas:

```
sig_nb1 <- coef(nb1)["sigma"]
sig_nb2 <- coef(nb2)["sigma"]
sig_ln <- coef(ln)["sigma"]
yb <- mean(trips$trips)
1 + sig_nb1
## sigma
## 3.394
1 + sig_nb2 * yb
## sigma
## 3.207
(1 + (exp(sig_ln ^ 2) - 1) * yb)
## sigma
## 3.707
```

Therefore, the three mixing models predict the same level of overdispersion, the conditional variance being about 3.5 times larger than the conditional expectation. This value should be compared to the ratio of the sample variance (24.4) and the sample mean (4.6), which is (5.4). Figure 12.4 presents the rootogram for the Poisson and the Negbin model. We can see that taking into account the overdispersion using a mixing model resolves most of the overdispersion problem.

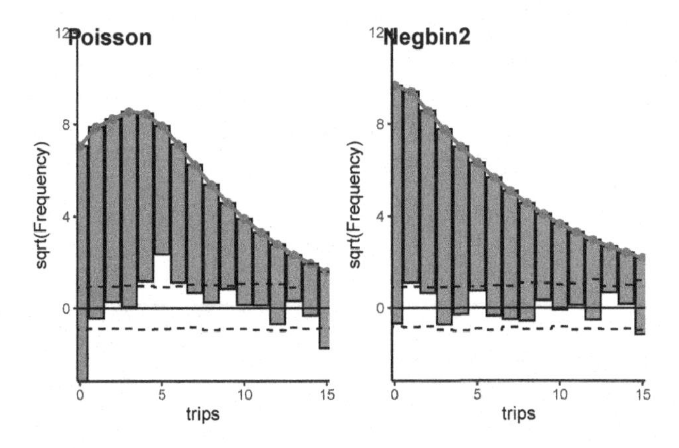

Figure 12.4: Rootograms for the Poisson and the Negbin model

12.3.2 Hurdle and ZIP models

As we have seen previously for the `trips` data set, it seems that the Poisson model is unable to model correctly the probability of $y = 0$. Different phenomena can explain zero values:

- some people may never (or almost never) take a trip out of the house because, for example, of a bad physical condition,
- some people may take a trip infrequently (for example, twice a week) so that a 0 value may be reported if the survey concerns a Tuesday as the individual has taken a trip on Monday and Thursday.

This question of zero vs. positive observations is closely related to the analysis of consumption expenditures based on individual data for which the value is 0 for a large part of the sample. For example, a large share of the individuals in the sample will report a null consumption of tobacco simply because they don't smoke. Some other individuals will report a null consumption of fish because fish is bought infrequently (once a month, for example) and the length of the survey is 2 weeks.

12.3.2.1 Hurdle models

For the first situation, Cragg (1971) proposed the hurdle model (presented in Section 11.6), for which the level of consumption of a good is taken into account by two different models:

- a binomial model that explains the fact that the expenditure is zero or positive,
- a model that explains the level of the expenditure if it is positive.

This hurdle model has been adapted to count data by Mullahy (1986). The first model returns the probability that $y = 0$. Therefore, the response is $d_n = \mathbf{1}(y_n > 0)$, which takes the two values of 0 (y_n is 0) and 1 (y_n is strictly positive). A set of covariates z_1 is used for this model, with corresponding coefficients denoted by γ_1. For example, using a probit model, we would have $\mathrm{P}(d_n = 1) = \Phi(\gamma_1^\top z_1)$. Another choice is to use a count distribution to modelize the probability that $d_n = 0$. For example, using the Poisson distribution ($\mathrm{P}(y) = e^{-\theta}\theta^y/y!$) and the log link ($\theta = \exp(\gamma_1^\top z_1)$ we have $\mathrm{P}(d_n = 0) = \exp(e^{-\gamma_1^\top z_1})$. Therefore the first model writes:

$$\left(\exp\left(-e^{\gamma_1^\top z_{n1}}\right)\right)^{d_n}\left(1-\exp\left(-e^{\gamma_1^\top z_{n1}}\right)\right)^{1-d_n} \tag{12.12}$$

The second model concerns the zero left-truncated sample, i.e., the subsample of individuals for which $y > 0$. Any count model can be used, but the probabilities returned by the chosen distribution for every positive value of y should be divided by their sum, so that they sum to 1. If once again, the Poisson distribution is used, the probabilities for positive values of y are:

$$\frac{\exp\left(-e^{\gamma_2^\top z_{n2}}\right)e^{y_n\gamma_2^\top z_{n2}}/y_n!}{1-\exp\left(-e^{\gamma_2^\top z_{n2}}\right)} \tag{12.13}$$

The hurdle model is a **finite mixture** obtained by combining the two distributions, one that generates the zero values and the other one that generates the positive values. As the two components of the hurdle model depend on a specific parameter's vector (γ_1 and γ_2), the estimation can be performed by independently fitting the two models. The general expression for the probability of one observation is the product of the two previous probabilities:

$$\left(\exp\left(-e^{\gamma_1^\top z_{n1}}\right)\right)^{1-d_n}\left(\frac{1-\exp\left(-e^{\gamma_1^\top z_{n1}}\right)}{1-\exp\left(-e^{\gamma_2^\top z_{n2}}\right)}\right)^{d_n}\left(\frac{\exp\left(-e^{\gamma_2^\top z_{n2}}\right)e^{y_n\gamma_2^\top z_{n2}}}{y_n!}\right)^{1-d_n}$$

This expression illustrates the interest of using the same distribution (here the Poisson distribution) for the two parts of the model. In this case, if $z_1 = z_2$, i.e., if the same set of covariates is used in the two models, the hurdle model is nested in the simple count model, the null hypothesis of the simple model being: $\mathrm{H}_0 : \gamma_1 = \gamma_2$. The **zero-inflated**

Poisson (or **ZIP** model),[6] proposed by Lambert (1992) starts from a count model, for example a Poisson with a set of covariates denoted by z_2, and corresponding coefficients γ_2. The probability of 0 ($e^{-\gamma_2^\top z_{n2}}$) is inflated using a further parameter ψ_n between 0 and 1:

$$P(Y=0|x_n) = \psi_n + (1-\psi_n)e^{-\gamma_2^\top z_{n2}} = e^{-\gamma_2^\top z_{n2}} + \psi_n\left(1-e^{-\gamma_2^\top z_{n2}}\right) > e^{-\gamma_2^\top z_{n2}}$$

The Poisson's probabilities for positive values of y are then deflated by $(1-\psi_n)$, so that the probabilities for all the possible values of y (0 and positive) sums to 1:

$$\mathrm{P}(y_n|z_{n2}) = (1-\psi_n)\frac{e^{-\mu_n}\mu_n^y}{y!} \text{ for } y_n > 0$$

ϕ_n can be a constant, or a function of covariates (z_1) that returns a value between 0 and 1. For example, a logistic specification can be used:

$$\psi_n = \frac{e^{\gamma_1^\top z_{n1}}}{1+e^{\gamma_1^\top z_{n1}}}$$

Hurdle and ZIP models are implemented in the **countreg** package.[7] `countreg::hurdle` and `countreg::zeroinfl` both have a `formula` and a `dist` argument. The first is a two-part formula where the first part describes x_1 and the second x_2. If a one-part formula is provided, the same set of covariates is used for the two parts of the model. The second a character indicating the count model family (`"poisson"` and `"negbin"` are the most common choices). `countreg::hurdle` has a `zero.dist` argument which is a character indicating the distribution for the binomial part of the model (`"poisson"`, `"negbin"` or `"binomial"`; in the latter case, the link can be indicated using the `link` argument). The link used for ψ_n is indicated in `countreg::zeroinfl` by the `link` argument.

```
hdl_pois <- countreg::hurdle(trips ~ workschl + size + dist + smsa +
                               fulltime + distnod + realinc + weekend +
                               car, dist = "poisson",
                             zero.dist = "poisson", trips)
hdl_nb2 <- update(hdl_pois, dist = "negbin", zero.dist = "negbin", trips)
zip_pois <- countreg::zeroinfl(trips ~ workschl + size + dist + smsa +
                               fulltime + distnod + realinc + weekend +
                               car, dist = "poisson", trips)
zip_nb2 <- update(zip_pois, dist = "negbin")
```

The rootograms for the Hurdle and the ZIP models are presented in Figure 12.5. Using the AIC criteria, we can now compare all the models estimated in this section. The AIC of the basic model is:

```
AIC(pois)
## [1] 3341
```

Adding either overdispersion (using a Negbin model) or an excess of 0 (using either a hurdle or a ZIP model) leads to:

[6] Also called with zeros model.

[7] The **countreg** package is not on CRAN, it can be installed using `install.packages("countreg", repos="http://R-Forge.R-project.org")`.

```
AIC(nb2)
## [1] 2780
AIC(hdl_pois)
## [1] 3009
AIC(zip_pois)
## [1] 3028
```

The performance of the hurdle model is slightly better than the ZIP model. The AIC is much less than the one of the Poisson model, but much larger than the one of the Negbin2 model.

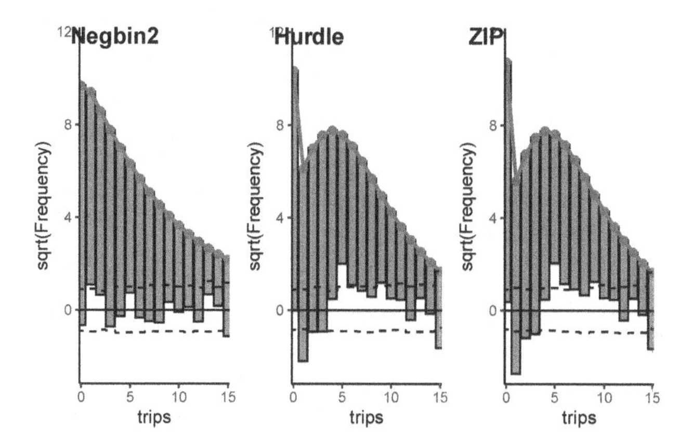

Figure 12.5: Rootograms for the Negbin2, Hurdle and ZIP models

Finally the Negbin2 hurdle or zip model can be compared either to the Negbin2 Poisson model (for which the treatment of excess of 0 is added) or to the hurdle or zip Poisson model (for which the treatment of overdispersion is added):

```
AIC(hdl_nb2)
## [1] 2643
AIC(zip_nb2)
## [1] 2699
```

The AICs of these two models are much lower than the ones of their particular cases (`hdl_pois` and `nb2` for the first one `zip_pois` and `nb2` for the second one). Therefore, the conclusion of this analysis is that either the ZIP or the hurdle Negbin model should be favored. The rootograms for these two models are presented on Figure 12.6.

12.4 Endogeneity and selection

12.4.1 Instrumental variable estimators for count data

We've seen in Section 7.3 that, denoting W the matrix of instruments, the IV estimator is $\hat{\gamma} = (Z^\top P_W Z)^{-1}(Z^\top P_W y)$ (Equation 7.9) and the GMM estimator is $\hat{\gamma} =$

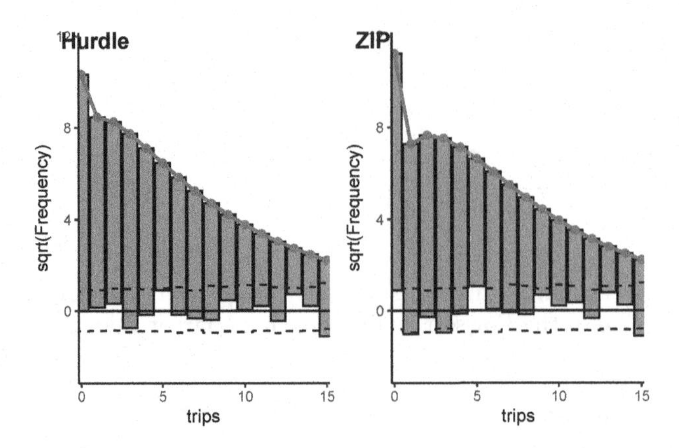

Figure 12.6: Rootograms for the Negbin Hurdle and ZIP models

$\left(Z^\top W\hat{S}^{-1}W^\top Z\right)^{-1} Z^\top W\hat{S}^{-1}W^\top y$ (Equation 7.10), with $\hat{S} = \sum_n \hat{\epsilon}_n^2 w_n w_n^\top$, $\hat{\epsilon}_n$ being the residual of a consistent estimation.

12.4.1.1 Exponential linear conditional mean model

The linear model is often inappropriate if the conditional distribution of y is asymmetric. In this case, a common solution is to use $\ln y$ instead of y as the response: $\ln y_n = \gamma^\top z_n + \epsilon$. This is of course possible only if $y_n > 0\,\forall n$, which is usually not the case with count data. An alternative is to use an exponential linear conditional mean model (Mullahy 1997), with additive ($y_n = e^{\gamma^\top z_n} + \epsilon_n$) or multiplicative errors ($y_n = e^{\gamma^\top z_n}\nu_n$). With additive errors, the empirical moments are then:

$$\frac{1}{N}\sum_{n=1}^N (y_n - e^{\gamma^\top z_n})w_n = W^\top(y - e^{Z\gamma})/N$$

If the errors are spherical, the variance of the empirical moments vector is: $\sigma_\epsilon^2(W^\top W)/N^2$ and the IV estimator minimizes:

$$\epsilon^\top W\left(\sigma^2 W^\top W\right)^{-1} W^\top \epsilon/\sigma^2 = \epsilon^\top P_W \epsilon/\sigma^2$$

Denoting $\hat{\epsilon}$ the residuals of this regression, the optimal weighting matrix $\hat{S} = \sum_{n=1}^N \hat{\epsilon}_n^2 w_n w_n^\top$ can be constructed and used in a second step to get the more efficient GMM estimator. With additive errors, the only difference with the linear case is that the minimization process results in a set of non linear equations, so that some numerical methods should be used. With multiplicative errors, we have: $\nu_n = y_n/e^{\gamma^\top z_n}$. The moment conditions are then: $\mathrm{E}\left((y_n/e^{\gamma^\top z_n} - 1)^\top w_n\right) = 0$ which leads to the following empirical moments:

$$\frac{1}{N}\sum_{n=1}^N (y_n/e^{\gamma^\top z_n} - 1)w_n = W^\top(y/e^{Z^\top\gamma} - 1)/N = Z^\top \tau_n/N$$

with $\tau_n = y_n/e^{\gamma^\top z_n} - 1$. Minimizing the quadratic form of these empirical moments with $(W^\top W)^{-1}$ or $\left(\sum_{n=1}^N \hat{\tau}_n^2 w_n w_n^\top\right)^{-1}$ leads respectively to the IV and the GMM estimators.

When the number of external instruments is greater that the number of endogenous variables, the Sargan test enables to test the hypothesis of exogeneity of all the instruments. The value of the objective function at convergence times the size of the sample is, under the null hypothesis that all the instruments are exogenous, a χ^2 with a number of degrees of freedom equal to the difference between the number of instruments and the number of covariates.

12.4.1.2 Cigarette smoking behavior

Mullahy (1997) estimates a demand function for cigarettes which depends on the stock of smoking habits. This variable is quite similar to a lagged dependent variable and is likely to be endogenous as the unobservable determinants of current smoking behavior should be correlated with the unobservable determinants of past smoking behavior. The data set, called `cigmales`, contains observations of 6160 males in 1979 and 1980 from the smoking supplement to the 1979 National Health Interview Survey. The response `cigarettes` is the number of cigarettes smoked daily. The covariates are the habit "stock" `habit`, the current state-level average per-pack price of cigarettes `price`, a dummy indicating whether there is in the state of residence a restriction on smoking in restaurants `restaurant`, the age `age` and the number of years of schooling `educ` and their squares, the number of family members `famsize`, and a dummy `race` which indicates whether the individual is white or not. The external instruments are cubic terms in `age` and `educ` and their interaction, the one-year lagged price of a pack of cigarettes `lagprice` and the number of years the state's restaurant smoking restrictions had been in place.

The starting point is a basic count model, i.e., a Poisson model with a log-link and robust standard errors:

```
cigmales <- cigmales %>%
    mutate(age2 = age ^ 2, educ2 = educ ^ 2,
           age3 = age ^ 3, educ3 = educ ^ 3,
           educage = educ * age)
pois_cig <- glm(cigarettes ~ habit + price + restaurant + income +
                    age + age2 + educ + educ2 + famsize + race,
                  data = cigmales, family = quasipoisson)
```

The IV and the GMM estimators are estimated using the `micsr::expreg` function. Its main argument is a two-part formula, the first part indicating the covariates and the second part the instruments. The `method` argument can be set to `"iv"` or `"gmm"` to estimate respectively the instrumental variable and the general method of moments estimators and the `error` argument can be equal to `"mult"` (the default) or `"add"` to select respectively multiplicative or additive errors.

```
iv_cig <- expreg(cigarettes ~ habit + price + restaurant + income +
                     age + age2 + educ + educ2 + famsize + race |
                     . - habit + age3 + educ3 + educage +
                     lagprice + reslgth,
                   data = cigmales, method = "iv")
gmm_cig <- update(iv_cig, method = "gmm")
```

Table 12.3: Cigarette consumption and smoking habits

	ML	IV	GMM
habit	0.0055	0.0031	0.0032
	(88.8791)	(1.2536)	(1.2009)
price	−0.0094	−0.0106	−0.0089
	(−3.5253)	(−2.3333)	(−2.0623)
restaurant	−0.0469	−0.0431	−0.0619
	(−1.4949)	(−0.7934)	(−1.0722)
income	−0.0028	−0.0076	−0.0064
	(−1.7915)	(−2.9016)	(−2.3976)
age	0.0087	0.0993	0.0928
	(1.5697)	(2.8577)	(2.3338)
age2	−0.0003	−0.0013	−0.0012
	(−5.0396)	(−3.6384)	(−2.9454)
educ	0.0353	0.1299	0.1408
	(1.8457)	(3.0713)	(3.3685)
educ2	−0.0031	−0.0088	−0.0093
	(−3.7454)	(−3.9537)	(−4.1349)
famsize	−0.0081	−0.0085	−0.0119
	(−1.0498)	(−0.6754)	(−0.9547)
racewhite	−0.0511	−0.0311	−0.0902
	(−1.2462)	(−0.4485)	(−1.3005)
Num.Obs.	6160	6160	6160
F	958.818		
RMSE	14.38		

The results are presented in Table 12.3. For the two flavors of instrumental variable estimators, the coefficient of `habit` is about half of the one of the Poisson model, indicating that this covariate is positively correlated with the unobserved determinants of smoking.

Computing the Sargan test:

```
# collapse: true
sargan(gmm_cig) %>% gaze
```

```
chisq = 7.469, df: 4, pval = 0.113
```

we conclude that the exogeneity of the instruments hypothesis is not rejected.

12.4.2 Sample selection and endogenous switching for count data

In two seminal papers, Heckman (1976, 1979) considered the case where two variables are jointly determined, a binomial variable and an outcome continuous variable. For example, the continuous variable can be the wage and the binomial variable the labor force participation. In this case, the wage is observed only for the subsample of the individuals who work. If the unobserved determinants of labor force participation are correlated with the unobserved determinants of wage, estimating the wage equation only on the subset of individuals who work will result in an inconsistent estimator. This case is called the **sample selection** model and has been presented in Section 11.6. Consider now the case where the binomial variable is a a dummy for private vs public sector. In this case the wage is observed for the whole sample (for the individuals in the public and in the private sector) but, once again, if the unobserved determinants of the chosen sector are correlated with those of the wage, estimating the wage equation only will lead to an inconsistent estimator. This case is called the **endogenous switching** model. Two consistent methods of estimation can be used in this context:

- a two-step method where, in a first step, a probit model for the binomial variable is estimated and, in a second step, the outcome equation is estimated by OLS with a supplementary covariate which is a function of the linear predictor of the probit,[8]
- the maximum likelihood estimation of the system of the two equations, assuming that the two errors are jointly normally distributed.

Let y be a count response (for the sake of simplicity a Poisson variable) and w a binomial variable. The value of w_n is given by the sign of $\gamma^\top z_{n1} + \nu_n$, where ν_n is a standard normal deviate, z_{n1} a vector of covariates and γ_1 the associated vector of unknown parameters. The distribution of y_n is Poisson with parameter λ_n, given by $\lambda_n = e^{\gamma_2^\top z_{n2}} + \epsilon_n$ where z_{n2} is a second set of covariates (which can overlap with z_{n1}), γ_2 is the corresponding set of unknown parameters and ϵ_n is a random normal deviate with 0 mean and a standard deviation equal to σ. ϵ and ν being potentially correlated, their joint distribution has to be considered:

$$\begin{aligned} f(\epsilon, \nu; \sigma, \rho) &= \frac{1}{2\pi\sqrt{1-\rho^2}\sigma} e^{-\frac{1}{2}\frac{\left(\frac{\epsilon}{\sigma}\right)^2 + \nu^2 - 2\rho\left(\frac{\epsilon}{\sigma}\right)\nu}{1-\rho^2}} \\ &= \frac{1}{\sqrt{2\pi}}\sigma e^{-\frac{1}{2}\left(\frac{\epsilon}{\sigma}\right)^2} \frac{1}{\sqrt{2\pi}\sqrt{1-\rho^2}} e^{-\frac{1}{2}\left(\frac{(\nu-\rho\epsilon/\sigma)}{\sqrt{1-\rho^2}}\right)^2} \end{aligned} \tag{12.14}$$

The second expression gives the joint distribution of ϵ and ν as the product of the marginal distribution of ϵ and the conditional distribution of ν. The conditional distribution of y, given ϵ is:

$$g(y_n \mid z_{n2}, \epsilon_n; \gamma_2) = \frac{e^{-\exp(\gamma_2^\top z_{n2} + \epsilon_n)} e^{y_n(\gamma_2^\top z_{n2} + \epsilon_n)}}{y_n!}$$

For $w = 1$, the unconditional distribution of y is obtained by integrating out g with the two random deviates ϵ and ν, using their joint distribution given by Equation 12.14:

$$\begin{aligned} P(y_n \mid z_{n1}, z_{n2}, w_n = 1) &= \int_{-\infty}^{+\infty} \int_{-\gamma_1^\top z_{n1}}^{+\infty} g(y_n \mid z_{n2}, \epsilon_n; \gamma_2) f(\epsilon, \nu; \sigma, \rho) d\epsilon d\nu \\ &= \int_{-\infty}^{+\infty} g(y_n \mid z_{n2}, \epsilon_n) \left(\int_{-\gamma_1^\top z_{n1}}^{+\infty} \frac{1}{\sqrt{2\pi}\sqrt{1-\rho^2}} e^{-\frac{1}{2}\left(\frac{(\nu-\rho\epsilon/\sigma)}{\sqrt{1-\rho^2}}\right)^2} d\nu \right) \\ &\times \frac{1}{\sqrt{2\pi}\sigma} e^{-\frac{1}{2}\left(\frac{\epsilon}{\sigma}\right)^2} d\epsilon \end{aligned}$$

[8] More precisely the inverse Mills ratio.

By symmetry of the normal distribution, the term in brackets is:

$$\Phi\left(\frac{\gamma_1^\top z_{n1} + \rho/\sigma\epsilon}{\sqrt{1-\rho^2}}\right)$$

which is the probability that $w = 1$ for a given value of ϵ. The density of y given that $w = 1$ is then:

$$\begin{array}{rcl} P(y_n \mid z_{n1}, z_{n2}, w_n = 1) & = & \int_{-\infty}^{+\infty} \frac{e^{-\exp(\gamma_2^\top z_{n2}+\epsilon_n)} e^{y_n(\gamma_2^\top z_{n2}+\epsilon_n)}}{y_n!} \\ & \times & \Phi\left(\frac{\gamma_1^\top z_{n1}+\rho/\sigma\epsilon}{\sqrt{1-\rho^2}}\right) \frac{1}{\sqrt{2\pi}\sigma} e^{-\frac{1}{2}\left(\frac{\epsilon}{\sigma}\right)^2} d\epsilon \end{array} \quad (12.15)$$

By symmetry, it is easily shown that $P(y_n \mid z_{n1}, z_{n2}, w_n = 0)$ is similar except that $\Phi\left((\gamma_1^\top z_{n1} + \rho/\sigma\epsilon)/\sqrt{1-\rho^2}\right)$ is replaced by $\Phi\left(-(\gamma_1^\top z_{n1} + \rho/\sigma\epsilon)/\sqrt{1-\rho^2}\right)$, so that a general formulation of the distribution of y_n is, denoting $q_n = 2w_n - 1$:

$$\begin{array}{rcl} P(y_n \mid z_{n1}, z_{n2}) & = & \int_{-\infty}^{+\infty} \frac{e^{-\exp(\gamma_2^\top z_{n2}+\epsilon_n)} e^{y_n(\gamma_2^\top z_{n2}+\epsilon_n)}}{y_n!} \\ & \times & \Phi\left(q_n \frac{\gamma_1^\top z_{n1}+\rho/\sigma\epsilon}{\sqrt{1-\rho^2}}\right) \frac{1}{\sqrt{2\pi}\sigma} e^{-\frac{1}{2}\left(\frac{\epsilon}{\sigma}\right)^2} d\epsilon \end{array} \quad (12.16)$$

There is no closed form for this integral but, using the change of variable $t = \epsilon/\sqrt{2}/\sigma$, we get:

$$\begin{array}{rcl} P(y_n \mid z_{n1}, z_{n2}) & = & \int_{-\infty}^{+\infty} \frac{e^{-\exp(\gamma_2^\top z_{n2}+\sqrt{2}\sigma t)} e^{y_n(\gamma_2^\top x_{n2}+\sqrt{2}\sigma t)}}{y_n!} \\ & \times & \Phi\left(q_n \frac{\gamma_1^\top z_{1n}+\sqrt{2}\rho t_n}{\sqrt{1-\rho^2}}\right) \frac{1}{\sqrt{\pi}} e^{-t^2} dt \end{array}$$

which can be approximated using Gauss-Hermite quadrature. Denoting t_r the nodes and ω_r the weights:

$$\begin{array}{rcl} P(y_n \mid z_{n1}, z_{n2}) & \approx & \sum_{r=1}^{R} \omega_r \frac{e^{-\exp(\gamma_2^\top z_{n2}+\sqrt{2}\sigma t_r)} e^{y_n(\gamma_2^\top z_{n2}+\sqrt{2}\sigma t_r)}}{y_n!} \\ & \times & \Phi\left(q_n \frac{\gamma_1^\top z_{n1}+\sqrt{2}\rho t_r}{\sqrt{1-\rho^2}}\right) \frac{1}{\sqrt{\pi}} e^{-t_r^2} \end{array}$$

For the exogenous switching model, the contribution of one observation to the likelihood is given by Equation 12.16. For the sample selection model, the contribution of one observation to the likelihood is given by Equation 12.15 if $w_n = 1$. If $w_n = 0$, y is unobserved and the contribution of such observations to the likelihood is the probability that $w_n = 0$, which is:

$$P(x_n = 0 \mid z_{n1}) = \int_{-\infty}^{+\infty} \Phi\left(q_n \frac{\gamma_1^\top z_{n1} + \rho/\sigma\epsilon}{\sqrt{1-\rho^2}}\right) \frac{1}{\sqrt{2\pi}\sigma} e^{-\frac{1}{2}\left(\frac{\epsilon}{\sigma}\right)^2} d\epsilon$$

The ML estimator is computing-intensive, as the integral has no closed form. One alternative is to use non-linear least squares, by first computing the expectation of y. Terza (1998) showed that this expectation is:

$$\mathrm{E}(y_n \mid z_{n1}, z_{n2}) = \exp\left(\gamma_2^\top z_{n2} + \ln \frac{\Phi\left(q_n(\gamma_1^\top z_{n1} + \theta)\right)}{\Phi\left(q_n(\gamma_1^\top z_{n1})\right)}\right)$$

for the endogenous switching case and:

$$\mathrm{E}(y_n \mid z_{n1}, z_{n2}, w_n = 1) = \exp\left(\gamma_2^\top z_{n2} + \ln \frac{\Phi(\gamma_1^\top z_{n1} + \theta)}{\Phi(\gamma_1^\top z_{n1})}\right)$$

for the sample selection case, where $\theta = \sigma\rho$. Greene (2001) noted that, taking a first order Taylor series of $\ln \frac{\Phi(\gamma_1^\top z_{n1} + \theta)}{\Phi(\gamma_1^\top z_{n1})}$ around $\theta = 0$ gives: $\theta\phi(\gamma_1^\top z_{n1})/\Phi(\gamma_1^\top z_{n1})$, which is the inverse mills ratio that is used in the linear model in order to correct the inconsistency due to sample selection. As γ_1 can be consistently estimated by a probit model, the NLS estimator is obtained by minimizing with respect to γ_2 and θ the sum of squares of the following residuals:

$$y_n - e^{\gamma_2^\top z_{n2} + \ln \frac{\Phi(\hat{\gamma}_1^\top z_{n1} + \theta)}{\Phi(\hat{\gamma}_1^\top z_{n1})}}$$

As it is customary for a two-step estimator, the covariance matrix of the estimators should take into account the fact that γ_1 has been estimated in the first step. Moreover, only $\theta = \rho\sigma$ is estimated. To retrieve an estimator of σ, Terza (1998) proposed to insert into the log-likelihood function the estimated values of γ_1, γ_2 and θ and then to maximize it with respect to σ.[9]

The `micsr::escount` function estimates the endogenous switching and the sample selection model for count data. The first is obtained by setting the `model` argument to `'es'` (the default) and the second to `'ss'`. The estimation method is selected using the `method` argument, which can be either `"twostep"` for the two-step, non-linear least squares model (the default) or `"ml"` for maximum likelihood. The model is described by a two-part formula. On the left side, the outcome and the binomial responses are indicated. On the right side, the first part contains the covariates for the outcome equation and the second part the covariates of the selection equation. Relevant only for the ML method, the `hessian` argument is a boolean: if `TRUE`, the covariance matrix of the coefficients is estimated using the numerical hessian, which is computed using the `hessian` function of the **numDeriv** package; otherwise, the outer product of the gradient is used and `R` is an integer that indicates the number of points used for the Gauss-Hermite quadrature method. `escount` returns an object of class `escount` which inherits from `micsr`.

12.4.2.1 Physician advice and alcohol consumption

Kenkel and Terza (2001) investigate the effect of physician's advice on alcohol consumption; the data set is called `drinks`. The outcome variable `drinks` is the number of drinks in the past 2 weeks and the selection variable `advice` is a dummy based on the respondents' answer to the question: "Have you ever been told by a physician to drink less". The unobserved part of the equation indicating the propensity to receive advice from the physician can obviously be correlated with the one of the alcohol consumption equation. The covariates are monthly income in thousands of dollars (`income`), `age` (a factor with six 10-year categories of age), education in years (`educ`), `race` (a factor with levels `white`, `black` and `other`), marital

[9] This once again requires the use of Gauss-Hermite quadrature, but the problem is considerably simpler as the likelihood is maximized with respect with only one parameter.

Table 12.4: Poisson and endogenous switching models for alcohol demand

	Poisson	two-step	ML
advice	0.477	−1.400	−0.848
	(0.011)	(0.338)	(0.022)
income	0.003	0.001	0.009
	(0.001)	(0.003)	(0.001)
educ	−0.026	−0.061	−0.055
	(0.002)	(0.012)	(0.003)
theta		1.142	
		(0.319)	
sigma			1.298
			(0.011)
rho			0.463
			(0.022)
Num.Obs.	2467	2467	2467
F	229.655		
RMSE	22.16		

status (`marital`, a factor with levels `single`, `married`, `widow`, `separated`), employment status (a factor `empstatus` with levels `other`, `emp` and `unemp`) and region (`region`, a factor with levels `west`, `northeast`, `midwest` and `south`). For the binomial part of the model, the same covariates are used (except of course `advice`) and 10 supplementary covariates indicating insurance coverage and health status are added.

Table 12.4 presents the results of a Poisson model and of the endogenous switching model, estimated by the two-step, non-linear least squares and by the maximum likelihood estimators.[10] The coefficient of `advice` in the alcohol demand equation is positive in the Poisson model, which would imply that a physician's advice have a positive effect on alcohol consumption. The estimation of the endogenous switching model shows that this positive coefficient is due to the positive correlation between the error terms of the two equations (the unobserved propensities to drink and to receive advice from a physician are positively correlated).

```
kt_pois <- glm(drinks ~ advice + income + age + educ + race + marital +
                 empstatus + region, data = drinks, family = poisson)
kt_ml <- escount(drinks + advice ~ advice + income + age + educ + race +
                   marital + empstatus + region | . - advice + medicare +
                   medicaid + champus + hlthins + regmed + dri + limits +
                   diabete + hearthcond + stroke,
                 data = drinks, method = "ml")
kt_2s <- update(kt_ml, method = "twostep")
```

[10]The coefficients of `marital`, `empstatus`, `income`, `race` and `region` are omitted to save place.

13

Duration models

When the response is a duration, specific models are used, which comes from the biostatistics (eg lifetime after an hearth transplant) and operations research (eg lifetime of a light bulb) literature. These data are called **duration** or **transition** data. The latter term indicates that we are interested in the time span of the transition from one state to another, the latter being called **death** and **failure**, respectively in the biostatistics and the operations research fields. For a given time, the observations for which the transition didn't take place are called **at-risk** observations.

Duration data have two main characteristics, that require the use of specific models. The first is that they are continuous-positive variables. The second is that they are often censored, especially right-censored; for example, trying to modelize the length of unemployment spells, we'll use survey data on employment realized in a given period (for example 2 years). At the end of the survey, some individuals are still unemployed because the spell is not finished, so that the duration of their spell is not observed, but only the fact that it is greater than the observed duration at the end of the survey.

Denote t as the duration response. Its distribution is given by a probability $\mbox{P}(T \leq t) = F(t)$ and a density function $f(t)$. The survival function is often used, which is $S(t) = P(T > t) = 1 - F(t)$.[1] For example, if T is measured in month and if $t = 12$, $F(12) = 0.8$ means that 80% of unemployment spells are less than 1 year, and $S(12) = 0.2$ means that 20% of unemployment spells are more than 1 year. For a small variation of the spell (say $dt = 1$ month), $f(t)dt$ is the probability that the length of the spell is between t and $t+dt$ (between 12 and 13 months in our example). Assume that $f(12) = 0.06$. It means that 6% of the spells ends between the 12$^{\mbox{th}}$ and the 13$^{\mbox{th}}$ month. If we divide by $S(12) = 0.2$, we get $f(12)/S(12) = 0.3$, which means that 30% of the spells of at least one year ends during the 13$^{\mbox{th}}$ month. This is called the **hazard function** and it plays a central role in duration analysis. More formally, the hazard function is defined by:

$$\theta(t) = \lim_{dt \rightarrow 0} \frac{\mbox{P}(t \leq T < t + dt \mid T > t)}{dt} = \lim_{dt \rightarrow 0} \frac{F(t + dt) - F(t)}{1 - F(t)} = \frac{f(t)}{S(t)} \quad (13.1)$$

The hazard function is also the derivative of the opposite of the logarithm of the survival function: $\frac{\partial \ln S}{\partial t} = \frac{-f(t)}{S(t)} = -\theta(t)$ Inversely, the log survival function is a primitive of the hazard, so that, using the fact that $S(0) = 1$, the **cumulative hazard** function $\Lambda(t)$ is:

$$\Lambda(t) = \int_0^t \theta(s)ds = -\ln S(t) \quad (13.2)$$

Now consider the area under the survival curve: $\int_0^{+\infty} S(t)dt$. Integrating by part, we get:

[1]The survival function is sometimes defined as $S(t) = P(T \geq t)$, see Cameron and Trivedi (2005).

$\int_0^{+\infty} S(t)dt = [tS(t)]_0^{+\infty} - \int_0^{+\infty} -f(t)tdt = \int_0^{+\infty} f(t)tdt$ as $\lim_{t\rightarrow +\infty} S(t)t = 0$. Therefore, this area is the expected value of the duration of the spell.

The hazard and survival functions can also be defined in terms of discrete time either because the duration is intrinsically discrete or because it is rounded to a given time unit (weeks or months, for example). Consider that the event is observed for K distinct times, denoted by $t_1, t_2, \ldots t_k, \ldots t_K$. If there are no ties, which means that all the events occur at a specific time, then $K = N$. The **discrete-time hazard** function is the probability of transition at time t_j given survival to time t_j:

$$\theta_j = P(T = t_j \mid T \geq t_j) = \frac{f^d(t_j)}{S^d(t_{j-})}$$

where $S^d(t_{j-})$ is the value of the survival function just before t_j. The **discrete-time survivor function** is obtained from the hazard function: $S^d(t) = P(T \geq t) = \prod_{j|t_j \leq t}(1 - \theta_j)$. For example, the probability of surviving at least to t_2 is: $S^d(t_2) = (1 - \theta_1)(1 - \theta_2)$. The first term is the probability of no transition at t_1, and the second one is the probability of no transition at t_2 conditional on surviving to just before t_2.

Section 13.1 is devoted to the non-parametric estimation of the survival function. Section 13.2 presents different parametric and semi-parametric estimators. Section 13.3 tackles the problem of heterogeneity in duration models. Finally, Section 13.4 presents miscellaneous duration models.

Duration analysis using **R** is performed using the **survival** package, which is a "recommended" package, which means that it comes with any **R** distribution.

```
library(survival)
```

13.1 Kaplan-Meier non-parametric estimator

The Kaplan-Meier estimator is a non-parametric estimator of the survival function. As such, it is very useful as a primary tool before developing more complicated parametric models.

13.1.1 Uncensored sample

To demonstrate this estimator, it is convenient to use first uncensored data. This is the case of the `oil` (Favero, Pesaran, and Sharma 1994) data set which indicates the number of months between the discovery of an oil field and the beginning of development. For each failure times t_k, we count the number of events d_k (the number of spells ending at time t_k) and the number of observations at-risk, i.e., the number of spells which haven't ended just before t_k, denoted by r_k. For the `oil` data set, we have 53 observations and the first ordered values of duration are:

```
oil %>% arrange(dur) %>% pull(dur) %>% head
## [1]  1  3  3  7  8 10
```

Therefore, $t_1 = 1$ and $r_1 = 53$ because all the spells are at risk just before time is equal to 1. The hazard function for t_1 is then estimated by $1/53 = 0.0189$, and the survival function by $1-1/53 = 0.981$, which means that the estimated probability that the spell is greater than 1 month is 98.1%. For $t_2 = 3$, we have $d_2 = 2$ and the number of observations at risk are the ongoing spells just before time is equal to 2, which is $r_2 = r_1 - d_1 = 52$. The hazard function for $t_2 = 3$ is then $d_2/r_2 = 2/52 = 0.038$, which means that the estimated probability of ending at time 3 for spells which are at least 1 month long is 3.8%. The survival function for time equal to 3 is therefore:$(1 - d_1/r_1) \times (1 - d_2/r_2) = 0.981 \times 0.961 = 0.943$. Note that as $r_1 = N$ and $r_2 = r_1 - d_1 = N - d_1$, the estimated survival probability for t_2 is $(1 - d_1/N)(1 - d_2/(N - d_1))$, which simplifies to $1 - (d_1 + d_2)/N$. For $t_3 = 7$, we have $d_3/r_3 = 1/50 = 0.02$ and the survival function can be either estimated by $0.943 \times (1-0.02)$ or by $(1 - 4/53) = 0.9245$.

The whole survival series can be easily obtained with the following code; we first create a frequency table of durations (the durations are then automatically arranged in an increasing order), the `n` column then contains the number of spells that end at each period. We then compute the cumulative sums of this counts `cumn` and the survival probabilities as the cumulative count divided by N.

```
surv_tb <- oil %>% count(dur) %>%
  mutate(cumn = cumsum(n), surv = 1 - cumn / 53)
surv_tb
```

```
# A tibble: 46 x 4
    dur     n  cumn  surv
  <int> <int> <int> <dbl>
1     1     1     1 0.981
2     3     2     3 0.943
3     7     1     4 0.925
4     8     1     5 0.906
5    10     2     7 0.868
# i 41 more rows
```

A more complex calculus (but which will prove to be useful if there are censored spells) consists of computing the number of spells at risk, the hazard and then the survival function as the cumulative product of 1 minus the hazard rate.

```
oil %>%
    count(dur) %>%
    mutate(cumn = cumsum(n),
           at_risk =  ifelse(dur == 1, 53, 53 - lag(cumn)),
           hazard = n / at_risk,
           surv = cumprod(1 - hazard))
```

```
# A tibble: 46 x 6
    dur     n  cumn at_risk hazard  surv
```

```
  <int> <int> <int>   <dbl>  <dbl> <dbl>
1     1     1     1      53 0.0189 0.981
2     3     2     3      52 0.0385 0.943
3     7     1     4      50 0.02   0.925
4     8     1     5      49 0.0204 0.906
5    10     2     7      48 0.0417 0.868
# i 41 more rows
```

The survival function can also be plotted. As it is a step function, using the **geom_step** function is particularly useful, see Figure 13.1.

```
km_oil <- surv_tb %>% ggplot(aes(dur, surv)) + geom_step()
km_oil
```

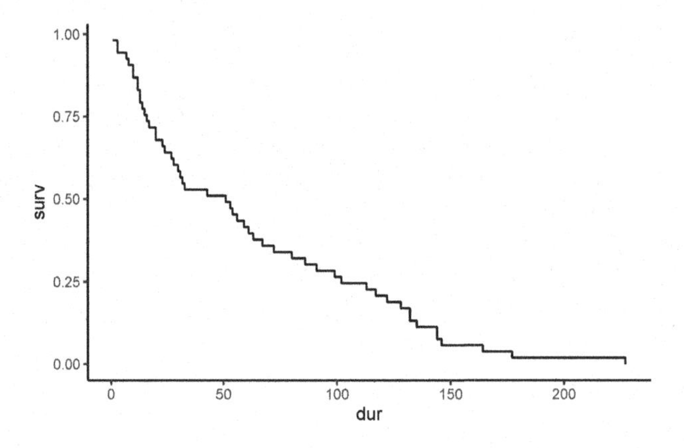

Figure 13.1: Survival curve for the oil data set

Reminding that $\frac{\partial \ln S(t)}{\partial t} = -\lambda(t)$, it is useful to use a logarithmic y-axis as, the hazard is then the slope (in absolute value) of the curve (see Figure 13.2). In particular, the constant hazard hypothesis would induce a roughly linear survival curve.

```
km_oil + coord_cartesian(xlim = c(0, 150)) +
    scale_y_continuous(trans = "log", breaks = c(0.01, 0.1, 1)) +
    geom_smooth(se = FALSE)
```

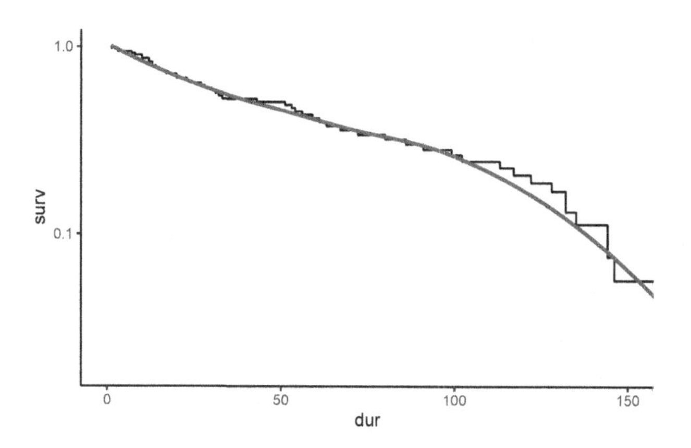

Figure 13.2: Survival curve for the oil data set with a log scale

We can see in this example that the hazard rate seems to be roughly constant for a wide range of values of duration (0-100 months) and then decreasing for high values of duration (> 100). The Kaplan-Meier estimator can also be computed using the `survreg::surfit` function, with the usual formula-data interface. The response must be a `Surv` object, and the right-hand side of the formula is just an intercept. A `Surv` object is obtained using the `Surv` function. With uncensored data, its only argument is the duration response. We then have:

```
sf <- survfit(Surv(dur) ~ 1, data = oil)
sf
```

```
Call: survfit(formula = Surv(dur) ~ 1, data = oil)

      n events median 0.95LCL 0.95UCL
[1,] 53     53     51      28      72
```

The `print` method returns the number of observations (`n`) the number of events (`events`), which is equal to the number of observations if all the spells are uncensored, the median spell duration and its 95% confidence interval. The `print.summary` function returns a table with one row for each time some spells end. The object returned by `survfit` is not a data frame, but the `broom:tidy` method coerces it easily to a tibble:

```
broom::tidy(sf)
```

```
# A tibble: 46 x 8
   time n.risk n.event n.censor estimate std.error conf.high conf.low
  <dbl>  <dbl>   <dbl>    <dbl>    <dbl>     <dbl>     <dbl>    <dbl>
1     1     53       1        0    0.981    0.0190     1        0.945
2     3     52       2        0    0.943    0.0336     1        0.883
3     7     50       1        0    0.925    0.0392     0.998    0.856
4     8     49       1        0    0.906    0.0443     0.988    0.830
5    10     48       2        0    0.868    0.0536     0.964    0.781
# i 41 more rows
```

There is a `plot` method for `survfit` objects that draws the survival curve and its confidence interval.

13.1.2 Censored sample

When there are censored observations, the computation is modified. Consider the `retirement` data set (An, Christensen, and Gupta 2004). It contains the joint duration until retirement for both spouses of 243 couples in Denmark.

```
retirement
```

```
# A tibble: 2,430 x 17
     id period  agem  agef dum84 unemp year0  durm  durf censm censf
  <dbl>  <dbl> <dbl> <dbl> <dbl> <dbl> <dbl> <dbl> <dbl> <dbl> <dbl>
1     1      1    52    54     0   9.2    81    10     6     0     1
2     1      2    53    55     0   9.8    81    10     6     0     1
3     1      3    54    56     0  10.5    81    10     6     0     1
4     1      4    55    57     1  10.1    81    10     6     0     1
5     1      5    56    58     1   9.1    81    10     6     0     1
# i 2,425 more rows
# i 6 more variables: skillm <dbl>, skillf <dbl>, owner <dbl>,
#   province <dbl>, year <dbl>, dum84bis <dbl>
```

`id` is the couple identifier, there are several annual observations for every couple (because some covariates are time-varying), the duration until retirement for males is `durm` and `censm` is a dummy which equals 1 if the observation is right-censored. The corresponding variables for women are `durf` and `censf`. We'll concentrate on women. To get one observation for each woman, we simply select these two variables, which are time-invariant, and we keep only the distinct rows:

```
ret <- retirement %>%
    select(id, duration = durf, censored = censf) %>%
    mutate(censored = ifelse(censored == 1, "yes", "no")) %>%
    distinct
```

As there is no transition before 7, we set duration to 6 for durations lower than 7. We then construct a table containing for each period the number of events and the number of censored spells; we first count the occurrence of any duration-censored combinations, we then pivot the tibble in order to have one line for each duration and two columns for the number of events (`"no"`) and the number of censored observations (`"yes"`) for which we give finally more relevant names.

```
ret_tbl <- ret %>%
  mutate(duration = ifelse(duration < 7, 6, duration)) %>%
  count(duration, censored) %>%
  pivot_wider(names_from = censored, values_from = n, values_fill = 0) %>%
  rename(event = no, censored = yes)
```

For each discrete time, we define the cumulative number of spells that exit the sample `cumn`, either because of transition or censoring. The number of observations at-risk just before t_j is equal to the total sample size minus the lag of `cumn`:

```
ret_tbl <- ret_tbl %>%
    mutate(cumn = cumsum(event + censored),
           atrisk = ifelse(duration == 6, 243, 243 - lag(cumn)))
```

Finally, we compute the hazard $\hat{\theta}_j$ as the ratio of the number of events at t_j and the number of observations at risk just before t_j and the survival $\hat{S}_j$ as the cumulative product of $(1-\hat{\theta}_j)$.

```
ret_tbl %>%
    mutate(hazard = event / atrisk,
           surv = cumprod(1 - hazard))
```

```
# A tibble: 5 x 7
  duration censored event  cumn atrisk hazard   surv
     <dbl>    <int> <int> <int>  <dbl>  <dbl>  <dbl>
1        6       25     0    25    243 0       1
2        7       11    21    57    218 0.0963 0.904
3        8       12    12    81    186 0.0645 0.845
4        9       10    17   108    162 0.105  0.757
5       10       14   121   243    135 0.896  0.0785
```

The same results are easily obtained using `survfit`. Once again, a formula-data interface is used, the response being a `Surv` object. In the context of a sample with right-censored observations, this object is obtained using the `Surv` function with two arguments:

- `time` : the duration (observed or right-censored),
- `event` : a dummy for observed duration (i.e., uncensored observations). It can be either a boolean, 0/1 or 1/2.

If the `Surv` function is used with two unnamed arguments, it is assumed that these two arguments are `time` and `event` in that order. We use in our example a boolean obtained from the `cens` variable:

```
surv_ret <- survfit(Surv(duration, censored == "no") ~ 1, data = ret)
broom::tidy(surv_ret) %>% slice(- (1:6))
```

```
# A tibble: 4 x 8
   time n.risk n.event n.censor estimate std.error conf.high conf.low
  <dbl>  <dbl>   <dbl>    <dbl>    <dbl>     <dbl>     <dbl>    <dbl>
1     7    218      21       11   0.904     0.0221     0.944   0.865
2     8    186      12       12   0.845     0.0293     0.895   0.798
3     9    162      17       10   0.757     0.0398     0.818   0.700
4    10    135     121       14   0.0785    0.256      0.130   0.0475
```

13.1.3 Different groups

When the observations belong to two or more groups, the survival curves can be estimated for each group.

We now use the **unemp_teachers** data set, initially used by Kastoryano and Klaauw (2022). It contains 3064 observations of unemployed Dutch teachers. Some of them performed a training; they are identified by a dummy called **training**. We restrict the sample to those who didn't perform a training and we measure the duration in years:

```
unteach <- unemp_teachers %>% filter(! training) %>%
  select(- training, - time_training) %>%
  mutate(time = time / 365.25)
head(unteach, 3)
```

```
# A tibble: 3 x 9
  gender lowsk   age  wage  time spell uidur month  year
  <fct>  <dbl> <dbl> <dbl> <dbl> <dbl> <dbl> <dbl> <dbl>
1 female     0  48.5 14.0  0.586     1   822     8  2007
2 female     0  53.3 12.4  0.151     1    91     6  2007
3 female     0  50.9  7.99 0.501     1   183     8  2006
```

We use **survfit**, this time with a factor, more precisely the gender factor to compute separate survival curves for males and females:

```
survfit_gender <- survfit(Surv(time, spell) ~ gender, unteach)
broom::tidy(survfit_gender) %>% print(n = 3)
```

```
# A tibble: 450 x 9
    time n.risk n.event n.censor estimate std.error conf.high conf.low
   <dbl>  <dbl>   <dbl>    <dbl>    <dbl>     <dbl>     <dbl>    <dbl>
1 0.0110   1805       4        0    0.998   0.00111     1.00     0.996
2 0.0137   1801       7        0    0.994   0.00184     0.998    0.990
3 0.0164   1794       6        0    0.991   0.00230     0.995    0.986
# i 447 more rows
# i 1 more variable: strata <chr>
```

The **survminer** package provides the **ggsurvplot** function, based on **ggplot2**, which enables to draw nice and highly customizable survival curves, as in Figure 13.3 where we set the **risk.table** argument to **TRUE** in order to have a table of observations at risk just below the figure.

```
survminer::ggsurvplot(survfit_gender, risk.table = TRUE, conf.int = TRUE)
```

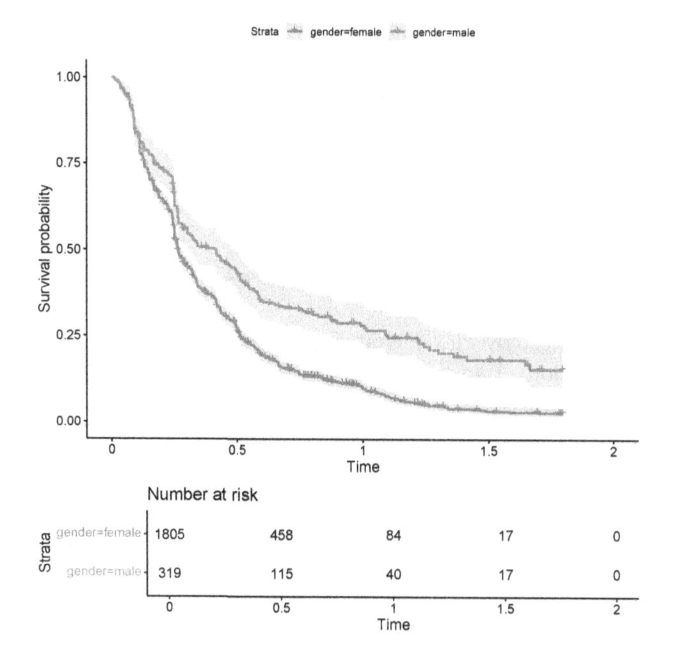

Figure 13.3: Survival curves for unemployment duration in the United States

The hypothesis of identical survival curves can be tested using a log-rank test. Denoting d_{tg} and r_{tg} as the number of failures and the number of observations at risk for the tth duration and the gth group, under the hypothesis of identical survival curves, the fitted number of failures is:

$$\hat{d}_{tg} = \frac{d_{t1} + d_{t2}}{r_{t1} + r_{t2}} r_{tg}$$

The test statistic is based on $\sum_t (d_{tg} - \hat{d}_{tg})$. Note that the value for the second group is necessarily the opposite of the one for the first group. The variance of this statistic is:

$$V = \frac{r_{t1} r_{t2} (r_t - d_t) d_t}{r_t^2 (r_t - 1)}$$

The statistic is then $\left(\sum_t (d_{tg} - \hat{d}_{tg})\right)^2 / V$ and is χ^2 with 1 degree of freedom under the hypothesis of same survival. This test can be computed using the `survival::surfdiff` function, with the same syntax as the one used for `survfit`:

```
survdiff(Surv(time, spell) ~ gender, unteach)
```

```
Call:
survdiff(formula = Surv(time, spell) ~ gender, data = unteach)

                 N Observed Expected (O-E)^2/E (O-E)^2/V
gender=female 1805     1600     1488      8.44      47.7
gender=male    319      222      334     37.60      47.7
```

```
Chisq= 47.7  on 1 degrees of freedom, p= 5e-12
```

The hypothesis of identical survival for males and females is highly rejected.

13.2 Parametric and semi-parametric estimators

In this section, we introduce covariates that may explain the length of the duration, or the value of the hazard function. Two kinds of models can be developed. The full parametric model specifies the whole distribution of the hazard function. On the contrary, in a semi-parametric approach, the hazard is the product of two component, the first depending on time and some parameters and the second depending only on the covariates and some other parameters. Then, only the second component is estimated, and therefore, the shape of the hazard doesn't need to be specified. We'll first consider the benchmark exponential model, for which the hazard is constant.[2]

13.2.1 Constant hazard and the exponential distribution

Consider the special case where the hazard is a constant θ. From Equation 13.2, the cumulative hazard is then: $\Lambda(t) = \int_0^t \theta ds = \theta t = -\ln S(t)$. The survival function is then $S(t) = e^{-\theta t}$ and the opposite of its derivative is the density function $f(t) = \theta e^{-\theta t}$. Therefore, if the hazard is constant, the distribution of T is the exponential distribution with parameter θ, which will be denoted by: $T \sim E(\theta)$.

The **moment generating function** of T for the exponential distribution is:

$$M(s) = \mathrm{E}\left(e^{sT}\right) = \int_0^{+\infty} e^{st}\theta e^{-\theta t} dt = \frac{\theta}{\theta - s}$$

for $\theta > s$. The **cumulant generating function** is the logarithm of the moment generating function:

$$K(s) = \ln M(s) = \ln\theta - \ln(\theta - s)$$

and the limit for $s \rightarrow 0$ of its first two derivatives with respect to s are the expected value and the variance of T. Therefore, $\mathrm{E}(T) = \frac{1}{\theta}$ and $\mathrm{V}(T) = \frac{1}{\theta^2}$. Therefore, for the exponential distribution, the mean equals the standard deviation. Moreover, the unit of measurement of θ is the inverse of the time unit, so that θdt is a number without units.

Denote $Z = \int_0^T \theta(s) ds$ as the cumulative hazard. This is a transformation of the time-scale such that $dZ = \frac{\partial Z}{\partial T} dT = \theta(T) dT$. With constant hazard, Z is simply proportional to T, as $dZ = \theta dT$, and the cumulative hazard is then θT. For example, if the constant hazard is 0.1 and the time is measured in days, a variation of 1 in the initial time scale (1 day) corresponds to a variation of 0.1 on the transformed time scale. As the hazard is positive, we have:

[2] This section relies heavily on Lancaster (1990). pp. 17-20.

$$\mathrm{P}(T \geq t) = \mathrm{P}\left(\int_0^T \theta(s)ds \geq \int_0^t \theta(s)ds\right) = \mathrm{P}(Z \geq z)$$

As $\mathrm{P}(T \geq t) = e^{-\int_0^t \theta(s)ds}$, we finally get: $\mathrm{P}(Z \geq z) = e^{-z}$, which means that Z, the cumulative hazard, is a unit exponential random variable: $Z \sim E(1)$.

Finally, consider the moment generating function of $\ln T$ with $T \sim E(\theta)$:

$$M(s)_{\ln T} = \mathrm{E}\left(e^{s \ln T}\right) = \mathrm{E}\left(T^s\right) = \int_0^{+\infty} t^s \theta e^{-\theta t} dt = \frac{\Gamma(1+s)}{\theta^s}$$

with $\Gamma(z) = \int_0^{+\infty} t^{z-1}e^{-t}dt$ the Gamma function. The corresponding cumulant generating function is: $K(s)_{\ln T} = \ln\Gamma(1+s) - s\ln\theta$, from which we obtain the expected value $\psi(1) - \ln\theta$ and the variance, $\psi'(1)$, where ψ and ψ' are the first and second derivatives of $\ln\Gamma$. ψ is called the digamma function and $\psi(1) = -\gamma$, $\gamma \approx 0.57721$ being the Euler-Mascheronni constant. ψ' is called the trigamma function and $\psi'(1) = \pi^2/6$. As $Z = \theta T \sim E(1)$, the density of Z is $g(Z) = e^{-Z}$. Consider now $U = \ln Z$. The density of U is:

$$f(U) = e^{-e^U}\left|\frac{dZ}{dU}\right| = e^{-e^U}e^U$$

which is the Gumbel distribution, also known as the extreme value distribution of type 1, for which $F(U) = 1 - e^{-e^U}$, $S(U) = e^{-e^U}$, $\mathrm{E}(U) = -\gamma$ and $\mathrm{V}(U) = \pi^2/6$.[3] As $U = \ln Z = \ln T + \ln\theta$, we can write:

$$\ln T = -\ln\theta + U$$

Now consider that the hazard depends on some covariates and denote θ_n the value of the hazard for individual n. Then we can write the exponential model in the following regression form:

$$\ln T_n = -\ln\theta_n + U_n = -(\gamma + \ln\theta_n) + (U_n + \gamma) = -(\gamma + \ln\theta_n) + W_n \tag{13.3}$$

Then $\mathrm{E}(\ln T_n \mid z_n) = -(\gamma + \ln\theta_n)$, $\mathrm{E}(W \mid x_n) = 0$ and $\mathrm{V}(W \mid x_n) = \pi^2/6$. θ can be parametrized as $e^{-\alpha + \beta^\top x_n}$, so that $\ln T_n = (\alpha - \gamma) + \beta^\top x_n + W_n$.

13.2.2 Estimation

Duration models are estimated by maximum likelihood. For censored observations, the contribution to the likelihood is the probability of surviving at the censoring time, which is $S(t_n \mid x_n)$. For uncensored observations, this is the density function: $f(t_n \mid x_n)$. Therefore, denoting d_n a dummy for complete (uncensored) observations, we get:

$$\ln L = \sum_{n=1}^N \left[d_n \ln f(t_n \mid x_n) + (1 - d_n)\ln S(t_n \mid x_n)\right]$$

[3]More precisely, this is the Gumbel distribution (minimum); for the Gumbel distribution (maximum), $F(U) = \exp(-\exp(-U))$.

The log-likelihood function can also be expressed in terms of the hazard function, as $\theta_n = f_n/S_n$:

$$\ln L = \sum_{n=1}^{N} \left[d_n \ln \theta(t_n \mid x_n) + \ln S(t_n \mid x_n)\right]$$

If there are no censored observations and if the distribution of time is exponential, Equation 13.3 can be estimated by least squares, which will be consistent, although inefficient. The `oil` data set doesn't have censored observations. Therefore, OLS estimation is consistent, even if it is less efficient than maximum likelihood if the distribution is really exponential. We estimated a model by maximum likelihood in Section 5.2.2, with `p98` and `varp98` as covariates. We reproduce this estimation using `surveg` and we compare the results with those obtained using `lm`. The `survreg` function has a `dist` argument that we set to `"exponential"`. The response should be generated using the `Surv` function, with the duration as the only argument in this example, as all the observations are uncensored:

```
survreg(Surv(dur) ~ p98 + varp98, oil, dist = "exponential") %>% summary
```

```
Call:
survreg(formula = Surv(dur) ~ p98 + varp98, data = oil, dist = "exponential")
            Value Std. Error    z      p
(Intercept) 1.440      0.551 2.61 0.0089
p98         0.830      0.493 1.68 0.0924
varp98      0.295      0.157 1.88 0.0597

Scale fixed at 1

Exponential distribution
Loglik(model)= -258.5   Loglik(intercept only)= -272.6
    Chisq= 28.2 on 2 degrees of freedom, p= 7.5e-07
Number of Newton-Raphson Iterations: 4
n= 53
```

```
lm(log(dur) ~ p98 + varp98, oil) %>% gaze
```

```
       Estimate Std. Error t value Pr(>|t|)
p98       0.948      0.446    2.12    0.039
varp98    0.226      0.146    1.55    0.127
```

As expected, the two sets of slopes are rather close, but the ML estimator doesn't seem to be more efficient in this example. β_k is the derivative of $\ln T_n$ with x_{nk}, so that $dT_n/T_n = \beta_k dx_{nk}$. Moreover, $d\theta_n/\theta_n = -\beta_k dx_{nk}$.

We then estimate the model for unemployment duration of Dutch teachers using as covariates the age, gender and (in log) last hourly wage:

```
un_exp <- survreg(Surv(time, spell) ~ age + gender + log(wage),
```

```
                  unteach, dist = "exponential")
coef(un_exp)
```

```
(Intercept)         age  gendermale    log(wage)
   -2.63528     0.02794     0.14972      0.36472
```

The results indicate that male unemployment spell duration are 15% longer than those for females. One more year of age increases the length of the spell by 2.79%. Finally, 1% of more previous wage increases the duration of the spell by 0.36%.

13.2.3 Accelerated failure time model

The previous model is a special case of a class of models called **accelerated failure time** (**AFT**), for which the duration can be written as: $T_n = Z/\lambda_n$, $\lambda_n = e^{\gamma^\top z_n}$, and Z is a random variable that doesn't depend on γ and x_n. λ_n therefore has a multiplicative effect on failure time, and failure time is decelerated if $\lambda < 1$ and accelerated if $\lambda > 1$. The exponential model is such a model, with $Z \sim E(1)$, or equivalently $U = \ln Z$ is an extreme value of type-1. Therefore, the AFT models are based on the following specification:

$$\ln T_n = -\ln \lambda_n + U_n$$

where λ_n can be parametrized as $e^{-\gamma^\top z_n}$, so that $\ln T_n = \gamma^\top z_n + U_n$. Different choices of the distribution of U_n lead to different flavors of AFT models. The basic exponential model is an AFT model with U following an extreme value type-1 distribution. Two natural extensions of this model consist of introducing a supplementary parameter $\sigma = 1/\delta$ or considering different distributions for U:

$$\ln T_n = -\ln \lambda_n + \frac{1}{\delta} U_n = -\ln \lambda_n + \sigma U_n$$

Keeping for the moment the hypothesis that U_n is type-1 extreme value, as $U = \delta \ln \lambda T$, the density of T is then:

$$f(T) = e^{\delta \ln(\lambda T)} e^{-e^{\delta \ln(\lambda T)}} \frac{dU}{dT} = (\lambda T)^\delta e^{-(\lambda T)^\delta} \frac{\delta}{T} = \delta \lambda^\delta T^{\delta - 1} e^{-(\lambda T)^\delta}$$

and the corresponding survival function is: $S(t) = e^{-(\lambda t)^\delta}$. This is a Weibull distribution, with a shape parameter equal to δ and a scale parameter equal to $\frac{1}{\lambda}$. The corresponding hazard function is: $\theta(t) = \delta \lambda^\delta t^{\delta - 1}$. This **Weibull model** is the most common model used in duration analysis. It is particularly appealing, as it is very easy to estimate and because it can deal with either a hazard function that increases ($\delta > 1$) or decrease ($\delta < 1$) with time. In the first case, there is a **positive time dependence** and in the second case, a **negative time dependence**. Obviously, the Weibull model reduces to the exponential model if $\delta = 1$. With $\lambda_n = e^{-\gamma z_n}$ and denoting $\epsilon_n = \ln T_n - \gamma^\top z_n$, the hazard and the survival functions are:

$$\theta(t) = \delta t^{\delta - 1} e^{-\delta \gamma^\top z} = \frac{\delta}{t} e^{\delta \epsilon} \text{ and } S(t) = e^{-t^\delta e^{-\delta \gamma^\top z}} \tag{13.4}$$

The log-likelihood function is then:

$$\ln L = \sum_{n=1}^{n} d_n \left[\ln \delta - \ln t + \delta \epsilon_n\right] - e^{\delta \epsilon}$$

With $U \sim \mathcal{N}(0,1)$, $\ln T = -\ln \lambda + \sigma U \sim \mathcal{N}(-\ln \lambda, \sigma)$ and therefore the time duration is log-normal and the hazard function is:

$$\theta(t) = \frac{1}{\sigma t} \frac{\phi\left(\frac{\ln t + \ln \lambda}{\sigma}\right)}{1 - \Phi\left(\frac{\ln t + \ln \lambda}{\sigma}\right)} = \frac{1}{\sigma t} r\left(\frac{\ln t + \ln \lambda}{\sigma}\right)$$

with ϕ and Φ the density and the probability functions of a standard normal variable and $r(z) = \phi(z)/\left(1 - \Phi(z)\right)$ is the inverse Mills' ratio. With, as previously, $\ln \lambda = -\gamma^\top z$, in terms of $\epsilon = \ln t - \gamma^\top z$, we have:

$$\theta(t) = \frac{1}{\sigma t} \frac{\phi(\epsilon/\sigma)}{1 - \Phi(\epsilon/\sigma)} \text{ and } S(t) = 1 - \Phi(\epsilon/\sigma)$$

The hazard function is then inverted U-shaped, whatever the value of σ.

$$\ln L = \sum_{n=1}^{n} d_n \left[-\ln \sigma - \ln t + \ln \phi(\epsilon_n/\sigma)\right] + (1 - d)\ln(1 - \Phi(\epsilon/\sigma))$$

Finally, consider that the distribution of U is standard logistic, so that $f(u) = \frac{e^u}{(1+e^u)^2}$. As $\delta \ln T\lambda = U$, $\frac{dU}{dT} = \delta/T$ and the density of T is then:

$$f(T) = \frac{(\lambda T)^\delta}{\left(1 + (\lambda T)^\delta\right)^2} \delta/T = \frac{\delta v T^{\delta-1}}{(1 + vT^\delta)^2}$$

with $v = \lambda^\delta$. The cumulative density and the survival function are $F(t) = \frac{vT^\delta}{1+vT^\delta}$ and $S(t) = \frac{1}{1+vT^\delta}$, so that the hazard function is:

$$\theta(t) = \delta \frac{vt^{\delta-1}}{1 + vt^\delta}$$

which is the **log-logistic model**. In terms of ϵ, we have:

$$\theta(t) = \frac{\delta}{t} \frac{e^{\delta\epsilon}}{1 + e^{\delta\epsilon}}, \; S(t) = \frac{1}{1 + e^{\delta\epsilon}}$$

and the following log-likelihood function:

$$\ln L = \sum_{n=1}^{n} d_n \left[\ln \delta - \ln t + \delta \epsilon_n\right] - (1 + d_n)\ln(1 + e^{\delta\epsilon})$$

If $\delta = 1$, the hazard reduces to $\theta(t) = v/(1 + vT)$ and is therefore decreasing from v for $T \to 0$ to 0 for $T \to +\infty$. The derivative of the hazard function is:

$$\theta'(t) = -\frac{\delta v T^{\delta-2}}{(1+vT^\delta)^2}\left[vT^\delta + (1-\delta)\right]$$

If $\delta < 1$ $\theta'(t)$ is negative for all T, so that the hazard function is decreasing, from $+\infty$ to 0. Finally, if $\delta > 1$, the hazard function is inverted U-shaped, with a maximum for $T = \left(\frac{\delta-1}{v}\right)^{1/\delta}$ and tends to 0 for $T \rightarrow 0$ and $T \rightarrow +\infty$.

All these models are estimated by updating the exponential model previously estimated:

```
un_wei <- update(un_exp, dist = "weibull")
un_lnorm <- update(un_exp, dist = "lognormal")
un_llog <- update(un_exp, dist = "loglogistic")
```

We first plot the hazard curves for a specific individual: we choose a male aged 35 with a last hourly wage of 10. We then compute $\hat{\lambda}$ for the different models:

```
x <- c(1, 35, 1, log(10))
l_llog <- exp(- sum(coef(un_llog) * x))
l_wei <- exp(- sum(coef(un_wei) * x))
l_lnorm <- exp(- sum(coef(un_lnorm) * x))
```

`survreg` objects have a `scale` object, which is the estimation of σ with our notations. As the Weibull and the log-logistic are expressed in term of δ, we take the inverse of this parameter for these two distributions.

```
s_llog <- un_llog$scale ^ (-1)
s_lnorm <- un_lnorm$scale
s_wei <- un_wei$scale ^ (-1)
c("log-log" = s_llog, "log-normal" = s_lnorm, "weibul" = s_wei)
##    log-log log-normal     weibul
##     1.7530     0.9948     1.1789
```

δ being larger than one for the log-logistic, we have an inverted U-shaped distribution for the hazard function. For the Weibull model, $\delta < 1$, the hazard curve is then increasing. We can then write functions for the hazards functions and plot them using `geom_function` (see Figure 13.4).

```
lnorm <- function(d) 1 / (s_lnorm * d) *
  dnorm( (log(d * l_lnorm) / s_lnorm)) /
  pnorm( (log(d * l_lnorm) / s_lnorm), lower.tail = FALSE)
loglog <- function(d) s_llog * l_llog ^ s_llog *
  d ^ (s_llog - 1) / (1 + (l_llog * d) ^ s_llog)
weib <- function(d) s_wei * l_wei ^ s_wei * d ^ (s_wei - 1)
ggplot() +
  geom_function(fun = lnorm, aes(linetype = "log-normal")) +
  geom_function(fun = loglog, aes(linetype = "log-log")) +
  geom_function(fun = weib, aes(linetype = "weibull")) +
  scale_x_continuous(limits = c(0, 2)) +
  labs(linetype = "distribution")
```

Table 13.1: Accelerated failure time model for unemployment duration

	exponential	weibull	log-logistic	log-normal
(Intercept)	−2.635	−2.511	−2.705	−2.634
	(0.149)	(0.128)	(0.138)	(0.140)
age	0.028	0.026	0.022	0.022
	(0.002)	(0.002)	(0.002)	(0.002)
gendermale	0.150	0.132	0.117	0.108
	(0.076)	(0.065)	(0.068)	(0.067)
log(wage)	0.365	0.350	0.310	0.270
	(0.061)	(0.051)	(0.057)	(0.058)
Log(scale)		−0.165	−0.561	−0.005
		(0.018)	(0.019)	(0.017)
Num.Obs.	2124	2124	2124	2124
AIC	442.5	369.4	294.6	294.4
BIC	465.1	397.7	322.9	322.7
RMSE	0.36	0.35	0.33	0.33

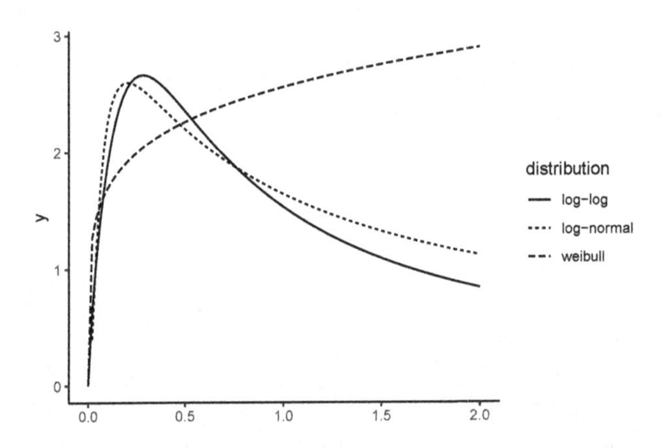

Figure 13.4: Hazard functions for employment duration

The results of the four AFT models are presented in Table 13.1.

13.2.4 Proportional hazard models

The **proportional hazard (PH)** models starts with the specification of the hazard function. More specifically, it is assumed that the hazard function can be written as the product of two terms:

$$\theta(t|x) = \lambda(t, \rho) f(x, \gamma)$$

- the first depends only on the duration and on some unknown parameters; it is called the baseline hazard,

- the second depends on some covariates and some unknown parameters, but not on duration; it is parametrized as $e^{\gamma^\top z} = e^{\alpha+\beta^\top x}$. Consider two individuals characterized by two sets of values for the covariates x_1 and x_2. The ratio of their hazard for a given duration is:

$$\frac{\theta(t|x = x_1)}{\theta(t|x = x_2)} = \frac{f(x_1, \gamma)}{f(x_2, \gamma)} = \frac{e^{\alpha+\beta^\top x_1}}{e^{\alpha+\beta^\top x_2}} = e^{\beta(x_1-x_2)}$$

and therefore doesn't depend on the duration. This model is called for this reason the **proportional hazard model**. The marginal effect of one covariate on the hazard is:

$$\frac{\partial\theta(t|x)}{\partial x_k} = \lambda(t, \rho)\frac{\partial f}{\partial x_k} = \lambda(t, \rho)\beta_k f(x, \gamma) = \beta_k\theta(t, x)$$

Therefore, the marginal effect is proportional to the hazard; more precisely, it is just β_k times the hazard. Note that the parametrization is the opposite of the one used for the AFT models. Therefore, the interpretation of the coefficients should be modified: a positive value of β_k means that the covariate has a positive effect on the hazard and therefore a negative effect on the duration. The most common used PH models are the exponential and the Weibull model. For the Weibull model, the hazard and the corresponding survival functions are generally parametrized as:

$$\theta(t_n) = \delta t_n^{\delta-1}e^{\gamma^\top z_n} \text{ and } S(t_n) = e^{-t_n^\delta e^{\gamma^\top z_n}} \tag{13.5}$$

This is exactly the same hazard function as Equation 13.4 with a different parametrization: starting from the PH coefficients, the AFT coefficients are $-\gamma/\delta$. The exponential model is obtained as the special case of Equation 13.5 where $\delta = 1$. In this case, the two vectors of coefficients of the AFT and the PH models are just opposite. The Weibull and the exponential models are the only two models which belong to the AFM and the PH family. Two more specifications of proportional hazard models are often used, namely the generalized Weibull and the Gompertz models.

The Cox proportional hazard model (D. R. Cox 1972, 1975) is semi-parametric, as only the $f()$ function is estimated and not the baseline hazard function. It is better understood using a simple example. Without lack of generality, consider that the observations are arranged in an increasing order of duration. Consider a sample of size 4, with $t = (2, 3, 5, 8)$. For time $t_1 = 2$, all the observations are at risk (none of them failed before). Consider the probability that one observation n fails at time $t_1 = 2$, given that it was at risk. This writes: $P(t_n = t_1|t_n \geq t_1) = \lambda(t_1)f(x_n, \gamma)$. The probability that an observation fails at $t_1 = 2$ is the sum of this expression for the four observations: $\lambda(t_1)\sum_{n=1}^4 f(x_n, \gamma)$. Finally, the probability that observation n fails given that one observation fails is:

$$\frac{\lambda(t_1)f(x_n, \gamma)}{\lambda(t_1)\sum_{n=1}^4 f(x_n, \gamma)} = \frac{f(x_n, \gamma)}{\sum_{n=1}^4 f(x_n, \gamma)}$$

This probability therefore doesn't depend on the baseline hazard function. Moreover, if $f(x_n, \gamma) = e^{\gamma^\top z_n}$, it has a logit form: $e^{\gamma^\top z_n}/\sum_n e^{\gamma^\top z_n}$. As this is the first observation that failed for $t_1 = 2$, the contribution of this observation is:

$$\frac{e^{\gamma^\top z_1}}{e^{\gamma^\top z_1} + e^{\gamma^\top z_2} + e^{\gamma^\top z_3} + e^{\gamma^\top z_4}}$$

The reasoning is similar for $t_2 = 3$. The set of observations at risk is the last three observations and the contribution of observation 2 is then:

$$\frac{e^{\gamma^\top z_2}}{e^{\gamma^\top z_2} + e^{\gamma^\top z_3} + e^{\gamma^\top z_4}}$$

For observation 3, the set of observations at risk reduces to observations 3 and 4:

$$\frac{e^{\gamma^\top z_3}}{e^{\gamma^\top z_3} + e^{\gamma^\top z_4}}$$

Finally, for the last observation, the set of observations at risk is only this observation, so that the contribution of observation 4 is:

$$\frac{e^{\gamma^\top z_4}}{e^{\gamma^\top z_4}} = 1$$

The **partial likelihood** for the whole sample is the product of these four contributions:

$$L = \frac{e^{\gamma^\top z_1}}{e^{\gamma^\top z_1} + e^{\gamma^\top z_2} + e^{\gamma^\top z_3} + e^{\gamma^\top z_4}} \times \frac{e^{\gamma^\top z_2}}{e^{\gamma^\top z_2} + e^{\gamma^\top z_3} + e^{\gamma^\top z_4}} \times \frac{e^{\gamma^\top z_3}}{e^{\gamma^\top z_3} + e^{\gamma^\top z_4}} \times 1$$

A more general expression is obtained by denoting R_n the set of observations for which $t_m \geq t_n$, i.e., the set of observations at risk at time t_n:

$$L = \prod_{n=1}^{N} \frac{e^{\gamma^\top z_n}}{\sum_{n \in R_n} e^{\gamma^\top z_n}}$$

and the partial log-likelihood is therefore:

$$\ln L = \sum_{n=1}^{N} \left(\gamma^\top z_n - \ln \sum_{n \in R_n} e^{\gamma^\top z_n} \right)$$

An interesting feature of Cox proportional hazard model is that the duration response is only used as a way to order the observations. For example, if $t = (1, 2, 3, 4)$ instead of $t = (2, 3, 5, 8)$, we would obtain exactly the same likelihood and therefore the same fitted parameters. Censored observations contribute only to the partial likelihood as elements of the set of observations at risk for some observations. For example, if observation 2 is censored, the likelihood is:

$$L = \frac{e^{\gamma^\top z_1}}{e^{\gamma^\top z_1} + e^{\gamma^\top z_2} + e^{\gamma^\top z_3} + e^{\gamma^\top z_4}} \times \frac{e^{\gamma^\top z_3}}{e^{\gamma^\top z_3} + e^{\gamma^\top z_4}} \times 1$$

and the general formula for the partial log-likelihood is, denoting d_n a dummy equal to 0 if observation n is censored and 1 otherwise:

$$\ln L = \sum_{n=1}^{N} d_n \left(\gamma^\top z_n - \ln \sum_{n \in R_n} e^{\gamma^\top z_n} \right)$$

Consider now that the sample includes ties, for example $t = (2, 4, 4, 8)$. Durations for $n = 2$ and $n = 3$ are equal because of rounding measures, and we don't know which observation failed first. If the event occured first for $n = 2$, the probability for observations 2 and 3 is:

$$\frac{e^{\gamma^\top z_2}}{e^{\gamma^\top z_2} + e^{\gamma^\top z_3} + e^{\gamma^\top z_4}} + \frac{e^{\gamma^\top z_3}}{e^{\gamma^\top z_3} + e^{\gamma^\top z_4}}$$

On the contrary, if the event occurs first for $n = 3$:

$$\frac{e^{\gamma^\top z_3}}{e^{\gamma^\top z_2} + e^{\gamma^\top z_3} + e^{\gamma^\top z_4}} + \frac{e^{\gamma^\top z_2}}{e^{\gamma^\top z_2} + e^{\gamma^\top z_4}}$$

The probability for observations 2 and 3 is the sum of these two probabilities and the partial likelihood is then:

$$\left(\frac{e^{\gamma^\top z_2}}{e^{\gamma^\top z_2} + e^{\gamma^\top z_3} + e^{\gamma^\top z_4}} + \frac{e^{\gamma^\top z_3}}{e^{\gamma^\top z_3} + e^{\gamma^\top z_4}} + \frac{e^{\gamma^\top z_3}}{e^{\gamma^\top z_2} + e^{\gamma^\top z_3} + e^{\gamma^\top z_4}} + \frac{e^{\gamma^\top z_2}}{e^{\gamma^\top z_2} + e^{\gamma^\top z_4}} \right)$$

Therefore, in the case of a tie with two observations, the product of two probabilities is replaced by the sum of four probabilities. If there were a tie with three observations, the product of three probabilities would be replaced by the sum of 18 probabilities. More generally, with a tie of k observations, the number of probabilities that should be computed is $k \times k!$. For example, with $k = 10$, more than 30 million probabilities should be computed. Therefore, this formula can only be applied when there are few and "thin" ties. Otherwise, approximations have been proposed by Breslow and Efron. The `survival::coxph` function fits the Cox proportional hazard regression model. We've already estimated the Weibull model using the `survival::survreg` function in its accelerated failure time parametrization. `micsr::weibreg` enables to estimate the Weibull model using either the AFT or the proportional hazard parametrization, using the `model` argument:

```
weib_aft <- weibreg(Surv(time, spell) ~ gender + age +
                      log(wage), unteach, model = "aft")
weib_ph <- update(weib_aft, model = "ph")
cox <- coxph(Surv(time, spell) ~ gender + age + log(wage),
             unteach)
```

The results are presented in Table 13.2. As stated previously, starting from the PH coefficient, the AFT is obtained as the opposite divided by δ; for example, for age: $-(-0.031/1.179) = 0.026$. Obviously, the value of the log-likelihood function is the same for the PH and the AFT versions of the Weibull model.

Table 13.2: Weibull model (AFT and PH) and Cox semi-parametric model for the unemployment data set

	Weibull (AFT)	Weibull (PH)	Cox model
(Intercept)	−2.511	2.961	
	(0.128)	(0.155)	
gendermale	0.132	−0.156	−0.137
	(0.065)	(0.076)	(0.076)
age	0.026	−0.031	−0.028
	(0.002)	(0.003)	(0.002)
log(wage)	0.350	−0.412	−0.371
	(0.051)	(0.061)	(0.061)
shape	1.179	1.179	
	(0.021)	(0.021)	
Num.Obs.	2124	2124	2124
AIC	369.4	369.4	24 745.2
BIC	397.7	397.7	24 762.2
Log.Lik.	−179.682	−179.682	
RMSE			0.91

13.3 Heterogeneity

Consider a population with two groups characterized by two different constant hazard rates, 0.4 for the first group and 0.1 for the second group.[4] If there are 100 individuals in both groups, the number of events (observations at risk) for the first group is $0.4 \times 100 = 40$ (60) for the first period, $0.4 \times 60 = 24$ (36) for the second and $0.4 \times 36 = 21.6$ (12.96) for the third. For the second group, we get $0.1 \times 100 = 10$ (90) for the first period, $0.1 \times 90 = 9$ (81), $0.1 \times 81 = 8.1$ (72.9) and $0.1 \times 72.9 = 7.29$ (65.61). Summing for the two groups, we get 50 (150), 33 (117), 22.5 (94.5). The hazard is then $50/200 = 0.25$ for the first period, $33/150 = 0.22$ for the second one and $22.5/117 = 0.1923$ for the third period. The hazard rate for the whole population is therefore decreasing, although it is constant for every individual. This result is easy to understand because, as times goes on, the proportion of individuals with the low hazard rate in the observations at risk is increasing. It is therefore important to detect the presence of heterogeneity and, in the presence of heterogeneity, to fit models that take heterogeneity into account.

13.3.1 Detecting heterogeneity

If heterogeneity is present, fitting a model that doesn't take it into account results in a wrong specification that can be tested using **generalized residuals**. In the context of duration models, integrated hazards can be considered as generalized residuals, as, if the specification is correct, their distribution is a unit exponential. Using the accelerating failure specification, remember that the integrated hazard is $\epsilon_n = \Lambda(t_n) = -\ln S(t_n) = e^{\delta(\ln t - \gamma^\top z_n)}$. The distribution of ϵ_n doesn't depend on x_n and , being a unit exponential distribution,

[4]This example is from Cameron and Trivedi (2005), p. 611.

its first four uncentered moments are 1, 2, 6, 24. To get an idea of the relevance of the specification, we can plot minus the log of the empirical survival (which is the number of observations m such that $\epsilon_m > \epsilon_n$ divided by the total number of observations) against the generalized residuals; the points should then be grossly aligned on the 45° line. For censored observations, the duration is not known and the integrated hazard is replaced by its expectation. Denoting L the censoring time, the expectation is:

$$\mathrm{E}(\epsilon(T) \mid T \geq L) = \int_{\epsilon(L)}^{+\infty} \frac{\epsilon f(\epsilon)}{S(\epsilon(L))} d\epsilon = \frac{1}{e^{-\epsilon(L)}} \int_{\epsilon(L)}^{+\infty} \epsilon e^{-\epsilon} d\epsilon = \epsilon(L) + 1$$

Figure 13.5 plots the generalized residuals, that has been computed using the `micsr::gres` function. The fit seems excellent for low values of the generalized residuals, and bad for values higher than 4 (about 2% of the sample).

```
weib <- weibreg(Surv(time, spell) ~ gender + age + log(wage),
                unteach)
un_gres <- gres(weib)
N <- nobs(weib)
int_haz <- tibble(gres = un_gres) %>% group_by(gres) %>%
  summarise(n = n()) %>% arrange(gres) %>%
  mutate(cumn = cumsum(n), esurv = 1 - cumn / N, ech = - log(esurv))
int_haz %>% ggplot(aes(gres, ech)) + geom_step() +
  geom_abline(intercept = 0, slope = 1, linetype = "dotted")
```

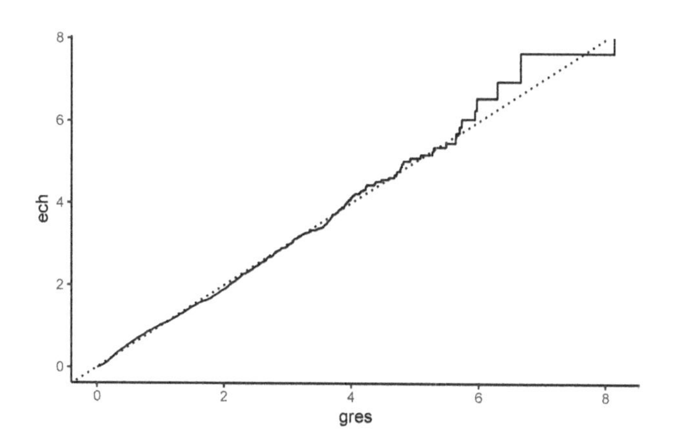

Figure 13.5: Generalized residuals for the Weibull model

The empirical moments are:

```
sapply(2:4, function(i) mean(un_gres ^ i))
## [1]  2.012  5.918 22.307
```

Conducting a conditional moment test, we get:

```
cmtest(weib, powers = 2:4, opg = FALSE) %>% gaze
## chisq = 3.860, df: 3, pval = 0.277
```

and the hypothesis of a correct specification is therefore not rejected.

13.3.2 Weibull-Gamma model

To take into account the heterogeneity, we can use mixing distributions, either finite (taking for example two categories of individuals as in the introductory example) or infinite. We'll present in this section a model very widely used for its simplicity, the Weibull-Gamma model. Assume that the survival function for individual n is:

$$S(t_n \mid x_n, \nu_n) = e^{-\mu_n \nu_n t_n^\delta} \tag{13.6}$$

where as usual, $\mu_n = e^{\alpha + \beta x_n}$. ν_n is an unobserved term that is specific to the individual. If it were observed, we would get the usual Weibull model. As it is unobserved, we have to consider the unconditional survival function, i.e., we have to integrate out Equation 13.6 with ν. Denoting f the density of ν, we get:

$$S(t_n \mid x_n) = \int_0^{+\infty} e^{-\mu_n \nu t_n^\delta} f(\nu) d\nu \tag{13.7}$$

If the regression contains an intercept, the mean of ν is not identified and can be set to 1. The one-parameter Gamma distribution is:

$$f(\nu) = \frac{\rho^\rho \nu^{\rho-1} e^{-\rho\nu}}{\Gamma(\rho)}$$

with $\mathrm{E}(\nu) = 1$ and $\mathrm{V}(\nu) = 1/\rho$. Using this mixing distribution, we get:

$$S(t_n \mid x_n) = \int_0^{+\infty} e^{-\mu_n \nu t_n^\delta} \frac{\rho^\rho \nu^{\rho-1} e^{-\rho\nu}}{\Gamma(\rho)} d\nu = \frac{\rho^\rho}{\Gamma(\rho)} \int_0^{+\infty} e^{-\nu(\mu_n \nu t_n^\delta + \rho)} \nu^{\rho-1} d\nu$$

With the change of variable: $u = \nu(\mu_n \nu t_n^\delta + \rho)$:

$$S(t_n \mid x_n) = \int_0^{+\infty} e^{-\mu_n \nu t_n^\delta} \frac{\rho^\rho \nu^{\rho-1} e^{-\rho\nu}}{\Gamma(\rho)} d\nu = \frac{\rho^\rho}{\Gamma(\rho)} \frac{1}{(\mu_n t_n^\delta + \rho)^\rho} \int_0^{+\infty} e^{-u} u^{\rho-1} d\nu$$

which simplifies to: $\left(\frac{\mu_n t_n^\delta}{\rho} + 1\right)^{-\rho}$. As ρ is the inverse of the variance, it is simpler to parametrize the survival function with $\chi = 1/\rho$, so that the hypothesis of homogeneity is simply $\chi = 0$:

$$S(t_n \mid x_n) = \frac{1}{(1 + \chi \mu_n t_n^\delta)^{\frac{1}{\chi}}}$$

The density is obtained as the opposite of the derivative of the survival function. The corresponding hazard function is:

$$\theta(t_n \mid x_n) = \frac{\delta\mu_n t_n^{\delta-1}}{1 + \chi\mu_n t_n^\delta}$$

and the log-likelihood function is then:

$$\ln L = \sum_{n=1}^{n} d_n \left[\ln\delta + (\delta - 1)\ln t + \ln\mu_n\right] - (1/\chi + d_n)\ln\left(1 + \chi\mu_n t_n^\delta\right)$$

as $\chi \to 0$, $\ln\left(1 + \chi\mu_n t_n^\delta\right) \to \chi\mu_n t_n^\delta$ so that the Weibull is obtained as the limiting case where the variance of the mixing distribution tends to 0. Remember from Table 13.1 that, for the Weibull AFT model, the scale parameter was $\sigma = 0.85$, which corresponds to $\delta = 1/0.85 = 1.18$ and therefore to an increasing hazard rate. The `micsr::weibreg` function estimates the Weibull-Gamma model when the `mixing` argument is set to `TRUE`:

```
weib <- weibreg(Surv(time, spell) ~ gender + age + log(wage),
                unteach)
weib %>% gaze(coef = "shape")
##        Estimate Std. Error z-value Pr(>|z|)
## shape    1.1789     0.0215      55   <2e-16
gweib <- update(weib, mixing = TRUE)
gweib %>% gaze(coef = c("shape", "var"))
##        Estimate Std. Error z-value Pr(>|z|)
## shape    1.5978     0.0635   25.15  < 2e-16
## var      0.7017     0.1039    6.75  1.5e-11
```

The supplementary parameter (χ) is called `"var"` and is highly significantly different from 0, so that the hypothesis of homogeneity is highly rejected. Note also that the shape parameter is much higher than previously.

13.4 Other models

13.4.1 Multi-state models

Consider now the case where several events are possible, for example two events denoted a and b. Let $\tilde{t}_n^a$ and $\tilde{t}_n^b$ be the duration for individual n for the two events a and b. The observed duration is t_n. Three cases should be considered:

- $t_n < \min(\tilde{t}_n^a, \tilde{t}_n^b)$ (which means that $t_n < \tilde{t}_n^a$ and $t_n < \tilde{t}_n^b$), the observation is then censored,
- $t_n = \min(\tilde{t}_n^a, \tilde{t}_n^b)$, the event is:
  - a if $\tilde{t}_n^a < \tilde{t}_n^b$,
  - b if $\tilde{t}_n^a > \tilde{t}_n^b$.

The **competing risk** model is a very simple multi-state model which extends the Cox proportional hazard model to the multi-state case. Consider for example state a. For an observation n, if the event a is observed, we have an observation of a complete spell. Otherwise, if the event is b, this means that $t_n = \tilde{t}_n^b < \tilde{t}_n^a$ and the observation is censored as far as we are interested in state a. We can therefore estimate the competing risk model with two independent Cox PH models:

- for the first, the event is a and censoring occurs either if the observation is censored or if the event is b,
- for the second, the event is b and censoring occurs either if the observation is censored or if the event is a,

Consider the `recall` data set, initially used by Katz (1986) and later by Sueyoshi (1995). It contains 1045 spells of unemployment in 1980.

```
recall %>% print(n = 3)
```

```
# A tibble: 1,045 x 16
     id spell end      duration   age sex    educ race      nb ui
  <dbl> <dbl> <fct>       <dbl> <int> <fct> <int> <fct> <int> <fct>
1     4     1 recall         17    26 male      5 white     3 yes
2     7     2 recall         39    24 male      8 white     4 no
3     7     3 censored       40    23 male      8 white     4 no
# i 1,042 more rows
# i 6 more variables: marital <fct>, unemp <int>, wifemp <fct>,
#   homeowner <fct>, occupation <fct>, industry <fct>
```

The analysis of Katz (1986) showed that an important transition from unemployment was a recall from the previous employer. The `end` covariate indicates the different modalities of transition:

```
recall %>% count(end)
```

```
# A tibble: 3 x 2
  end          n
  <fct>    <int>
1 new-job    218
2 recall     593
3 censored   234
```

We consider the `new-job` and `recall` modalities as an indication of a complete spell. The model for unemployment spells is estimated using as covariates the age, sex, years of education, number of dependents (`nb`), a dummy for unemployment insurance during the spell (`ui`), marital status, county unemployment rate (`unemp`), wife's employment status, a dummy for home ownership and factors for occupation and industry. The competing risk model can then be estimated using the following two calls to `coxph`:

```
un_recall <- coxph(Surv(duration, end == "recall") ~ age + sex + educ +
                       race + unemp + wifemp + homeowner +
                       occupation + industry, recall)
```

```
un_new <- coxph(Surv(duration, end == "new-job") ~ age + sex + educ +
                  race + unemp + wifemp + homeowner +
                  occupation + industry, recall)
```

where the **event** argument is a dummy equal to **TRUE** when **end** equals **"recall"** for the first regression and **"new-job"** for the second regression. The competing risk model can be estimated more simply with a simple call to **coxph** when the **event** argument is a factor for which the first level indicates censored observations. In the **recall** data set, the first level is **"new-job"**, so that we need to change the order of the levels using **relevel**:

```
recall %>% pull(end) %>% levels
## [1] "new-job"  "recall"   "censored"
recall <- recall %>% mutate(end = relevel(end, "censored"))
recall %>% pull(end) %>% levels
## [1] "censored" "new-job"  "recall"
```

Moreover, the **id** argument is mandatory. This is the individual identifier and it is really useful only when there are several observations for the same individuals, which is not the case here. We therefore create an **idspell** variable before proceeding to the estimation. To save space, we use a very limited subset of covariates.

```
recall <- recall %>%
  add_column(id_spell = 1:nrow(recall), .before = 1)
coxph(Surv(duration, end) ~ age + sex, data = recall, id = id_spell)
```

```
Call:
coxph(formula = Surv(duration, end) ~ age + sex, data = recall,
    id = id_spell)


1:2           coef exp(coef) se(coef) robust se  z    p
  age       -0.017     0.983    0.008     0.008 -2 0.02
  sexfemale -0.381     0.683    0.189     0.190 -2 0.04


1:3           coef exp(coef) se(coef) robust se  z     p
  age        0.022     1.022    0.004     0.004  6 2e-09
  sexfemale -0.386     0.680    0.119     0.122 -3 0.002

 States:  1= (s0), 2= new-job, 3= recall

Likelihood ratio test=56  on 4 df, p=2e-11
n= 1045, unique id= 1045, number of events= 811
```

13.4.2 Grouped data

Duration is often recorded not as a continuous variable, but as a set of intervals. For example, unemployment duration may not be measured in days, but in months or years. In this case, the model must be written in terms of discrete time. Denote $t_0, t_1, t_2, ...t_T$ (with $t_0 = 0$) and a the interval defined by $t_{a-1} \leq t < t_a$. The survival at time t_a is such that $-\ln S(t_a) = \int_0^{t_a} \theta(s) ds$. Therefore, for the ath interval, we have:

$$-\ln S(t_a) + \ln S(t_{a-1}) = -\ln \frac{S(t_a)}{S(t_{a-1})} = \int_{t_{a-1}}^{t_a} \theta(s) ds$$

The hazard in the ath interval is the probability of an event in this interval (which is the difference in survival) divided by the initial value of the survival:

$$\theta_a = \frac{S(t_{a-1}) - S(t_a)}{S(t_{a-1})} = 1 - e^{-\int_{t_{a-1}}^{t_a} \theta(s) ds}$$

Now assume a proportional hazard: $\theta(s) = \lambda(s) e^{\gamma^\top z_s}$. Denoting $\delta_a = \ln \int_{t_{a-1}}^{t_a} \lambda(s) d_s$, we get:

$$\theta_a = 1 - e^{-e^{\delta_a + \gamma^\top z_{t_{a-1}}}} = F(\delta_a + \gamma^\top z_{t_{a-1}})$$

Where F is the CDF of the extreme value type 1 distribution. Note that the covariates may be time-varying from one interval to another. In a given interval, values of the covariates are measured at the beginning of the interval. The discrete survival function at the beginning of this interval is:

$$S_{t_{a-1}} = \prod_{s=1}^{a-1} (1 - \theta_s) = \prod_{s=1}^{a-1} (1 - F(\delta_s + \beta^\top x_{t_{s-1}}))$$

The probability of a transition in the ath interval is the product of the discrete hazard for interval a and the value of the survival function at the beginning of this interval (i.e., t_{a-1}). For observation n, the contribution to the likelihood is then:

$$F(\delta_{a_n} + \gamma^\top z_{nt_{a-1}}) \prod_{s=1}^{a_n - 1} (1 - F(\delta_s + \gamma^\top z_{nt_{s-1}}))$$

Finally, the log-likelihood is, denoting $F_{ns} = F(\delta_s + \gamma^\top z_{nt_{s-1}})$ and y_{ns} a dummy equal to 1 if n has a transition in the sth interval and 0 otherwise:

$$\begin{array}{rcl} \ln L & = & \sum_{n=1}^N \left[\sum_{s=1}^{a_{n-1}} \ln(1 - F_{ns}) + \ln F_{na_n} \right] \\ & = & \sum_{n=1}^N \sum_{s=1}^{a_n} \left[(1 - y_{ns}) \ln(1 - F_{ns}) + y_{ns} \ln F_{ns} \right] \end{array} \quad (13.8)$$

This is an unbalanced pooled binomial model, the number of observation for each individual being the index of the interval of its transition. Equation 13.8 deals correctly with right-censored observation as, in this case, $y_{na_n} = 0$ and the correct term $(1 - F_{na_n})$ is introduced in the log-likelihood. A specific set of δ_s parameters can be estimated, which enables to retrieve from the estimation the shape of the hazard function.

An alternative to this binomial model is to start with the proportional hazard formulation:

$\theta(t) = \lambda(t)e^{-\gamma^\top z}$[5] and to compute the integrated hazard: $\int_0^t \theta(s)ds = e^{-\gamma^\top z_t}\int_0^t \lambda(s)ds = e^{\delta_t-\gamma^\top z_t}$ where γ_t is defined as: $\ln \int_0^t \lambda(s)ds$. As the integrated hazard is the opposite of the logarithm of the survival function, we have: $S(t) = e^{-e^{\delta_t-\gamma^\top z_t}}$. The probability of transition in the $a^{\mbox{th}}$ interval is $S(t_{a-1}) - S(t_a) = F(t_a) - F(t_{a-1})$, with $F(z) = 1 - e^{-e^z}$ the CDF of the extreme value type 1 distribution. Therefore, the log-likelihood is:

$$\ln L = \sum_{n=1}^N \sum_{s=1}^A y_{ns} \ln\left[F(\delta_{t_s} - \gamma^\top z_{nt_s}) - F(\delta_{t_{s-1}} - \gamma^\top z_{nt_{s-1}})\right]$$

which is the ordered model (see Equation 10.17), δ_t being a set of threshold. With censored observations, denoting d_n a dummy for complete observations, the log-likelihood becomes:

$$\begin{array}{rl}\ln L = \sum_{n=1}^N y_{ns} \sum_{s=1}^A & \left(d_n \ln\left[F(\delta_{t_{a_n}} - \gamma^\top z_n) - F(\delta_{t_{a_n-1}} - \gamma^\top z_n)\right]\right. \\ & \left. + (1-d_n)\ln\left[1 - F(\delta_{t_{a_n}} - \gamma^\top z_n)\right]\right)\end{array}$$

With the `micsr::ordreg` function, the response can be a `Surv` objects in order to take into account censored observations.

```
form_ord <- Surv(duration, end == "recall") ~ age + sex + educ + race +
  nb + ui + marital + unemp + wifemp + homeowner + occupation + industry
recall_ord <- ordreg(form_ord, recall, link = "cloglog")
```

To estimate the binomial model, we first need to construct the pooled sample, i.e., to get one line for one period for every spell. We first create a data frame with only the identifiers for the individual and the spell and the duration. We then nest the data frame by the two identifiers. This results in a new column called `data` that contains, for each observation a one-line one-column tibble containing the duration of the spell. Using `mutate`, we then create a new column in `data` containing integers from 1 to the given value of `duration` for each observation. We then ungroup and remove the duration column.

```
durations <- recall$duration %>% unique
pooled <- recall %>%
  select(spell, id, duration) %>%
  nest_by(spell, id) %>%
  mutate(period = list(1:data$duration)) %>%
  unnest(cols = c(data, period)) %>%
  ungroup %>%
  select(- duration)
pooled %>% print(n = 3)
```

```
# A tibble: 14,646 x 3
  spell    id period
  <dbl> <dbl>  <int>
1     1     4      1
2     1     4      2
3     1     4      3
```

[5] Note that we used $-\gamma$ instead of γ, like for the AFT models, for a reason that will be clear later in this section.

Table 13.3: Estimation of the grouped duration model for the recall data set

	ordered cloglog	binomial cloglog	probit
age	−0.014 (0.004)	0.014 (0.004)	0.006 (0.002)
female	0.014 (0.155)	−0.016 (0.155)	0.006 (0.073)
educ	0.032 (0.022)	−0.030 (0.022)	−0.016 (0.011)
non-white	0.240 (0.097)	−0.240 (0.097)	−0.117 (0.048)
No of dependents	0.002 (0.029)	−0.002 (0.029)	−0.005 (0.015)
UI receipt	0.191 (0.096)	−0.187 (0.096)	−0.084 (0.047)
married	−0.044 (0.146)	0.044 (0.146)	0.036 (0.071)
Area unemploy.	0.007 (0.017)	−0.007 (0.017)	−0.003 (0.008)
Wife works	−0.123 (0.100)	0.127 (0.100)	0.065 (0.051)
Homeowner	−0.387 (0.099)	0.391 (0.098)	0.206 (0.049)
Industry indicators	Yes	Yes	Yes
Occup. indicators	Yes	Yes	Yes
Num.Obs.	1045	13 246	13 246
AIC	4431.5	4443.6	4449.0
BIC	4693.9	4848.1	4853.5
Log.Lik.	−2162.733	−2167.804	−2170.479
F		7.861	7.661
RMSE		0.20	0.20

```
# i 14,643 more rows
```

There are some periods without observations of an event or a censoring. Observations for these periods should be removed:

```
durations <- recall %>% pull(duration) %>% unique
pooled <- pooled %>% filter(period %in% durations)
```

Finally we join this tibble with the initial tibble, and we create the event variable (`end` should be `"recall"` and the period should be equal to the duration).

```
recall_pooled <- pooled %>%
  left_join(recall, by = c("spell", "id")) %>%
  mutate(event = (end == "recall" & period == duration))
```

We then estimate different flavors of binomial models, using the same formula than for the ordered model, except that the response is now `event` and that dummies for periods are added:

```
form_binom <- update(form_ord, event ~ . + factor(period))
recall_binom <- glm(form_binom, family = binomial(link = "cloglog"),
                    data = recall_pooled)
recall_probit <- update(recall_binom, family = binomial(link = "probit"))
recall_logit <- update(recall_probit, family = binomial(link = "logit"))
```

Results are presented in Table 13.3.

13.4.3 Interval regression

Sometimes, the exact value of the response is not known, but only an interval that contains the value. For example, income is often indicated this way in individual surveys:

- less than \$10,000,
- from \$10 to \$20,000 dollars,
- from \$20 to \$50,000 dollars,
- more than \$50,000 dollars.

Note that, in this case, there is no bound for the last category, so that we also have a right-censoring problem. Denote μ_l and μ_u as the lower and the upper bounds of the interval and f the density of the distribution of the variable of interest. The three different cases are presented in Figure 13.6. The variable of interest is represented on the horizontal axis and its distribution by the density curve. The lower and upper bounds are respectively denoted μ_l and μ_u. For Figure 13.6 (a), we have $\mu_l = 0$, so that the interval is left-censored. In Figure 13.6 (b), both the lower and the upper bounds are observed. Finally, in Figure 13.6 (c), only the lower bound is observed and the interval is therefore right-censored. Denoting $F(y_n; \theta_n)$ the cumulative density function for y, the general formula for the probability that enters the likelihood function for one observation is: $F(\mu_n^u; \theta_n) - F(\mu_n^l; \theta_n)$ which simplifies respectively to $F(\mu_n^l)$ and $1 - F(\mu_n^u)$ for the cases where $\mu_l = 0$ (left-censored interval) and $\mu_u = \infty$ (right-censored interval).

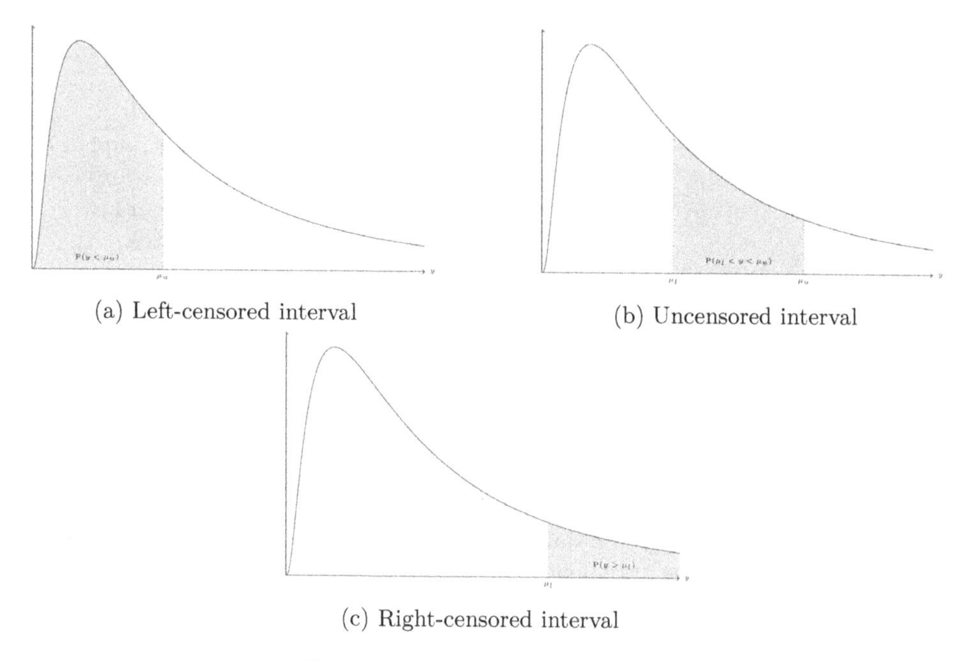

(a) Left-censored interval

(b) Uncensored interval

(c) Right-censored interval

Figure 13.6: Interval regression

Popular choices of distribution functions are the log-normal and the Weibull distribution. For the first, a simple parametrization is to set the mean of $\ln y$ equal to $\gamma^\top z_n$ and its standard deviation to σ. The probability is then:

$$\Phi\left(\frac{\ln \mu_n^u - \gamma^\top z_n}{\sigma}\right) - \Phi\left(\frac{\ln \mu_n^l - \gamma^\top z_n}{\sigma}\right)$$

For the Weibull distribution, the cumulative density function is: $F(y; \lambda, \delta) = 1 - e^{-\left(\frac{y}{\lambda}\right)^\delta}$ and the expected value is proportional to λ ($\mbox{E}(y) = \lambda\Gamma(1 + 1/\delta)$). Setting $\lambda_n = e^{\gamma^\top z_n}$, we get:

$$e^{-\delta e^{(\ln \mu_n^l - \gamma^\top z_n)}} - e^{-\delta e^{(\ln \mu_n^u - \gamma^\top z_n)}}$$

An interesting case of interval response occurs in ecological economics, for which one of the methods used to estimate the value of non-market resources is contingent valuation. Consider for example the `kakadu` data set, which was used by Carson, Wilks, and Imber (1994) and Werner (1999). The study concerns the Kakadu Conservation Zone, a part of the Kakadu National Park, and the survey aimed to measure the willingness to pay to preserve this zone from mining. The double-bounded, discrete-choice elicitation method of Hanemann (1991) is used; a respondent is asked whether they are willing to pay a pre-chosen randomly assigned amount to preserve the zone. If the answer is yes (no), they are asked whether he is willing to pay a higher (lower) pre-chosen amount. Four sets of dollar amounts were used ($[A: 100, 250, 50], [B: 50, 100, 20], [C: 20, 50, 5]$, and $[D: 5, 20, 2]$). The response is therefore a factor with four modalities: (no, yes), (yes, no), (no, no) and (yes, yes). For the first two levels, the interval is closed as the willingness to pay is in the interval formed by the two successive amounts. For (no, no), the willing to pay is lower than the second amount and for (yes, yes), it is greater than the second amount. In the `kakadu` data set, the lower and the upper bonds are respectively `lower` and `upper`, and `NA`s indicate unbounded willingness to pay. The interval regression model can be estimated using the `survival::survreg` function. The response should be a `Surv` object with the `time` and `time2` arguments set to the lower and to the upper bounds and the `type` argument set to `"interval2"`. The set of covariates is the one used by Werner (1999), table 4, p. 484. The results are presented in table Table 13.4.

```
kakadu <- kakadu %>% mutate(upper = ifelse(upper == 999, NA, upper),
                            lower = ifelse(lower == 0, NA, lower))
kak_weil <- survreg(Surv(time = log(lower), time2 = log(upper),
                         type = "interval2") ~ jobs + lowrisk +
                      aboriginal + finben + mineparks + moreparks +
                      envcon + age + income + major,
                    kakadu, dist = "weibull")
kak_ln <- update(kak_weil, dist = "lognormal")
modelsummary::msummary(list(Weibull = kak_weil, `Log-normal` = kak_ln),
                       output = "kableExtra")
```

In this example, the log-normal model fits slightly better the data than the Weibull model. The `predict` methods enables to obtain the fitted values (or the predicted values if a new data frame is provided). The argument `type` can be set to its default value `"response"` to get fitted value of the response or to `"link"` to obtain the linear predictor. For example, the mean and the median values of the willingness to pay are, for the two models:

```
c(mean(predict(kak_weil)), median(predict(kak_weil)))
## [1] 13.212  7.714
c(mean(predict(kak_ln)), median(predict(kak_ln)))
## [1] 9.418 4.909
```

Table 13.4: Interval regression for the Kakadu data set

	Weibull	Log-normal
(Intercept)	2.026	1.428
	(0.269)	(0.308)
jobs	−0.191	−0.206
	(0.033)	(0.040)
lowrisk	−0.271	−0.294
	(0.035)	(0.038)
aboriginal	0.064	0.073
	(0.031)	(0.036)
finben	−0.245	−0.256
	(0.036)	(0.040)
mineparks	0.329	0.384
	(0.034)	(0.039)
moreparks	0.142	0.182
	(0.032)	(0.039)
envconyes	0.221	0.204
	(0.078)	(0.086)
age	−0.010	−0.015
	(0.002)	(0.003)
income	0.006	0.007
	(0.002)	(0.003)
majoryes	0.186	0.205
	(0.076)	(0.084)
Log(scale)	0.034	0.279
	(0.051)	(0.049)
Num.Obs.	1827	1827
RMSE	16.99	12.41

14

Discrete choice models

Consider the case where the response is the choice of an alternative among a set of mutually exclusive alternatives. This choice can be modeled in a utility maximization framework, which means that we hypothesize that the individual chooses the alternative which corresponds to the maximum level of utility.[1] Namely, denoting $j = 1 \ldots J$ the set of alternatives, we'll denote U_{nj} the level of utility of individual n if he chooses alternative j and the j^{th}alternative will be chosen if $U_{nj} > U_{nk}\ \forall k \neq j$. The level of utility will depends on some observable covariates and on some other unobservables whose effect will be summarized in a variable ϵ_{nj} that will be considered, from the researcher's point of view, as the realization of a random variable. Two key questions for these models are how the covariates enter the utility function and what is the assumed distribution of the error. Section 14.1 will deal with the first point, namely how to deal with covariates in discrete choice model. Section 14.2 will present the landmark model in this field, the multinomial logit model. Section 14.3 will go beyond the hypothesis of iid errors to present two extensions of the basic model, the nested and the heteroscedastic logit models. Section 14.4 will introduce the rich field of random parameters logit models. Finally, Section 14.5 will be devoted to the multinomial probit model.[2]

Throughout this chapter, we'll use the **mlogit** package which is devoted to the estimation of discrete choice models:

```
library(mlogit)
```

14.1 Data management and model description

14.1.1 Data management

Data sets used for discrete choice models estimation concern some individuals, who make one or a sequential choice of one alternative among a set of mutually exclusive alternatives. The determinants of these choices are covariates that can depend on the alternative and the choice situation, only on the alternative or only on the choice situation. Data sets can have two different shapes: a *wide* shape (one row for each choice situation) or a *long* shape (one row for each alternative and, therefore, as many rows as there are alternatives for each choice situation). **mlogit** deals with both formats. It depends on the **dfidx** package which takes as first argument a `data.frame` and returns a `dfidx` object, which is a `data.frame` in "long" format with a special data frame column which contains the indexes. The second

[1] We've already encountered this model in the special case where the response is binomial in Section 10.2.2.
[2] This chapter is largely based on Croissant (2020).

argument is called `idx`. In its simple use, it should be a list (or a vector) of two characters containing the choice situation and the alternative indexes.

14.1.1.1 Wide format

`dutch_railways`[3] is an example of a *wide* data set:

```
dutch_railways %>% print(n = 4)
```

```
# A tibble: 2,929 x 11
     id choiceid choice price_A price_B time_A time_B change_A
  <int>    <int> <fct>    <dbl>   <dbl>  <dbl>  <dbl>    <int>
1     1        1 A         10.9    18.2  2.5     2.5         0
2     1        2 A         10.9    14.5  2.5     2.17        0
3     1        3 A         10.9    18.2  1.92    1.92        0
4     1        4 B         18.2    14.5  2.17    2.5         0
# i 2,925 more rows
# i 3 more variables: change_B <int>, comfort_A <int>,
#   comfort_B <int>
```

This data set contains data about a stated preference survey in the Netherlands in 1987. Each individual has responded to several (up to 16) scenarios. For every scenario, two train trips are proposed to the user, with different combinations of four attributes: `price` (the price in euros), `time` (travel time in minutes), `change` (the number of changes) and `comfort` (the class of comfort, 0, 1 or 2, 0 being the most comfortable class). This "wide" format is suitable to store choice situation (or individual specific) variables because, in this case, they are stored only once in the data. It is cumbersome for alternative-specific variables because there are as many columns for such variables as there are alternatives.

For such a wide data set, the `shape` argument of `dfidx` is mandatory, as its default value is `"long"`. The alternative-specific variables are indicated with the `varying` argument which is a numeric vector that indicates their position in the data frame. This argument is then passed to `stats::reshape` that coerced the original `data.frame` in "long" format. Further arguments may be passed to `reshape`. For example, as the names of the variables are of the form `price_A`, one must add `sep = "_"` (the default value being `"."`). The `choice` argument is also mandatory because the response has to be transformed in a logical value in the long format. In "wide" format, there is no alternative index. The choice situation index is not mandatory, as there is one line for each choice situation. In this data set, there is a choice situation index called `id`, and it is nested in the individual index called `choiceid`. To take the panel dimension into account, `idx` is a list of length 1 (the choice situation) containing a vector of length 2 with `choiceid` and `id`. The `idnames` is used to give a relevant name for the second index, the `NA` in first position indicating that the name of the first index is unchanged.

```
Tr <- dfidx(dutch_railways, shape = "wide", varying = 4:11, sep = "_",
            idx = list(c("choiceid", "id")), idnames = c(NA, "alt"),
            opposite = c("price", "time", "change", "comfort"))
```

[3] Used by Ben-Akiva, Bolduc, and Bradley (1993) and Meijer and Rouwendal (2006).

Note the use of the `opposite` argument for the four covariates: we expect negative coefficients for all of them, taking the opposite of the covariates will lead to expected positive coefficients.

```
Tr %>% print(n = 4)
```

```
# A tibble: 5,858 x 6
# Index:     2929 (choiceid) x 2 (alt)
# Balanced: yes
# Nesting:  choiceid (id)
  idx   choice price  time change comfort
  <idx> <fct>  <dbl> <dbl>  <int>   <int>
1 1:A   A      -10.9 -2.5       0      -1
2 1:B   A      -18.2 -2.5       0      -1
3 2:A   A      -10.9 -2.5       0      -1
4 2:B   A      -14.5 -2.17      0      -1
5 3:A   A      -10.9 -1.92      0      -1
# i 5,853 more rows
```

An `idx` column is added to the data, which contains the three relevant indexes: `choiceid` is the choice situation index, `alt` the alternative index and `id` the individual index. This column can be extracted using the `idx` function:

```
idx(Tr)
```

14.1.1.2 Long format

`toronto_montreal`,[4] is an example of a data set in long format. It presents the choice of individuals for a transport mode for the Toronto-Montreal corridor in 1989:

```
toronto_montreal %>% print(n = 5)
```

```
# A tibble: 15,520 x 11
   case alt   choice  dist  cost   ivt   ovt  freq income urban noalt
  <int> <fct>  <int> <int> <dbl> <int> <int> <int>  <int> <int> <int>
1     1 train      0    83  28.2    50    66     4     45     0     2
2     1 car        1    83  15.8    61     0     0     45     0     2
3     2 train      0    83  28.2    50    66     4     25     0     2
4     2 car        1    83  15.8    61     0     0     25     0     2
5     3 train      0    83  28.2    50    66     4     70     0     2
# i 15,515 more rows
```

There are four transport modes (`air`, `train`, `bus` and `car`) and most of the variables are alternative-specific (`cost` for monetary cost, `ivt` for in-vehicle time, `ovt` for out-vehicle time, `freq` for frequency). The only choice situation-specific variables are `dist` (distance of the trip), `income` (household income), `urban` (a dummy for trips which have a large city at the

[4]Used in particular by Forinash and Koppelman (1993), Bhat (1995), Koppelman and Wen (1998) and Koppelman and Wen (2000).

origin or the destination) and **noalt** (the number of available alternatives). The advantage of this shape is that there are much fewer columns than in the wide format, the caveat being that values of **dist**, **income** and **urban** are repeated up to four times. For data in "long" format, the **shape** and the **choice** arguments are no longer mandatory. To replicate published results later in the text, we'll use only a subset of the choice situations, namely those for which the four alternatives are available. This can be done using the **subset** function with the **subset** argument set to **noalt == 4** while estimating the model. This can also be done within **dfidx**, using the **subset** argument.

The information about the structure of the data can be explicitly indicated using choice situations and alternative indexes (respectively **case** and **alt** in this data set) or, in part, guessed by the **dfidx** function. Here, after subsetting, we have 2779 choice situations with 4 alternatives, and the rows are ordered first by choice situation and then by alternative (**train**, **air**, **bus**, and **car** in this order). The first way to read correctly this data frame is to ignore completely the two index variables. In this case, the only supplementary argument to provide is the **alt.levels** argument, which is a character vector that contains the name of the alternatives in their order of appearance:

```
MC <- dfidx(toronto_montreal, subset = noalt == 4,
            alt.levels = c("train", "air", "bus", "car"))
```

Note that this can only be used if the data set is "balanced", which means that the same set of alternatives is available for all choice situations. It is also possible to provide the name of the variable that contains the alternatives through the argument **idx**:

```
MC <- dfidx(toronto_montreal, subset = noalt == 4, idx = list(NA, "alt"))
```

The name of the variable that contains the information about the choice situations can also be indicated through the argument **idx**:

```
MC <- dfidx(toronto_montreal, subset = noalt == 4, idx = "case",
            alt.levels = c("train", "air", "bus", "car"))
```

Both alternative and choice situation variables can also be provided:

```
MC <- dfidx(toronto_montreal, subset = noalt == 4,
            idx = c("case", "alt"))
```

More simply, as the two indexes are stored in the first two columns of the original data frame, the **idx** argument can be unset:

```
MC <- dfidx(toronto_montreal, subset = noalt == 4)
```

and the indexes can be kept as standalone series if the **drop.index** argument is set to **FALSE**:

```
MC <- dfidx(toronto_montreal, subset = noalt == 4,
            idx = c("case", "alt"), drop.index = FALSE)
MC %>% print(n = 5)
```

```
# A tibble: 11,116 x 12
# Index:    2779 (case) x 4 (alt)
# Balanced: yes
  idx          case alt   choice  dist  cost   ivt   ovt  freq income
  <idx>       <int> <fct>  <int> <int> <dbl> <int> <int> <int>  <int>
1 109:train     109 train      0   377  58.2   215    74     4     45
2 109:air       109 air        1   377 143.     56    85     9     45
3 109:bus       109 bus        0   377  27.5   301    63     8     45
4 109:car       109 car        0   377  71.6   262     0     0     45
5 110:train     110 train      0   377  58.2   215    74     4     70
# i 11,111 more rows
# i 2 more variables: urban <int>, noalt <int>
```

14.1.2 Model description

Standard `formulas` are not very practical to describe random utility models, as these models may use different sets of covariates. Actually, working with random utility models, one has to consider at most three sets of covariates:

1. alternative- and choice situation-specific covariates x_{nj} with generic coefficients β and alternative-specific covariates t_j with a generic coefficient ν,
2. choice situation-specific covariates z_n with alternative-specific coefficients γ_j,
3. alternative- and choice situation-specific covariates w_{nj} with alternative-specific coefficients δ_j.

The covariates enter the observable part of the utility which can be written, for alternative j:

$$V_{nj} = \alpha_j + \beta x_{nj} + \nu t_j + \gamma_j z_n + \delta_j w_{nj}$$

As the absolute value of utility is irrelevant, only utility differences are useful to modelize the choice for one alternative. For two alternatives j and l, we obtain:

$$V_{nj} - V_{nl} = (\alpha_j - \alpha_l) + \beta(x_{nj} - x_{nl}) + \nu(t_j - t_l) + (\gamma_j - \gamma_l)z_n + (\delta_j w_{nj} - \delta_k w_{nl})$$

It is clear from the previous expression that coefficients of choice situation-specific variables (the intercept being one of those) should be alternative-specific; otherwise they would disappear in the differentiation. Moreover, only differences of these coefficients are relevant and can be identified. For example, with three alternatives 1, 2 and 3, the three coefficients $\gamma_1, \gamma_2, \gamma_3$ associated with a choice situation-specific variable cannot be identified, but only two linear combinations. Therefore, one has to make a choice of normalization, and the simplest one is just to set $\gamma_1 = 0$.

Coefficients for alternative and choice situation-specific variables may (or may not) be alternative-specific. For example, transport time is alternative-specific, but 10 mn in public transport may not have the same impact on utility than 10 mn in a car. In this case, alternative-specific coefficients are relevant. Monetary cost is also alternative-specific, but in this case, one can consider than $1 is $1 however it is spent for the use of a car or in public transports. In this case, a generic coefficient is relevant. The treatment of alternative-specific variables doesn't differ much from the alternative and choice situation-specific variables with a generic coefficient. However, if some of these variables are introduced, the ν parameter can only be estimated in a model without intercepts to avoid perfect multicolinearity.

A logit model with only choice situation-specific variables is sometimes called a *multinomial logit model*, one with only alternative-specific variables, a *conditional logit model*, and one with both kinds of variables, a *mixed logit model*. This is seriously misleading: *conditional logit model* is also a logit model for longitudinal data in the statistical literature, and *mixed logit* is one of the names of a logit model with random parameters. Therefore, in what follows, we'll use the name *multinomial logit model* for the model we've just described whatever the nature of the explanatory variables used.

The **mlogit** package provides objects of class `mFormula` which are built upon `Formula` objects provided by the **Formula** package. To illustrate the use of `mFormula` objects, we use again the `toronto_montreal` data set and consider three sets of covariates that will be indicated in a three-part formula, which refers to the three items at the beginning of this section.

- `cost` (monetary cost) is an alternative-specific covariate with a generic coefficient (part 1),
- `income` and `urban` are choice situation-specific covariates (part 2),
- `ivt` (in-vehicle travel time) is alternative-specific and alternative-specific coefficients are expected (part 3).

```
library(Formula)
f <- Formula(choice ~ cost | income + urban | ivt)
```

Some parts of the formula may be omitted when there is no ambiguity. For example, the following sets of `formulas` are identical:

```
f2 <- Formula(choice ~ cost + ivt | income + urban)
f2 <- Formula(choice ~ cost + ivt | income + urban | 0)
```

```
f3 <- Formula(choice ~ 0 | income | 0)
f3 <- Formula(choice ~ 0 | income)
```

```
f4 <- Formula(choice ~ cost + ivt)
f4 <- Formula(choice ~ cost + ivt | 1)
f4 <- Formula(choice ~ cost + ivt | 1 | 0)
```

By default, an intercept is added to the model; it can be removed by using `+ 0` or `- 1` in the second part.

```
f5 <- Formula(choice ~ cost | income + 0 | ivt)
f5 <- Formula(choice ~ cost | income - 1 | ivt)
```

A `model.frame` method is provided for `dfidx` objects. It differs from the `formula` method by the fact that the returned object is an object of class `dfidx` and not an ordinary data frame, which means that the information about the structure of the data is not lost. Defining a specific `model.frame` method for `dfidx` objects implies that the first argument of the function should be a **dfidx** object, which results in an unusual order of the arguments in the function (the data first, and then the formula). Moreover, as the model matrix for random utility models has specific features, we add a supplementary argument called **pkg** to the **dfidx** function so that the returned object has a specific class (and inherits the **dfidx** class):

```
MC <- dfidx(toronto_montreal, subset = noalt == 4, pkg = "mlogit")
class(MC)
## [1] "dfidx_mlogit" "dfidx"          "tbl_df"         "tbl"
## [5] "data.frame"
f <- Formula(choice ~ cost | income  | ivt)
mf <- model.frame(MC, f)
class(mf)
## [1] "dfidx_mlogit" "dfidx"          "tbl_df"         "tbl"
## [5] "data.frame"
```

Using **mf** as the argument of **model.matrix** enables the construction of the relevant model matrix for random utility model, as a specific **model.matrix** method for **dfidx_mlogit** objects is provided.

```
head(model.matrix(mf), 4)
```

```
  (Intercept):air (Intercept):bus (Intercept):car   cost income:air
1               0               0               0  58.25          0
2               1               0               0 142.80         45
3               0               1               0  27.52          0
4               0               0               1  71.63          0
  income:bus income:car ivt:train ivt:air ivt:bus ivt:car
1          0          0       215       0       0       0
2          0          0         0      56       0       0
3         45          0         0       0     301       0
4          0         45         0       0       0     262
```

The model matrix contains $J-1$ columns for every choice situation-specific variables (**income** and the intercept), which means that the coefficient associated with the first alternative (**train**) is set to 0. It contains only one column for **cost** because we want a generic coefficient for this variable. It contains J columns for **ivt**, because it is an alternative specific variable for which we want alternative specific coefficients.

14.2 Random utility model and multinomial logit model

14.2.1 Random utility model

The utility for alternative l is written as: $U_l = V_l + \epsilon_l$ where V_l is a function of some observable covariates and unknown parameters to be estimated, and ϵ_l is a random deviate which contains all the unobserved determinants of the utility. Alternative l is therefore chosen if $\epsilon_j < (V_l - V_j) + \epsilon_l \; \forall \; j \neq l$ and the probability of choosing this alternative is then:

$$\mathrm{P}(\epsilon_1 < V_l - V_1 + \epsilon_l, \epsilon_2 < V_l - V_2 + \epsilon_l, ..., \epsilon_J < V_l - V_J + \epsilon_l).$$

Denoting F_{-l} as the cumulative density function of all the ϵs except ϵ_l, this probability is:

$$(\mathrm{P}_l \mid \epsilon_l) = F_{-l}(V_l - V_1 + \epsilon_l, ..., V_l - V_J + \epsilon_l).$$

Note that this probability is conditional on the value of ϵ_l. The unconditional probability (which depends only on β and on the value of the observed covariates) is obtained by integrating out the conditional probability using the marginal density of ϵ_l, denoted by f_l:

$$\mathrm{P}_l = \int F_{-l}(V_l - V_1 + \epsilon_l, ..., V_l - V_J) + \epsilon_l) f_l(\epsilon_l) d\epsilon_l.$$

The conditional probability is an integral of dimension $J-1$, and the computation of the unconditional probability adds one more dimension of integration.

14.2.2 Distribution of the error terms

The multinomial logit model (McFadden 1974) is a special case of the model developed in the previous section. It is based on three hypotheses. The first hypothesis is the independence of the errors. In this case, the univariate distribution of the errors can be used, which leads to the following conditional and unconditional probabilities:

$$(\mathrm{P}_l \mid \epsilon_l) = \prod_{j \neq l} F_j(V_l - V_j + \epsilon_l) \text{ and } \mathrm{P}_l = \int \prod_{j \neq l} F_j(V_l - V_j + \epsilon_l) \; f_l(\epsilon_l) \; d\epsilon_l,$$

which means that the conditional probability is the product of $J-1$ univariate cumulative density functions, and the evaluation of only a one-dimensional integral is required to compute the unconditional probability. The second hypothesis is that each ϵ follows a Gumbel (maximum) distribution, whose density and probability functions are respectively:

$$f(z) = \frac{1}{\theta} e^{-\frac{z-\mu}{\theta}} e^{-e^{-\frac{z-\mu}{\theta}}} \text{ and } F(z) = \int_{-\infty}^{z} f(t) dt = e^{-e^{-\frac{z-\mu}{\theta}}},$$

where μ is the location parameter and θ the scale parameter. The first two moments of the Gumbel distribution are $\mathrm{E}(z) = \mu + \theta\gamma$, where γ is the Euler-Mascheroni constant (≈ 0.57721) and $\mathrm{V}(z) = \frac{\pi^2}{6}\theta^2$. The mean of ϵ_j is not identified if V_j contains an intercept. We can then, without loss of generality suppose that $\mu_j = 0, \; \forall j$. Moreover, the overall scale of utility is not identified. Therefore, only $J-1$ scale parameters may be identified, and a natural choice of normalization is to impose that one of the θ_j is equal to 1. The last hypothesis is that the errors are identically distributed. As the location parameter is not identified for any error term, this hypothesis is essentially a homoskedasticity hypothesis, which means that the scale parameter of the Gumbel distribution is the same for all the alternatives. As one of them has been previously set to 1, we can therefore suppose that, without loss of generality, $\theta_j = 1, \ \forall j \in 1...J$. The conditional and unconditional probabilities then further simplify to:

$$(\mathrm{P}_l \mid \epsilon_l) = \prod_{j \neq l} e^{-e^{-(V_l - V_j + \epsilon_l)}} \text{ and } \mathrm{P}_l = \int_{-\infty}^{+\infty} \prod_{j \neq l} e^{-e^{-(V_l - V_j + t)}} e^{-t} e^{-e^{-t}} dt.$$

The probabilities have then very simple, closed forms, which correspond to the logit transformation of the deterministic part of the utility.[5]

$$P_l = \frac{e^{V_l}}{\sum_{j=1}^{J} e^{V_j}}.$$

14.2.3 IIA property

If we consider the probabilities of choice for two alternatives l and m, we have $P_l = e^{V_l} / \sum_j e^{V_j}$ and $P_m = e^{V_m} / \sum_j e^{V_j}$. The ratio of these two probabilities is:

$$\frac{P_l}{P_m} = \frac{e^{V_l}}{e^{V_m}} = e^{V_l - V_m}.$$

This probability ratio for the two alternatives depends only on the characteristics of these two alternatives and not on those of other alternatives. This is called the independence of irrelevant alternatives (**IIA**) property. IIA relies on the hypothesis that the errors are identical and independent. It is not a problem in itself and may even be considered as a useful feature for a well-specified model. However, this hypothesis may be in practice violated, especially if some important variables are omitted.

14.2.4 Interpretation

14.2.4.1 Marginal effects

The marginal effects are the derivatives of the probabilities with respect to the covariates, which can be choice situation-specific (z_n) or alternative-specific (x_{nj}):

$$\begin{aligned}
\frac{\partial P_{nl}}{\partial z_n} &= P_{nl}\left(\beta_l - \sum_j P_{nj}\beta_j\right) \\
\frac{\partial P_{nl}}{\partial x_{nl}} &= \gamma P_{nl}(1 - P_{nl}) \\
\frac{\partial P_{nl}}{\partial x_{nk}} &= -\gamma P_{nl} P_{nk}.
\end{aligned}$$

- For a choice situation-specific variable, the sign of the marginal effect is not necessarily the sign of the coefficient. Actually, the sign of the marginal effect is given by $\left(\beta_l - \sum_j P_{nj}\beta_j\right)$, which is positive if the coefficient for alternative l is greater than a weighted average of the coefficients for all the alternatives, the weights being the probabilities of choosing

[5] See Train (2009), pp. 74-75.

the alternatives. In this case, the sign of the marginal effect can be established with no ambiguity only for the alternatives with the lowest and the greatest coefficients.

- For an alternative-specific variable, the sign of the coefficient can be directly interpreted. The marginal effect is obtained by multiplying the coefficient by the product of two probabilities which is at most 0.25. The rule of thumb is therefore to divide the coefficient by 4 in order to have an upper bound of the marginal effect.

Note that the last equation can be rewritten: $\frac{\mathrm{d}P_{nl}/P_{nl}}{\mathrm{d}x_{nk}} = -\gamma P_{nk}$. Therefore, when a characteristic of alternative k changes, the relative changes of the probabilities for every alternative except k are the same, which is a consequence of the IIA property.

14.2.4.2 Marginal rates of substitution

Coefficients are marginal utilities, which cannot be interpreted. However, ratios of coefficients are marginal rates of substitution. For example, if the observable part of utility is: $V = \beta_o + \beta_1 x_1 + \beta x_2 + \beta x_3$, joint variations of x_1 and x_2 which ensure the same level of utility are such that: $dV = \beta_1 dx_1 + \beta_2 dx_2 = 0$ so that:

$$-\frac{dx_2}{dx_1}\mid_{dV=0} = \frac{\beta_1}{\beta_2}.$$

For example, if x_2 is transport cost (in dollars), x_1 transport time (in hours), $\beta_1 = 1.5$ and $\beta_2 = 0.05$, $\frac{\beta_1}{\beta_2} = 30$ is the marginal rate of substitution of time in terms of dollars and the value of 30 means that, to reduce the travel time of 1 hour, the individual is willing to pay at most \$30 more. Stated more simply, time value is \$30 per hour.

14.2.4.3 Consumer surplus

Consumer's surplus has a very simple expression for multinomial logit models, which was first derived by Small and Rosen (1981). The level of utility attained by an individual is $U_j = V_j + \epsilon_j$, j being the chosen alternative. The expected utility, from the searcher's point of view is then: $\mathrm{E}(\max_j U_j)$, where the expectation is taken over the values of all the error terms. Its expression is simply, up to an additive unknown constant, the log of the denominator of the logit probabilities, often called the "log-sum":

$$\mathrm{E}(U) = \ln \sum_{j=1}^{J} e^{V_j} + C.$$

If the marginal utility of income (α) is known and constant, the expected surplus is simply $\frac{\mathrm{E}(U)}{\alpha}$.

14.2.5 Application

Random utility models are fitted using the `mlogit` function. Basically, only two arguments are mandatory, `formula` and `data`, if an `dfidx` object (and not an ordinary `data.frame`) is provided. We use the `toronto_montreal` data set, which was already coerced to a `dfidx` object (called `MC`) in the previous section. The same model can then be estimated using as `data` argument this `dfidx` object:

```
MC <- dfidx(toronto_montreal, subset = noalt == 4)
ml.MC1 <- mlogit(choice ~ cost + freq + ovt | income | ivt, MC)
```

or a `data.frame`. In this latter case, further arguments that will be passed to `dfidx` should be indicated:

```
ml.MC1b <- mlogit(choice ~ cost + freq + ovt | income | ivt,
                  toronto_montreal, subset = noalt == 4,
                  idx = c("case", "alt"))
```

`mlogit` provides two further useful arguments:

- `reflevel` indicates which alternative is the "reference" alternative, i.e., the one for which the coefficients of choice situation-specific covariates are set to 0,
- `alt.subset` indicates a subset of alternatives on which the estimation has to be performed; in this case, only the lines that correspond to the selected alternatives are used, and all the choice situations where unselected alternatives have been chosen are removed.

We estimate the model on the subset of three alternatives (we exclude `bus` whose market share is negligible in our sample) and we set `car` as the reference alternative. Moreover, we use a total transport time variable computed as the sum of the in-vehicle and out-vehicle time variables.

```
MC <- MC %>% mutate(time = ivt + ovt)
ml.MC1 <- mlogit(choice ~ cost + freq | income | time, MC,
                 alt.subset = c("car", "train", "air"), reflevel = "car")
```

The main results of the model are computed and displayed using the `summary` method:

```
summary(ml.MC1)
```

```

Call:
mlogit(formula = choice ~ cost + freq | income | time, data = MC,
    alt.subset = c("car", "train", "air"), reflevel = "car",
    method = "nr")

Frequencies of alternatives:choice
  car train   air
0.458 0.167 0.375

nr method
6 iterations, 0h:0m:0s
g'(-H)^-1g = 6.94E-06
successive function values within tolerance limits

Coefficients :
```

```
                     Estimate Std. Error z-value Pr(>|z|)
(Intercept):train -0.970344   0.265131   -3.66  0.00025 ***
(Intercept):air   -1.898566   0.684143   -2.78  0.00552 **
cost              -0.028497   0.006559   -4.34  1.4e-05 ***
freq               0.074029   0.004733   15.64  < 2e-16 ***
income:train      -0.006469   0.003104   -2.08  0.03713 *
income:air         0.028246   0.003654    7.73  1.1e-14 ***
time:car          -0.014024   0.001380  -10.16  < 2e-16 ***
time:train        -0.010969   0.000818  -13.40  < 2e-16 ***
time:air          -0.017551   0.003992   -4.40  1.1e-05 ***
---
Signif. codes:  0 '***' 0.001 '**' 0.01 '*' 0.05 '.' 0.1 ' ' 1

Log-Likelihood: -1950
McFadden R^2:  0.312
Likelihood ratio test : chisq = 1770 (p.value = <2e-16)
```

The frequencies of the different alternatives in the sample are first indicated. Next, some information about the optimization is displayed: the Newton-Raphson method (with analytical gradient and hessian) is used, as it is the most efficient method for this simple model for which the log-likelihood function is globally concave. Note that very few iterations and computing times are required to estimate this model. Then the usual table of coefficients is displayed, followed by some goodness-of-fit measures: the value of the log-likelihood function, which is compared to the value when only intercepts are introduced, which leads to the computation of the McFadden R^2 and to the likelihood ratio test. The `fitted` method can be used either to obtain the probability of actual choices (`type = "outcome"`) or the probabilities for all the alternatives (`type = "probabilities"`).

```
head(fitted(ml.MC1, type = "outcome"))
```

```
   109    110    111    112    113    114
0.1909 0.3400 0.1471 0.3400 0.3400 0.2440
```

```
head(fitted(ml.MC1, type = "probabilities"), 4)
```

```
       car  train    air
109 0.4206 0.3884 0.1909
110 0.3696 0.2904 0.3400
111 0.4297 0.4233 0.1471
112 0.3696 0.2904 0.3400
```

Note that the log-likelihood is the sum of the log of the fitted outcome probabilities and that, as the model contains intercepts, the average fitted probabilities for every alternative equals the market shares of the alternatives in the sample.

```
sum(log(fitted(ml.MC1, type = "outcome")))
## [1] -1951
```

```
logLik(ml.MC1)
## 'log Lik.' -1951 (df=9)
apply(fitted(ml.MC1, type = "probabilities"), 2, mean)
##    car  train    air
## 0.4576 0.1672 0.3752
```

Predictions can be made using the `predict` method. If no data is provided, predictions are made for the sample mean values of the covariates.

```
predict(ml.MC1)
##    car  train    air
## 0.5066 0.2117 0.2817
```

Assume, for example, that we wish to predict the effect of a reduction of train transport time of 20%. We first create a new `data.frame` simply by multiplying train transport time by 0.8 and then using the `predict` method with this new `data.frame`.

```
NMC <- MC
NMC <- NMC %>% mutate(time = ifelse(idx$alt == "train", 0.8 * time, time))
Oprob <- fitted(ml.MC1, type = "probabilities")
Nprob <- predict(ml.MC1, newdata = NMC)
rbind(old = apply(Oprob, 2, mean), new = apply(Nprob, 2, mean))
```

```
       car  train    air
old 0.4576 0.1672 0.3752
new 0.4045 0.2636 0.3319
```

If, for the first individuals in the sample, we compute the ratio of the probabilities of the air and the car mode, we obtain:

```
head(Nprob[, "air"] / Nprob[, "car"])
##    109    110    111    112    113    114
## 0.4539 0.9198 0.3422 0.9198 0.9198 0.6021
head(Oprob[, "air"] / Oprob[, "car"])
##    109    110    111    112    113    114
## 0.4539 0.9198 0.3422 0.9198 0.9198 0.6021
```

which is an illustration of the IIA property. If train time changes, it changes the probabilities of choosing air and car, but not their ratio. We next compute the surplus for individuals of the sample induced by train time reduction. This requires the computation of the log-sum term (also called inclusive value or inclusive utility) for every choice situation, which is:

$$\mathrm{iv}_n = \ln \sum_{j=1}^{J} e^{\beta^\top x_{nj}}.$$

For this purpose, we use the `logsum` function, which works on a vector of coefficients and a model matrix. The basic use of `logsum` consists of providing as unique argument (called `coef`) a `mlogit` object. In this case, the `model.matrix` and the `coef` are extracted from the same model:

```
ivbefore <- logsum(ml.MC1)
```

To compute the log-sum after train time reduction, we must provide a model matrix which is not the one corresponding to the fitted model. This can be done using the `X` argument which is a matrix or an object from which a model matrix can be extracted. This can also be done by filling the `data` argument (a data frame or an object from which a data frame can be extracted using `model.frame`), and eventually the `formula` argument (a formula or an object for which the `formula` method can be applied). If no formula is provided, but if `data` is a `dfidx` object, the formula is extracted from it.

```
ivafter <- logsum(ml.MC1, data = NMC)
```

Surplus variation is then computed as the difference of the log-sums divided by the opposite of the cost coefficient which can be interpreted as the marginal utility of income:

```
surplus <- - (ivafter - ivbefore) / coef(ml.MC1)["cost"]
summary(surplus)
##    Min. 1st Qu.  Median    Mean 3rd Qu.    Max.
##   0.585   2.844   3.900   4.697   5.844  31.391
```

Consumer surplus variations range from 0.6 to 31 Canadian dollars, with a median value of about $4. Marginal effects are computed using the `effects` method. By default, they are computed at the sample mean, but a `data` argument can be provided. The variation of the probability and the covariate can be either absolute or relative. This is indicated with the `type` argument which is a combination of two `a` (as absolute) and `r` (as relative) characters. For example, `type = "ar"` means that what is measured is an absolute variation of the probability for a relative variation of the covariate.

```
effects(ml.MC1, covariate = "income", type = "ar")
##     car   train     air
## -0.1822 -0.1509  0.3331
```

The results indicate that, for a 100% increase of income, the probability of choosing `air` increases by 33 percentage points, as the probabilities of choosing `car` and `train` decrease by 18 and 15 percentage points.

For an alternative specific covariate, a matrix of marginal effects is displayed.

```
effects(ml.MC1, covariate = "cost", type = "rr")
```

```
          car   train     air
car   -0.9131  0.9377  0.9377
train  0.3358 -1.2505  0.3358
air    1.2317  1.2317 -3.1410
```

The cell in the lth row and the cth column indicates the change of the probability of choosing alternative c when the cost of alternative l changes. As `type = "rr"`, elasticities are computed. For example, a 10% change of train cost increases the probabilities of choosing car and air by 3.36%. Note that the relative changes of the probabilities of choosing one of these two modes are equal, which is a consequence of the IIA property. Finally, in order to compute travel time valuation, we divide the coefficients of travel times (in minutes) by the coefficient of monetary cost (in dollars).

```
coef(ml.MC1)[grep("time", names(coef(ml.MC1)))] /
    coef(ml.MC1)["cost"] * 60
##    time:car time:train    time:air
##       29.53      23.09       36.95
```

The value of travel time ranges from 23 for a train to 37 Canadian dollars per hour for a plane.

14.3 Logit models relaxing the iid hypothesis

In the previous section, we assumed that the error terms were iid (identically and independently distributed), i.e., uncorrelated and homoskedastic. Extensions of the basic multinomial logit model have been proposed by relaxing one of these two hypotheses while maintaining the hypothesis of a Gumbel distribution.

14.3.1 Heteroskedastic logit model

The heteroskedastic logit model was proposed by Bhat (1995). The probability that $U_l > U_j$ is:

$$P(\epsilon_j < V_l - V_j + \epsilon_l) = e^{-e^{-\frac{(V_l - V_j + \epsilon_l)}{\theta_j}}},$$

which implies the following conditional and unconditional probabilities

$$(P_l \mid \epsilon_l) = \prod_{j \neq l} e^{-e^{-\frac{(V_l - V_j + \epsilon_l)}{\theta_j}}},$$

$$\begin{aligned} P_l &= \int_{-\infty}^{+\infty} \prod_{j \neq l} \left(e^{-e^{-\frac{(V_l - V_j + t)}{\theta_j}}} \right) \frac{1}{\theta_l} e^{-\frac{t}{\theta_l}} e^{-e^{-\frac{t}{\theta_l}}} dt \\ &= \int_0^{+\infty} \left(e^{-\sum_{j \neq l} e^{-\frac{V_l - V_j - \theta_l \ln t}{\theta_j}}} \right) e^{-t} dt. \end{aligned} \quad (14.1)$$

There is no closed form for this integral, but it can be efficiently computed using a Gauss quadrature method, and more precisely the Gauss-Laguerre quadrature method. Bhat (1995) estimated the heteroskedastic logit model on the `toronto_montreal` data set. Using `mlogit`, the heteroskedastic logit model is obtained by setting the `heterosc` argument to `TRUE`:

```
ml.MC <- mlogit(choice ~ freq + cost + ivt + ovt |
                  urban + income, MC, reflevel = 'car',
                alt.subset = c("car", "train", "air"))
hl.MC <- mlogit(choice ~ freq + cost + ivt + ovt |
                  urban + income, MC, reflevel = 'car',
                alt.subset = c("car", "train", "air"),
                heterosc = TRUE)
hl.MC %>% gaze(coef = 11:12)
```

```
         Estimate Std. Error z-value Pr(>|z|)
sp.train    1.237      0.110   11.20  < 2e-16
sp.air      0.540      0.112    4.83  1.4e-06
```

Two supplementary coefficients (**sp.train** and **sp.air**) are estimated (θ_j in Equation 14.1), the third for the reference modality being set to 1. The variance of the error terms of train and air are respectively higher and lower than the variance of the error term of car (set to 1). Note that the z-values and p-values of the output are not particularly meaningful, as the hypothesis that the coefficient is zero (and not 1) is tested. The homoskedasticity hypothesis can be tested using any of the three tests. For the likelihood ratio and the Wald test, one can use only the fitted heteroskedastic model as argument. In this case, it is guessed that the hypothesis that the user wants to test is the homoskedasticity hypothesis.

```
lrtest(hl.MC, ml.MC) %>% gaze
## Chisq = 6.888, df: 2, pval = 0.032
waldtest(hl.MC, heterosc = FALSE) %>% gaze
## chisq = 25.196, df: 2, pval = 0.000
```

or, more simply:

```
lrtest(hl.MC)
waldtest(hl.MC)
```

The Wald test can also be computed using the **linearHypothesis** function from the **car** package:

```
car::linearHypothesis(hl.MC, c('sp.air = 1', 'sp.train = 1')) %>% gaze
## Chisq = 25.196, df: 2, pval = 0.000
```

For the score test, we provide the constrained model as argument, which is the standard multinomial logit model and the supplementary argument which defines the unconstrained model, which is in this case **heterosc = TRUE**.

```
scoretest(ml.MC, heterosc = TRUE) %>% gaze
## chisq = 9.488, df: 2, pval = 0.009
```

The homoskedasticity hypothesis is therefore strongly rejected using the Wald test, but only at the 1 and 5% level for, respectively, the score and the likelihood ratio tests.

14.3.2 Nested logit model

The nested logit model was first proposed by McFadden (1978). It is a generalization of the multinomial logit model that is based on the idea that some alternatives may be joined in several groups (called nests). The error terms may then present some correlation in the same nest, whereas error terms of different nests are still uncorrelated. Denoting $m = 1...M$ the nests and B_m the set of alternatives belonging to nest m, the cumulative distribution of the errors is:

$$\exp\left(-\sum_{m=1}^{M}\left(\sum_{j\in B_m} e^{-\epsilon_j/\lambda_m}\right)^{\lambda_m}\right).$$

The marginal distributions of the ϵs are still univariate extreme values, but there is now some correlation within nests. $1 - \lambda_m$ is a measure of the correlation, i.e., $\lambda_m = 1$ implies no correlation. In the special case where $\lambda_m = 1\ \forall m$, the errors are iid Gumbel errors and the nested logit model reduce to the multinomial logit model. It can then be shown that the probability of choosing alternative j that belongs to nest l is:

$$P_j = \frac{e^{V_j/\lambda_l}\left(\sum_{k\in B_l} e^{V_k/\lambda_l}\right)^{\lambda_l - 1}}{\sum_{m=1}^{M}\left(\sum_{k\in B_m} e^{V_k/\lambda_m}\right)^{\lambda_m}},$$

and that this model is a random utility model if all the λ parameters are in the $0-1$ interval.[6] Let us now write the deterministic part of the utility of alternative j as the sum of two terms: the first (Z_j) being specific to the alternative and the second (W_l) to the nest it belongs to:

$$V_j = Z_j + W_l.$$

We can then rewrite the probabilities as follow:

$$\begin{aligned} P_j &= \frac{e^{(Z_j+W_l)/\lambda_l}}{\sum_{k\in B_l} e^{(Z_k+W_l)/\lambda_l}} \times \frac{\left(\sum_{k\in B_l} e^{(Z_k+W_l)/\lambda_l}\right)^{\lambda_l}}{\sum_{m=1}^{M}\left(\sum_{k\in B_m} e^{(Z_k+W_m)/\lambda_m}\right)^{\lambda_m}} \\ &= \frac{e^{Z_j/\lambda_l}}{\sum_{k\in B_l} e^{Z_k/\lambda_l}} \times \frac{\left(e^{W_l/\lambda_l}\sum_{k\in B_l} e^{Z_k/\lambda_l}\right)^{\lambda_l}}{\sum_{m=1}^{M}\left(e^{W_m/\lambda_m}\sum_{k\in B_m} e^{Z_k/\lambda_m}\right)^{\lambda_m}}. \end{aligned}$$

Then denote $I_l = \ln\sum_{k\in B_l} e^{Z_k/\lambda_l}$ which is often called the log-sum, the inclusive value or the inclusive utility.[7] We then can write the probability of choosing alternative j as:

$$P_j = \frac{e^{Z_j/\lambda_l}}{\sum_{k\in B_l} e^{Z_k/\lambda_l}} \times \frac{e^{W_l+\lambda_l I_l}}{\sum_{m=1}^{M} e^{W_m+\lambda_m I_m}}.$$

[6] A slightly different version of the nested logit model (Daly 1987) is often used, but is not compatible with the random utility maximization hypothesis. Its difference from the previous expression is that the deterministic parts of the utility for each alternative are not divided by the nest elasticity. The differences between the two versions have been discussed in Koppelman and Wen (1998), Heiss (2002) and Hensher and Greene (2002).

[7] We've already encountered this expression in Section 14.2.4.3.

The first term $P_{j|l}$ is the conditional probability of choosing alternative j if nest l is chosen. It is often referred to as the *lower model*. The second term P_l is the marginal probability of choosing nest l and is referred to as the *upper model*. $W_l + \lambda_l I_l$ can be interpreted as the expected utility of choosing the best alternative in l, W_l being the expected utility of choosing an alternative in this nest (whatever this alternative is) and $\lambda_l I_l$ being the expected extra utility gained by being able to choose the best alternative in the nest. The inclusive values link the two models. It is then straightforward to show that IIA applies within nests, but not for two alternatives in different nests.

A consistent but inefficient way of estimating the nested logit model is to estimate separately its two components. The coefficients of the lower model are first estimated, which enables the computation of the inclusive values I_l. The coefficients of the upper model are then estimated, using I_l as covariates. Maximizing directly the likelihood function of the nested model leads to a more efficient estimator.

To illustrate the estimation of the nested logit model, we use the `telephone` data set, used by Train, McFadden, and Ben-Akiva (1987) and Walker, Ben-Akiva, and Bolduc (2007).

```
telephone %>% print(n = 5)
```

```
# A tibble: 2,170 x 4
  choice service  household  cost
  <lgl>  <fct>        <int> <dbl>
1 FALSE  budget           1  1.76
2 FALSE  extended         1 13.8
3 FALSE  local            1  2.55
4 FALSE  metro            1  3.15
5 TRUE   standard         1  1.75
# i 2,165 more rows
```

A total of 428 households were surveyed in 1984 about their choice of a local telephone service, which typically involves the choice between a flat service (a fixed monthly charge for an unlimited calls within a specified geographical area) and a measured (a reduced fixed monthly charge for a limited number of calls plus usage charges for additional calls) service. Households had the choice between five services:

- *budget measured* (`budget`): no fixed monthly charge; usage charges apply to each call made,
- *standard measured* (`standard`): a fixed monthly charge covers up to a specified dollar amount (greater than the fixed charge) of local calling, after which usage charges apply to each call made,
- *local flat* (`local`): a greater monthly charge that may depend upon residential location; unlimited free calling within a local calling area; usage charges apply to calls made outside local calling area,
- *extended area flat* (`extended`): a further increase in the fixed monthly charge to permit unlimited free calling within an extended area,
- *metro area flat* (`metro`): the greatest fixed monthly charge that permits unlimited free calling within the entire metropolitan area.

The first two services are measured, and the last three are flat services. There is therefore an obvious nesting structure for this example. We first estimate the multinomial logit model, with the log of cost as the unique covariate:

```
ml_tel <- mlogit(choice ~ log(cost), telephone,
                 idx = c("household", "service"))
```

We then update this model in order to introduce nests, using the **nests** argument. It is a list of characters that contains the alternatives for the different nests. It is advisable to use a named list (we use here **"measured"** and **"flat"** as names of the nests):

```
nl_tel <- mlogit(choice ~ cost, telephone,
                 idx = c("household", "service"),
                 nests = list(measured = c("budget", "standard"),
                              flat = c("local", "metro", "extended")))
coef(nl_tel)
## (Intercept):extended     (Intercept):local     (Intercept):metro
##               1.2255                1.2716                1.7837
## (Intercept):standard                  cost           iv:measured
##               0.3782               -1.4900                0.4848
##              iv:flat
##               0.4362
```

Two supplementary coefficients are estimated, `iv:measured` and `iv:flat`. The two values are in the 0-1 interval and close to each other. The `un.nest.el` argument enables to estimate a unique supplementary coefficient for the two nests:

```
nl_tel_u <- update(nl_tel, un.nest.el = TRUE)
```

We then have three nested models and the hypothesis of a unique parameter can be tested using any of the three tests:

```
scoretest(nl_tel_u, un.nest.el = FALSE) %>% gaze
## chisq = 0.109, df: 1, pval = 0.741
lrtest(nl_tel, nl_tel_u) %>% gaze
## Chisq = 0.137, df: 1, pval = 0.712
waldtest(nl_tel, un.nest.el = TRUE) %>% gaze
## chisq = 0.160, df: 1, pval = 0.689
```

The three tests conclude that a unique parameter can be estimated. Then, we can test whether this parameter is 1, in which case the nested logit model reduces to the multinomial logit model:

```
scoretest(ml_tel,
          nests = list(measured = c("budget", "standard"),
                       flat = c("local", "metro", "extended"))) %>% gaze
## chisq = 5.685, df: 2, pval = 0.058
lrtest(nl_tel_u, ml_tel) %>% gaze
## Chisq = 17.177, df: 1, pval = 0.000
```

```
waldtest(nl_tel_u, nests = NULL) %>% gaze
## chisq = 25.541, df: 1, pval = 0.000
car::linearHypothesis(nl_tel_u, "iv = 1") %>% gaze
## Chisq = 25.541, df: 1, pval = 0.000
```

Based on the Wald and the likelihood ratio, the preferred specification is the nested logit model (but note that the p-value for the score test is slightly higher than 5%).

14.4 Random parameters (or mixed) logit model

14.4.1 Derivation of the model

A **mixed logit** model (or random parameters logit model) is a logit model whose parameters are assumed to vary from one individual to another. It is therefore a model that takes the heterogeneity of the population into account. For the standard logit model, the probability that individual n chooses alternative j is:

$$P_{nl} = \frac{e^{\beta' x_{nl}}}{\sum_j e^{\beta' x_{nj}}}.$$

Suppose now that the coefficients are individual-specific. The probabilities are then:

$$P_{nl} = \frac{e^{\beta_n' x_{nl}}}{\sum_j e^{\beta_n' x_{nj}}}.$$

A first approach consists of estimating the parameters for every individual. However, these parameters are identified and can be consistently estimated only if a large number of choice situations per individual is available, which is scarcely the case. A more appealing approach consists of considering β_n as random draws on a distribution whose parameters are estimated, which leads to the mixed logit model. The probability that individual n will choose alternative l, for a given value of β_n is:

$$P_{nl} \mid \beta_n = \frac{e^{\beta_n' x_{nl}}}{\sum_j e^{\beta_n' x_{nj}}}.$$

To get the unconditional probability, we have to integrate out this conditional probability, using the density function of β. Suppose that $V_{nl} = \beta_n x_{nl}$, i.e., that there is only one individual-specific coefficient and that the density of β_n is $f(\beta, \theta)$, θ being the vector of the parameters of the distribution of β. The unconditional probability is then:

$$P_{nl} = \mathrm{E}(P_{nl} \mid \beta_n) = \int_\beta (P_{nl} \mid \beta) f(\beta, \theta) d\beta = \int_\beta \frac{e^{\beta^\top x_{nl}}}{\sum_j e^{\beta^\top x_{nj}}} f(\beta, \theta) d\beta,$$

which is a one-dimensional integral that can be efficiently estimated by quadrature methods.

If $V_{nl} = \beta_n^\top x_{nl}$ where β_n is a vector of length K and $f(\beta, \theta)$ is the joint density of the K individual-specific coefficients, the unconditional probability is:

$$P_{nl} = \mathrm{E}(P_{nl} \mid \beta_n) = \int_{\beta_1} \int_{\beta_2} \ldots \int_{\beta_K} (P_{nl} \mid \beta) f(\beta, \theta) d\beta_1 d\beta_2 \ldots d\beta_K.$$

This is a K-dimensional integral which cannot easily be estimated by quadrature methods. The only practical method is then to use simulations. More precisely, R draws of the parameters are taken from the distribution of β, the probability is computed for every draw and the unconditional probability, which is the expected value of the conditional probabilities is estimated by the average of the R probabilities.

The expected value of a random coefficient ($\mathrm{E}(\beta)$) is simply estimated by the mean of the R draws on its distribution: $\bar{\beta} = \sum_{r=1}^R \beta_r$. Individual parameters are obtained by first computing the probabilities of the observed choice of n for every value of β_r:

$$P_{nr} = \frac{\sum_j y_{nj} e^{\beta_r' x_{nj}}}{\sum_j e^{\beta_r' x_{nj}}},$$

where y_{nj} is a dummy equal to 1 if n has chosen alternative j. The expected value of the parameter for an individual is then estimated by using these probabilities to weight the R β values:

$$\hat{\beta}_n = \frac{\sum_r P_{nr} \beta_r}{\sum_r P_{nr}}.$$

If there are repeated observations for the same individuals, the longitudinal dimension of the data can be taken into account in the mixed logit model, assuming that the random parameters of individual n are the same for all their choice situations. Denoting y_{ntl} a dummy equal to 1 if n chooses alternative l for the tth choice situation, the probability of the observed choice is:

$$P_{nt} \mid \beta_n = \prod_j \frac{\sum_j y_{ntj} e^{\beta_n^\top x_{ntj}}}{\sum_j e^{\beta_n^\top x_{ntj}}}.$$

The joint probability for the T observations of individual n is then:

$$P_n \mid \beta_n = \prod_t \prod_j \frac{\sum_j y_{ntj} e^{\beta_n^\top x_{ntj}}}{\sum_j e^{\beta_n^\top x_{ntj}}}$$

14.4.2 Application

The random parameter logit model is estimated by providing a `rpar` argument to `mlogit`. This argument is a named vector, the names being the random coefficients and the acronyms for the law of distribution. Currently, the normal (`"n"`), log-normal (`"ln"`), zero-censored normal (`"cn"`), uniform (`"u"`) and triangular (`"t"`) distributions are available. For these distributions, two parameters are estimated which are, for normal related distributions, the mean and the standard-deviation of the underlying normal distribution and for the uniform and triangular distribution, the mean and the half-range of the distribution. For these last two distributions, zero-bounded variants are also provided (`"zbt"` and `"zbu"`). These two distributions are defined by only one parameter (the mean) and their definition domain varies from 0 to twice the mean.

Several considerations may lead to the choice of a specific distribution:

- if correlated coefficients are required, the natural choice is a (transformed-) normal distribution, `"n"`, `"ln"`, `"tn"` and `"cn"`,
- it's often the case that one wants to impose that the distribution of a random parameter takes only positive or negative values. For example, the price coefficient should be negative for every individual. In this case, `"zbt"` and `"zbu"` can be used. The use of `"ln"` and `"cn"` can also be relevant but, in this case, if only negative values are expected, one should consider the distribution of the opposite of the random price coefficient. This can easily be done using the `opposite` argument of `dfidx`,[8]
- the use of unbounded distributions often leads to implausible values of some statistics of the random parameters, especially the mean. This is particularly the case of the log-normal distribution, which has an heavy right tail. In this case, the use of bounded distribution like the uniform and the triangular distributions can be used.

`R` is the number of draws, `halton` indicates whether Halton draws (see Train 2009, , chapter 9) should be used (`NA` and `NULL` indicate respectively that default Halton draws are used and that pseudo-random numbers are used), `panel` is a boolean which indicates if the panel data version of the log-likelihood should be used.

Correlations between random parameters can be introduced only for normal-related distributed random parameters, using the `correlation` argument. If `TRUE`, all the normal-related random parameters are correlated. The `correlation` argument can also be a character vector indicating a subset of the random parameters that are assumed to be correlated.

We use the `dutch_railways` data set, previously coerced to a `dfidx` object called `Tr`. We first estimate the multinomial model: both alternatives being virtual train trips, it is relevant to use only generic coefficients and to remove the intercept:

```
Tr <- dfidx(dutch_railways, choice = "choice", varying = 4:11, sep = "_",
            opposite = c("price", "comfort", "time", "change"),
            idx = list(c("choiceid", "id")), idnames = c("chid", "alt"))
Train.ml <- mlogit(choice ~ price + time + change + comfort | - 1, Tr)
Train.ml %>% gaze
```

```
        Estimate Std. Error z-value Pr(>|z|)
price     0.3271     0.0165   19.85  < 2e-16
time      1.7206     0.1604   10.73  < 2e-16
change    0.3263     0.0595    5.49  4.1e-08
comfort   0.9457     0.0649   14.56  < 2e-16
```

[8]See Section 14.1.1.1.

All the coefficients are highly significant and have the predicted positive sign (remember than an increase in the variable `comfort` implies using a less comfortable class). The coefficients can't be directly interpreted, but dividing them by the price coefficient, we get monetary values:

```
coef(Train.ml)[- 1] / coef(Train.ml)[1]
##    time  change comfort
##  5.2598  0.9976  2.8911
```

```
mv <- coef(Train.ml)[- 1] / coef(Train.ml)[1]
```

We obtain the value of 5.3 euros for an hour of traveling, 1 euro for a change and 2.9 euros to travel in a more comfortable class.[9] We then estimate a model with three random parameters, `time`, `change` and `comfort`. We first estimate the uncorrelated mixed logit model:

```
Train.mxlu <- mlogit(choice ~ price + time + change + comfort | - 1, Tr,
                     rpar = c(time = "n", change = "n", comfort = "n"),
                     R = 100, panel = TRUE, correlation = FALSE,
                     halton = NA, method = "bhhh")
names(coef(Train.mxlu))
## [1] "price"      "time"        "change"      "comfort"     "sd.time"
## [6] "sd.change"  "sd.comfort"
```

Compared to the multinomial logit model, there are now three more coefficients which are the standard deviations of the distribution of the three random parameters. The correlated model is obtained by setting the `correlation` argument to `TRUE`.

```
Train.mxlc <- update(Train.mxlu, correlation = TRUE)
names(coef(Train.mxlc))
```

```
 [1] "price"                 "time"
 [3] "change"                "comfort"
 [5] "chol.time:time"        "chol.time:change"
 [7] "chol.change:change"    "chol.time:comfort"
 [9] "chol.change:comfort"   "chol.comfort:comfort"
```

There are now six parameters which are the elements of the Choleski decomposition of the covariance matrix of the three random parameters. These six parameters are therefore the elements of the following matrix:

$$C = \begin{pmatrix} c_{11} & c_{12} & c_{13} \\ 0 & c_{22} & c_{23} \\ 0 & 0 & c_{33} \end{pmatrix}$$

such that:

[9] Remember that the survey took place in 1987. The values should be multiplied by 1.73 to get the value of 1 euro in 2024.

$$C^\top C = \begin{pmatrix} c_{11}^2 & c_{11}c_{12} & c_{11}c_{13} \\ c_{11}c_{12} & c_{12}^2 + c_{22}^2 & c_{12}c_{23} + c_{22}c_{23} \\ c_{11}c_{13} & c_{12}c_3 + c_{22}c_{23} & c_{13}^2 + c_{23}^2 c_{33}^2 \end{pmatrix} = \begin{pmatrix} \sigma_1^2 & \sigma_{12} & \sigma_{13} \\ \sigma_{12} & \sigma_2^2 & \sigma_{23} \\ \sigma_{13} & \sigma_{23} & \sigma_3^2 \end{pmatrix}$$

where σ_k^2 and σ_{kl} are respectively the variance of the random parameter k and the covariance between two random parameters k and l. Therefore, the first estimated parameter can be simply interpreted as the standard deviation of the first random parameter, but the five other can't be interpreted easily. Random parameters may be extracted using the function `rpar` which takes as first argument a `mlogit` object and as second argument `par` the parameter(s) to be extracted. This function returns a `rpar` object and a `summary` method is provided to describe it:

```
marg.ut.time <- rpar(Train.mxlc, "time")
summary(marg.ut.time)
##      Min. 1st Qu.  Median    Mean 3rd Qu.    Max.
##      -Inf   1.284   4.894   4.894   8.504     Inf
```

The estimated random parameter is in the "preference space", which means that it is the marginal utility of time. Parameters in the "willingness to pay" (WTP) space are more easy to interpret. They can be estimated directly (a feature not supported by `mlogit`) or can be obtained from the marginal utility by dividing by the coefficient of a covariate expressed in monetary value (a price for example), taken as a non-random parameter. The ratio can then be interpreted as a monetary value (or willingness to pay). To obtain the distribution of the random parameters in the WTP space, one can use the `norm` argument of `rpar`:

```
wtp.time <- rpar(Train.mxlc, "time", norm = "price")
summary(wtp.time)
##      Min. 1st Qu.  Median    Mean 3rd Qu.    Max.
##      -Inf   1.802   6.871   6.871  11.939     Inf
```

The median value (and the mean value as the distribution is symmetric) of transport time is about 33 euros. Several methods/functions are provided to extract the individual statistics (`mean`, `med` and `stdev` respectively for the mean, median and standard deviation):

```
mean(rpar(Train.mxlc, "time", norm = "price"))
## [1] 6.871
med(rpar(Train.mxlc, "time", norm = "price"))
## [1] 6.871
stdev(rpar(Train.mxlc, "time", norm = "price"))
## [1] 7.515
```

In case of correlated random parameters, as the estimated parameters can't be directly interpreted, a `vcov` method for `mlogit` objects is provided. It has a `what` argument whose default value is `coefficient`. In this case the usual covariance matrix of the coefficients is return. If `what = "rpar"`, the covariance matrix of the correlated random parameters is returned if `type = "cov"` (the default) and the correlation matrix (with standard deviations on the diagonal) is returned if `type = "cor"`. The object is of class `vcov.mlogit` and a `summary` method for this object is provided which computes, using the delta method, the standard errors of the parameters of the covariance or the correlation matrix.

```
vcov(Train.mxlc, what = "rpar")
```

```
           time  change comfort
time    28.6460 -0.2788   5.558
change  -0.2788  3.1047   1.232
comfort  5.5579  1.2325   7.896
```

```
vcov(Train.mxlc, what = "rpar", type = "cor")
```

```
            time   change comfort
time     5.35220 -0.02956  0.3696
change  -0.02956  1.76203  0.2489
comfort  0.36956  0.24893  2.8099
```

```
summary(vcov(Train.mxlc, what = "rpar", type = "cor"))
```

```
                    Estimate Std. Error z-value Pr(>|z|)
sd.time               5.3522     0.3811   14.04   <2e-16 ***
sd.change             1.7620     0.1446   12.19   <2e-16 ***
sd.comfort            2.8099     0.1783   15.76   <2e-16 ***
cor.time:change      -0.0296     0.2324   -0.13   0.8988
cor.time:comfort      0.3696     0.1141    3.24   0.0012 **
cor.change:comfort    0.2489     0.1103    2.26   0.0240 *
---
Signif. codes:  0 '***' 0.001 '**' 0.01 '*' 0.05 '.' 0.1 ' ' 1
```

The correlation can be restricted to a subset of random parameters by filling the `correlation` argument with a character vector indicating the corresponding covariates:

```
Train.mxlc2 <- update(Train.mxlc, correlation = c("time", "comfort"))
vcov(Train.mxlc2, what = "rpar", type = "cor")
```

```
          time comfort
time    5.5726  0.3909
comfort 0.3909  3.0631
```

The presence of random coefficients and their correlation can be investigated using any of the three tests. Actually, three nested models can be considered, a model with no random effects, a model with random but uncorrelated effects and a model with random and correlated effects. We first present the three tests of no correlated random effects:

```
lrtest(Train.mxlc, Train.ml) %>% gaze
## Chisq = 388.057, df: 6, pval = 0.000
waldtest(Train.mxlc) %>% gaze
```

```
## chisq = 288.287, df: 6, pval = 0.000
car::linearHypothesis(Train.mxlc,
                      c("chol.time:time = 0", "chol.time:change =  0",
                        "chol.time:comfort = 0", "chol.change:change = 0",
                        "chol.change:comfort = 0",
                        "chol.comfort:comfort = 0")) %>% gaze
## Chisq = 288.287, df: 6, pval = 0.000
scoretest(Train.ml, rpar = c(time = "n", change = "n", comfort = "n"),
          R = 100, correlation = TRUE, halton = NA,
          panel = TRUE) %>% gaze
## chisq = 208.765, df: 6, pval = 0.000
```

The hypothesis of no correlated random parameters is strongly rejected. We then present the three tests of no correlation, the existence of random parameters being maintained.

```
lrtest(Train.mxlc, Train.mxlu) %>% gaze
## Chisq = 42.621, df: 3, pval = 0.000
car::linearHypothesis(Train.mxlc,
                      c("chol.time:change = 0","chol.time:comfort = 0",
                        "chol.change:comfort = 0")) %>% gaze
## Chisq = 103.195, df: 3, pval = 0.000
waldtest(Train.mxlc, correlation = FALSE) %>% gaze
## chisq = 103.195, df: 3, pval = 0.000
scoretest(Train.mxlu, correlation = TRUE) %>% gaze
## chisq = 10.483, df: 3, pval = 0.015
```

The hypothesis of no correlation is strongly rejected with the Wald and the likelihood ratio test, only at the 5% level for the score test.

14.5 Multinomial probit

14.5.1 The model

The multinomial probit is obtained with the same modeling that we used while presenting the random utility model. The utility of an alternative is still the sum of two components : $U_j = V_j + \epsilon_j$ but the joint distribution of the error terms is now a multivariate normal with mean 0 and with a matrix of covariance denoted by Ω.[10] Alternative l is chosen if:

$$\left\{\begin{array}{rcl} U_1 - U_l & = & (V_1 - V_l) + (\epsilon_1 - \epsilon_l) < 0 \\ U_2 - U_l & = & (V_2 - V_l) + (\epsilon_2 - \epsilon_l) < 0 \\ & \vdots & \\ U_J - U_l & = & (V_J - V_l) + (\epsilon_J - \epsilon_l) < 0 \end{array}\right.$$

wich implies, denoting $V_j^l = V_j - V_l$:

[10] See Hausman and Wise (1978) and Daganzo (1979).

$$\left\{\begin{array}{lllll} \epsilon_1^l & = & (\epsilon_1 - \epsilon_l) & < & -V_1^l \\ \epsilon_2^l & = & (\epsilon_2 - \epsilon_l) & < & -V_2^l \\ \vdots & & & & \vdots \\ \epsilon_J^l & = & (\epsilon_J - \epsilon_l) & < & -V_J^l \end{array}\right.$$

The initial vector of errors ϵ are transformed using the following transformation: $\epsilon^l = M^l\epsilon$, where the transformation matrix M^l is a $(J-1) \times J$ matrix obtained by inserting in an identity matrix a lth column of -1. For example, if $J = 4$ and $l = 3$:

$$M^3 = \begin{pmatrix} 1 & 0 & -1 & 0 \\ 0 & 1 & -1 & 0 \\ 0 & 0 & -1 & 1 \end{pmatrix}$$

The covariance matrix of the error differences is:

$$\mathrm{V}\left(\epsilon^l\right) = \mathrm{V}\left(M^l\epsilon\right) = M^l\mathrm{V}\left(\epsilon\right)M^{l^\top} = M^l\Omega M^{l^\top} \tag{14.2}$$

The probability of choosing l is then:

$$P_l = \mathrm{P}(\epsilon_1^l < -V_1^l \ \&\ \epsilon_2^l < -V_2^l \ \&\ \ldots\ \epsilon_J^l < -V_J^l) \tag{14.3}$$

with the hypothesis of normal distribution, this writes:

$$P_l = \int_{-\infty}^{-V_1^l}\int_{-\infty}^{-V_2^l} \ldots \int_{-\infty}^{-V_J^l} \phi(\epsilon^l)d\epsilon_1^l d\epsilon_2^l \ldots d_J^l$$

with :

$$\phi\left(\epsilon^l\right) = \frac{1}{(2\pi)^{(J-1)/2} \mid \Omega^l \mid^{1/2}} e^{-\frac{1}{2}\epsilon^l \Omega^{l^{-1}} \epsilon^l}$$

Two problems arise with this model:

- the identified parameters are the elements of Ω^l and not of Ω. We must then carefully investigate the meanings of these elements,
- the probability is a $J-1$ integral, which should be numerically computed. The relevant strategy in this context is to use simulations.

14.5.2 Identification

The meaningful parameters are those of the covariance matrix of the error Ω. For example, with $J = 3$:

$$\Omega = \begin{pmatrix} \sigma_1^2 & \sigma_{12} & \sigma_{13} \\ \sigma_{21} & \sigma_2^2 & \sigma_{23} \\ \sigma_{31} & \sigma_{32} & \sigma_3^3 \end{pmatrix}$$

Computing Equation 14.2 for $l = 1$, we get:

$$\Omega^1 = M^1 \Omega M^{1^\top} = \begin{pmatrix} \sigma_1^2 + \sigma_2^2 - 2\sigma_{12} & \sigma_1^2 + \sigma_{23} - \sigma_{12} - \sigma_{13} \\ \sigma_1^2 + \sigma_{23} - \sigma_{12} - \sigma_{13} & \sigma_1^2 + \sigma_3^2 - 2\sigma_{13} \end{pmatrix}$$

The overall scale of utility being unidentified, one has to impose the value of one of the variance, for example the first one is set to 1. We then have :

$$\Omega^1 = \begin{pmatrix} 1 & \frac{\sigma_1^2 + \sigma_{23} - \sigma_{12} - \sigma_{13}}{\sigma_1^2 + \sigma_2^2 - 2\sigma_{12}} \\ \frac{\sigma_1^2 + \sigma_{23} - \sigma_{12} - \sigma_{13}}{\sigma_1^2 + \sigma_2^2 - 2\sigma_{12}} & \frac{\sigma_1^2 + \sigma_3^2 - 2\sigma_{13}}{\sigma_1^2 + \sigma_2^2 - 2\sigma_{12}} \end{pmatrix}$$

Therefore, out of the six structural parameters of the covariance matrix, only three can be identified. Moreover, it's almost impossible to interpret these parameters. More generally, with J alternatives, the number of the parameters of the covariance matrix is $(J+1) \times J/2$ and the number of identified parameters is $J \times (J-1)/2 - 1$.

14.5.3 Simulations

Let C^l be the Choleski decomposition of the covariance matrix of the error differences:

$$\Omega^C = C^{l^\top} C^l$$

This matrix is an upper triangular matrix of dimension $(J-1)$:

$$C^l = \begin{pmatrix} l_{11} & l_{12} & l_{13} & \cdots & l_{1(J-1)} \\ 0 & l_{22} & l_{23} & \cdots & l_{2(J-1)} \\ 0 & 0 & l_{33} & \cdots & l_{3(J-1)} \\ \vdots & \vdots & \vdots & \ddots & \vdots \\ 0 & 0 & 0 & \cdots & l_{(J-1)(J-1)} \end{pmatrix}$$

Let ν be a vector of standard normal deviates: $\nu \sim \mathcal{N}(0, I)$

Therefore, we have :

$$\mathrm{V}\left(C^{l^\top}\nu\right) = C^{l^\top} V(\nu) C^l = C^{l^\top} I C^l = \Omega^l$$

Therefore, if we draw a vector of standard normal deviates ν and apply to it this transformation, we get a realization of ϵ^l. This probability of choosing l given by Equation 14.3 can be written as a product of conditional and marginal probabilities:

$$\begin{array}{rcl} P_l & = & \mathrm{P}(\epsilon_1^l < -V_1^l \;\&\; \epsilon_2^l < -V_2^l \;\&\; \ldots \;\&\; \epsilon_J^l < -V_J^l)) \\ & = & \mathrm{P}(\epsilon_1^l < -V_1^l)) \\ & \times & \mathrm{P}(\epsilon_2^l < -V_2^l \mid \epsilon_1^l < -V_1^l) \\ & \times & \mathrm{P}(\epsilon_3^l < -V_3^l \mid \epsilon_1^l < -V_1^l \;\&\; \epsilon_2^l < -V_2^l) \\ & \vdots & \\ & \times & \mathrm{P}(\epsilon_J^l < -V_J^l \mid \epsilon_1^l < -V_1^l \;\&\; \ldots \;\&\; \epsilon_{J-1}^l < -V_{J-1}^l)) \end{array}$$

The vector of error difference deviates is:

$$\begin{pmatrix} \epsilon_1^l \\ \epsilon_2^l \\ \epsilon_3^l \\ \vdots \\ \epsilon_J^l \end{pmatrix} = C^{l\top} \nu = \begin{pmatrix} l_{11} & 0 & 0 & \dots & 0 \\ l_{12} & l_{22} & 0 & \dots & 0 \\ l_{13} & l_{23} & l_{33} & \dots & 0 \\ \vdots & \vdots & \vdots & \ddots & \vdots \\ l_{1(J-1)} & l_{2(J-1)} & l_{3(J-1)} & \dots & l_{(J-1)(J-1)} \end{pmatrix} \times \begin{pmatrix} \nu_1 \\ \nu_2 \\ \nu_3 \\ \vdots \\ \nu_J \end{pmatrix}$$

$$\begin{pmatrix} \epsilon_1^l \\ \epsilon_2^l \\ \epsilon_3^l \\ \vdots \\ \epsilon_J^l \end{pmatrix} = \begin{pmatrix} l_{11}\nu_1 \\ l_{12}\nu_1 + l_{22}\nu_2 \\ l_{13}\nu_1 + l_{23}\nu_2 + l_{33}\nu_3 \\ \vdots \\ l_{1(J-1)}\nu_1 + l_{2(J-1)}\nu_2 + \dots + l_{(J-1)(J-1)}\nu_{J-1} \end{pmatrix}$$

Let's now investigate the marginal and conditional probabilities:

- the first is simply the marginal probability for a standard normal deviate, therefore we have: $\mathrm{P}(\epsilon_1^l < -V_1^l) = \Phi\left(-\frac{V_1^l}{l_{11}}\right)$
- the second is, for a given value of ν_1 equal to $\Phi\left(-\frac{V_2^l + l_{21}\nu_1}{l_{22}}\right)$. We then have to compute the mean of this expression for any value of ν_1 lower than $-\frac{V_1^l}{l_{11}}$. We then have, denoting $\bar{\phi}_1$ the truncated normal density:

$$\mathrm{P}(\epsilon_2^l < -V_2^l) = \int_{-\infty}^{-\frac{V_1^l}{l_{11}}} \Phi\left(-\frac{V_2^l + l_{21}\nu_1}{l_{22}}\right) \bar{\phi}_1(\nu_1) d\nu_1$$

- the third is, for given values of ν_1 and ν_2 equal to: $\Phi\left(-\frac{V_3^l + l_{31}\nu_1 + l_{32}\nu_2}{l_{33}}\right)$. We then have:

$$\mathrm{P}(\epsilon_3^l < -V_3^l) = \int_{-\infty}^{-\frac{V_1^l}{l_{11}}} \int_{-\infty}^{-\frac{V_2^l + l_{21}\nu_1}{l_{22}}} \Phi\left(-\frac{V_3^l + l_{31}\nu_1 + l_{32}\nu_2}{l_{33}}\right) \bar{\phi}_1(\nu_1)\bar{\phi}_2(\nu_2) d\nu_1 d\nu_2$$

- and so on.

These probabilities can easily be simulated by drawing numbers from a truncated normal distribution. This so called GHK (for Geweke, Hajivassiliou and Keane) algorithm (see for example Geweke, Keane, and Runkle 1994) can be described as follow:

1. compute $\Phi\left(-\frac{V_1^l}{l_{11}}\right)$,
2. draw a number called ν_1^r from a standard normal distribution upper-truncated at $-\frac{V_1^l}{l_{11}}$ and compute $\Phi\left(-\frac{V_2^l + l_{12}\nu_1^r}{l_{22}}\right)$,
3. draw a number called ν_2^r from a standard normal distribution upper-truncated at $-\frac{V_2^l + l_{12}\nu_1^r}{l_{22}}$ and compute $\Phi\left(-\frac{V_3^l + l_{13}\nu_1^r + l_{23}\nu_2^r}{l_{33}}\right)$,
4. ... draw a number called ν_{J-1}^r from a standard normal distribution upper-truncated at $-\frac{V_{J-1}^l + l_{1(J-1)}\nu_1^r + \dots l_{(J-2)(J-1)}\nu_{J-2}^r}{l_{(J-1)(J-1)}}$,
5. multiply all these probabilities and get a realization of the probability of choosing l called P_l^r,
6. repeat all these steps many times and average all these probabilities; this average is an estimation of the probability: $\bar{P}_l = \sum_{r=1}^{R} P_l^r / R$.

Several points should be noted concerning this algorithm:

- the utility differences should be computed respective to the chosen alternative for each individual,
- the Choleski decomposition used should rely on the same covariance matrix of the errors. One method to attained this goal is to start from a given difference, e.g., the difference respective with the first alternative. The vector of error difference is then ϵ^1 and its covariance matrix is $\Omega^1 = L^{1^\top} L^1$. To apply a difference with another alternative l, we construct a matrix called S^l which is obtained by using a $J-2$ identity matrix, adding a first row of 0 and inserting a column of -1 at the $l-1$th position. For example, with four alternatives and $l=3$, we have:

$$S^3 = \begin{pmatrix} 0 & -1 & 0 \\ 1 & -1 & 0 \\ 0 & -1 & 1 \end{pmatrix}$$

  The elements of the Choleski decomposition of the covariance matrix is then obtained as follow:

$$\Omega^l = S^l \Omega^1 S^{l^\top} = L^l L^{l^\top}$$

- to compute draws from a normal distribution truncated at a, the following trick is used : take a draw μ from a uniform distribution (between 0 and 1); then $\nu = \Phi^{-1}\left(\mu\Phi(a)\right)$ is a draw from a normal distribution truncated at a.

14.5.4 Application

In Section 14.2.5, we estimated a multinomial logit for mode choice in the Toronto-Montreal corridor, using `cost` and `freq` as alternative specific covariates with a generic coefficient, `income` as a choice situation specific covariate and `time` as an alternative specific covariate with a specific coefficient. We fit the same model using this time a probit. This is simply done by setting the `probit` argument to `TRUE`.

```
pbt <- mlogit(formula = choice ~ cost + freq | income | time, data = MC,
              probit = TRUE, alt.subset = c("car", "train", "air"),
              reflevel = "car")
```

As previously, we want to predict the effect of a reduction of 20% of a train's fare:

```
Oprob <- fitted(pbt, type = "probabilities")
Nprob <- predict(pbt, newdata = NMC)
rbind(old = apply(Oprob, 2, mean), new = apply(Nprob, 2, mean))
```

```
       car  train    air
old 0.4606 0.1676 0.3721
new 0.4194 0.2457 0.3353
```

With this model, the IIA property is not operative, as the ratio of probabilities of choosing air and car is no longer the same before and after the change of a train's fare.

```
head(Nprob[, "air"] / Nprob[, "car"])
## [1] 0.4549 0.8923 0.3402 0.8972 0.9090 0.6034
head(Oprob[, "air"] / Oprob[, "car"])
## [1] 0.4737 0.8897 0.3585 0.9016 0.9288 0.6299
```

References

Abadie, Alberto. 2021. "Using Synthetic Controls: Feasibility, Data Requirements, and Methodological Aspects." *Journal of Economic Literature* 59 (2): 391–425. https://doi.org/10.1257/jel.20191450.

Abadie, Alberto, Alexis Diamond, and Jens Hainmueller. 2010. "Synthetic Control Methods for Comparative Case Studies: Estimating the Effect of California's Tobacco Control Program." *Journal of the American Statistical Association* 105 (490): 493–505. https://ideas.repec.org/a/bes/jnlasa/v105i490y2010p493-505.html.

———. 2011. "Synth: An R Package for Synthetic Control Methods in Comparative Case Studies." *Journal of Statistical Software* 42 (i13). https://doi.org/http://hdl.handle.net/10.

———. 2015. "Comparative Politics and the Synthetic Control Method." *American Journal of Political Science* 59 (2): 495–510. https://doi.org/10.1111/ajps.12116.

Abadie, Alberto, and Javier Gardeazabal. 2003. "The Economic Costs of Conflict: A Case Study of the Basque Country." *American Economic Review* 93 (1): 113–32. https://ideas.repec.org/a/aea/aecrev/v93y2003i1p113-132.html.

Abiad, Abdul, and Ashoka Mody. 2005. "Financial Reform: What Shakes It? What Shapes It?" *American Economic Review* 95 (1): 66–88. https://doi.org/10.1257/0002828053828699.

Adkins, Lee C. 2012. "Testing Parameter Significance in Instrumental Variables Probit Estimators: Some Simulation." *Journal of Statistical Computation and Simulation* 82: 1415–36. https://api.semanticscholar.org/CorpusID:18158285.

Adkins, Lee C., David A. Carter, and W. Gary Simpson. 2007. "Managerial Incentives And The Use Of Foreign-Exchange Derivatives By Banks." *Journal of Financial Research* 30 (3): 399–413. https://doi.org/10.1111/j.1475-6803.2007.

Aldrich, J. H., and F. D. Nelson. 1984. *Linear Probability, Logit, and Probit Models.* Sage University Press.

Amemiya, Takeshi. 1971. "The Estimation of the Variances in a Variance–Components Model." *International Economic Review* 12: 1–13.

———. 1978. "The Estimation of a Simultaneous Equation Generalized Probit Model." *Econometrica* 46 (5): 1193–1205. https://EconPapers.repec.org/RePEc:ecm:emetrp:v:46:y:1978:i:5:p:1193-1205.

———. 1981. "Qualitative Response Models: A Survey." *Journal of Economic Literature* 19 (4): 1483–1536. https://ideas.repec.org/a/aea/jeclit/v19y1981i4p1483-1536.html.

———. 1984. "Tobit Models: A Survey." *Journal of Econometrics* 24 (1): 3–61. https://doi.org/10.1016/0304-4076(84)90074-5.

———. 1985. *Advanced Econometrics.* Harvard University Press.

An, Mark Y., Bent Jesper Christensen, and Nabanita Datta Gupta. 2004. "Multivariate Mixed Proportional Hazard Modelling of the Joint Retirement of Married Couples." *Journal of Applied Econometrics* 19 (6): 687–704. https://doi.org/10.1002/jae.783.

Ananat, Elizabeth Oltmans. 2011. "The Wrong Side(s) of the Tracks: The Causal Effects of Racial Segregation on Urban Poverty and Inequality." *American Economic Journal: Applied Economics* 3 (2): 34–66. https://doi.org/10.1257/app.3.2.34.

Angrist, Joshua D. 1990. "Lifetime Earnings and the Vietnam Era Draft Lottery: Evidence from Social Security Administrative Records." *The American Economic Review* 80 (3): 313–36. http://www.jstor.org/stable/2006669.

Angrist, Joshua D., Eric Bettinger, Erik Bloom, Elizabeth King, and Michael Kremer. 2002. "Vouchers for Private Schooling in Colombia: Evidence from a Randomized Natural Experiment." *American Economic Review* 92 (5): 1535–58. https://doi.org/10.1257/000282802762024629.

Angrist, Joshua D., Guido W. Imbens, and Donald B. Rubin. 1996. "Identification of Causal Effects Using Instrumental Variables." *Journal of the American Statistical Association* 91 (434): 444–55. http://www.jstor.org/stable/2291629.

Anselin, Luc. 1995. "Local Indicators of Spatial Association—LISA." *Geographical Analysis* 27 (2): 93–115. https://doi.org/10.1111/j.1538-4632.1995.tb00338.x.

Anselin, Luc, Anil K. Bera, Raymond Florax, and Mann J. Yoon. 1996. "Simple Diagnostic Tests for Spatial Dependence." *Regional Science and Urban Economics* 26 (1): 77–104. https://doi.org/10.1016/0166-0462(95)02111-6.

Arel-Bundock, Vincent. 2023. *Marginaleffects: Predictions, Comparisons, Slopes, Marginal Means, and Hypothesis Tests.* https://CRAN.R-project.org/package=marginaleffects.

Becker, Martin, and Stefan Klosner. 2017. "Estimating the Economic Costs of Organized Crime by Synthetic Control Methods." *Journal of Applied Econometrics* 32 (7): 1367–69. https://EconPapers.repec.org/RePEc:wly:japmet:v:32:y:2017:i:7:p:1367-1369.

Ben-Akiva, M., D. Bolduc, and M. Bradley. 1993. "Estimation of Travel Choice Models with Randomly Distributed Values of Time." *Transportation Research Record* 1413: 88–97.

Berndt, Ernst R. 1991. *The Practice of Econometrics, Classic and Contemporary.* Addison-Wesley Publishing Company.

Berndt, Ernst R., Bronwyn Hall, Robert Hall, and Jerry Hausman. 1974. "Estimation and Inference in Nonlinear Structural Models." In *Annals of Economic and Social Measurement, Volume 3, Number 4*, 653–65. National Bureau of Economic Research, Inc. https://EconPapers.repec.org/RePEc:nbr:nberch:10206.

Bhat, C. R. 1995. "A Heterocedastic Extreme Value Model of Intercity Travel Mode Choice." *Transportation Research B* 29 (6): 471–83.

Bivand, Roger S., and David W. S. Wong. 2018. "Comparing Implementations of Global and Local Indicators of Spatial Association." *Test* 27: 716–48.

Bivand, Roger, Giovanni Millo, and Gianfranco Piras. 2021. "A Review of Software for Spatial Econometrics in r." *Mathematics* 9 (11). https://doi.org/10.3390/math9111276.

Bohn, Sarah, Magnus Lofstrom, and Steven Raphael. 2014. "Did the 2007 Legal Arizona Workers Act Reduce the State's Unauthorized Immigrant Population?" *The Review of Economics and Statistics* 96 (2): 258–69. https://ideas.repec.org/a/tpr/restat/v96y2014i2p258-269.html.

Bonjour, Dorothe, Lynn F Cherkas, Jonathan E Haskel, Denise D Hawkes, and Tim D Spector. 2003. "Returns to Education: Evidence from U.K. Twins." *American Economic Review* 93 (5): 1799–1812. https://doi.org/10.1257/000282803322655554.

Bonnel, Patrick. 2004. *Prévoir La Demande de Transports.* Presses de l'école nationale des ponts et chaussées.

Bound, John, David A. Jaeger, and Regina M. Baker. 1995. "Problems with Instrumental Variables Estimation When the Correlation Between the Instruments and the Endogeneous Explanatory Variable Is Weak." *Journal of the American Statistical Association* 90 (430): 443–50. http://www.jstor.org/stable/2291055.

Breusch, T. S., and A. R. Pagan. 1979. "A Simple Test for Heteroscedasticity and Random Coefficient Variation." *Econometrica* 47 (5): 1287. https://doi.org/10.2307/1911963.

———. 1980. "The Lagrange Multiplier Test and Its Applications to Model Specification in Econometrics." *The Review of Economic Studies* 47 (1): 239. https://doi.org/10.2307/2297111.

Burde, Dana, and Leigh L. Linden. 2013. "Bringing Education to Afghan Girls: A Randomized Controlled Trial of Village-Based Schools." *American Economic Journal: Applied Economics* 5 (3): 27–40. http://www.jstor.org/stable/43189440.

Buser, Thomas. 2015. "The Effect of Income on Religiousness." *American Economic Journal: Applied Economics* 7 (3): 178–95. https://doi.org/10.1257/app.20140162.

Calonico, Sebastian, Matias D. Cattaneo, and Rocío Titiunik. 2015. "rdrobust: An R Package for Robust Nonparametric Inference in Regression-Discontinuity Designs." *The R Journal* 7 (1): 38–51. https://doi.org/10.32614/RJ-2015-004.

Cameron, A. Colin, and Per Johansson. 1997. "Count Data Regression Using Series Expansions: With Applications." *Journal of Applied Econometrics* 12 (3): 203–23. https://doi.org/10.1002/(SICI)1099-1255(199705)12:3%3C203::AID-JAE446%3E3.0.CO;2-2.

Cameron, A. Colin, and Pravin K. Trivedi. 1986. "Econometric Models Based on Count Data. Comparisons and Applications of Some Estimators and Tests." *Journal of Applied Econometrics* 1 (1): 29–53. https://doi.org/10.1002/jae.3950010104.

———. 2005. *Microeconometrics.* Cambridge University Press. https://EconPapers.repec.org/RePEc:cup:cbooks:9780521848053.

———. 2013. *Regression Analysis of Count Data.* 2nd ed. 53. Cambridge: Cambridge University Press.

Carlevaro, Fabrizio, and Yves Croissant. 2023. "Hurdle Regression Models: An Application to Consumer Behavior in the United States." In *Applied Econometrics Analysis Using Cross Section and Panel Data*, edited by Deep Mukherjee, 227–68. Contributions to Economics. Singapour: Springer.

Carson, Richard T., Leanne Wilks, and David Imber. 1994. "Valuing the Preservation of Australia's Kakadu Conservation Zone." *Oxford Economic Papers* 46: 727–49. http://www.jstor.org/stable/2663496.

Costanigro, Marco, Ron C. Mittelhammer, and Jill J. McCluskey. 2009. "Estimating Class-Specific Parametric Models Under Class Uncertainty: Local Polynomial Regression Clustering in an Hedonic Analysis of Wine Markets." *Journal of Applied Econometrics* 24 (7): 1117–35. https://doi.org/10.1002/jae.1094.

Cox, D. R. 1972. "Regression Models and Life-Tables." *Journal of the Royal Statistical Society. Series B (Methodological)* 34 (2): 187–220. http://www.jstor.org/stable/2985181.

———. 1975. "Partial Likelihood." *Biometrika* 62 (2): 269–76. http://www.jstor.org/stable/2335362.

Cox, D. R, and E. J. Snell. 1989. *Analysis of Binary Data.* 2nd ed. Monographs on statistics; applied probability.

Cragg, John G. 1971. "Some Statistical Models for Limited Dependent Variables with Applications for the Demand for Durable Goods." *Econometrica* 39 (5): 829–44.

Cragg, John G., and Russell S. Uhler. 1970. "The Demand for Automobiles." *Canadian Journal of Economics* 3 (3): 386–406. https://ideas.repec.org/a/cje/issued/v3y1970i3p386-406.html.

Croissant, Yves. 2020. "Estimation of Random Utility Models in r: The Mlogit Package"." *Journal of Statistical Software* 95 (11): 1–41. https://doi.org/10.18637/jss.v095.i11.

Croissant, Yves, and Giovanni Millo. 2008. "Panel Data Econometrics in R: The plm Package." *Journal of Statistical Software* 27 (2): 1–43. https://doi.org/10.18637/jss.v027.i02.

———. 2018. *Panel Data Econometrics with R.* Wiley.

Daganzo, C. 1979. *Multinomial Probit: The Theory and Its Application to Demand Forecasting.* Academic Press, New York.

Daly, A. 1987. "Estimating 'Tree' Logit Models." *Transportation Research B*, 251–67.

Davidson, Russell, and James G. MacKinnon. 1993. *Estimation and Inference in Econometrics*. New-York: Oxford University Press.

———. 2004. *Econometric Theory and Methods*. Oxford University Press.

De Crombrugghe, Denis, Franz C. Palm, and Jean-Pierre Urbain. 1997. "Statistical Demand Functions for Food in the USA and the Netherlands." *Journal of Applied Econometrics* 12 (5): 615–45. https://doi.org/10.1002/(SICI)1099-1255(199709/10)12:5%3C615::AID-JAE455%3E3.0.CO;2-L.

Dean, C. B. 1992. "Testing for Overdispersion in Poisson and Binomial Regression Models." *Journal of the American Statistical Association* 87 (418): 451–57. http://www.jstor.org/stable/2290276.

Deb, Partha, and Pravin K. Trivedi. 1997. "Demand for Medical Care by the Elderly: A Finite Mixture Approach." *Journal of Applied Econometrics* 12 (3): 313–36. http://www.jstor.org/stable/2285252.

———. 2002. "The Structure of Demand for Health Care: Latent Class Versus Two-Part Models." *Journal of Health Economics* 21 (4): 601–25. https://doi.org/10.1016/S0167-6296(02)00008-5.

Dehejia, Rajeev H., and Sadek Wahba. 1999. "Causal Effects in Nonexperimental Studies: Reevaluating the Evaluation of Training Programs." *Journal of the American Statistical Association* 94 (448): 1053–62. http://www.jstor.org/stable/2669919.

———. 2002. "Propensity Score-Matching Methods for Nonexperimental Causal Studies." *The Review of Economics and Statistics* 84 (1): 151–61. https://doi.org/10.1162/003465302317331982.

Di Tella, Rafael, and Ernesto Schargrodsky. 2004. "Do Police Reduce Crime? Estimates Using the Allocation of Police Forces After a Terrorist Attack." *The American Economic Review* 94 (1): 115–33. http://www.jstor.org/stable/3592772.

Douch, Mustapha, and Terence Huw Edwards. 2022. "The bilateral trade effects of announcement shocks: Brexit as a natural field experiment." *Journal of Applied Econometrics* 37 (2): 305–29. https://doi.org/10.1002/jae.2878.

Dow, William H., and Edward C. Norton. 2003. "Choosing Between and Interpreting the Heckit and Two-Part Models for Corner Solutions." *Health Services and Outcomes Research Methodology* 4 (1): 5–18.

Dunford, Eric. 2021. *Tidysynth: A Tidy Implementation of the Synthetic Control Method.* https://CRAN.R-project.org/package=tidysynth.

Duranton, Gilles, and Diego Puga. 2020. "The Economics of Urban Density." *Journal of Economic Perspectives* 34 (3): 3–26. https://doi.org/10.1257/jep.34.3.3.

Edelman, Benjamin, Michael Luca, and Dan Svirsky. 2017. "Racial Discrimination in the Sharing Economy: Evidence from a Field Experiment." *American Economic Journal: Applied Economics* 9 (2): 1–22. https://doi.org/10.1257/app.20160213.

Efron, Bradley. 1978. "Regression and ANOVA with Zero-One Data: Measures of Residual Variation." *Journal of the American Statistical Association* 73 (361): 113–21. http://www.jstor.org/stable/2286531.

Ellison, Glenn, and Ashley Swanson. 2016. "Do Schools Matter for High Math Achievement? Evidence from the American Mathematics Competitions." *American Economic Review* 106 (6): 1244–77. https://doi.org/10.1257/aer.20140308.

Ertur, Cem, and Wilfried Koch. 2007. "Growth, Technological Interdependence and Spatial Externalities: Theory and Evidence." *Journal of Applied Econometrics* 22 (6): 1033–62. http://www.jstor.org/stable/25146563.

Estrella, Arturo. 1998. "A New Measure of Fit for Equations with Dichotomous Dependent Variables." *Journal of Business & Economic Statistics* 16 (2): 198–205.

Favero, Carlo A., M. Hashem Pesaran, and Sunil Sharma. 1994. "A Duration Model of Irreversible Oil Investment: Theory and Empirical Evidence." *Journal of Applied Econometrics* 9: S95–112. http://www.jstor.org/stable/2285225.

Forinash, C. V., and F. S. Koppelman. 1993. "Application and Interpretation of Nested Logit Models and Intercity Mode Choice." *Transportation Record* 1413: 98–106.

Geil, Peter, Andreas Million, Ralph Rotte, and Klaus F. Zimmermann. 1997. "Economic Incentives and Hospitalization in Germany." *Journal of Applied Econometrics* 12 (3): 295–311. http://www.jstor.org/stable/2285251.

Geweke, J., M. Keane, and D. Runkle. 1994. "Alternative Computational Approaches to Inference in the Multinomial Probit Model." *Review of Economics and Statistics* 76: 609–32.

Giraud, Timothée. 2023. *Mapsf: Thematic Cartography.* https://CRAN.R-project.org/package=mapsf.

Goldberg, Pinelopi Koujianou, and Giovanni Maggi. 1999. "Protection for Sale: An Empirical Investigation." *American Economic Review* 89 (5): 1135–55. https://doi.org/10.1257/aer.89.5.1135.

Goldberger, A. S. 1964. *Econometric Theory.* Wiley, New-York.

———. 1981. "Linear Regression After Regression." *Journal of Econometrics* 15: 357–66.

Gourieroux, Christian, Alain Monfort, Eric Renault, and Alain Trognon. 1987. "Generalised Residuals." *Journal of Econometrics* 34 (1): 5–32. https://doi.org/10.1016/0304-4076(87)90065-0.

Greene, William H. 1981. "On the Asymptotic Bias of the Ordinary Least Squares Estimator of the Tobit Model." *Econometrica* 49: 505–13.

———. 2001. "Fiml Estimation of Sample Selection Models for Count Data." In *Economic Theory, Dynamics and Markets: Essays in Honor of Ryuzo Sato*, edited by Takashi Negishi, Rama V. Ramachandran, and Kazuo Mino, 73–91. Boston, MA: Springer US. https://doi.org/10.1007/978-1-4615-1677-4_6.

———. 2003. *Econometrics Analysis.* 5th ed. Pearson.

———. 2018. *Econometrics Analysis.* 8th ed. Pearson.

Gronau, Reuben. 1973. "The Effect of Children on the Housewife's Value of Time." *Journal of Political Economy* 81 (2): S168–99. http://www.jstor.org/stable/1840419.

Grossman, Gene M., and Elhanan Helpman. 1994. "Protection for Sale." *The American Economic Review* 84 (4): 833–50. http://www.jstor.org/stable/2118033.

Gurmu, Shiferaw. 1997. "Semi-Parametric Estimation of Hurdle Regression Models with an Application to Medicaid Utilization." *Journal of Applied Econometrics* 12 (3): 225–42. http://www.jstor.org/stable/2285247.

Hanemann, W. Michael. 1991. "Willingness to Pay and Willingness to Accept: How Much Can They Differ?" *The American Economic Review* 81 (3): 635–47. http://www.jstor.org/stable/2006525.

Hausman, Jerry. 1978. "Specification Tests in Econometrics." *Econometrica* 46 (6): 1251–71. http://www.jstor.org/stable/1913827.

Hausman, Jerry, and D. Wise. 1978. "A Conditional Probit Model for Qualitative Choice: Discrete Decisions Recognizing Interdemendence and Heterogeneous Preferences." *Econometrica* 48: 403–29.

Hausman, Jerry, and David Wise. 1976. "The Evaluation of Results from Truncated Samples: The New Jersey Income Maintenance Experiment," 421–45. https://EconPapers.repec.org/RePEc:nbr:nberch:10489.

———. 1977. "Social Experimentation, Truncated Distributions, and Efficient Estimation." *Econometrica* 45 (4): 919–38. https://ideas.repec.org/a/ecm/emetrp/v45y1977i4p919-38.html.

Heckman, James J. 1976. "The Common Structure of Statistical Models of Truncation, Sample Selection and Limited Dependent Variables and a Simple Estimator for Such Models," 475–92. https://EconPapers.repec.org/RePEc:nbr:nberch:10491.

———. 1979. "Sample Selection Bias as a Specification Error." *Econometrica* 47 (1): 153–61. http://www.jstor.org/stable/1912352.

———. 2000. "Causal Parameters and Policy Analysis in Economics: A Twentieth Century Retrospective." *The Quarterly Journal of Economics* 115 (1): 45–97. http://www.jstor.org/stable/2586935.

Heiss, Florian. 2002. "Structural Choice Analysis with Nested Logit Models." *The Stata Journal* 2 (3): 227–52.

Henningsen, Arne, and Jeff D. Hamann. 2007. "Systemfit: A Package for Estimating Systems of Simultaneous Equations in r." *Journal of Statistical Software* 23 (4): 1–40. https://www.jstatsoft.org/v23/i04/.

Henningsen, Arne, and Ott Toomet. 2011. "maxLik: A Package for Maximum Likelihood Estimation in R." *Computational Statistics* 26 (3): 443–58. https://doi.org/10.1007/s00180-010-0217-1.

Hensher, David A., and William H. Greene. 2002. "Specification and Estimation of the Nested Logit Model: Alternative Normalisations." *Transportation Research Part B* 36: 1–17.

Hijmans, Robert J. 2023. *Terra: Spatial Data Analysis.* https://CRAN.R-project.org/package=terra.

Hochguertel, Stefan. 2003. "Precautionary Motives and Portfolio Decisions." *Journal of Applied Econometrics* 18 (1): 61–77. https://doi.org/10.1002/jae.658.

Hong, Seung-Hyun. 2013. "Measuring the Effect of Napster on Recorded Music Sales: Difference-in-Differences Estimates Under Compositional Changes." *Journal of Applied Econometrics* 28 (2): 297–324. http://www.jstor.org/stable/23355919.

Horowitz, Joel L. 1993. "Semiparametric Estimation of a Work-Trip Mode Choice Model." *Journal of Econometrics* 58 (1-2): 49–70. https://EconPapers.repec.org/RePEc:eee:econom:v:58:y:1993:i:1-2:p:49-70.

Houthakker, H. S. 1951. "Some Calculations on Electricity Consumption in Great Britain." *Journal of the Royal Statistical Society. Series A (General)* 114 (3): 359–71. http://www.jstor.org/stable/2980781.

Ichino, Andrea. 2002. "Estimation of Average Treatment Effects Based on Propensity Scores." *Stata Journal* 2 (4): 358–377(20). https://www.stata-journal.com/article.html?article=st0026.

Ichino, Andrea, Fabrizia Mealli, and Tommaso Nannicini. 2008. "From Temporary Help Jobs to Permanent Employment: What Can We Learn from Matching Estimators and Their Sensitivity?" *Journal of Applied Econometrics* 23 (3): 305–27. http://www.jstor.org/stable/25144550.

Imbens, Guido W., and Joshua D. Angrist. 1994. "Identification and Estimation of Local Average Treatment Effects." *Econometrica* 62 (2): 467–75. http://www.jstor.org/stable/2951620.

Ivaldi, Marc, Norbert Ladoux, Hervé Ossard, and Michel Simioni. 1996. "Comparing Fourier and Translog Specifications of Multiproduct Technology: Evidence from an Incomplete Panel of French Farmers." *Journal of Applied Econometrics* 11 (6): 649–67. https://doi.org/10.1002/(sici)1099-1255(199611)11:6%3C649::aid-jae416%3E3.0.co;2-4.

Jones, Andrew. 2000. "Health Econometrics." In *Handbook of Health Economics*, edited by A. J. Culyer and J. P. Newhouse, 1st ed., 1:265–344. Elsevier. https://EconPapers.repec.org/RePEc:eee:heachp:1-06.

Kastoryano, Stephen, and Bas van der Klaauw. 2022. "Dynamic Evaluation of Job Search Assistance." *Journal of Applied Econometrics* 37 (2): 227–41. https://doi.org/10.1002/jae.2866.

Katz, Lawrence F. 1986. "Layoffs, Recall and the Duration of Unemployment." NBER Working Papers 1825. National Bureau of Economic Research, Inc. https://ideas.repec.org/p/nbr/nberwo/1825.html.

Kenkel, Donald S., and Joseph V. Terza. 2001. "The Effect of Physician Advice on Alcohol Consumption: Count Regression with an Endogenous Treatment Effect." *Journal of Applied Econometrics* 16 (2): 165–84. https://doi.org/10.1002/jae.596.

Kennan, John. 1985. "The Duration of Contract Strikes in u.s. Manufacturing." *Journal of Econometrics* 28 (1): 5–28. https://doi.org/10.1016/0304-4076(85)90064-8.

Kiefer, Nicholas M. 1988. "Economic Duration Data and Hazard Functions." *Journal of Economic Literature* 26 (2): 646–79. http://www.jstor.org/stable/2726365.

Koenker, Roger. 1981. "A Note on Studentizing a Test for Heteroscedasticity." *Journal of Econometrics* 17: 107–12.

Koenker, Roger, and Achim Zeileis. 2009. "On Reproducible Econometric Research." *Journal of Applied Econometrics* 24 (5): 833–47. https://doi.org/10.1002/jae.1083.

Koop, Gary, Dale J. Poirier, and Justin Tobias. 2005. "Semiparametric Bayesian Inference in Multiple Equation Models." *Journal of Applied Econometrics* 20 (6): 723–47. https://doi.org/10.1002/jae.810.

Koppelman, Franck S., and Chieh-Hua Wen. 1998. "Alternative Nested Logit Models: Structure, Properties and Estimation." *Transportation Research B* 32 (5): 289–98.

———. 2000. "The Paired Combinatorial Logit Model: Properties, Estimation and Application." *Transportation Research B* 34: 75–89.

Kumar, Anil. 2018. "Do Restrictions on Home Equity Extraction Contribute to Lower Mortgage Defaults? Evidence from a Policy Discontinuity at the Texas Border." *American Economic Journal: Economic Policy* 10 (1): 268–97. https://www.jstor.org/stable/26529015.

Kuznets, Simon. 1955. "Economic Growth and Income Inequality." *The American Economic Review* 45 (1): 1–28. http://www.jstor.org/stable/1811581.

LaLonde, Robert. 1986. "Evaluating the Econometric Evaluations of Training Programs with Experimental Data." *American Economic Review* 76 (4): 604–20. https://EconPapers.repec.org/RePEc:aea:aecrev:v:76:y:1986:i:4:p:604-20.

Lambert, Diane. 1992. "Zero-Inflated Poisson Regression, with an Application to Defects in Manufacturing." *Technometrics* 34 (1): 1–14. http://www.jstor.org/stable/1269547.

Lancaster, Tony. 1990. In *The Econometric Analysis of Transition Data.* Econometric Society Monographs. Cambridge University Press.

Lave, Charles A. 1970. "The Demand for Urban Mass Transportation." *The Review of Economics and Statistics* 52 (3): 320–23. http://www.jstor.org/stable/1926301.

Lindo, Jason M., Nicholas J. Sanders, and Philip Oreopoulos. 2010. "Ability, Gender, and Performance Standards: Evidence from Academic Probation." *American Economic Journal: Applied Economics* 2 (2): 95–117. http://www.jstor.org/stable/25760207.

Long, John Scott. 1997. *Regression Models for Categorical and Limited Dependent Variables.* Sage Publications.

Maddala, G. S. 1983. *Limited-Dependent and Qualitative Variables in Econometrics.* Econometric Society Monographs. Cambridge University Press. https://doi.org/10.1017/CBO9780511810176.

Maddala, G. S., and Kajal Lahiri. 2009. *Introduction to Econometrics.* 4th ed. Wiley.

Magee, Lonnie. 1990. "R2 Measures Based on Wald and Likelihood Ratio Joint Significance Tests." *The American Statistician* 44 (3): 250–53. http://www.jstor.org/stable/2685352.

Makowsky, Michael D., and Thomas Stratmann. 2009. "Political Economy at Any Speed: What Determines Traffic Citations?" *American Economic Review* 99 (1): 509–27. https://doi.org/10.1257/aer.99.1.509.

Mankiw, N. Gregory, David Romer, and David N. Weil. 1992. "A Contribution to the Empirics of Economic Growth." *The Quarterly Journal of Economics* 107 (2): 407–37. https://ideas.repec.org/a/oup/qjecon/v107y1992i2p407-437..html.

Matschke, Xenia, and Shane M. Sherlund. 2006. "Do Labor Issues Matter in the Determination of u.s. Trade Policy? An Empirical Reevaluation." *The American Economic Review* 96 (1): 405–21. http://www.jstor.org/stable/30034374.

McCrary, Justin. 2008. "Manipulation of the Running Variable in the Regression Discontinuity Design: A Density Test." *Journal of Econometrics* 142 (2): 698–714. https://doi.org/10.1016/j.jeconom.2007.05.005.

McCullagh, P., and J. A. Nelder. 1989. *Generalized Linear Models.* 2nd ed. London: Chapman; Hall.

McDonald, John F, and Robert A Moffitt. 1980. "The Uses of Tobit Analysis." *The Review of Economics and Statistics* 62 (2): 318–21. https://ideas.repec.org/a/tpr/restat/v62y1980i2p318-21.html.

McFadden, Daniel. 1973. "Conditional Logit Analysis of Qualitative Choice Behaviour." In *Frontiers in Econometrics*, edited by P. Zarembka, 105–42. Academic Press New York.

———. 1974. "The Measurement of Urban Travel Demand." *Journal of Public Economics* 3: 303–28.

———. 1978. "Spatial Interaction Theory and Planning Models." In *Modeling the Choice of Residential Location*, edited by A. Karlqvist, 75–96. North-Holland, Amsterdam.

McKelvey, Richard D., and William Zavoina. 1975. "A Statistical Model for the Analysis of Ordinal Level Dependent Variables." *The Journal of Mathematical Sociology* 4 (1): 103–20. https://doi.org/10.1080/0022250X.1975.9989847.

Meijer, Erik, and Jan Rouwendal. 2006. "Measuring Welfare Effects in Models with Random Coefficients." *Journal of Applied Econometrics* 21 (2): 227–44. https://doi.org/10.1002/jae.841.

Messner, Jakob W., Georg J. Mayr, Achim Zeileis, and Daniel S. Wilks. 2014. "Heteroscedastic Extended Logistic Regression for Postprocessing of Ensemble Guidance." *Monthly Weather Review* 142 (1): 448–56. https://doi.org/10.1175/MWR-D-13-00271.1.

Miller, Nathan H. 2009. "Strategic Leniency and Cartel Enforcement." *American Economic Review* 99 (3): 750–68. https://doi.org/10.1257/aer.99.3.750.

Mullahy, John. 1986. "Specification and Testing of Some Modified Count Data Models." *Journal of Econometrics* 33 (3): 341–65. https://doi.org/10.1016/0304-4076(86)90002-3.

———. 1997. "Instrumental-Variable Estimation of Count Data Models: Applications to Models of Cigarette Smoking Behavior." *The Review of Economics and Statistics* 79 (4): 586–93. http://www.jstor.org/stable/2951410.

———. 1998. "Much Ado about Two: Reconsidering Retransformation and the Two-Part Model in Health Econometrics." *Journal of Health Economics* 17 (3): 247–81. https://doi.org/10.1016/S0167-6296(98)00030-7.

Nagelkerke, Nico J. D. 1991. "A Note on a General Definition of the Coefficient of Determination." *Biometrika* 78: 691–92.

Nerlove, M. 1971. "Further Evidence on the Estimation of Dynamic Economic Relations from a Time–Series of Cross–Sections." *Econometrica* 39: 359–82.

Newey, Whitney K. 1985. "Maximum Likelihood Specification Testing and Conditional Moment Tests." *Econometrica* 53 (5): 1047–70. http://www.jstor.org/stable/1911011.

———. 1987. "Efficient Estimation of Limited Dependent Variable Models with Endogenous Explanatory Variables." *Journal of Econometrics* 36 (3): 231–50. https://doi.org/10.1016/0304-4076(87)90001-7.

Nunn, Nathan. 2008. "The Long-Term Effects of Africa's Slave Trades." *The Quarterly Journal of Economics* 123 (1): 139–76. https://EconPapers.repec.org/RePEc:oup:qjecon:v:123:y:2008:i:1:p:139-176.

Olsen, Randall J. 1978. "Note on the Uniqueness of the Maximum Likelihood Estimator for the Tobit Model." *Econometrica* 46 (5): 1211–15. https://EconPapers.repec.org/RePEc:ecm:emetrp:v:46:y:1978:i:5:p:1211-15.

Pebesma, Edzer, and Roger S. Bivand. 2023. *Spatial Data Science with Applications in R.* Chapman & Hall. https://r-spatial.org/book/.

Pinotti, Paolo. 2015. "The Economic Costs of Organised Crime: Evidence from Southern Italy." *The Economic Journal* 125 (586): F203–32. https://doi.org/10.1111/ecoj.12235.

Plug, Erik. 2004. "Estimating the Effect of Mother's Schooling on Children's Schooling Using a Sample of Adoptees." *American Economic Review* 94 (1): 358–68. https://doi.org/10.1257/000282804322970850.

Powell, J. 1986. "Symmetrically Trimed Least Squares Estimators for Tobit Models." *Econometrica* 54: 1435–60.

Reynaerts, Jo, and Jakob Vanschoonbeek. 2022. "The economics of state fragmentation: Assessing the economic impact of secession." *Journal of Applied Econometrics* 37 (1): 82–115. https://doi.org/10.1002/jae.2857.

Rivers, Douglas, and Quang H. Vuong. 1988. "Limited Information Estimators and Exogeneity Tests for Simultaneous Probit Models." *Journal of Econometrics* 39 (3): 347–66. https://EconPapers.repec.org/RePEc:eee:econom:v:39:y:1988:i:3:p:347-366.

Robinson, Sherman. 1976. "A Note on the u Hypothesis Relating Income Inequality and Economic Development." *The American Economic Review* 66 (3): 437–40. http://www.jstor.org/stable/1828182.

Romer, Paul M. 1986. "Increasing Returns and Long-Run Growth." *Journal of Political Economy* 94 (5): 1002–37. http://www.jstor.org/stable/1833190.

Rosenbaum, Paul R., and Donald B. Rubin. 1983. "The Central Role of the Propensity Score in Observational Studies for Causal Effects." *Biometrika* 70 (1): 41–55. http://www.jstor.org/stable/2335942.

Santos Silva, João M. C, and Frank Windmeijer. 2001. "Two-Part Multiple Spell Models for Health Care Demand." *Journal of Econometrics* 104 (1): 67–89. https://doi.org/10.1016/S0304-4076(01)00059-8.

Schaller, Huntley. 1990. "A Re-Examination of the q Theory of Investment Using u.s. Firm Data." *Journal of Applied Econometrics* 5 (4): 309–25. http://www.jstor.org/stable/2096476.

Seller, Christine, John R. Stoll, and Jean-Paul Chavas. 1985. "Validation of Empirical Measures of Welfare Change: A Comparison of Nonmarket Techniques." *Land Economics* 61 (2): 156–75. http://www.jstor.org/stable/3145808.

Shi, Xiaoxia. 2015. "A Nondegenerate Vuong Test." *Quantitative Economics*, 85–121.

Sjoberg, Daniel D., Karissa Whiting, Michael Curry, Jessica A. Lavery, and Joseph Larmarange. 2021. "Reproducible Summary Tables with the Gtsummary Package." *The R Journal* 13: 570–80. https://doi.org/10.32614/RJ-2021-053.

Skeels, Christopher L., and Francis Vella. 1999. "A Monte Carlo Investigation of the Sampling Behavior of Conditional Moment Tests in Tobit and Probit Models." *Journal of Econometrics* 92 (2): 275–94. https://doi.org/10.1016/S0304-4076(98)00092-X.

Small, K. A., and H. S. Rosen. 1981. "Applied Welfare Economics with Discrete Choice Models." *Econometrica* 49: 105–30.

Smith, Richard J., and Richard W. Blundell. 1986. "An Exogeneity Test for a Simultaneous Equation Tobit Model with an Application to Labor Supply." *Econometrica* 54 (3): 679–85. http://www.jstor.org/stable/1911314.

Stewart, Kenneth. 2019. "Suits' Watermelon Model: The Missing Simultaneous Equations Empirical Application." *Journal of Economics Teaching* 4 (2): 115–39. https://EconPapers.repec.org/RePEc:jtc:journl:v:4:y:2019:i:2:p:115-139.

Stock, James H., and Mark W. Watson. 2015. *Introduction to Econometrics*. Pearson.

Sudhir, Anand, and S. M. R. Kanbur. 1993. "The Kuznets Process and the Inequality-Development Relationship." *Journal of Development Economics* 40: 25–52.

Sueyoshi, Glenn T. 1995. "A Class of Binary Response Models for Grouped Duration Data." *Journal of Applied Econometrics* 10 (4): 411–31. https://ideas.repec.org/a/jae/japmet/v10y1995i4p411-31.html.

Suits, Daniel B. 1955. "An Econometric Model of the Watermelon Market." *American Journal of Agricultural Economics* 37 (2): 237–51. https://doi.org/10.2307/1233923.

Swamy, P. A. V. B., and S. S Arora. 1972. "The Exact Finite Sample Properties of the Estimators of Coefficients in the Error Components Regression Models." *Econometrica* 40: 261–75.

Tauchen, George. 1985. "Diagnostic Testing and Evaluation of Maximum Likelihood Models." *Journal of Econometrics* 30 (1): 415–43. https://doi.org/10.1016/0304-4076(85)90149-6.

Tennekes, Martijn. 2018. "tmap: Thematic Maps in R." *Journal of Statistical Software* 84 (6): 1–39. https://doi.org/10.18637/jss.v084.i06.

Terza, Joseph V. 1998. "Estimating Count Data Models with Endogenous Switching: Sample Selection and Endogenous Treatment Effects." *Journal of Econometrics* 84 (1): 129–54. https://doi.org/10.1016/S0304-4076(97)00082-1.

Tjur, Tue. 2009. "Coefficients of Determination in Logistic Regression Models—a New Proposal: The Coefficient of Discrimination." *The American Statistician* 63 (4): 366–72. http://www.jstor.org/stable/25652317.

Tobin, James. 1958. "Estimation of Relationships for Limited Dependent Variables." *Econometrica* 26 (1): 24–36.

Toomet, Ott, and Arne Henningsen. 2008. "Sample Selection Models in R: Package sampleSelection." *Journal of Statistical Software* 27 (7). https://www.jstatsoft.org/v27/i07/.

Train, Kenneth. 2009. *Discrete Choice Methods with Simulation*. Cambridge University Press. https://EconPapers.repec.org/RePEc:cup:cbooks:9780521766555.

Train, Kenneth, Daniel McFadden, and Moshe Ben-Akiva. 1987. "The Demand for Local Telephone Service: A Fully Discrete Model of Residential Calling Patterns and Service Choices." *The RAND Journal of Economics* 18 (1): 109–23. http://www.jstor.org/stable/2555538.

Vandaele, Walter. 1981. "Wald, Likelihood Ratio, and Lagrange Multiplier Tests as an f Test." *Economics Letters* 8 (4): 361–65. https://doi.org/10.1016/0165-1765(81)90026-4.

Veall, Michael R., and Klaus F. Zimmermann. 1996. "Pseudo-R2 Measures for Some Common Limited Dependent Variable Models." *Journal of Economic Surveys* 10 (3): 241–59. https://doi.org/10.1111/j.1467-6419.1996.tb00013.x.

Vuong, Quang H. 1989. "Likelihood Ratio Tests for Selection and Non-Nested Hypotheses." *Econometrica* 57 (2): 307–33.

Walker, Joan L., Moshe Ben-Akiva, and Denis Bolduc. 2007. "Identification of Parameters in Normal Error Component Logit-Mixture (NECLM) Models." *Journal of Applied Econometrics* 22 (6): 1095–125. http://www.jstor.org/stable/25146565.

Wallace, T. D., and A. Hussain. 1969. "The Use of Error Components Models in Combining Cross Section with Time Series Data." *Econometrica* 37 (1): 55–72.

Werner, Megan. 1999. "Allowing for Zeros in Dichotomous-Choice Contingent-Valuation Models." *Journal of Business & Economic Statistics* 17 (4): 479–86. http://www.jstor.org/stable/1392405.

Wheeler, Christopher H. 2003. "Evidence on Agglomeration Economies, Diseconomies, and Growth." *Journal of Applied Econometrics* 18 (1): 79–104. http://www.jstor.org/stable/30035189.

White, Halbert. 1980. "A Heteroskedasticity-Consistent Covariance Matrix Estimator and a Direct Test for Heteroskedasticity." *Econometrica* 48 (4): 817. https://doi.org/10.2307/1912934.

Wickham, Hadley, Mine Cetinkaya-Rundel, and Garrett Grolemund. 2023. *R for Data Science: Import, Tidy, Transform, Visualize, and Model Data.* O'Reilly.

Wilhelm, Mark Ottoni. 2008. "Practical Considerations for Choosing Between Tobit and SCLS or CLAD Estimators for Censored Regression Models with an Application to Charitable Giving." *Oxford Bulletin of Economics and Statistics* 70 (4): 559–82. https://doi.org/10.1111/j.1468-0084.2008.00506.x.

Windmeijer, Frank. 1995. "Goodness-of-Fit Measures in Binary Choice Models." *Econometric Reviews* 14 (February): 101–16. https://doi.org/10.1080/07474939508800306.

Winkelmann, Rainer. 2004. "Health Care Reform and the Number of Doctor Visits: An Econometric Analysis." *Journal of Applied Econometrics* 19 (4): 455–72. http://www.jstor.org/stable/25146297.

Wold, H. O. A. 1958. "A Case Study of Interdependent Versus Causal Chain Systems." *Revue de l'Institut International de Statistique / Review of the International Statistical Institute* 26 (1/3): 5–25. http://www.jstor.org/stable/1401568.

Wooldridge, Jeffrey M. 2010. *Econometric Analysis of Cross Section and Panel Data.* Vol. 1. MIT Press Books 0262232588. The MIT Press. https://ideas.repec.org/b/mtp/titles/0262232588.html.

Zeileis, Achim. 2004. "Econometric Computing with HC and HAC Covariance Matrix Estimators." *Journal of Statistical Software* 11 (10): 1–17. https://doi.org/10.18637/jss.v011.i10.

———. 2006. "Object-Oriented Computation of Sandwich Estimators." *Journal of Statistical Software* 16 (9): 1–16. https://doi.org/10.18637/jss.v016.i09.

Zeileis, Achim, and Yves Croissant. 2010. "Extended Model Formulas in R: Multiple Parts and Multiple Responses." *Journal of Statistical Software* 34 (1): 1–13. https://www.jstatsoft.org/v34/i01/.

Zeileis, Achim, and Torsten Hothorn. 2002. "Diagnostic Checking in Regression Relationships." *R News* 2 (3): 7–10. https://CRAN.R-project.org/doc/Rnews/.

Zeileis, Achim, Susanne Köll, and Nathaniel Graham. 2020. "Various Versatile Variances: An Object-Oriented Implementation of Clustered Covariances in r." *Journal of Statistical Software* 95 (1): 1–36. https://doi.org/10.18637/jss.v095.i01.

Zellner, Arnold. 1962. "An Efficient Method of Estimating Seemingly Unrelated Regressions and Tests of Aggregation Bias." *Journal of the American Statistical Association* 57: 500–509.

Zellner, Arnold, and N. S. Revankar. 1969. "Generalized Production Functions." *Review of Economic Studies* 36 (2): 241–50. https://ideas.repec.org/a/oup/restud/v36y1969i2p241-250..html.

Zellner, Arnold, and H. Theil. 1962. "Three-Stage Least Squares: Simultaneous Estimation of Simultaneous Equations." *Econometrica* 30 (1): 54–78. http://www.jstor.org/stable/1911287.

Indexes

General index

accelerated failure time model, 413–415
    exponential model, 413
    log-logistic model, 414–415
    log-normal model, 414
    weibul model, 413–414
Akaike information criteria, 326, 392–393
asymptotic distribution, 52
    maximum likelihood estimator, 141
asymptotic equality, 146
asymptotic hessian, 145
asymptotic information matrix, 145
asymptotic properties, 49–55
    multiple linear regression model, 85
at-risk observations, 401
attenuation bias, 204
average treatment effect of the treated, 265

bandwidth, 253–254
Bayes information criteria, 326
best linear unbiased estimator
    multiple linear regression model, 84
    simple linear regression model, 45–49
between matrix, 176
bias, 33, 36
    errors in variables, 204
    instrumental variable estimator, 211
    omitted variable, 204–206
    simultaneity, 206–209
    truncated response, 350
binomial model
    duration data, 426
    minimum chi-squared estimator, 335
    two-step estimator, 333–335
Breusch Pagan test
    heteroskedasticity, 178–180
    individual effects, 180–183
    inter-equation correlation, 183–184

caliper matching, 264
censored regression model
    two-sided, 360
censoring, 345
central-limit theorem, 52–53, 141
chi-square test, 243–244
Choleski decomposition, 455, 460, 462
coefficient of determination, 18–19
    adjusted, 85
    binomial model, 326–329
    pseudo-three tests principles, 158–159, 163
common factor hypothesis, 301
competing risk model, 423–424
compliers, 247
concentrated log-likelihood function, 151
conditional expectation, 5–10, 34
conditional logit model, 438
conditional moment test, 163–167
    binomial, 331–332
    tobit1, 362–363
    Weibull model, 421
confidence interval, 63, 64
    multiple linear regression model, 86–90
    prediction, 63–66
    R, 62
    simple linear regression model, 59
consistency, 49–55
    maximum likelihood estimator, 138–139
constant hazard hypothesis, 404, 410, 420
constrained least squares, 103
constrained model, 92
contiguity matrix, 292–295
convergence

in distribution, 52–53
in mean square, 50
in probability, 49–51
corner solution, 343–345
correlated errors, 174–176
random utility model, 449–450
counterfactual, 241, 247
covariance, 5–10
covariance matrix
binomial model, 323
Poisson
information based, 381
Poisson model
quasi-Poisson, 382
sandwich, 381–383
two-step estimator, 353
covariance matrix estimation
binomial model, 324
maximum likelihood estimator
gradient based, 141, 146
hessian based, 142, 147
information based, 141, 146
sandwich, 142, 147
multiple linear regression
sandwich, 188, 191
sandwich, 184–191
simple linear regression
sandwich, 185–187
critical value, 56–57
cumulant generating function, 410, 411
cumulative hazard, 401

data generator process, 23
delta method, 62–63, 305–306, 456
deviance, 318, 325
binomial model, 319
null, 319
scaled, 318
difference in differences, 256–259
digamma function, 411
discrete-time hazard, 402
discrete-time survivor functions, 402
Durbin's model, 301

efficiency, 33
endogeneity
probit, 332–335
tobit1, 363, 364
endogenous switching
Poisson model, 396–399
maximum likelihood, 397–398
two-step estimator, 398–399
error component model, 175, 194–198
errors, 14
errors in variables, 203–204
estimating function, 188
Euler-Mascheronni constant, 411
excess of zero, 376
exclusion restriction, 210, 247
experimental data, 241
exponential distribution, 133, 136–138, 144, 147–411
exponential linear conditional mean model, 393–395
exponential model, 410–411
maximum likelihood estimator, 411–412
extreme value distribution of type 1, 411

F statistic, 96–98, 157
individual effects, 237
weak instruments, 237
first stage regression, 217
first-difference estimator, 233
fixed effects model, 231–233
forcing variable, 249, 255
Frisch-Waugh theorem, 76–77, 232

Gamma function, 411
gaussian quadrature, 388, 398, 447
Geary test, 295–296
local, 297–298
general method of moments estimator, 222–223
Poisson model, 394
generalized least squares, 173, 192–201
error component model, 194–198
seemingly unrelated regression, 199–201
weighted least squares, 193–194
generalized linear model, 317–319
generalized production function, 153–156
generalized residuals, 323
binomial model, 322
censored regression model, 354–355
Weibull model, 420–421
geometry of least squares
multiple linear regression model, 75–76
simple linear regression model, 17–18
goodness-of-fit indicators

R, 62
grouped data, 426–428
binomial model, 426
ordered model, 426–427
Gumbel distribution, 440

Hausman test, 236–237
hazard function, 401
constant, 410–411
heckit model, 368–370
heteroskedastic logit model, 447
heteroskedasticity, 174, 348
random utility model, 447
homoskedasticity, 37, 42, 82
random utility model, 440
hurdle model
censored regression model, 367–368
count model, 391

idempotent matrix, 176
identification, 34, 209
indempotent matrix, 72
independence of irrelevant alternatives hypothesis, 441, 445
index function, 311
indirect least squares, 209
information matrix equality, 139–141, 145–146
instrumental variable estimator, 209–226
general, 221–223
general method of moments estimator, 222–223
just identified, 209–217
Poisson model, 393–395
regression discontinuity, 254
three-stage least squares, 226–228
treatment effect, 246–249
two stage, 222–223
two-stage least squares, 211
Wald estimator, 217–221
intention to treat, 248
interval regession model, 429–430
inverse mills ratio, 323
inverse of a partitioned matrix, 103, 166

Jacobian, 151
generalized production function, 154

Kaplan-Meier estimator, 402–410
censored sample, 406–407
different groups, 408–410
uncensored sample, 402–406
Kronecker product, 175, 227
Kullback-Leibler information criterion, 168

Lagrange multiplier test
nested logit, 451
latent variable, 316, 323
ordered model, 337
law of large numbers, 49–51
least squares dummy variable estimator, 232
likelihood ratio test, 93, 95–96
binomial, 330
generalized production function, 161
heteroskedastic logit model, 448
linear gaussian model, 157
linear gaussian model, 150–151
linear predictor, 311
linear probability model, 311
link, 317
canonical, 379
log, 379
local average treatment effect, 248
log-normal model, 151–153
log-rank test, 409
log-sum, 442
logit, 312–313
lower model, 450

manipulation test, 255
marginal effects, 122–128
at the mean, 125, 128
average marginal effects, 125, 127, 128
matching, 259
caliper, 264
maximum likelihood estimator
exponential model, 411–412
minimum chi-squared estimator
binomial model, 335
mixed logit model, 438, *see* random parameter logit model
mixed model
multinomial logit, 452–453
mixing distribution
Poisson model, 386–388
gamma, 387–388
normal, 388
moment generating function, 410, 411
monotonicity hypothesis, 247
Moran test, 295–296

local, 297–298
multinomial logit model, 438
multinomial probit model, 458–462

negative binomial distribution, 387
negbin, 383
NB1, 383, 384, 388
NB2, 383, 384, 387
nested logit model, 449–450
two-step estimation, 450
Newton-Raphson algorithm, 147–148
non-linear least squares, 352–399
normal distribution, 25, 43–44, 150–151
null model, 319

observational data, 241
offset, 153
omitted variable bias, 204–206
ordered model, 337–340
duration data, 426–427
overdispersion, 376, 383
test, 384

panel data, 174–176
error component model, 194–198
first-difference estimator, 233
fixed effects model, 231–233
least squares dummy variable, 232
within estimator, 232
partial likelihood, 418
Pearson statistic, 319
Pearson's chi-square test, 243–244
placebo tests, 252
Poisson distribution, 133–144
Poisson model, 378–383
potential outcome, 241
prediction interval, 64
probability value, 57–59
probit, 312–313, 353
propensity score, 259
proportional hazard model, 416–419

QR decomposition, 80–81
quasi-Poisson model, 382

random parameter logit model, 452–453
random utility model, 316–317, 439–440
randomized experiment, 242–246
regression discontinuity, 249
fuzzy, 250, 254
manipulation test, 255
placebo tests, 252
sharp, 250
spatial, 281–286
reparametrization
Cobb-Douglas, 152
generalized production function, 153, 159
Solow model, 291
residual standard error, 43
R, 62
residuals, 14
deviance, 319
generalized, 322, 323
censored regression model, 354–355
Pearson's, 319
Poisson
Pearson's, 382
response, 382
response, 319, 323

sample selection, 345
Poisson model, 396–399
maximum likelihood, 397–398
two-step estimator, 398–399
sampling error, 33
sandwich, 173, 184–191
cluster, 185–186, 189
heteroskedastic-consistent, 185, 188–189
maximum likelihood estimator, 142, 147
multiple linear regression, 82
Poisson model, 381–383
Sargan test, 237, 395
saturated model, 318
score test, 93, 96
binomial, 330
generalized production function, 161–162
heteroskedastic logit model, 448
linear gaussian model, 157
nested logit, 451
seemingly unrelated regression, 199–201
semi-parametric estimator
tobit-1, 355–356
simulations
convergence of the OLS estimator, 54–55
simple linear regression model, 25–31

variance of the OLS estimator, 38–41
Vuong test, 170
simultaneity bias, 206–209
small sample properties
instrumental variable estimator, 211
spatial autoregressive model, 299–301
Durbin, 301
spatial correlation
tests, 295–298
spatial error model, 299–301
Durbin, 301
stated preference survey, 434
Student t distribution, 44, 60
synthetic control, 267–274
system estimation, 98–104, 177
seemingly unrelated regression, 199–201
simultaneity bias, 206–209
three-stage least squares, 226–228

t-test, 242–243, 258
Welch, 242
Taylor expansion, 140–141, 149
exact, 141, 146
Taylor series, 399
test
heteroskedasticity, 178–180
individual effects, 180–183
inter-equation correlation, 183–184
multiple linear regression model, 91–92
simple linear model, 56–59
three test principles
linear gaussian model, 157–158
linear regression models, 92–93, 98
maximum likelihood estimator, 156–163
three-stage least squares, 226–228
tobit-1
semi-parametric estimator, 355–356
two-step estimator, 353
tobit-2, 366–370
top-coding, 345
treatment effect
average treatment effect of the treated, 265
local average treatment effect, 248
trigamma function, 411
truncated normal distribution, 346
two-part models, 367
two-stage least squares estimator, 211
two-step estimator, 370
binomial model, 333–335
nested logit, 450
Poisson model, 398–399
selection models, 370
tobit-1, 353

unbiasedness, 50
multiple linear regression model, 81–82
simple linear regression model, 35
unconstrained model, 92
uncorrelated errors
random utility model, 440
uncorrelation, 37, 42, 82
uniform distribution, 13, 25
upper model, 450

variance decomposition, 387
Vuong test, 168–172
generalized production function, 170
translog production function, 170

Wald estimator, 217–221
Wald test, 92–93, 95
binomial, 330
generalized production function, 159–161
heteroskedastic logit model, 448
linear gaussian model, 157
nested logit, 451
weak instruments test, 237
Weibull distribution, 413
Weibull-Gamma model, 422–423
weighted least squares, 193–194
Welch's t-test, 242
within estimator, 232
within matrix, 176

zero inflated Poisson, 391–392

Authors

Abadie, 267, 268
Abiad, 339, 340
Adams, xxi
Adkins, 335
Aldrich, 158
Allaire, xix
Amemiya, xix, 50, 196, 313–315, 335, 346, 353
An, 406
Ananat, 216
Anand, 120
Angrist, 218, 219, 246–249
Anselin, 297, 301
Arel-Bundock, 123
Arora, 196
Ashenfelter, xxi

Baker, 237
Becker, 259, 261, 267
Ben-Akiva, 434, 450
Bera, 301
Berndt, 146, 174
Bettinger, 248, 249
Bhat, 435, 447
Bivand, 277, 292, 295, 299, 303
Blundell, 334, 335
Bohn, 267
Bolduc, 434, 450
Bonjour, 174, 181
Bonnel, 10
Bound, 237
Bradley, 434
Breusch, 178–180, 182, 183
Buser, 250

Calonico, 252
Cameron, xix, 164, 210, 222, 237, 241, 376, 384, 401, 420
Carlevaro, 368
Carson, 430
Carter, 335
Cattaneo, 252
Cetinkaya-Rundel, xix
Chavas, 376
Cherkas, 174, 181
Christensen, 406
Costanigro, 105
Cox, 158, 417
Cragg, 158, 345, 367, 391
Croissant, 101, 182, 368, 433
Curry, 244

Daganzo, 458
Daly, 449
Davidson, xix, 17, 50, 52, 54, 80, 146, 211
De Crombrugghe, 345
Dean, 384
Deb, 376
Dehejia Wahba, 259
Di Tella, 256
Diamond, 267, 268
Douch, 267
Dow, 370
Dunford, 268
Dunvely, xxi
Duranton, 186

Edelman, 314
Edwards, 267
Efron, 326
Ellison, 376
Ertur, 286, 288, 290, 296, 301, 302, 304, 306
Estrella, 327

Favero, 136, 402
Florax, 301
Forinash, 435

Gardeazabal, 267
Geil, 376
Geweke, 461
Giraud, 280
Goldberg, 364
Goldberger, 345, 350
Goldin, xxi
Gourieroux, 322
Graddy, xxi
Graham, 189, 190
Greene, 103, 166, 193, 350, 376, 399, 449
Grolemund, xix
Gronau, 345, 367
Grossman, 364
Gupta, 406
Gurmu, 376

Hainmueller, 267, 268
Hall, Browny, 146
Hall, Robert, 146
Ham, xxi
Hamann, 104, 154
Hanemann, 430
Haskel, 174, 181
Hausman, xxi, 146, 236, 345, 458
Hawkes, 174, 181
Heckman, 210, 353, 367, 396
Heiss, 449
Helpman, 364
Henningsen, 104, 154
Hensher, 449
Hong, 258
Horowitz, 313
Hothorn, 330
Houthakker, 174, 179
Hussain, 196

Ichino, 259–261
Imbens, 246–248
Imber, 430
Ivaldi, 100

Jaeger, 237
Johansson, 376
Jones, xxi, 370

Kambur, 120
Kastoryano, 408

Katz, xxi, 424
Keane, 461
Kenkel, 399
Kennan, 376
Kiefer, xix
King, 248, 249
Klosner, 267
Koch, 286, 288, 290, 296, 301, 302, 304, 306
Koenker, 180, 376
Koop, 205
Koppelman, xxi, 435, 449
Kotlikoff, xxi
Kremer, 248, 249
Krishnan, xxi
Kumar, 282
Kuznets, 120
Köll, 189, 190

Ladoux, 100
Lahiri, 206
Lalonde, 259
Lambert, 392
Lancaster, xix, 410
Larmarange, 244
Lave, 326
Lavery, 244
Linden, 242
Lindo, 250
Lofstrom, 267
Luca, 314

MacKinnon, xix, xxi
Maddala, xix, 158, 206
Magee, 158, 326
Maggi, 364
Makowsky, 370, 371
Mankiw, 69, 83
Matschke, 364–366
Mayr, 360
McCluskey, 105
McCrary, 255
McCullagh, 319
McDonald, xxi, 352
McFadden, 329, 440, 449, 450
McKelvey, 329
McKinnon, 17, 50, 52, 54, 80, 146, 211
Mealli, 260
Meijer, 434
Messner, 360
Miller, 133, 376
Million, 376
Millo, 182, 299
Mittelhammer, 105
Mody, 339, 340
Moffitt, xxi, 352
Monfort, 322
Mullahy, xxi, 376, 391, 394, 395
Mullbauer, 6

Nagelkerke, 158
Nannicini, 260
Nelder, 319
Nelson, 158
Nerlove, 196
Newey, 164, 335
Norton, 370
Nunn, 223, 224

Olsen, 355
Oreopoulos, 250
Ossard, 100

Pagan, xxi, 178–180, 182, 183
Palm, 345
Pebesma, 277, 292, 303
Pesaran, 136, 402
Pinotti, 267
Piras, 299
Plug, 6
Pohlmeier, xxi
Poirier, 205
Powell, 355
Puga, 186

Raphael, 267
Renault, 322
Revankar, 153
Reynaerts, 267
Rivers, 334
Robinson, 120
Romer, 69, 83, 302
Rosen, 442
Rosenbaum, 259
Rosenzweig, xxi
Rotte, 376
Rouwendal, 434
Rubin, 246, 248, 259
Runkle, 461

Sanders, 250
Santos Silva, 376
Schaller, 174, 182, 234
Schargrodsky, 256
Seller, 376
Sharma, 136, 402
Sherlund, 364–366
Shi, 170
Simioni, 100
Simpson, 335
Sjoberg, 244
Skeels, xxi, 165
Small, 442
Smith, xxi, 334, 335
Snell, 158
Spector, 174, 181
Stewart, 228
Stock, 90
Stoll, 376
Stratmann, 370, 371
Sueyoshi, 424
Suit, 228
Svirsky, 314
Swamy, 196
Swanson, 376

Tauchen, 163
Tennekes, 280
Terza, xxi, 376, 377, 398, 399
Theil, 227
Titiunik, 252
Tjur, 327
Tobias, 205
Tobin, 345
Train, xix, 450, 454
Trivedi, xix, 164, 210, 222, 237, 241, 376, 384, 401, 420
Trognon, 322

Uhler, 158
Urbain, 345

van de Ven, xxi
van der Klaauw, 408
Vandaele, 158
Vanschoonbeek, 267

Veall, 159, 326
Vella, xxi, 165
Vuong, 156, 168, 334

Wahba, 259
Walker, 450
Wallace, 196
Watson, 90
Weil, 69, 83
Wen, 435, 449
Werner, 430
Wheeler, 286, 287
White, 179, 185, 353
Whiting, 244
Wickham, xix
Wilhelm, xxi, 357, 358, 362
Wilks, 360, 430
Windmeijer, 326, 376
Winkelman, xxi, 376
Wise, 345, 458
Wold, 228
Wong, 295
Wooldridge, 344

Yoon, 301

Zavoina, 329
Zeileis, 101, 189, 190, 330, 360, 376
Zellner, 153, 199, 227
Zimmermann, 159, 326, 376

Data Sets

`micsr.data::adoptees`, 6, 8
`micsr.data::afghan_girls`, 242–246
`micsr.data::agglo_growth`, 286–288, 292–298
`micsr.data::airbnb`, 314–315
`micsr.count::amc12`, 375
`micsr::apples`, 100, 151–153, 155–156, 159–163, 166–167, 171–172, 183–184, 199–201
`micsr.count::asymptotic`, 375, 376

`micsr.data::basque_country`, 268
`micsr.data::bdh`, 250–251, 254
`micsr.count::bids`, 375, 377
`micsr::birthwt`, 6

`micsr.data::car_thefts`, 256
`micsr.count::cartels`, 375, 377
`micsr::charitable`, 357–358, 362–363
`micsr::cigarettes`, 6
`micsr::cigmales`, 376, 377, 395–396

`micsr.count::doctor_aus`, 375
`micsr.count::doctor_cal`, 375
`micsr.count::doctor_ger`, 375
`micsr.count::doctor_us`, 375
`micsr::drinks`, 399–400
`micsr.data::dutch_railways`, 434–435, 454–458

`micsr.count::elderly`, 375

`micsr::federiv`, 335–337
`micsr::fin_reform`, 339–340
`micsr::food`, 358–359

`micsr.data::growth`, 69, 83

`micsr.count::health_ins`, 375
`micsr.count::health_reform`, 375
`micsr.count::hospitalization`, 375

`micsr.data::kakadu`, 430
`micsr.data::kuznets`, 120–125

`micsr.data::loan_market`, 207–209

`micsr.count::majordrg`, 376, 377
`micsr.data::mode_choice`, 313, 314, 324–330, 332

`micsr.data::napster`, 258–259

`micsr.data::oil`, 402–406, 412

`micsr.data::paces`, 248
`micsr::portfolio`, 360–362
`micsr.data::price_time`, 10
`micsr.data::probation`, 250–255
`micsr.count::publications`, 376

`micsr.data::recall`, 424–425, 427–428
`micsr.data::retirement`, 406–407

`micsr.data::sibling_educ`, 205–206
`necountries::slave_trade`, 223–226, 238–239
`micsr.count::somerville`, 376
`necountries::sp_solow`, 288–296, 302–306
`micsr.count::strikes`, 376

micsr::telephone, 450–452
micsr.data::tobinq, 175, 182–183, 198, 235–238
micsr.data::toronto_montreal, 435–448, 462–463
micsr.data::tracks_side, 216–217
micsr::trade_protection, 364–366
micsr.data::traffic_citations, 370–372
micsr::trips, 376–380, 382–386, 388–390, 392–393
micsr.data::twa, 260–267
micsr.data::twins, 174, 181–182, 190–191, 196–198, 233–234
micsr.data::uk_elec, 174, 179–180, 189–190, 194
micsr.data::unemp_teachers, 408–410, 412, 415–416, 419, 421–423
micsr.data::urban_gradient, 186–187
micsr.data::vietnam, 218–221
micsr.data::watermelon, 228–231
micsr.data::wine, 105–119, 126–128

Functions

tibble::add_column, 182, 187, 212, 225, 284, 296, 298, 425
gtsummary::add_p, 244
dplyr::add_row, 65
ggplot2::aes, 170, 186, 214, 220, 223, 244, 251, 280, 283–285, 288, 296, 385, 404, 408, 415, 421, 430
ggplot2::after_stat, 251
stats::AIC, 326, 392, 393
gtsummary::all_continuous, 244
base::apply, 121, 148, 161, 163, 167, 190, 200, 300, 444, 445, 462
dplyr::arrange, 271, 402, 421
Formula::as.Formula, 365, 366
base::as.matrix, 199, 229, 230
base::as.numeric, 152, 161, 163, 171, 199, 230, 250, 254, 296, 298, 330
kableExtra::as_kable_extra, 244
tibble::as_tibble, 65, 124–128, 212, 214, 296–298
base::attr, 161, 163, 166
broom::augment, 107, 108
ggplot2::autoplot, 378, 385, 390, 393
marginaleffects::avg_slopes, 125, 127
Matrix::bdiag, 101, 199
dplyr::bind_cols, 102, 199
stats::binomial, 320, 428
micsr::binomreg, 313, 315, 324, 330, 336
lmtest::bptest, 180
sandwich::bread, 189
linearHypothesis::car, 330
base::cat, 149
base::cbind, 167, 200, 226, 230, 300, 305
censReg::censReg, 357
necountries::check_join, 290
stats::chisq.test, 244
base::chol, 199, 212, 230
base::class, 439
micsr::clm, 104, 199, 200
dplyr::closest, 263
micsr::cmtest, 332, 362, 363, 421
stats::coef, 21, 26, 62, 63, 78, 84, 87, 88, 91, 94, 95, 98, 102, 104, 106–113, 115–119, 122, 123, 126, 148, 149, 152, 153, 155, 156, 159–161, 163, 166, 171, 181, 200, 213, 217, 230, 234, 248, 287, 291, 305, 314, 365, 380, 389, 412, 415, 427, 446, 447, 451, 455
lmtest::coeftest, 191
base::colnames, 199, 212
kableExtra::column_spec, 244
stats::confint, 62, 87
ggplot2::coord_cartesian, 215, 378, 385, 390, 393, 404, 430
stats::cor, 206, 217
dplyr::count, 114, 250, 263, 264, 403, 406, 424
necountries::countries, 290

survival::coxph, 419, 424, 425
crch::crch, 361
base::crossprod, 79–81, 96, 104, 121, 148, 149, 162, 167, 183, 189, 190, 199, 226, 230, 382
base::cumprod, 403, 407
base::cumsum, 403, 407, 421
base::cut, 287, 296, 298

utils::data, 288
base::data.frame, 170
marginaleffects::datagrid, 128
dplyr::desc, 271
stats::deviance, 21, 96, 321
stats::df.residual, 45, 61, 64, 88, 91, 321, 382
dfidx::dfidx, 434, 436, 439, 454
base::diag, 45, 163, 200, 300, 305, 325
base::diff, 234
base::dimnames, 200
dplyr::distinct, 257, 406
spdep::dnearneigh, 296
stats::dnorm, 371, 415
stats::dpois, 377
base::drop, 81, 104, 148, 159, 160, 162, 163, 167, 226

stats::effects, 446
micsr::endogtest, 336, 366
plm::ercomp, 197, 198
spatialreg::errorsarlm, 303
micsr::escount, 399, 400
sandwich::estfun, 189
dplyr::everything, 214
micsr::expreg, 395

ggplot2::facet_grid, 220
ggplot2::facet_wrap, 244
base::factor, 212, 234, 258, 428
forcats::fct_lump, 114
forcats::fct_lump_min, 224
forcats::fct_relevel, 116
forcats::fct_rev, 298
dplyr::filter, 27, 65, 83, 100, 126, 127, 151, 199, 218–220, 229, 251, 260, 262, 264, 271, 282–285, 287, 297, 315, 359, 408, 428
stats::fitted, 21, 64, 106, 162, 179, 225, 230, 321, 382, 384, 385, 444, 445, 462
plm::fixef, 235, 236

base::format, 279
Formula::Formula, 101, 199, 225, 438, 439
stats::formula, 390, 393
micsr::ftest, 97, 98, 217
dplyr::full_join, 285
base::function, 29, 148, 149, 191

geodata::gadm, 279
micsr::gaze, 95, 97, 98, 160, 182, 186, 197, 205–208, 216, 217, 224, 225, 233, 238, 243, 244, 246, 248, 249, 253–255, 257, 258, 287, 291, 295, 296, 303, 304, 313, 330, 332, 336, 359, 360, 362, 363, 366, 384, 396, 412, 421, 423, 447, 448, 451, 454, 457, 458
spdep::geary.test, 296
tidysynth::generate_control, 270
tidysynth::generate_predictor, 269
tidysynth::generate_weights, 270
ggplot2::geom_abline, 421
micsr::geom_binmeans, 251, 285
ggplot2::geom_boxplot, 244
ggplot2::geom_density, 170, 214, 430
ggplot2::geom_errorbar, 64
ggplot2::geom_function, 170, 415
ggplot2::geom_hline, 220
ggrepel::geom_label_repel, 223
ggplot2::geom_line, 408
ggplot2::geom_point, 64, 70, 288, 385
ggplot2::geom_ribbon, 220
ggplot2::geom_sf, 280, 282–284, 287, 293, 296, 298
ggplot2::geom_sf_label, 283
ggplot2::geom_smooth, 64, 70, 186, 220, 223, 251, 285, 288, 385, 404
ggplot2::geom_step, 404, 421
ggpubr::ggarrange, 385, 390, 393
ggplot2::ggplot, 64, 70, 170, 186, 214, 223, 244, 251, 280, 282–285, 287, 288, 293, 296, 298, 385, 404, 408, 415, 421, 430
survminer::ggsurvplot, 408
stats::glm, 317, 320, 371, 379, 390, 395, 400, 428
MASS::glm.nb, 390, 393
tidysynth::grab_balance_table, 270
tidysynth::grab_loss, 272
tidysynth::grab_predictor_weights, 271

tidysynth::grab_significance, 273
tidysynth::grab_unit_weights, 271
micsr::gres, 421
dplyr::group_by, 28, 40, 113, 115, 116, 182, 187, 234, 250, 257, 260, 265, 282, 339, 340, 421
ggplot2::guide_legend, 291
ggplot2::guides, 291

utils::head, 78, 79, 101, 121, 123, 170, 234–236, 284, 292, 294, 321, 322, 382, 383, 402, 408, 424, 439, 444, 445, 462
countreg::hurdle, 392

base::I, 118, 119, 121, 171, 254
dfidx::idx, 435
base::ifelse, 250, 257, 258, 260, 263, 266, 285, 290, 294, 315, 340, 403, 406, 407, 445
base::is.na, 315, 340
ivreg::ivreg, 226, 238, 249, 254

kableExtra::kable_styling, 244

ggplot2::labs, 271, 274, 291, 415
dplyr::lag, 229, 339, 340, 403, 407
spatialreg::lagsarlm, 303
dplyr::left_join, 220, 257, 263, 264, 266, 290, 340, 428
base::length, 44, 135, 144, 148, 162, 167, 171, 182, 187, 264, 292, 294
base::levels, 114, 115
base::lfactorial, 135
car::linearHypothesis, 95, 160, 303, 336, 366, 448, 451, 457, 458
base::list, 149, 200, 214, 430, 451
stats::lm, 20, 24, 26, 27, 44, 47, 60, 78, 83, 87, 88, 93, 94, 96, 98, 102, 106, 108–113, 115–119, 121–124, 126, 148, 149, 152, 162, 163, 179, 181, 186, 194, 198–200, 205–208, 216, 217, 224, 225, 230, 233, 234, 243, 246, 248, 249, 253, 254, 257, 258, 266, 287, 291, 296, 302, 372, 379, 384, 412
stats::lm.fit, 213
spdep::lm.LMtests, 303
spdep::lm.morantest, 296
spdep::localmoran, 296, 297
stats::logLik, 152, 153, 161, 163, 326, 330, 444
micsr::loglm, 152, 153, 161, 171
mlogit::logsum, 446
lmtest::lrtest, 303, 304, 330, 448, 451, 457, 458

purrr::map, 30, 31, 214
MatchIt::match.data, 266
MatchIt::matchit, 265, 266
base::matrix, 63, 95, 103, 159, 167, 183, 199, 200, 212, 300, 305
maxLik::maxLik, 155, 156
sandwich::meat, 189
mlogit::med, 456
stats::median, 430
mlogit::mlogit, 443, 447, 451, 454, 455, 462
stats::model.frame, 78, 91, 96, 101, 148, 155, 199, 225, 382, 384, 439
stats::model.matrix, 78, 80, 96, 101, 104, 148, 155, 189, 190, 199, 225, 229, 382, 439
Formula::model.part, 102, 199, 225
stats::model.response, 79, 148, 155, 199, 382, 384
spdep::moran.plot, 296
spdep::moran.test, 295
rmapshaper::ms_simplify, 279
modelsummary::msummary, 430
dplyr::mutate, 27, 49, 64, 65, 70, 93, 106, 113, 114, 118, 121, 126, 186, 196, 205, 212, 218–220, 229, 233, 234, 250, 254, 257, 258, 260, 263, 264, 266, 284, 285, 287, 289, 290, 298, 302, 313, 315, 321, 339, 340, 357, 360, 364, 371, 395, 403, 406–408, 421, 425, 428, 430, 443, 445

stats::na.omit, 294
base::names, 21, 97, 212, 365, 447, 455
spdep::nb2listwdist, 295, 296
tidyr::nest, 29, 31, 214
dplyr::nest_by, 427
stats::nobs, 21, 45, 63, 64, 126, 179, 183, 189, 199, 421
base::nrow, 78, 126, 143, 229, 244, 425

utils::object.size, 279
stats::offset, 161
micsr::ordreg, 340, 427
base::outer, 183, 200, 243

base::paste, 101, 149, 199
stats::pchisq, 94, 180, 184
plm::pdim, 197
stats::pf, 94
plm::pFtest, 238
pglm::pglm, 197
plm::phtest, 238
tidyr::pivot_longer, 102, 199, 214, 264, 266, 430
tidyr::pivot_wider, 214, 218, 219, 250, 258, 406
plm::plm, 191, 196, 198, 235
plm::plmtest, 182
tidysynth::plot_differences, 272
tidysynth::plot_mspe_ratio, 274
tidysynth::plot_placebos, 274
tidysynth::plot_trends, 271
tidysynth::plot_weights, 271
stats::pnorm, 58, 305, 371, 415
micsr::poisreg, 388
stats::poisson, 379
MASS::polr, 340
stats::poly, 121–124, 205, 206, 287, 288, 296
spdep::poly2nb, 292, 293
stats::predict, 64, 65, 322, 430, 445, 462
base::prop.table, 243
micsr::pscore, 260
stats::pt, 88
dplyr::pull, 24, 26, 29, 30, 44, 48, 91, 114, 115, 182, 187, 206, 217, 225, 255, 264, 284, 296, 298, 402, 425, 428

stats::qchisq, 91
stats::qf, 91
stats::qnorm, 57
base::qr, 80
base::qr.Q, 80
base::qr.R, 80
stats::qt, 61, 64
micsr::quad_form, 159, 160, 162, 167
stats::quasipoisson, 382, 395

base::rbind, 111, 190, 191, 198–201, 230, 380, 445, 462
rddensity::rddensity, 255
rdrobust::rdplot, 252, 253
rddensity::rdplotdensity, 255
rdrobust::rdrobust, 254, 255
stats::relevel, 425
dplyr::rename, 70, 263, 289, 297, 364, 406
base::rep, 28, 162, 163, 167
stats::resid, 21, 44, 96, 98, 106, 107, 109, 179, 182, 183, 187, 189, 190, 199, 226, 230, 296, 321, 382, 384, 385
dplyr::right_join, 285, 287
stats::rnorm, 25, 26, 28, 212
countreg::rootogram, 378, 385, 390, 393
base::round, 59, 62, 144
mlogit::rpar, 456
micsr::rsq, 97, 162, 163, 167, 179, 217, 225, 326, 327, 329
stats::runif, 25

spatialreg::sacsarlm, 303
base::sapply, 149, 292–294, 421
micsr::sargan, 396
ggplot2::scale_fill_brewer, 287, 296
ggplot2::scale_fill_grey, 284, 298
ggplot2::scale_shape_manual, 186
ggplot2::scale_size_continuous, 291
ggplot2::scale_x_continuous, 214, 288, 415
ggplot2::scale_x_log10, 223
ggplot2::scale_y_continuous, 404, 408
ggplot2::scale_y_log10, 223
micsr::scoretest, 330
mlogit::scoretest, 448, 451, 457, 458
stats::sd, 25, 28, 31, 44, 137
dplyr::select, 64, 65, 106, 126–128, 218–220, 229, 244, 258, 262–264, 266, 279, 290, 406, 408, 427
sampleSelection::selection, 372
tidyr::separate, 214, 264
base::set.seed, 25, 214
units::set_units, 281, 284
stats::sigma, 45, 63, 64, 96
dplyr::slice, 263, 407
marginaleffects::slopes, 124–128
base::solve, 79, 95, 96, 104, 148, 149, 162, 167, 189, 190, 199, 226, 230, 300, 382
dnearneigh::spdep, 294
base::sqrt, 45, 49, 88
sf::st_area, 281
sf::st_as_sf, 279

sf::st_centroid, 284
sf::st_distance, 280, 284
sf::st_geometry, 279, 280, 283, 284
sf::st_intersection, 283
micsr::st_nb, 293
sf::st_point, 278
sf::st_set_geometry, 294, 296
sf::st_sf, 278, 285
sf::st_sfc, 278
sf::st_union, 283
ggplot2::stat_ellipse, 70
micsr::stder, 45, 149, 183, 189–191, 194, 198, 226, 325, 382, 383
mlogit::stdev, 456
gtsummary::style_pvalue, 244
dplyr::summarise, 27, 28, 31, 40, 48, 49, 91, 113, 115, 116, 137, 182, 187, 196, 206, 214, 217, 234, 257, 260, 265, 282, 340, 359, 377, 421
survival::Surv, 405, 407–409, 412, 419, 421, 423–425, 427, 430
survival::survdiff, 409
survival::survfit, 405, 407, 408
survival::survreg, 412, 430
tidysynth::synthetic_control, 269
systemfit::systemfit, 104, 200, 230

stats::t.test, 243, 246, 258
base::table, 243
base::tapply, 190
gtsummary::tbl_summary, 244
ggplot2::theme, 291
tibble::tibble, 26–29, 106, 125, 194, 213, 278, 385, 421, 430
broom::tidy, 405, 407, 408
AER::tobit, 357
micsr::tobit1, 357, 359–361, 365
dplyr::top_n, 264
dplyr::transmute, 30, 31, 48, 100, 125, 151, 199, 207, 214

dplyr::ungroup, 182, 339, 427
base::unique, 182, 187, 264, 427, 428
base::unname, 22, 63, 88, 115, 116, 122, 123, 153, 208, 213, 248, 291, 295, 305
tidyr::unnest, 30, 214, 427
stats::update, 47, 265, 303, 313, 315, 321, 324, 336, 340, 358, 371, 372, 382, 388, 392, 395, 400, 415, 419, 423, 428, 430, 451, 455, 457
base::upper.tri, 184

stats::var, 377
stats::vcov, 45, 83, 88, 95, 123, 159, 160, 163, 190, 191, 194, 198, 305, 325, 456, 457
sandwich::vcovCL, 190, 191, 198
sandwich::vcovHC, 189–191, 194, 325, 383
micsr::vuong_sim, 170

lmtest::waldtest, 330, 366, 448, 451, 457, 458
micsr::weibreg, 419, 421, 423
base::with, 252, 255

micsr::zellner_revankar, 161, 163, 166, 171
countreg::zeroinfl, 392

Libraries

AER
- tobit, 357

Formula
- Formula, 101, 199, 225, 438, 439
- as.Formula, 365, 366
- model.part, 102, 199, 225

MASS
- glm.nb, 390, 393
- polr, 340

MatchIt
- match.data, 266
- matchit, 265, 266

Matrix
- bdiag, 101, 199

base
- I, 118, 119, 121, 171, 254
- apply, 121, 148, 161, 163, 167, 190, 200, 300, 444, 445, 462
- as.matrix, 199, 229, 230
- as.numeric, 152, 161, 163, 171, 199, 230, 250, 254, 296, 298, 330
- attr, 161, 163, 166
- cat, 149
- cbind, 167, 200, 226, 230, 300, 305
- chol, 199, 212, 230
- class, 439
- colnames, 199, 212
- crossprod, 79–81, 96, 104, 121, 148, 149, 162, 167, 183, 189, 190, 199, 226, 230, 382
- cumprod, 403, 407
- cumsum, 403, 407, 421
- cut, 287, 296, 298
- data.frame, 170
- diag, 45, 163, 200, 300, 305, 325
- diff, 234
- dimnames, 200
- drop, 81, 104, 148, 159, 160, 162, 163, 167, 226
- factor, 212, 234, 258, 428
- format, 279
- function, 29, 148, 149, 191
- ifelse, 250, 257, 258, 260, 263, 266, 285, 290, 294, 315, 340, 403, 406, 407, 445
- is.na, 315, 340
- length, 44, 135, 144, 148, 162, 167, 171, 182, 187, 264, 292, 294
- levels, 114, 115
- lfactorial, 135
- list, 149, 200, 214, 430, 451
- matrix, 63, 95, 103, 159, 167, 183, 199, 200, 212, 300, 305
- names, 21, 97, 212, 365, 447, 455
- nrow, 78, 126, 143, 229, 244, 425
- outer, 183, 200, 243
- paste, 101, 149, 199
- prop.table, 243
- qr.Q, 80
- qr.R, 80
- qr, 80
- rbind, 111, 190, 191, 198–201, 230, 380, 445, 462
- rep, 28, 162, 163, 167
- round, 59, 62, 144
- sapply, 149, 292–294, 421
- set.seed, 25, 214
- solve, 79, 95, 96, 104, 148, 149, 162, 167, 189, 190, 199, 226, 230, 300, 382
- sqrt, 45, 49, 88
- table, 243
- tapply, 190
- unique, 182, 187, 264, 427, 428
- unname, 22, 63, 88, 115, 116, 122, 123, 153, 208, 213, 248, 291, 295, 305
- upper.tri, 184
- with, 252, 255

broom
- augment, 107, 108
- tidy, 405, 407, 408

car
- linearHypothesis, 95, 160, 303, 336, 366, 448, 451, 457, 458

censReg
- censReg, 357

countreg
- hurdle, 392
- rootogram, 378, 385, 390, 393
- zeroinfl, 392

crch

crch, 361
dfidx
dfidx, 434, 436, 439, 454
idx, 435
dnearneigh
spdep, 294
dplyr
add_row, 65
arrange, 271, 402, 421
bind_cols, 102, 199
closest, 263
count, 114, 250, 263, 264, 403, 406, 424
desc, 271
distinct, 257, 406
everything, 214
filter, 27, 65, 83, 100, 126, 127, 151, 199, 218–220, 229, 251, 260, 262, 264, 271, 282–285, 287, 297, 315, 359, 408, 428
full_join, 285
group_by, 28, 40, 113, 115, 116, 182, 187, 234, 250, 257, 260, 265, 282, 339, 340, 421
lag, 229, 339, 340, 403, 407
left_join, 220, 257, 263, 264, 266, 290, 340, 428
mutate, 27, 49, 64, 65, 70, 93, 106, 113, 114, 118, 121, 126, 186, 196, 205, 212, 218–220, 229, 233, 234, 250, 254, 257, 258, 260, 263, 264, 266, 284, 285, 287, 289, 290, 298, 302, 313, 315, 321, 339, 340, 357, 360, 364, 371, 395, 403, 406–408, 421, 425, 428, 430, 443, 445
nest_by, 427
pull, 24, 26, 29, 30, 44, 48, 91, 114, 115, 182, 187, 206, 217, 225, 255, 264, 284, 296, 298, 402, 425, 428
rename, 70, 263, 289, 297, 364, 406
right_join, 285, 287
select, 64, 65, 106, 126–128, 218–220, 229, 244, 258, 262–264, 266, 279, 290, 406, 408, 427
slice, 263, 407
summarise, 27, 28, 31, 40, 48, 49, 91, 113, 115, 116, 137, 182, 187, 196, 206, 214, 217, 234, 257, 260, 265, 282, 340, 359, 377, 421
top_n, 264
transmute, 30, 31, 48, 100, 125, 151, 199, 207, 214
ungroup, 182, 339, 427
forcats
fct_lump_min, 224
fct_lump, 114
fct_relevel, 116
fct_rev, 298
geodata
gadm, 279
ggplot2
aes, 170, 186, 214, 220, 223, 244, 251, 280, 283–285, 288, 296, 385, 404, 408, 415, 421, 430
after_stat, 251
autoplot, 378, 385, 390, 393
coord_cartesian, 215, 378, 385, 390, 393, 404, 430
facet_grid, 220
facet_wrap, 244
geom_abline, 421
geom_boxplot, 244
geom_density, 170, 214, 430
geom_errorbar, 64
geom_function, 170, 415
geom_hline, 220
geom_line, 408
geom_point, 64, 70, 288, 385
geom_ribbon, 220
geom_sf_label, 283
geom_sf, 280, 282–284, 287, 293, 296, 298
geom_smooth, 64, 70, 186, 220, 223, 251, 285, 288, 385, 404
geom_step, 404, 421
ggplot, 64, 70, 170, 186, 214, 223, 244, 251, 280, 282–285, 287, 288, 293, 296, 298, 385, 404, 408, 415, 421, 430
guide_legend, 291
guides, 291
labs, 271, 274, 291, 415
scale_fill_brewer, 287, 296
scale_fill_grey, 284, 298
scale_shape_manual, 186
scale_size_continuous, 291
scale_x_continuous, 214, 288, 415
scale_x_log10, 223
scale_y_continuous, 404, 408
scale_y_log10, 223

stat_ellipse, 70
theme, 291
ggpubr
ggarrange, 385, 390, 393
ggrepel
geom_label_repel, 223
gtsummary
add_p, 244
all_continuous, 244
style_pvalue, 244
tbl_summary, 244
ivreg
ivreg, 226, 238, 249, 254
kableExtra
as_kable_extra, 244
column_spec, 244
kable_styling, 244
linearHypothesis
car, 330
lmtest
bptest, 180
coeftest, 191
lrtest, 303, 304, 330, 448, 451, 457, 458
waldtest, 330, 366, 448, 451, 457, 458
marginaleffects
avg_slopes, 125, 127
datagrid, 128
slopes, 124–128
maxLik
maxLik, 155, 156
micsr
binomreg, 313, 315, 324, 330, 336
clm, 104, 199, 200
cmtest, 332, 362, 363, 421
endogtest, 336, 366
escount, 399, 400
expreg, 395
ftest, 97, 98, 217
gaze, 95, 97, 98, 160, 182, 186, 197, 205–208, 216, 217, 224, 225, 233, 238, 243, 244, 246, 248, 249, 253–255, 257, 258, 287, 291, 295, 296, 303, 304, 313, 330, 332, 336, 359, 360, 362, 363, 366, 384, 396, 412, 421, 423, 447, 448, 451, 454, 457, 458
geom_binmeans, 251, 285
gres, 421
loglm, 152, 153, 161, 171
ordreg, 340, 427
poisreg, 388
pscore, 260
quad_form, 159, 160, 162, 167
rsq, 97, 162, 163, 167, 179, 217, 225, 326, 327, 329
sargan, 396
scoretest, 330
st_nb, 293
stder, 45, 149, 183, 189–191, 194, 198, 226, 325, 382, 383
tobit1, 357, 359–361, 365
vuong_sim, 170
weibreg, 419, 421, 423
zellner_revankar, 161, 163, 166, 171
mlogit
logsum, 446
med, 456
mlogit, 443, 447, 451, 454, 455, 462
rpar, 456
scoretest, 448, 451, 457, 458
stdev, 456
modelsummary
msummary, 430
necountries
check_join, 290
countries, 290
pglm
pglm, 197
plm
ercomp, 197, 198
fixef, 235, 236
pFtest, 238
pdim, 197
phtest, 238
plmtest, 182
plm, 191, 196, 198, 235
purrr
map, 30, 31, 214
rddensity
rddensity, 255
rdplotdensity, 255
rdrobust
rdplot, 252, 253
rdrobust, 254, 255
rmapshaper
ms_simplify, 279
sampleSelection
selection, 372
sandwich

bread, 189
estfun, 189
meat, 189
vcovCL, 190, 191, 198
vcovHC, 189–191, 194, 325, 383

sf
st_area, 281
st_as_sf, 279
st_centroid, 284
st_distance, 280, 284
st_geometry, 279, 280, 283, 284
st_intersection, 283
st_point, 278
st_set_geometry, 294, 296
st_sfc, 278
st_sf, 278, 285
st_union, 283

spatialreg
errorsarlm, 303
lagsarlm, 303
sacsarlm, 303

spdep
dnearneigh, 296
geary.test, 296
lm.LMtests, 303
lm.morantest, 296
localmoran, 296, 297
moran.plot, 296
moran.test, 295
nb2listwdist, 295, 296
poly2nb, 292, 293

stats
AIC, 326, 392, 393
binomial, 320, 428
chisq.test, 244
coef, 21, 26, 62, 63, 78, 84, 87, 88, 91, 94, 95, 98, 102, 104, 106–113, 115–119, 122, 123, 126, 148, 149, 152, 153, 155, 156, 159–161, 163, 166, 171, 181, 200, 213, 217, 230, 234, 248, 287, 291, 305, 314, 365, 380, 389, 412, 415, 427, 446, 447, 451, 455
confint, 62, 87
cor, 206, 217
deviance, 21, 96, 321
df.residual, 45, 61, 64, 88, 91, 321, 382
dnorm, 371, 415
dpois, 377
effects, 446
fitted, 21, 64, 106, 162, 179, 225, 230, 321, 382, 384, 385, 444, 445, 462
formula, 390, 393
glm, 317, 320, 371, 379, 390, 395, 400, 428
lm.fit, 213
lm, 20, 24, 26, 27, 44, 47, 60, 78, 83, 87, 88, 93, 94, 96, 98, 102, 106, 108–113, 115–119, 121–124, 126, 148, 149, 152, 162, 163, 179, 181, 186, 194, 198–200, 205–208, 216, 217, 224, 225, 230, 233, 234, 243, 246, 248, 249, 253, 254, 257, 258, 266, 287, 291, 296, 302, 372, 379, 384, 412
logLik, 152, 153, 161, 163, 326, 330, 444
median, 430
model.frame, 78, 91, 96, 101, 148, 155, 199, 225, 382, 384, 439
model.matrix, 78, 80, 96, 101, 104, 148, 155, 189, 190, 199, 225, 229, 382, 439
model.response, 79, 148, 155, 199, 382, 384
na.omit, 294
nobs, 21, 45, 63, 64, 126, 179, 183, 189, 199, 421
offset, 161
pchisq, 94, 180, 184
pf, 94
pnorm, 58, 305, 371, 415
poisson, 379
poly, 121–124, 205, 206, 287, 288, 296
predict, 64, 65, 322, 430, 445, 462
pt, 88
qchisq, 91
qf, 91
qnorm, 57
qt, 61, 64
quasipoisson, 382, 395
relevel, 425
resid, 21, 44, 96, 98, 106, 107, 109, 179, 182, 183, 187, 189, 190, 199, 226, 230, 296, 321, 382, 384, 385
rnorm, 25, 26, 28, 212
runif, 25
sd, 25, 28, 31, 44, 137

sigma, 45, 63, 64, 96
t.test, 243, 246, 258
update, 47, 265, 303, 313, 315, 321, 324, 336, 340, 358, 371, 372, 382, 388, 392, 395, 400, 415, 419, 423, 428, 430, 451, 455, 457
var, 377
vcov, 45, 83, 88, 95, 123, 159, 160, 163, 190, 191, 194, 198, 305, 325, 456, 457

survival
Surv, 405, 407–409, 412, 419, 421, 423–425, 427, 430
coxph, 419, 424, 425
survdiff, 409
survfit, 405, 407, 408
survreg, 412, 430

survminer
ggsurvplot, 408

systemfit
systemfit, 104, 200, 230

tibble
add_column, 182, 187, 212, 225, 284, 296, 298, 425
as_tibble, 65, 124–128, 212, 214, 296–298
tibble, 26–29, 106, 125, 194, 213, 278, 385, 421, 430

tidyr
nest, 29, 31, 214
pivot_longer, 102, 199, 214, 264, 266, 430
pivot_wider, 214, 218, 219, 250, 258, 406
separate, 214, 264
unnest, 30, 214, 427

tidysynth
generate_control, 270
generate_predictor, 269
generate_weights, 270
grab_balance_table, 270
grab_loss, 272
grab_predictor_weights, 271
grab_significance, 273
grab_unit_weights, 271
plot_differences, 272
plot_mspe_ratio, 274
plot_placebos, 274
plot_trends, 271
plot_weights, 271
synthetic_control, 269

units
set_units, 281, 284

utils
data, 288
head, 78, 79, 101, 121, 123, 170, 234–236, 284, 292, 294, 321, 322, 382, 383, 402, 408, 424, 439, 444, 445, 462
object.size, 279

Printed in the United States
by Baker & Taylor Publisher Services